REACTION MECHANISMS
IN ORGANIC
ANALYTICAL CHEMISTRY

REACTION MECHANISMS IN ORGANIC ANALYTICAL CHEMISTRY

KENNETH A. CONNORS

Professor of Pharmacy
The University of Wisconsin

A WILEY-INTERSCIENCE PUBLICATION

JOHN WILEY & SONS, New York • London • Sydney • Toronto

Library of Congress Cataloging in Publication Data:

Connors, Kenneth Antonio, 1932-
Reaction mechanisms in organic analytical chemistry.

1. Chemistry, Analytic. 2. Chemistry, Physical
organic.   I. Title. [DNLM: 1. Chemistry, Analytical.
2. Chemistry, Organic. QD 73 C752r 1972]

QD271.C68   547'.1'39   72–5845
ISBN 0-471-16845-9

Printed in the United States of America

10 9 8 7 6 5 4 3 2 1

To my grandfather

ANTONIO GIOIA

PREFACE

Conventional discussions of organic analysis are organized by analytical method or technique (absorption spectroscopy, polarography, partition chromatography, mass spectrometry, etc.) or by functional group (amines, esters, olefins, and so on); much attention has also been given to classical areas of analytical interest, such as acid-base reactions, metal-ligand coordination equilibria, and solubility phenomena. This book expresses a different point of view by looking at organic analytical reactions on the basis of reaction type and mechanism. Such a shift of viewpoint sometimes allows connections to be made between ideas or observations that were formerly thought to be unrelated (if indeed they were thought of at all in the same context). In this manner it may stimulate analytical research. Moreover, this way of looking at organic analysis seems more effective, and more enjoyable, in teaching and learning the subject.

In this book I have organized and discussed analytical reactions within the framework that was developed for organic chemistry through the 1920's to 1950's, and that during the 1960's was applied to the study and systematization of bioorganic reactions. Students of analytical chemistry probably are not, as a class, well prepared for a thorough exposure to organic reaction mechanisms. This book is also, therefore, an introduction to many of the methods and concepts of physical organic chemistry. One of its contributions may be to direct the reading of the analytical chemist toward those subjects that he will find helpful in his analytical studies. I hope that no serious student of analytical chemistry will be satisfied to limit his pursuit of mechanistic studies to this book, but will follow my leads to other sources. This is a major purpose in providing a selection of literature citations. One advantage in reading elsewhere is that different points of view may be available, or even different ways of expressing the same idea, which may illuminate a difficult concept. Another advantage is the detail of experimental support and mechanistic interpretation that is present in the original literature but that had to be abbreviated or omitted in a review of this scope.

Despite the ultimate goal of relating the mechanistic viewpoint to analytical reactions, I have tried not to be too narrowly limited by present analytical practice. Restriction of the mechanistic treatment to reactions of present analytical interest might have resulted in a spotty coverage of some topics and would not have served to stimulate analytical development by revealing possible gaps and weaknesses in the traditional emphasis. Nevertheless, my selection of topics is related ultimately to analytical reactions, and in this way I may have provided a means for the analytical student who wishes to learn more organic chemistry to focus on those areas and authors that are most apt to be of professional interest to him.

I have omitted discussions of many important reaction types, among them free radical reactions, oxidation/reductions, photochemical reactions, rearrangements, and molecular complex formation. The treatment of the topics included is systematic, but does not pretend to be complete. This is especially so with the sections titled Analytical Reactions, which summarize, within each reaction class, the reactions that have been applied in analysis. The reader may find it amusing and instructive to make additions or reassignments. Since few analytical methods have been described in mechanistic terms by their developers, it is not easy to conduct a literature survey in these terms. Many interesting analytical reactions have probably been overlooked.

Few analytical methods have been developed or improved primarily by the use of systematic studies guided by mechanistic principles. For small-scale efforts involving the adaptation of known methods to specific problems the traditional trial-and-error adjustment of conditions, aided by experience and chemical intuition, is probably superior to a serious mechanistic study. But it is precisely in developing a reliable insight into the nature of the controlling variables that the mechanistic approach taken by this book should be analytically useful.

I wish to acknowledge with thanks my indebtedness to J. S. Fritz and G. S. Hammond, whose book on organic analysis, published in 1957 while I was a graduate student, influenced my approach to analytical chemistry; to my teachers Takeru Higuchi and Myron L. Bender; to my secretary Billie Hubacher; and to my colleagues Pasupati Mukerjee and Joseph R. Robinson.

Kenneth A. Connors

Madison, Wisconsin
May 1972

CONTENTS

REACTION MECHANISMS IN ORGANIC ANALYTICAL CHEMISTRY

Chapter 1.   INTRODUCTION

1.1   The Nature of Organic Analytical Methods

The term <u>organic analytical chemistry</u> obviously includes
a greater part of chemistry than can be bound in one book.
Some selection is necessary in treating the theoretical
foundations of the subject. The distinguishing feature of
the organic analytical methods in this book is the occurrence of a chemical reaction. In a broad sense, therefore, this book describes the principles of functional
group analysis, this phrase being interpreted more freely
than is conventional. Functional group analysis is based
upon the characteristic chemical behavior of an atom or
group of atoms (the functional group) in a molecule.
    Functional group methods consist of two parts: (1)
the chemical reaction or reactions that the sample compound is made to undergo; (2) the final measurement (the
"finish"), which is usually quantitative. Sometimes these
reaction and measurement components are merged together,
as in a simple acidimetric titration of an amine; more
frequently they are discrete, as when an ester is converted to the ferric complex of the corresponding hydroxamic
acid (the reaction portion), whose concentration is then
determined spectrophotometrically (the finish). Usually
these methods are designed to give quantitative information (amount or purity) about a sample of known identity,
but sometimes the same steps are utilized in qualitative
analysis, as in classical derivatization procedures, spot
tests, or derivatization prior to mass spectrometry.
    It is assumed that the reader is familiar with the
common organic analytical methods [1]. Detailed exposi-
tions are available in monographs [2] and review volumes
[3]. These analytical methods can be discussed by considering the generalized reaction

        sample + reagent ⇌ product A + product B   (1.1)

Two general phenomena control the extent of this reaction.
One of these is the <u>equilibrium</u> configuration of the system; for our purposes this is best described by the meth-

ods of thermodynamics. Chapter 2 considers some important
concepts of chemical equilibria. The second controlling
feature is the rate of reaction; this is treated in Chap.
3.

Table 1.I   Analytically Useful Examples of Reaction (1.1)

Sample	+	Reagent	$\rightleftharpoons$	Product A	+	Product B
R-OH		$(R'CO)_2O$		$R'COOR$		$R'COOH$
				N-OH		
R-CHO		$NH_3OH^+C\ell^-$		R-$\overset{\text{‖}}{C}$-H		$H_3O^+C\ell^-$
RCOOR'		OH^-		$R-COO^-$		$R'-OH$
Ar-OH		$Ar-N=N^+C\ell^-$		$Ar-N=N-Ar-OH$		$HC\ell$
$2RNH_3C\ell$		$Hg(OAc)_2$ [a]		$HgC\ell_2$		$2RNH_3OAc$
R-OH		CH_3MgI		CH_4		$R-OMgI$
R-O-R		2HI		2R-I		H_2O
R-CH=CH-R		Br_2		RCHBrCHBrR		--
R-SH		$AgNO_3$		R-SAg		HNO_3

[a] Ac represents the acetyl group, CH_3CO, so OAc is acetoxy
    and OAc$^-$ is acetate.

     Table 1.I lists a few reactions that have been devel-
oped into analytical methods. Now consider how Eq. (1.1)
might be utilized. An excess of reagent may be added to
the sample. After the reaction is "complete," the actual
quantitative finish is made, in this case by one of three
routes: (1) the unreacted reagent can be determined, and
this quantity subtracted from the total reagent added;
this gives the amount of sample; (2) the amount of product
A can be determined; (3) the amount of product B can be
determined. The determination of excess reagent, of A or
of B, may itself require further reactions. The selection
of the final analytical route will be determined by features specific to the analysis. Many examples appear in
later chapters.
     Many analytical methods employ titrimetric or spectro-
photometric finishes. The calculations required for a
titrimetric determination are readily grasped with the aid
of titration diagrams. As an example consider the determination of an ester by saponification (alkaline hydrolysis).

$$R-COOR' + OH^- \rightarrow R-COO^- + R'-OH \qquad (1.2)$$

It would appear that it is only necessary to add excess
standard alkali, heat the solution until the ester is completely hydrolyzed, and back-titrate with standard acid.
The difference between equivalents of alkali added and
equivalents of acid consumed in the back-titration is
equal to the equivalents of ester in the sample. This
procedure is adequate for many samples. It can be re-
fined, however, by including a blank determination. The
procedure just outlined assumes that alkali is consumed
only by the ester. This may not be a valid assumption,
especially if the alkali concentration is rather low or
if the reaction time is long, for alkali could be consumed
by the glass surface and by carbon dioxide absorbed from
the atmosphere. These types of error are minimized by
performing a blank analysis. Figure 1.1 shows a titration
diagram for the sample and blank determinations. The
lengths of the lines are proportional to milliequivalents
of reactants.

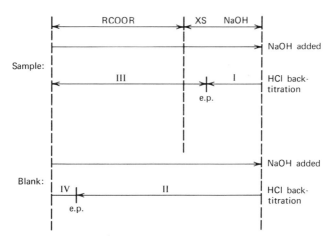

Fig. 1.1. Titration diagram for the saponification of an
ester. The lengths of the lines are proportional to milliequivalents of reactants (from Ref. [1], p. 454).

If no alkali were consumed by extraneous materials,
then milliequivalents of ester would equal III. This,
however, is not so. Instead we find, by inspection of the
diagram,

milliequivalents of ester = III - IV

which accounts for the consumption of alkali by substances
other than the ester [4]. In order to calculate III and

IV, the volumes and normalities of both the alkali and the
acid are required.
    An alternate, and simpler, calculation is available.
Figure 1.1 shows that

              milliequivalents of ester = II - I

The quantities II and I can be calculated from a knowledge
of the volumes and of the normality of the acid used in
the back-titrations; the normality of the alkali is unnecessary as long as the same volume of the same alkali
solution is used in the sample and in the blank determinations. (In effect, the blank determination is a standard-
ization of the alkali under the conditions of the assay.)
The approach illustrated by this example is widely used.
Another titration diagram is shown in Figure 13.7.
    Spectrophotometric finishes nearly always utilize the
"working curve" technique, in which several samples of an
authentic specimen, graded in amount or concentration, are
subjected to the same reaction conditions as the test sample. Usually a blank determination is also carried out,
with the sample compound being omitted but all other conditions being duplicated. The absorbances of the known
solutions, relative to the reagent blank, are plotted
against sample size or concentration to give a smooth,
usually straight, line (the working curve, or standard
curve), from which the unknown result can be obtained by
interpolation. Calculation of the original sample content
or purity then requires the appropriate dilution factors.
    This method possesses the weakness that it cannot
readily reveal systematic errors or deviations from the
anticipated stoichiometry, because the known and unknown
samples are treated alike, as far as is possible. Thus
if the reaction takes place to the extent of only 95% of
the theoretical, this discrepancy will occur for both the
standard and the test samples, and so it will be compen-
sated for. Often the working curve method is abbreviated
by using a single standard sample instead of a series
bracketing the unknown concentration; then deviations from
Beer's law or from quantitative reaction cannot be detected, although the analysis may be successful if the
concentrations of standard and unknown are similar. Although the inability of the standard curve method to detect nonquantitative reaction has been counted a weakness
here [5], this same feature can be a strength in that adequate analyses can be conducted using systems that do not
undergo quantitative reaction. In fact, the literature
contains examples of such assays in which less than 20%
of the sample is reacted; as long as the reaction conditions are duplicated for standard and unknown, reproduci-

ble results can be obtained (though with less than the
theoretical sensitivity). Unfortunately, when a reaction
is incomplete in an analytical system, the controlling
variables are seldom well understood, and so duplication
of the essential conditions may not be achieved. Poor
reproducibility or even failure of the analysis may result.
For this reason a worthy goal in the development of spectrophotometric finishes is reliance on previously established spectral properties of a product; in effect an authentic specimen of the product serves as the primary
standard for the sample determination. This method allows
study of the completeness of the functional group reaction.
Alternatively, a full understanding of the reasons for
incomplete reaction may give the analyst the necessary
control over reaction conditions to establish confidence
in the method.

These discussions of titrimetric and spectrophotometric finishes can be generalized to include most other
techniques. Many kinds of end-point detection are applicable in titrimetry, but the basic relationships are all
the same. Spectrophotometry is simply one example of a
technique that is based upon a relationship between a
solution property and solute concentration; in absorption
spectroscopy this relationship is Beer's law. Analogous
functions describe other useful techniques of measurement,
such as fluorimetry, polarimetry, and gas chromatographic
detection.

1.2   Plan of the Book

Descriptions of organic analytical methods have tradi-
tionally been organized around the function groups [2]--
hydroxy, amino, carbonyl, carboxyl, and so on. This is a
practical approach, because it starts with what the analyst already knows about his sample. Another way to
arrange a treatment of organic analysis is by methods--
absorption spectroscopy, acid-base titration, etc. [1,6].
Fritz and Hammond [7] showed how a consideration of the
fundamental chemistry, in particular of relative rates
and equilibria, might lead to the development of improved
methods. Schenk has extended their treatment to several
functional group determinations [8]. The present book is
a further and more systematic essay into the theory of
organic analytical methods. The organization is based on
type of reaction.

The broadest classification of reactions is into the
categories of heterolytic and homolytic reactions. In
homolytic (free radical) reactions, bond cleavage occurs
with one electron remaining with each atom, as in

$$R:X \xrightarrow{\text{homolytic}} R\cdot + \cdot X \qquad (1.3)$$

Free radical reactions are not discussed except as they appear incidentally to other processes. Heterolytic reactions occur with both electrons remaining with one of the atoms.

$$R:X \xrightarrow{\text{heterolytic}} R^+ + :X^- \qquad (1.4)$$

Heterolytic reactions are sometimes called ionic reactions.

A further classification is into these four categories:

1. Substitutions (displacements)
2. Additions
3. Eliminations
4. Rearrangements

Usually at least two reactants are present, and it is convenient to refer to one of these as the substrate and the other as the reagent. The distinction is arbitrary and conventional (see Sec. 4.2), and leads to a further classification in terms of reagent type. Reagents are nucleophiles (nucleus lovers) if they have an unshared electron pair and seek electron-deficient sites. Nucleophiles are either bases (in the Brønsted sense) or reducing agents; in a very general interpretation these classes are practically synonymous, for all such reagents function by donating electrons [9]. Electrophiles, which may be acids (including Lewis acids) or oxidizing agents, seek sites of high electron density.

Reactions are classified by specifying the class and the reagent type; thus a nucleophilic substitution (S_N) is a substitution reaction by a nucleophilic reagent, as in

$$I^- + C_2H_5\text{-}Br \longrightarrow C_2H_5\text{-}I + Br^- \qquad (1.5)$$

In this reaction iodide is the nucleophile. Since the substrate is aliphatic, the reaction is called an aliphatic nucleophilic substitution. An aromatic electrophilic substitution (S_E) is exemplified by Eq. (1.6), the nitronium ion being the electrophile.

Addition (Ad) reactions of multiple bonds also take place with electrophiles and nucleophiles; evidently if a reagent that is an electrophile adds to a double bond, the unsaturated bond must be functioning as a nucleophile. Eq. (1.7) is an Ad_E reaction.

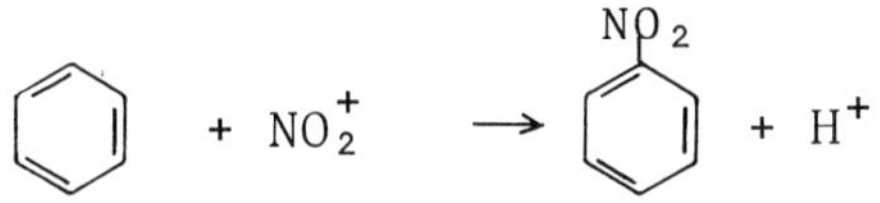

$$\text{⬡} + NO_2^+ \rightarrow \text{⬡-}NO_2 + H^+ \qquad (1.6)$$

$$HC\ell + CH_2=CH_2 \rightarrow CH_3CH_2C\ell \qquad (1.7)$$

[An ambiguity appears in that it is conceivable that the nucleophilic chloride ion initiated the attack, but much chemical experience supports the classification of (1.7) as an electrophilic reaction.] Eq. (1.8) shows an Ad_N reaction.

$$CN^- + CH_3\overset{O}{\overset{\|}{C}}CH_3 \rightarrow CH_3\overset{O^-}{\underset{CN}{\overset{|}{\underset{|}{C}}}}CH_3 \qquad (1.8)$$

Eliminations (E) often take the form of

$$R\text{-}CH_2\text{-}XY \rightarrow R\text{-}CH=X + HY \qquad (1.9)$$

X may be a heteroatom or a carbon atom, with valencies satisfied as appropriate, leading to carbonyl (for example) and olefin products, respectively. Rearrangements are reactions in which the positions of atoms in a molecule undergo change; keto-enol tautomerization is an example.

Multistep reactions include more than one of the types of reaction, as in the ester hydrolysis shown in

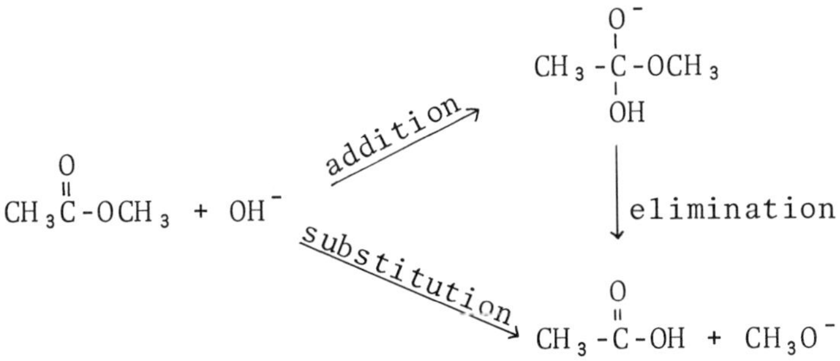

$$CH_3\overset{O}{\overset{\|}{C}}\text{-}OCH_3 + OH^- \xrightarrow{\text{addition}} CH_3\text{-}\overset{O^-}{\underset{OH}{\overset{|}{\underset{|}{C}}}}\text{-}OCH_3 \xrightarrow{\text{elimination}}$$

$$\xrightarrow{\text{substitution}} CH_3\text{-}\overset{O}{\overset{\|}{C}}\text{-}OH + CH_3O^- \qquad (1.10)$$

The overall reaction is a nucleophilic substitution of CH_3O^- (the leaving group) by OH^-. This occurs, however, via an initial nucleophilic addition to the carboxyl double bond, followed by an elimination. Such reactions

provide obstacles to tidy organization schemes, but they
obviously fit into the general classification.
    Chapters 2, 3, and 4 treat the physical chemistry
pertinent to the study of organic reactions, and Chaps. 5
and 6 apply these principles to some features--selectivity
and sensitivity--of particular analytical interest. The
chapters following treat many of the important classes of
heterolytic organic reactions from a mechanistic point of
view. In each of these chapters the basic organic chemistry of a reaction type is first presented, and then many
reactions of this type that are used in organic analysis
are introduced as examples, some of them being described
in detail.

                         Problems

1. Sketch a titration diagram, and show the method of
    calculation, for the conventional Kjeldahl technique
    of analyzing for nitrogen content, in which the suc-
    cessive steps are: (1) acid digestion of the sample
    to produce ammonium ion; (2) addition of excess alka-
    li and steam-distillation of the ammonia into excess
    acid; (3) back-titration of acid with standard alkali.

2. Sketch a titration diagram, and show the method of
    calculation, for the modified Kjeldahl analysis in
    which the ammonia produced is determined directly in
    the digestion mixture by redox titration, the succes-
    sive steps being as follows: (1) acid digestion of
    the sample to produce ammonium ion; (2) addition of
    excess sodium bicarbonate; (3) titration of ammonia
    with standard hypochlorite until a slight excess of
    hypochlorite is present; (4) addition of excess of
    standard arsenic trioxide to reduce the unreacted hy-
    pochlorite; (5) titration of the unreacted arsenic
    trioxide with hypochlorite to a final endpoint.

    Answer: See Ref. [1], pp. 395-396.

3. Classify the analytical reactions in Table 1.I with
    respect to reagent type and reaction class.

4. (a) Derive the equation $c_s = c_r(A_s/A_r)$, where c_s and
    c_r are concentrations of an absorbing species in sam-
    ple and reference standard solutions, respectively,
    and A_s, A_r are the corresponding absorbances; assume
    Beer's law is obeyed.
    (b) Show that this equation holds even when Beer's
    law is not obeyed, in the special case $A_s = A_r$.

References

1. See, for example, K. A. Connors, A Textbook of Pharma-
    ceutical Analysis, Wiley, New York, 1967, pp. 413-497.
2. S. Siggia, Quantitative Organic Analysis via Function-
    al Groups, 3rd ed., Wiley, New York, 1963; N. D.
    Cheronis and T. S. Ma, Organic Functional Group Ana-
    lysis by Micro and Semimicro Methods, Wiley (Inter-
    science), New York, 1964; F. E. Critchfield, Organic
    Functional Group Analysis, Macmillan, New York, 1963;
    Treatise on Analytical Chemistry (I. M. Kolthoff and
    P. J. Elving, eds.) Part II, Vol. 13, Wiley (Inter-
    science), New York, 1966; Comprehensive Analytical
    Chemistry (C. L. Wilson and D. W. Wilson, eds.) Vol.
    1B, Elsevier, Amsterdam, 1960, Chap. VIII; Instrumen-
    tal Methods of Organic Functional Group Analysis (S.
    Siggia, ed.), Wiley (Interscience), New York, 1972.
3. Organic Analysis, Vol. 1 (1953), Vol. 2 (1954), Vol.
    3 (1956), Vol. 4 (1960), Wiley (Interscience), New
    York.
4. The assumption that the blank exactly matches the
    sample in this regard is difficult to sustain, but if
    the correction is small, a moderate error can be
    tolerated.
5. Another weakness is that it, in effect, employs as a
    "primary standard" a specimen of the same compound
    that is to be determined.
6. S. Siggia and H. J. Stolten, An Introduction to Mod-
    ern Organic Analysis, Wiley (Interscience), New York,
    1956.
7. J. S. Fritz and G. S. Hammond, Quantitative Organic
    Analysis, Wiley, New York, 1957, Chaps. 2 and 9.
8. G. H. Schenk, Organic Functional Group Analysis:
    Theory and Development, Pergamon, Oxford, 1968.
9. C. K. Ingold, Structure and Mechanism in Organic Chem-
    istry, Cornell Univ. Press, Ithaca, 1953, Chap. V.
    Basicity generally refers to equilibrium affinity,
    usually for protons, whereas nucleophilicity refers
    to kinetic reactivity, usually toward carbon. See
    also Sec. 4.3.

Chapter 2.   CHEMICAL EQUILIBRIA

The macroscopic behavior of chemical systems is usefully
described by thermodynamics. From the analytical point
of view, dilute solutions of reacting species are of special interest, and this chapter reviews some important
properties of chemical solution equilibria. A general
familiarity with thermodynamics is assumed. Some topics
are treated because of their overwhelming importance, even
though they are very familiar. Other subjects, such as
unitary thermodynamic quantities, are discussed because
of their relative neglect. Further treatments can be
pursued in the many monographs on thermodynamics [1].

2.1   Equilibrium

Let G represent the Gibbs free energy of a homogeneous
system containing n_1, n_2,..., n_i moles each of i components, at temperature T and pressure P. Then Eq. (2.1)
defines the partial molar free energy $\overline{G}_i$ of substance i,
or, synonymously, its chemical potential μ_i.

$$\left(\frac{\partial G}{\partial n_i}\right)_{T,P,n_j} = \overline{G}_i = \mu_i \qquad (2.1)$$

The chemical potential is related to the activity a_i by

$$\mu_i = \mu_i^0 + RT \ln a_i \qquad (2.2)$$

in which μ_i^0 is the chemical potential of substance i in
an arbitrarily selected standard state.
    For the generalized process

$$aA + bB + \cdots \rightleftharpoons mM + nN + \cdots$$

the function given in Eq. (2.3) is a constant (at constant
temperature and pressure) when the system is at equilibrium.

10

$$\frac{a_M^m \, a_N^n \, \ldots}{a_A^a \, a_B^b \, \ldots} = K \qquad (2.3)$$

K is the equilibrium constant, which is a function of
temperature and pressure. An important result of thermodynamics is that the standard free energy change for the
process is related to K by

$$\Delta \overline{G}^0 = -RT \, \ln K \qquad (2.4)$$

$\Delta \overline{G}^0$ is the change in free energy, per mole, accompanying
the conversion of reactants into products when all substances are in their standard states. For reactions in
solution it is correct to consider that this free energy
change is really a change in partial molar free energies.
It is conventional, however, to omit the bar over the
symbol.
      The van't Hoff equation, Eq. (2.5), describes the
variation of the equilibrium constant with temperature.

$$\frac{d \ln K}{dT} = \frac{\Delta H^0}{RT^2} \qquad (2.5)$$

Constant pressure is understood. The quantity ΔH^0 is the
standard enthalpy change for the process. In the special
case that ΔH^0 is not a function of temperature, integration of Eq. (2.5) leads to

$$\ln K = -\frac{\Delta H^0}{RT} + \text{constant} \qquad (2.6)$$

or, by carrying out the integration between the temperatures T_1 and T_2,

$$\ln \frac{K_2}{K_1} = \frac{\Delta H^0}{R} \left(\frac{T_2 - T_1}{T_1 T_2} \right) \qquad (2.7)$$

      It is remarkable that the simple equation (2.6) ap-
pears to describe the temperature effect for many reactions, but this is mainly because the temperature range
studied is usually fairly limited. In the general case
it must be recognized that ΔH^0 is itself a function of
temperature, and this dependence must be taken into account. This can be done in terms of heat capacities, or
rather the change in heat capacities, ΔC_p^0. Heat capacities are also temperature dependent, and empirical relationships can be established between experimental heat
capacity data and temperature. The observation that ΔH^0
is not a function of temperature [applicability of

Eq. (2.6)] is equivalent to stating that $\Delta C_p^0 = 0$ over
the temperature range studied. If this condition does
not describe the system, it is usual to postulate that
ΔC_p^0 is a nonzero but constant quantity, independent of
temperature. These two levels of approximation appear to
account for most of the systems of interest in solution
chemistry. The dissociation equilibria of weak electrolytes do not, in general, yield ΔC_p^0 values of zero. The
temperature variation curves of dissociation constants
for these equilibria often pass through a maximum. All
experimental data can be fitted with the empirical assumption that the enthalpy change is a polynomial in temperature,

$$\Delta H^0 = A - BT - CT^2 \qquad (2.8)$$

with no higher terms being required. This expression
leads to a rather intractable relationship between ΔG^0 and
temperature, so it has been simplified by assuming that
either B or C is zero. Both of the simpler forms also
yield descriptions that are within experimental error, and
the usual practice is to assume B = 0 [2]. Then

$$\Delta H^0 = A - CT^2 \qquad (2.9)$$

Combining this with Eq. (2.5),

$$\Delta G^0 = -RT \ln K = A - DT + CT^2 \qquad (2.10)$$

These equations are used by fitting the data to a least-
squares equation of the form

$$\log K = -\frac{A'}{T} - C'T + D' \qquad (2.11)$$

where $A = 2.303RA'$, $C = 2.303RC'$, and $D = 2.303RD'$.

2.2   Standard States and Reference States

Conventional Standard States. Several concentration
scales are in common use in solution chemistry, and the
selection of a concentration scale can have an important
bearing on the magnitudes of quantities determined experimentally. The familiar concentration scales are the mole
fraction, molality, molarity, formality, and normality
scales. The mole ratio y_i of substance i is defined by

$$y_i = \frac{n_i}{n_1} \qquad (2.12)$$

where n_1 is the number of moles of solvent. In very dilute solution the mole ratio is very nearly equal to the mole fraction.

Expressions can be derived relating the concentration of a solute in one scale to that in another [3]. Some of these are employed in the later treatment. The relationship between the mole fraction x and molarity c of a constituent is given by

$$x_i = \frac{c_i}{1000\rho} \cdot \frac{\Sigma n_r M_r}{\Sigma n_r} \qquad (2.13)$$

where ρ is the density of the solution, Σn_r is the sum of the number of moles, and M_r is the molecular weight of constituent r; thus $\Sigma n_r M_r$ is the total mass of the solution. Equation (2.14) relates the mole fraction and molality m of a constituent.

$$x_i = \frac{m_i}{1000} \cdot \frac{n_1 M_1}{\Sigma n_r} \qquad (2.14)$$

The thermodynamic quantity most accessible in analytical systems is the equilibrium constant, which is defined in terms of equilibrium activities of reactants and products. These activities are related to chemical potentials by Eq. (2.2), which can be written in the form $\ln a = (\mu - \mu^0)/RT$, showing that the activity is not fully defined until the constant μ^0 is specified. μ^0 can be interpreted as the chemical potential of the substance when its activity is unity. There is some arbitrariness in specifying the state of the system that shall have unit activity, and certain states have been found to be particularly convenient for this purpose. We use the term standard state to mean that state of the system in which the activity of the substance under consideration is unity. The reference state is that state of the system in which the ratio of the activity to the concentration of the substance is unity. This ratio is called the activity coefficient, γ.

The following standard states are widely employed [4]. For a liquid solvent the usual standard state is the pure liquid at 1 atm pressure, at any temperature. That is, $a_1 = x_1 = 1$ in the pure liquid. It follows that the reference state and standard state are identical, since in the pure liquid $\gamma_x = a_1/x_1 = 1$. In an ideal solution a_1 would be equal to x_1 at all values of x_1; in real solutions departures from this equality are to be expected, and these are reflected in deviations of γ_x from unity.

The standard states for solutes depend upon the type of solution. For liquid solutes, especially those totally miscible with the solvent and those used over a wide range in composition, the standard state is selected to be (as with the solvent) the pure liquid. Again, the reference and standard states are identical. Clearly no distinction is made between solvent and solute for these kinds of solutions.

When low solute concentrations are used, a different reference state is usually chosen. This reference state is the infinitely dilute solution, and the standard state then depends upon the manner in which the solute concentration is specified. If the solute concentration is given in mole fraction, the reference state is defined to be the infinitely dilute solution, so that $a_2/x_2 \to 1$ as $x_2 \to 0$. (Compare this reference state with that in the preceding paragraph.) This fixes the standard state, which is hypothetical, for it is the state in which $x_2 = 1$, but some of the solute properties (e.g., partial molar enthalpy [4]) are those of the reference state. Since the standard state is hypothetical, it is common to define it, as here, in terms of the reference state.

If solute concentration is given in molality, the reference state is the infinitely dilute solution with the activity coefficient defined so that $\gamma_m = a_2/m_2 \to 1$ as $m_2 \to 0$. The standard state is a hypothetical 1 m solution with some of the properties of the infinitely dilute solution. For solute concentrations in molarity, the reference state is again the infinitely dilute solution with the activity coefficient defined by $\gamma_c = a_2/c_2 \to 1$ as $c_2 \to 0$. Again the standard state is hypothetical; it is a 1 M solution with some properties of the reference state.

The activity coefficients γ_x, γ_m, and γ_c are different. Glasstone [4] gives relationships connecting them.

It is not always clear to users of these concepts exactly how they are exercising their choice of standard state. Because of the conventions described, the standard state is, in effect, determined when the experimentalist chooses his concentration scale. A common example illustrates the use of the reference state definition. For the dissociation of the weak acid BH^+ in aqueous solution either of the equations

$$BH^+ \rightleftharpoons B + H^+$$

$$BH^+ + H_2O \rightleftharpoons B + H_3O^+$$

may be written. According to the first of these, the equilibrium constant is defined

$$K_a = \frac{a_B \; a_{H^+}}{a_{BH^+}}$$

One way to determine K_a is to evaluate the mixed constant
[5] K_a' as a function of solute concentration, and to ex-
trapolate K_a' to zero solute concentration. This extrapolation in effect defines the reference state to be the
infinitely dilute solution, and gives K_a as the extrapolated value of K_a'.

Note that K_a could have been defined from the second
of the formulations as

$$K_a = \frac{a_B \; a_{H_3O^+}}{a_{BH^+} \; a_{H_2O}}$$

but the same result would be obtained from the extrapolation, because the infinitely dilute solution is also the
standard state for the solvent ($a_{H_2O} = 1$).

Often the reference state is selected to be the infinitely dilute solution (with respect to the solute) in an
experimental solvent that is itself a more or less complex solution. This solvent might be a binary mixture,
or it might contain buffer components, or its ionic
strength might be maintained at some constant value other
than zero, etc. Such reference states are selected because it may not be possible to achieve a rigorous reference state applicable to all members of a set of comparative measurements, or at least not without great effort
[6].

Standard State Transformations. Since activity depends
upon the choice of standard state, and the thermodynamic
equilibrium constant is defined in terms of activities of
reactants (i.e., the participants on the left-hand side
of the equation) and products (those on the right-hand
side), the numerical value of the equilibrium constant
may depend upon the standard states used. But the standard states are defined, according to usual practice, when
the concentration scales are specified. This means that
an equilibrium constant may have different values when
calculated on the basis of different concentration scales,
and it is helpful to be able to convert an equilibrium
constant directly from one concentration scale to another.
This becomes easy to do if we restrict ourselves to extremely dilute solutions, which is often practicable.

Equilibrium constants evaluated experimentally are
often referred to as "concentration constants" to distinguish them from the thermodynamic constant defined in
terms of activities. The concentration constants on the

several concentration scales are symbolized K_x, K_y, K_m, and K_c, whereas K represents the thermodynamic constant on the molality scale. At extreme dilution, K_y becomes essentially equal to K_x and K_m becomes equal to K. We will find a relationship between K_y and K_m.

The molality m_i and mole ratio y_i of i can be related through the definition of mole ratio, $y_i = n_i/n_1$. Let M* be the number of moles of solvent contained in 1000 g of solvent. Then

$$M* = \frac{1000}{\text{molecular weight of solvent}} = \frac{1000}{M_1} \qquad (2.15)$$

M* can be substituted for n_1, and then n_i becomes equal to m_i by definition, giving

$$m_i = M*y_i \qquad (2.16)$$

The constants K_m and K_y are defined generally by

$$K_m = \frac{\overset{q+\Delta q}{\Pi} m_p}{\overset{q}{\Pi} m_r} \qquad (2.17)$$

$$K_y = \frac{\overset{q+\Delta q}{\Pi} y_p}{\overset{q}{\Pi} y_r} \qquad (2.18)$$

for the general reaction in which q particles of reactants (r) go to q + Δq particles of products (p). Taking logarithms of both equations, substituting from Eq. (2.16) for the m terms, and rearranging leads to [7]

$$\frac{K_m}{K_y} = (M*)^{\Delta q} \qquad (2.19)$$

In extreme dilution this can be written

$$\frac{K}{K_x} = (M*)^{\Delta q} \qquad (2.20)$$

Δq is determined by the definition of the equilibrium constant, and not by the way the reaction happens to have been written.

The relationship between K_c and K_y can be similarly

derived. Equations (2.13) and (2.14) give Eq. (2.21), relating molality and molarity in very dilute solution [9],

$$m_i = \frac{c_i}{\rho_0} \qquad (2.21)$$

where ρ_0 is the density of the solvent. Carrying through the earlier development:

$$\frac{K_c}{K_y} = (M^* \rho_0)^{\Delta q} \qquad (2.22)$$

These equations show that if $\Delta q = 0$ (as in isomerization reactions and liquid-liquid distribution), the equilibrium constant has the same value no matter which standard state is selected; more generally, however, Δq is not zero, and then the magnitude of the equilibrium constant [and therefore the standard free energy change--see Eq. (2.4)] depends upon the choice of standard state.

Unitary Quantities. We must now face the problem of the meaning of thermodynamic quantities whose values can be altered by changing the concentration scale upon which they are evaluated. What is the relationship of these quantities to the process under study, and which of the numerical values for a given property is the one pertinent to an interpretation of the process? Gurney has made the closest analysis of this problem [8]. Using a statistical mechanical argument, he has shown that Eq. (2.23) describes an ideal dilute system containing n_B solute particles (molecules) and n_S solvent particles.

$$\frac{\partial G}{\partial n_B} = \text{constant} + kT \ln x_B \qquad (2.23)$$

That is, the free energy change for the introduction of a few B molecules into the dilute solution has two components: one of these is dependent upon the composition of the solution, and one is independent of the composition. Gurney has called the composition-dependent portion the cratic term [9] and the composition-independent portion the unitary term. The cratic term evidently depends only upon the numbers of molecules, whereas the unitary term depends upon the kinds of molecules. It is the unitary free energy change that is characteristic of the specific process, and that is therefore of chemical interest. Letting U represent the unitary portion of the free energy change per molecule, Eq. (2.23) becomes

$$\frac{\partial G}{\partial n_B} = U + \underline{k}T \; \ell n \; x_B \qquad (2.24)$$

The condition for equilibrium is that the change in free energy for the system be zero, or $\partial G/\partial n_B = 0$; therefore

$$U = -\underline{k}T \; \ell n \; x_B \qquad (2.25)$$

On a molar basis we may write ΔG_{un} for the unitary free energy change

$$\Delta G_{un} = -RT \; \ell n \; x_B \qquad (2.26)$$

The unitary free energy change at equilibrium is equal to the cratic free energy change ΔG_{cr}, taken with opposite sign; $\Delta G = \Delta G_{un} + \Delta G_{cr}$. Differentiation of free energy with respect to temperature gives entropy, so we find also $\Delta S = \Delta S_{un} + \Delta S_{cr}$, with the unitary entropy change being independent of composition. The situation is different in the enthalpy term. Consider a reversible process in a solution of given composition; then ΔG_{cr} will be equal to RT times a term governed only by mole fractions, which under this condition are constant; thus

$$\Delta G_{cr} = constant \times T \qquad (2.27)$$

Differentiating with respect to T,

$$\frac{d\Delta G_{cr}}{dT} = constant$$

or

$$T \; \frac{d\Delta G_{cr}}{dT} = constant \times T = \Delta G_{cr}$$

so

$$\Delta G_{cr} - T \; \frac{d\Delta G_{cr}}{dT} = 0 \qquad (2.28)$$

But $\Delta H = \Delta G - T(d\Delta G/dT)$. Comparison of this with Eq. (2.28) shows that the cratic portion of the free energy change makes no contribution to the enthalpy change; thus ΔH is a unitary quantity.

Now let us obtain the unitary free energy change for the typical equilibrium

$$A + B \rightleftharpoons AB \qquad (2.29)$$

When one A particle is transferred from the left to the right side of the equation, n_A decreases by one to $(n_A - 1)$, n_B goes to $(n_B - 1)$, and n_{AB} goes to $(n_{AB} + 1)$. The corresponding changes in the cratic term are $-kT \ln x_A$, $-kT \ln x_B$, and $+kT \ln x_{AB}$. Therefore the cratic contribution to the free energy change at equilibrium, per mole, is

$$RT \ln \frac{x_{AB}}{x_A x_B}$$

so the unitary free energy change per mole is the negative of this,

$$\Delta G^0_{un} = -RT \ln \frac{x_{AB}}{x_A x_B} = -RT \ln K_x \qquad (2.30)$$

and we see that we have obtained the familiar equation relating standard free energy change to the equilibrium constant on the mole fraction scale.

Since most equilibrium constants for solution equilibria have been determined on the molar or molal scales, it is convenient to obtain expressions for the direct calculation of standard unitary free energy changes from these conventional equilibrium constants. With the stipulation of extremely dilute solutions, this is done by combining Eq. (2.30) with either (2.20) or (2.22). For example, for the molar concentration scale,

$$\Delta G^0_{un} = -RT \ln K_c + \Delta q \, RT \ln (M^* \rho_0) \qquad (2.31)$$

For Eq. (2.29), $\Delta q = -1$. If the reaction is carried out in (for example) dilute aqueous solution at 25°C, Eq. (2.31) becomes

$$
\begin{aligned}
\Delta G^0_{un} &= -(2.303)(1.987)(298.16)\log K_c \\
&\quad -(4.576)(298.16) \times \log 55.5 \\
&= -1364.4 \log K_c - 2379.5 \text{ (cal/mole)}
\end{aligned}
$$

From the relationship $\Delta G = \Delta H - T\Delta S$ and the knowledge that ΔH does not have a cratic portion, the equation $\Delta S_{un} = \Delta S + (\Delta G - \Delta G_{un})/T$ is obtained, where ΔS and ΔG refer to the conventional (molar or molal) scales. This is combined with (2.31) for the molar concentration scale to give

$$\Delta S_{un} = \Delta S - \Delta q \, R \ln (M^* \rho_0) \qquad (2.32)$$

Thus, in the preceding example (dilute aqueous solution, 25°C, $\Delta q = -1$), Eq. (2.32) becomes (in entropy units) [10],

$$\Delta S_{un} = \Delta S + 7.98 \tag{2.33}$$

In real solutions, and when the solute concentration is not restricted to the extremely dilute range, the free energy change will be a function of composition and of the sizes and shapes of the solute and solvent molecules. Eq. (2.34), which may be compared with Eq. (2.23), then applies.

$$\frac{\partial G}{\partial n_B} = \text{constant} + \underline{k}T \ln f_B + \underline{k}T \ln x_B$$

$$= U + \underline{k}T \ln f_B x_B \tag{2.34}$$

The quantity f_B evidently has the character of an activity coefficient.

These considerations become relevant in interpreting equilibrium constants measured at high concentrations, because in this situation the several concentration scales are not proportional to each other. Equation (2.13) can be transformed to [11]

$$x_i = \frac{c_i}{1000} \frac{\Sigma n_r V_r}{\Sigma n_r} \tag{2.35}$$

For any constituent i in a solution containing high concentrations of solute B in solvent S, this becomes

$$x_i = \frac{c_i}{1000} [V_S + x_B(V_B - V_S)] \tag{2.36}$$

Thus, for Reaction (2.29), K_x and K_c are related by

$$K_x = \frac{1000 \, K_c}{V_S + x_B(V_B - V_S)} \tag{2.37}$$

Except for the special case when $V_B = V_S$, Eq. (2.37) shows that K_x and K_c cannot both be constant quotients over any appreciable range of B concentration. The preceding argument demonstrated that K_x will vary with composition. For certain systems of hydrogen-bonded molecular complexes, it has been experimentally demonstrated that K_c is essentially constant over a considerable range of the concentration of B when $V_S \neq V_B$, and that K_x depends upon the

concentration of B [12]. The molar concentration scale
is therefore experimentally convenient for these systems
[13]; but it does not necessarily follow that K_c is the
superior measure of the reaction free energy change [14].

2.3 Determination of Thermodynamic Quantities

Classical titrimetric and gravimetric methods of analysis
are firmly based upon the principles of chemical equilibrium. Acid-base titrations, complexometric titrations,
precipitation titrations and gravimetric analyses, and
oxidation-reduction titrations are the fundamental methods.

Table 2.I   Some Equilibria of Analytical Importance

Process or property	Equilibrium measure	Equilibration time
Acid-base reactions	Dissociation, ioni-zation, salt forma-tion constants	< μsec
Solubility of electrolytes	Solubility product	h-days
Solubility of nonelectrolytes	Solubility	h-days
Metal-ligand complex formation	Stability (or instability) constants	< msec
Inorganic redox reactions	Standard potential	< μsec
Molecular complex formation	Stability constant	< msec
Liquid-liquid distribution	Partition coefficient	sec-h
Covalent bond changes	Equilibrium constant	μsec-years

The analytical behavior of these systems can be quantitatively accounted for with a knowledge of the relevant
equilibrium constants and activity coefficients. (For
redox reactions, the standard potential is customarily
employed instead of the standard free energy change, to
which it is proportional.) Extensive discussions are
available [15]. Equilibrium constants, especially acid-
base dissociation constants, can be found in several

important collections [16]. Often, however, the analyst
will find it necessary to determine equilibrium constants
and associated thermodynamic properties.

Measurement of Equilibrium Constants. Table 2.I lists
types of equilibria that are of analytical interest.
Proton-transfers (acid-base reactions and tautomerizations)
occur in both aqueous and nonaqueous media, and the study
of acid-base strength is primarily based upon measurements
of the equilibrium constant for reaction with a reference
acid or base, usually the solvent. In aqueous systems
the relevant quantities are the acid dissociation constant
K_a and the base dissociation constant K_b. Relative acid-
base strengths are commonly discussed in terms of the
corresponding quantities pK_a = -log K_a and pK_b = -log K_b,
since pK values are directly proportional to standard free
energy changes. K_a is determined on the molarity or molality scales, and K_b for the conjugate base is calculated
from the relationship $K_w = K_a K_b$, which holds for a conjugate acid-base pair related by the equilibrium

$$\text{Acid} \rightleftharpoons \text{base} + H^+$$

In solvents of intermediate or low dielectric constant, interpretation of acid-base strength is more ambig-
uous. This is because ionization (production of ions) and
dissociation (separation of species) become recognizably
different processes in solvents with low dielectric
constant.

$$AB \; \underset{\longleftarrow}{\xrightarrow{\text{ionization}}} \; A^+B^- \; \underset{\longleftarrow}{\xrightarrow{\text{dissociation}}} \; A^+ + B^-$$

That is, the ion-pair A^+B^- can be detected in these solutions. For example, a base B will react in glacial acetic
acid according to

$$B + HOAc \rightleftharpoons BH^+OAc^- \rightleftharpoons BH^+ + OAc^- \qquad (2.38)$$

for which the ionization constant K_i and dissociation
constant K_d can be defined.

$$K_i = \frac{a_{BH^+OAc^-}}{a_B}$$

$$K_d = \frac{a_{BH^+} \, a_{OAc^-}}{a_{BH^+OAc^-}}$$

An overall dissociation constant K_B may be defined by
$K_B = a_{BH^+} \, a_{OAc^-}/(a_B + a_{BH^+OAc^-})$. The constants K_i, K_d, K_B

provide three different measures of base strength relative to the reference acid HOAc. Sometimes another acid is added to the system to function as a reference; this may be an indicator acid, selected to expedite the equilibrium constant measurement, or it may be a strong acid (such as perchloric acid), selected so that a wide range of bases may be measured. In either case a salt formation constant K_f^{BHA} can be defined,

$$K_f^{BHA} = \frac{a_{BHA}}{a_B \; a_{HA}}$$

where the activities represent the sums of ionized and un-ionized species. Acid-base equilibria in nonaqueous systems provide a major area for analytical research [17].

The solubility product of a slightly soluble electrolyte B_mA_n is defined by

$$K_{sp} = \frac{a_B^m \; a_A^n}{a_{B_mA_n}}$$

The standard state of B_mA_n is taken to be the crystalline solid, so $a_{B_mA_n}$ is equal to unity. Solubility products can be determined from direct analyses of solution concentrations, though these may be difficult for very slightly soluble substances. Radiochemical techniques are advantageous for analyzing such very dilute solutions. Electrochemical measurements of suitable cells can yield solubility products. The dissolution of a nonelectrolyte is an equilibrium between the solute in the crystal environment and in the solution. The solubility has the character of an equilibrium constant.

Metal ion coordination compounds are formed in facile equilibria described by stability constants, which are equilibrium constants written for the formation of the complexes. Since the central cations generally have coordination numbers greater than one, multiple complexes are formed in stepwise fashion (unless a multidentate ligand can satisfy all of the binding capabilities of the cation in a 1:1 stoichiometry, as with ethylenediamine-tetracetic acid). Proton acids can be viewed as special cases of these complexes. Rossotti and Rossotti [18] have reviewed methods for stability constant measurements.

Oxidation-reduction reactions can be considered as electron transfers. Their study is facilitated by electrochemical techniques (potentiometry and polarographic methods), and their equilibrium properties are usually presented in terms of standard electrode potentials.

Inorganic systems have much analytical importance. Organic electron-transfer processes are particularly studied
for their biochemical interest [19].
    Partition coefficients are equilibrium constants de-
scribing this process:

        Solute in phase I ⇌ Solute in phase II

This general equation includes many analytical processes.
When both phases are liquids, the techniques of solvent
extraction, countercurrent distribution, and the several
forms of partition chromatography are generated by this
equilibrium. If one phase is a gas and the other is a
liquid, as in gas-liquid chromatography, the partition
coefficient is essentially a Henry's law constant. When
one phase is a solid surface, as in adsorption chromatography, it is common to speak of a distribution coefficient.
    Reactions in which covalent bonds are made or broken
include most of those that interest the organic analyst.
Not all of these are reversible, and relatively few equilibrium constants have been measured.
    This is not the place for a detailed treatment of
experimental methods for the measurement of all of these
kinds of equilibrium constants. Some generally applicable
comments may be useful. The third column of Table 2.I
gives a very rough guide to the time required to reach
the equilibrium state. These quantities, which should
not be taken in a quantitative sense, may help the experimentalist. The great range specified for processes involving two phases reflects the many variables upon which
the rates depend; these include temperature (as with all
equilibria), mode and vigor of agitation, and area of
contact between the phases. The classification of covalent changes is too broad for useful generalizations.
Thermodynamics is concerned with differences between the
initial and final states of a system, and not with the
speed of attainment of the equilibrium. This velocity
may be markedly increased by substances (catalysts) that
do not alter the position of equilibrium. Many examples
are shown in later chapters.
    Equilibrium constants (perhaps it is better to speak
of equilibrium quotients when concentrations rather than
activities are concerned) can be estimated by two general
approaches: (1) direct determination of the concentrations of all species, or of some species combined with
restricting relationships (stoichiometry, mass balance);
(2) measurement of physical or chemical properties that
differ for reactants and products. Sometimes the two
approaches are combined. The measurement of the equili-

brium quotient for ethyl acetate formation in aqueous
solution illustrates the direct method [20]; the initial
concentrations of ethanol and acetic acid were known, and
the concentration of ethyl acetate at equilibrium was
measured by the ferric hydroxamate method. The reaction
was carried out at several concentrations of hydrochloric
acid, and the equilibrium quotients were extrapolated to
zero ionic strength and HCl concentration. The second,
indirect method is exemplified by spectrophotometric
determination of acid dissociation constants.

According to the definition of activity as a ratio of
fugacities, activity is a dimensionless quantity, and so
the thermodynamic equilibrium constant must be dimensionless. Consistency then requires that the activity coefficient have the dimension of reciprocal concentration.
Equilibrium quotients evaluated in terms of concentrations
may have dimensions depending upon the particular concentration scale and type of reaction. For the reaction
$A \rightleftharpoons B$, with $\Delta q = 0$, the equilibrium quotient is dimensionless if the same concentration scale is used for both A
and B; for the equilibrium $A + B \rightleftharpoons C$, the equilibrium
quotient on the molar concentration scale has the units
M^{-1}. On the mole fraction scale any equilibrium quotient
is dimensionless because the mole fraction is dimensionless, but some authors assign units [e.g., (mole fraction)$^{-1}$ for the preceding example] to indicate the concentration scale that was used.

Many workers prefer to consider the activity coefficient a dimensionless quantity, thus forcing the activity
to have the dimensions and units of the concentration.
In this view even the thermodynamic equilibrium constant
may have units. It is often required to take the logarithm of an equilibrium constant or of other thermodynamic
quantities. One can take the logarithm of a pure number,
but not of a dimension. This dilemma can be resolved by
adopting the view that activities are dimensionless, or
in general, that the logarithm of a ratio is always taken
with the denominator having the value unity. Another
resolution of the difficulty is adopted by Guggenheim and
Prue [21], who treat the dimensions in an algebraic sense,
and therefore can always generate dimensionless measures.
Thus we may write $K = 15 \ M^{-1}$, which is equivalent to
$K/M^{-1} = 15$, and then we can take the logarithm of K/M^{-1},
for this is a pure number. This notation is convenient
for table headings and graph axes, though it is cumber-
some in equations.

Free Energy Changes. The term "free energy of reaction"
is often used synonymously with "free energy change"; of
course, only the change in the free energy can be meas-

ured. A goal of much thermodynamic research, experimental
and theoretical, is the estimation of standard free energy
changes for chemical reactions; then with the relationship
ΔG^0 = -RT $\ln$ K, the equilibrium constant can be calculated. For reactions in the gas phase considerable progress
has been made. Reactions in solution, however, are less
successfully treated by these methods, and it is easier
and more accurate to measure equilibrium constants than
to calculate them. The equilibrium constant is the basic
experimental quantity in studies of solution equilibria.
Since the numerical value of the equilibrium constant,
and therefore of the standard free energy change, may depend upon the concentration scale used in its evaluation
(see Sec. 2.2), it is essential to provide this information when reporting K or ΔG^0 values.

The standard free energy of formation of a compound,
ΔG_f^0, is the free energy change upon formation of one mole
of the compound in its standard state from its constituent
elements in their standard states. It is this quantity
that is tabulated in compilations of thermodynamic data
for organic compounds [22]. Effective empirical methods
have been developed to estimate thermodynamic quantities.
These are based upon correlations between increments in
thermodynamic functions and structural modifications.
For example, the increment in ΔG_f^0 (at 25°) for the CH_3,
CH_2, and CH groups are found from such analyses to be
-4.14, +2.05, and +7.46 kcal/mole, respectively, in hydrocarbons. This "method of group contributions" is based
upon the additivity of these group increments. (When the
additivity fails, the usual practice is to redefine the
group.) Group contributions have been tabulated for enthalpies, entropies, and heat capacities as well as for
free energies.

Whether ΔG_f^0 values are obtained experimentally or are
estimated empirically or theoretically, their availability
provides more than a measure of the free energy change for
the formation of a compound from its elements. The standard free energy of formation of any element is obviously
zero; therefore the standard free energy change for a
chemical reaction is given by

$$\Delta G^0 = \sum_{\text{prod.}} \Delta G_f^0 - \sum_{\text{react.}} \Delta G_f^0 \qquad (2.39)$$

For example, for the addition of ammonia to propene to
give n-propylamine, with all substances in the ideal gas
state,

$$CH_3CH=CH_2(g) + NH_3(g) \rightleftharpoons CH_3CH_2CH_2NH_2(g)$$

the standard free energy change is

$$\Delta G^0 = \Delta G_f^0 \text{ (propylamine)} - \Delta G_f^0 \text{ (ammonia)} - \Delta G_f^0 \text{ (propene)}$$

The standard free energies of formation, at 25^0, are found in Stull, Westrum, and Sinke [23].

$$\Delta G^0 = 9.51 - (-3.86) - (14.99) \text{ kcal/mole} = -1.62 \text{ kcal/mole}$$

For reactions in solution, it is necessary to convert ΔG_f^0 from the usual tabulated standard state (the gas at 1 atm pressure, or the pure liquid) to an appropriate solution standard state (such as the 1 $\underline{M}$ solution). Edsall and Wyman [24] discuss this transformation, and give standard free energies of formation in aqueous solution for some important biochemicals.

Free energy changes can also be obtained by "adding" reactions, and their corresponding free energy changes, such that the desired reaction is obtained. Jencks et al. [25], measured the equilibrium constants for the isomerization of mercaptopropyl acetate, Eq. (2.40), and for its formation, Eq. (2.41).

$$CH_3COS(CH_2)_3OH \rightleftharpoons CH_3COO(CH_2)_3SH \qquad (2.40)$$

$$CH_3COOH + HO(CH_2)_3SH \rightleftharpoons CH_3COO(CH_2)_3SH + H_2O \quad (2.41)$$

Table 2.II   Equilibrium Constants and Standard Free Energy Changes for Ester Reactions[a]

Reaction[b]	K	ΔG^0(kcal/mole)
$CH_3COSR\text{-}OH \rightleftharpoons CH_3COOR\text{-}SH$	56	-2.50
$CH_3COOR\text{-}SH + H_2O \rightleftharpoons CH_3COOH + HO\text{-}R\text{-}SH$	21.4	-1.90

[a]At 39^0C, ionic strength 0.3. K values based on molar concentrations of solutes. From Ref. [25].
[b]$R \equiv (CH_2)_3$.

Table 2.II gives the equilibrium constants and standard free energy changes for the isomerization of the thiol ester to the oxygen ester, and for the hydrolysis of the oxygen ester (which is the reverse of its formation). Addition of these two reactions gives the hydrolysis of the thiol ester,

$$CH_3COS(CH_2)_3OH + H_2O \rightleftharpoons CH_3COOH + HS(CH_2)_3OH \quad (2.42)$$

The standard free energy change of (2.42) is the sum of those for the reactions in Table 2.II; or $\Delta G^0_{2.42} = \Delta G^0_{2.40} - \Delta G^0_{2.41} = -2.50 - 1.90$ kcal/mole $= -4.40$ kcal/mole. The equilibrium constant for (2.42) is therefore 1.2×10^3 M.

Biochemists have evolved the concepts of "high-energy bonds" or "energy-rich compounds" to concentrate attention on those biochemically important substances that react to yield a relatively large amount of free energy. The present use of the term energy-rich compound is to signify a compound that hydrolyzes to liberate at least 7 kcal/mole of free energy at physiological conditions (usually pH 7). Specification of the pH introduces a complication in the calculation, and it is an exceedingly important one. The point can be demonstrated with a simple example of ester hydrolysis. Equation (2.43) shows this reaction written with the usual standard states of the pure liquid solvent and the solutes at unit activity (molar or molal scales).

$$RCOOR'(a = 1) + H_2O(\ell) \rightleftharpoons RCOOH(a = 1) + R'OH(a = 1)$$

$$(2.43)$$

Carpenter [26] calls this the nonionized compound convention. The standard free energy change for this reaction, ΔG^0_h, is given by $\Delta G^0_h = -RT \ln K$. The free energy change at any other concentrations of the reactants as written is given by the reaction isotherm:

$$\Delta G_h = \Delta G^0_h + RT \ln \frac{\Pi a_{prod.}}{\Pi a_{react.}} \quad (2.44)$$

Another convention represents the reaction carried out at constant specified pH:

$$RCOOR(a = 1) + H_2O(\ell) \rightleftharpoons RCOO^-(\Sigma a_{pH} = 1)$$

$$+ R'OH(a = 1) + H^+(a_{pH}) \quad (2.45)$$

The symbolism $(\Sigma a_{pH} = 1)$ signifies that the sum of the activities of the un-ionized and ionized forms (RCOOH, RCOO$^-$) is unity. The standard free energy change for Eq. (2.45), $\Delta G^0_{h,pH}$, is evaluated by measuring the equilibrium constant at the specified pH, and the partition of RCOOH into two forms is neglected. Note that the equilibrium constant for (2.43) is pH independent, whereas that for (2.45) is pH dependent [27].

The free energy changes for Eqs. (2.43) and (2.45)

are related by

$$\Delta G^0_{hpH} = \Delta G^0_h + \Delta G^0_{ipH} \qquad (2.46)$$

where ΔG^0_{ipH} is the free energy of ionization and dilution to the concentration determined by the specified pH and the pK_a of RCOOH. For the simple example given, this is found as the sum of the free energy changes for these processes:

Process	ΔG
$RCOOH(a = 1) \rightleftharpoons H^+(a = 1) + RCOO^-(a = 1)$	$-RT \ln K_a$
$H^+(a = 1) \rightleftharpoons H^+(a_{pH})$	$RT \ln a_{H^+}$
$RCOO^-(a = 1) \rightleftharpoons RCOO^-(\Sigma a = 1)$	$RT \ln a_{RCOO^-}$

Since $a_{RCOOH} + a_{RCOO^-} = 1$, $a_{RCOO^-} = K_a/(K_a + a_{H^+})$, and

$$\Delta G^0_{ipH} = RT \ln \frac{a_{H^+}}{a_{H^+} + K_a} \qquad (2.47)$$

If, for example, the ester hydrolysis is carried out at pH 7 and 25°, and if $pK_a = 5$ for RCOOH, then $\Delta G^0_{ipH} = -2730$ kcal/mole. Carpenter [26] has analyzed polyfunctional acids, which occur widely in physiological systems.

   It appears that a major portion of the free energy liberated upon hydrolysis of "energy-rich" compounds is a consequence of ionization reactions. The correction to physiological conditions alters the standard free energy change for oxygen acetate hydrolysis from -1800 to about -5100 cal/mole; for thiol acetates, the corresponding values are -4400 and -7700 cal/mole. This places thiol esters in the energy-rich class [25]. Note that the difference between acetates and thiol acetates is in ΔG^0_h. For the hydrolysis of a number of peptide (amide) bonds, ΔG^0_h is roughly constant at +4.7 kcal/mole [26]; this implies net synthesis of peptides. Incorporation of the ionization correction converts these values to negative quantities in the range -0.5 to -4.2 kcal/mole; this treatment therefore reveals that the thermodynamic reason for amide hydrolysis at pH 7 is in the release of the acidic and basic groups.

   Equilibrium constants or standard free energy changes can be useful qualitative guides, as in comparing relative strengths of a series of acids. Another application, this one quantitative, is in making calculations of product yields or reactant conversion at equilibrium. The specific form of the calculation depends upon the form of the equilibrium constant, on the boundary conditions (initial concentrations), and on the accuracy desired. Consider, as an example, the hydrolysis of ethyl acetate.

$$AcOEt + H_2O \rightleftharpoons HOAc + EtOH$$

The equilibrium constant for this reaction, at 39^0 in pure water, is 18.5 M, where the activity of the water is taken as unity [25]. Let c be the initial ester concentration, x the equilibrium concentration of acetic acid, and suppose the initial concentrations of acetic acid and ethanol are zero. Then we can write $K = x^2/(c - x)$, and x can be found by application of the quadratic formula. The percent of ester hydrolyzed at equilibrium is 100 x/c. If, for this system, c = 0.1 M, we then find x = 0.09412 M; 94.12% of the ester is hydrolyzed at equilibrium.

Enthalpies and Entropies. The indirect method of determining the standard enthalpy change ΔH^0 (also called the standard heat of reaction) requires evaluation of the equilibrium constant as a function of temperature. If a plot of log K against 1/T is linear (within experimental error), it can be concluded that ΔH^0 is constant ($\Delta Cp^0 = 0$) over the temperature range studied. Equation (2.6) describes the system; it can be written

$$\log K = - \frac{\Delta H^0}{4.576 \ T} + constant \qquad (2.48)$$

From the slope of this van't Hoff plot, ΔH^0 is obtained:

$$slope = \frac{-\Delta H^0}{4.576} \qquad (2.49)$$

ΔH^0 will have the units calories per mole. If ΔH^0 is negative, the reaction is exothermic; an endothermic reaction has a positive ΔH^0.

From the relation $\Delta G^0 = -RT \ \ell n \ K$, or $\Delta G^0 = -4.576 \ T$ log K, the standard free energy change is obtained at each temperature. Finally the standard entropy change is calculated:

$$\Delta S^0 = \frac{\Delta H^0 - \Delta G^0}{T} \qquad (2.50)$$

For reactions that do not involve the creation or neutralization of electrical charge, ΔH^0 is found to be essentially constant over moderate temperature ranges [28].

Table 2.III gives equilibrium constants for a gas-phase dissociation,

$$F_2CHCH_2NH_2 \cdot B(CH_3)_3 \rightleftharpoons F_2CHCH_2NH_2 + B(CH_3)_3 \qquad (2.51)$$

chosen because of the unusually large number of equilibrium constant-temperature points [29]. The table includes the calculated values of 1/T and log K, and the van't Hoff plot is shown in Fig. 2.1. The slope of the straight line is -3.37×10^3 deg, so ΔH^0 = 15.4 kcal/mole. The calculation of ΔS^0 can be illustrated by the 100°C value of the equilibrium constant. This yields ΔG^0 = -(4.576)(373.16) (0.433) cal/mole = -0.74 kcal/mole. Equation (2.50) then gives ΔS^0 = +43 cal/mole-deg, or +43 entropy units (e.u.).

Table 2.III   Equilibrium Constants for Reaction (2.51)[29]

$t/°C$	$10^3(deg)/T$	K/atm	$log(K/atm)$
65.0	2.957	0.315	-0.502
70.0	2.914	0.444	-0.353
75.0	2.872	0.608	-0.216
80.0	2.832	0.833	-0.079
85.0	2.792	1.13	0.053
90.0	2.754	1.53	0.185
95.0	2.716	2.05	0.312
100.0	2.680	2.17	0.433

If the plot of log K against 1/T is not linear, which may happen for ionic reactions, then ΔH^0 is a function of temperature. This problem was discussed in Sec. 2.1. The quantity log K is fitted to an equation of the form

$$\log K = -\frac{A'}{T} - C'T + D'$$

by the method of least-squares [30], and the quantities ΔH^0 and ΔS^0 are calculated using Eq. (2.9) and the equation given in Problem 1.

Heats of reaction can be determined directly by calorimetry [31], which is capable of great accuracy. For most analytical applications, the technique of thermometric titrimetry [32] probably is a more practical direct method for measurements of ΔH^0. This method involves the titration of a dilute solution of the sample with a relatively concentrated titrant under essentially adiabatic conditions; the temperature of the solution is recorded as a function of titrant volume. The thermometric titration curve of temperature against volume may be analyzed to yield the titration end point and the heat of reaction [33].

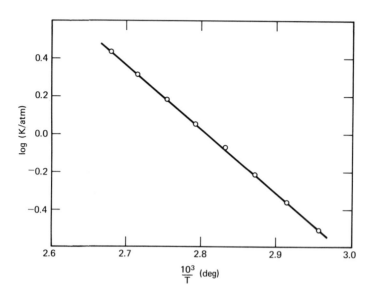

Fig. 2.1.   van't Hoff plot for the dissociation of a
            gaseous trimethylboron-difluoropropylamine
            complex [29].

     Table 2.IV gives standard enthalpy and entropy changes
for some reactions in solution; these values illustrate
the diversity in magnitude that may be observed. It should
be noted that the position of equilibrium in a reaction is
controlled by ΔG^0, and since $\Delta G^0 = \Delta H^0 - T\Delta S^0$, it is the
signs and sizes of the ΔH^0 and $T\Delta S^0$ terms that determine
the extent of reaction at equilibrium.

<u>Medium and Substituent Effects</u>. We have used the operator
Δ to signify a change in a quantity associated with a
chemical reaction. It is convenient to define different
symbols to describe changes caused by alterations in the
medium and in structure. The delta operators of Leffler
and Grunwald are employed for this purpose [34]. A change
caused by a change in the medium is indicated by the symbol δ_M, and a change caused by a structural alteration by
δ_R. Thus $\delta_M\bar{G}$ is a change in partial molar free energy
caused by a change in medium.
     These operators can be combined to describe changes
in changes. For example, $\delta_R\Delta H^0$ represents a change in
heat of reaction caused by a structural change. It can be
demonstrated that all of these operators commute [34], so
that we can write $\delta_R\Delta G = \Delta\delta_R G$, and so on [35].

Table 2.IV Standard Enthalpy and Entropy Changes for Some Solution Reactions

Reaction	Conditions	ΔH°/kcal-mole^{-1}	ΔS°/e.u.[a]	Ref.
$H_2O \rightleftharpoons H^+ + OH^-$	I = 0	+13.34		b
$CH_3COOH \rightleftharpoons CH_3COO^- + H^+$	aq	-0.01		c
$TRISH^+ \rightleftharpoons TRIS^e + H^+$	aq, I = 0.01	+11.39	+1.5	c,d
$IMH^+ \rightleftharpoons IM^f + H^+$	aq, I = 0.01	+8.78	-2.5	c
$PYRH^+ \rightleftharpoons PYR^g + H^+$	aq, I = 0.01	+4.98	-7.0	c
$C\ell_3CCOOH \rightleftharpoons C\ell_3CCOO^- + H^+$	aq, I = 0	+0.28	-1.4	h
$H_3BO_3 + OH^- \rightleftharpoons HBO_2^- + H_2O$	aq, 0.01 M	-10.6		i
$Ca^{2+} + Y^{4-} \rightleftharpoons CaY^{2-}$ ^{j}	aq	-5.7	+31	i
$Mg^{2+} + Y^{4-} \rightleftharpoons MgY^{2-}$ ^{j}	aq	+5.5	+60	i
$C_6H_5OH + Et_2O \rightleftharpoons C_6H_5OH{:}OEt_2$	hexane solvent	-3.8	-18	k
Fumaric acid + $H_2O \rightleftharpoons$ Malic acid	aq HCℓ	-4.6	-7.7	ℓ

aEntropy changes based on solute molar or molal scales

bJ. D. Hale, R. M. Izatt, and J. J. Christensen, Proc. Chem. Soc., 1963, 240.

cJ. J. Christensen, D. P. Wrathall, and R. M. Izatt, Anal. Chem., 40, 175 (1968).

dE. W. Wilson, Jr., and D. F. Smith, Anal. Chem., 41, 1903 (1969).

eTRIS ≡ tris-(hydroxymethyl)-aminomethane.

fIM ≡ imidazole.

gPYR ≡ pyridine.

hJ. L. Kurz and J. M. Farrar, J. Am. Chem. Soc., 91, 6057 (1969).

iJ. Jordan, Ref. [32].

jY⁴⁻ ≡ ethylenediaminetetraacetate.

kG. C. Pimentel and A. L. McClellan, The Hydrogen Bond, Freeman, San Francisco, 1960, Chap. 7.

ℓL. T. Rozelle and R. A. Alberty, J. Phys. Chem., 61, 1637 (1957); the original paper gives ΔS° = -16 e.u., which may be based on water concentration expressed in molar units.

Table 2.V   Thermodynamic Quantities for Dissociation of
            Protonated Amines [29]

Amine	pK_a	$\Delta G^0 (25^0)/$ kcal-mole^{-1}	ΔH^0/kcal-mole^{-1}	ΔS^0/e.u.[a]
CH_3ONH_2	4.60	6.27	7.63	+4.6
$F_3CCH_2NH_2$	5.59	7.65	8.30	+2.2
$F_2CHCH_2NH_2$	7.09	9.67	9.77	+0.3
$F(CH_2)_2NH_2$	8.79	11.99	11.96	0.0
$HO(CH_2)_2NH_2$	9.50	12.96	12.08	-2.9
$C_2H_5NH_2$	10.63	14.50	13.58	-3.1
$n-C_3H_7NH_2$	10.53	14.41	13.67	-2.5

[a]Based on molar concentration scale.

The description of substituent or medium effects in
these terms can be demonstrated with the data in Table
2.V, which are for the aqueous dissociation of the conjugate acids of some primary amines:

$$RNH_3^+ \overset{K_a}{\rightleftharpoons} H^+ + RNH_2$$

For this series only the group R is changed.   The corres-
ponding changes in ΔG^0, ΔH^0, and ΔS^0 are therefore symbolized $\delta_R\Delta G^0$, $\delta_R\Delta H^0$, and $\delta_R\Delta S^0$.   In order to calculate
these quantities a reference compound must be selected.
Suppose n-propylamine is chosen as the reference.   Then,
as an example,

$$\delta_R\Delta G^0_{CH_3O} = \Delta G^0_{CH_3O} - \Delta G^0_{C_3H_7}$$

$$= 6.27 - 14.41 \text{ kcal/mole}$$

$$= -8.14 \text{ kcal/mole}$$

In Table 2.VI these quantities are tabulated for the
data in Table 2.V.
From the relation $\Delta G^0 = \Delta H^0 - T\Delta S^0$, we can write

$$\delta_R\Delta G^0 = \delta_R\Delta H^0 - T\delta_R\Delta S^0$$

and so on.   In studying substituent or solvent effects on
equilibria (that is, on free energy changes), it is not
sufficient to investigate solely the ΔG^0 factor, because
this term is determined by both an enthalpy and an entropy

term. Four types of control can be recognized [36]:

1. Cooperative effects, with $\delta\Delta H^0$ and $\delta\Delta S^0$ having opposite signs. Then the $\delta\Delta H^0$ and $T\delta\Delta S^0$ terms will be additive.

2. Enthalpy-controlled effects, in which the two terms are opposed, but the enthalpy term is larger.

3. Entropy-controlled effects, in which the two terms oppose, but the $T\Delta S^0$ term is larger.

4. Compensating effects, in which the two terms are opposed and are nearly equal. A change in temperature may interconvert categories 2, 3, and 4.

Table 2.VI   Substituent Effects on Thermodynamic
              Quantities in Table 2.V [29]

R	$\delta_R\Delta G^0$/kcal-mole^{-1}	$\delta_R\Delta H^0$/kcal-mole^{-1}	$\delta_R\Delta S^0$/e.u.
n-C$_3$H$_7$	(0.00)	(0.00)	(0.00)
C$_2$H$_5$	+0.09	-0.09	-0.6
HO(CH$_2$)$_2$	-1.45	-1.59	-0.4
F(CH$_2$)$_2$	-2.42	-1.71	+2.5
F$_2$CHCH$_2$	-4.74	-3.90	+2.8
F$_3$CCH$_2$	-6.76	-5.37	+4.7
CH$_3$O	-8.14	-6.04	+7.1

Table 2.VI shows that $\delta_R\Delta G^0$ follows $\delta_R\Delta H^0$ fairly well for this series, but this is not always so. For the formic acid-acetic acid pair these effects are observed for the dissociation in water at 25^0 [36]: $\delta_R\Delta G^0$ = -1369 cal/mole, $\delta_R\Delta H^0$ = +69 cal/mole, $T\delta_R\Delta S^0$ = +1438 cal/mole-deg. (Acetic acid is the reference acid.) It is expected that electron release by the methyl group, relative to hydrogen, will weaken acetic acid relative to formic acid by an effect on ΔH^0 (a "potential energy" effect). In fact, however, although acetic acid is weaker than formic acid, as expected, the change in ΔH^0 is in the opposite direction, and it is the entropy term that controls the substituent effect. It is especially dangerous to compare small $\delta\Delta G^0$ values when heat capacity changes are significant [37].

Interpretation of the thermodynamic quantities requires the invention of specific models, which may be primarily physical in nature (for example, a description of electrolytic dissociation as a pure coulombic interaction) or chemical (as in the resonance theory of the

propagation of substituent effects). These are not introduced here, but are used when appropriate to discussion
of specific types of reactions.

Problems

1. Using the relationship $(\partial \Delta G/\partial T)_p = -\Delta S$, find an equa-
    tion for ΔS^0 in terms of the parameters of Eq. (2.10).

    Answer: $\Delta S^0 = D - 2CT$

2. Find a relationship between the overall dissociation
    constant K_B, the ionization constant K_i, and the dis-
    sociation constant K_d for an electrolyte in a low
    dielectric constant solvent.

3. From the data in Table 2.IV, obtain the heat of ioni-
    zation of boric acid.

    Answer: $\Delta H^0_i = +2.7$ kcal/mole

4. Develop an equation to calculate ΔG^0 directly from
    pK_a values.

    Answer: $\Delta G^0 = 4.576\ T\ pK_a$

5. The heat of ionization of $CHF_2CH_2NH_3^+$ is 9.77 kcal/mole
    [29] and its pK_a is 7.09 at 25^0. Calculate ΔS^0.

    Answer: $\Delta S^0 = +0.3$ e.u.

6. These stability constants were evaluated for 1:1 mole-
    cular complex formation between methyl trans-cinnamate
    and theophylline in water (J. L. Cohen, Ph.D. Disser-
    tation, University of Wisconsin, Madison, 1969):   at
    2.0^0C, $K = 39.4$ M^{-1}; 15.0^0C, 33.2 M^{-1}; 25.0^0C, 26.1
    M^{-1}; 35.0^0C, 22.9 M^{-1}.
    (a) Calculate ΔH^0 and ΔS^0.
    (b) What is the unitary free energy change at 25^0?
    (c) Predict K at 75^0C.

    Answer: (a) $\Delta H^0 = -2.8$ kcal/mole, $\Delta S^0 = -3.0$ e.u.;
              (b) $\Delta G^0_{un} = -4.3$ kcal/mole;
              (c) $K(75^0) = 13.4$ M^{-1}.

7. The reaction

$$CH_3I + LiC\ell \rightleftharpoons CH_3C\ell + LiI$$

was allowed to proceed to equilibrium at 25.2°C in acetone. The initial concentrations of methyl iodide and lithium chloride were 0.05725 M and 0.02960 M, respectively. At equilibrium the reaction was 94.2% complete. (E. D. Hughes, C. Ingold, and A. J. Parker, J. Chem. Soc., 1960, 4400.) Calculate the equilibrium constant.

Answer: K = 15.4

8. Benzaldehyde forms an addition compound with bisulfite according to this equation:

$$C_6H_5CHO + HSO_3^- \rightleftharpoons C_6H_5CH(OH)SO_3^-$$

The equilibrium constant for this reaction is 1×10^4 at 21° and pH 5.0. For a solution prepared by mixing 10 ml of 0.01 M benzaldehyde and 15 ml of 0.01 M bisulfite, calculate the concentration of benzaldehyde remaining unreacted at equilibrium, and the percent error introduced into a determination of the aldehyde if reaction is assumed to be complete. (It is acceptable to assume that the concentration of unreacted bisulfite is equal to its initial concentration minus the aldehyde concentration, thus simplifying the calculation.)

Answer: 2×10^{-4} M; 5% error.

References

1. See, for example, S. Glasstone, Thermodynamics for Chemists, Van Nostrand, Princeton, N. J., 1947; G. N. Lewis and M. Randall, Thermodynamics (revised by K. S. Pitzer and L. Brewer), 2nd ed., McGraw-Hill, New York, 1961.
2. H. S. Harned and R. A. Robinson, Trans. Faraday Soc., 36, 973 (1940); H. S. Harned and B. B. Owen, The Physical Chemistry of Electrolytic Solutions, 3rd ed., Reinhold, New York, 1958, p. 665.
3. S. Glasstone, Ref. [1], pp. 279-281.
4. S. Glasstone, Ref. [1], pp. 350-355.
5. The activity of hydrogen ion may be evaluated electrometrically, whereas most measurements of B and BH^+ yield concentrations. See A. Albert and E. P. Serjeant, Ionization Constants of Acids and Bases, Methuen, London, 1962.
6. J. E. Leffler and E. Grunwald, Rates and Equilibria of Organic Reactions, Wiley, New York, 1963, pp. 20-22.

7. R. W. Gurney, Ionic Processes in Solution, McGraw-Hill, New York, 1953, p. 105 (Dover Publications edition, 1962).
8. R. W. Gurney, Ref. [7], Chaps. 5, 6 (Dover Publications edition, 1962).
9. R. W. Gurney, Ref. [7], pp. 88-92. This word is pronounced with a long a, as in crater.
10. For some applications of unitary quantities see W. Kauzmann, Advances in Protein Chemistry, Vol. XIV, Academic, New York, 1959, pp. 34-49; C. Tanford, Accounts Chem. Res., $\underline{1}$, 161 (1968); W. P. Jencks, Catalysis in Chemistry and Enzymology, McGraw-Hill, New York, 1969, pp. 15-17, 373-390, 422; K. A. Connors, M. H. Infeld, and B. J. Kline, J. Am. Chem. Soc., $\underline{91}$, 3597, 5697 (1969).
11. The use of molar volumes V_r is not restrictive, because if the solution is not ideal, these can be replaced with partial molar volumes, $\overline{V}_r$.
12. I. D. Kuntz, Jr., F. P. Gasparro, M. D. Johnston, Jr., and R. P. Taylor, J. Am. Chem. Soc., $\underline{90}$, 4778 (1968); H. Buchowski, J. Devaure, P. V. Huong, and J. Lascombe, Bull. Soc. Chim. France, 1966, 2532; P. J. Trotter and M. W. Hanna, J. Am. Chem. Soc., $\underline{88}$, 3724 (1966); S. K. Alley, Jr., and R. L. Scott, J. Phys. Chem., $\underline{67}$, 1182 (1963).
13. For some treatments of the measurement of complex stability constants see G. Briegleb, Elektronen-Donator-Acceptor-Komplexe, Springer-Verlag, Berlin, 1961; L. J. Andrews and R. M. Keefer, Molecular Complexes in Organic Chemistry, Holden-Day, San Francisco, 1964; R. Foster, Organic Charge-Transfer Complexes, Academic, Inc., New York, 1969; K. A. Connors and J. A. Mollica, Jr., J. Pharm. Sci., $\underline{55}$, 772 (1966); D. A. Deranleau, J. Am. Chem. Soc., $\underline{91}$, 4044, 4050 (1969).
14. See also K. Denbigh, The Principles of Chemical Equilibrium, 2nd ed., Cambridge Univ. Press, Cambridge, 1966, p. 298.
15. I. M. Kolthoff and V. A. Stenger, Volumetric Analysis, 2nd ed., Vol. I, Wiley (Interscience), New York, 1942; J. E. Ricci, Hydrogen Ion Concentration, Princeton Univ. Press, Princeton, 1952; Treatise on Analytical Chemistry (I. M. Kolthoff and P. J. Elving, eds.), Part I, Vol. 1, (Wiley) Interscience, New York, 1959; Comprehensive Analytical Chemistry (C. L. Wilson and D. W. Wilson, eds.), Vol. I, Part B, Elsevier, Amsterdam, 1959; J. N. Butler, Ionic Equilibrium, Addison-Wesley, Reading, Mass., 1964; H. Freiser and Q. Fernando, Ionic Equilibria in Analytical Chemistry, Wiley, New York, 1963.

16. H. C. Brown, D. H. McDaniel, and O. Häfliger, Deter-
    mination of Organic Structures by Physical Methods
    (E. A. Braude and F. C. Nachod, eds.), Academic, New
    York, 1955, Chap. 14; A. Albert and E. P. Serjeant,
    Ionization Constants of Acids and Bases, Wiley, New
    York, 1962; G. Kortum, W. Vogel, and K. Andrussow,
    Dissociation Constants of Organic Acids in Aqueous
    Solution, Butterworths, London, 1961; D. D. Perrin,
    Dissociation Constants of Organic Bases in Aqueous
    Solution, Butterworths, London, 1965; K. B. Yatsimir-
    skii and V. P. Vasil'ev, Instability Constants of
    Complex Compounds, Van Nostrand, Princeton, 1966; J.
    Bjerrum, G. Schwarzenbach, and L. G. Sillen, Stability
    Constants: I. Organic Ligands, Spec. Pub. No. 6
    (1957), and Stability Constants: II. Inorganic
    Ligands, Spec. Pub. No. 7 (1958), The Chemical Socie-
    ty, London.
17. I. M. Kolthoff and S. Bruckenstein, Treatise on Ana-
    lytical Chemistry (I. M. Kolthoff and P. J. Elving,
    eds.), Part I, Vol. 1, Wiley (Interscience), New
    York, 1959, Chap. 130.
18. F. J. C. Rossotti and H. Rossotti, The Determination
    of Stability Constants in Solution, McGraw-Hill, New
    York, 1961.
19. W. M. Clark, Oxidation-Reduction Potentials of Orga-
    nic Systems, Williams and Wilkins, Baltimore, 1960.
20. W. P. Jencks, S. Cordes, and J. Carriuolo, J. Biol.
    Chem., $\underline{235}$, 3608 (1960).
21. E. A. Guggenheim and J. E. Prue, Physicochemical Cal-
    culations, North-Holland, Amsterdam, 1956, pp. 1-6;
    see also J. E. Briggs, J. Chem. Educ., $\underline{35}$, 30 (1958);
    G. N. Copley, ibid., $\underline{35}$, 366 (1958).
22. G. J. Janz, Thermodynamic Properties of Organic Com-
    Pounds, 2nd ed., Academic, New York, 1967; D. R.
    Stull, E. F. Westrum, Jr., and G. C. Sinke, The
    Chemical Thermodynamics of Organic Compounds, Wiley,
    New York, 1969.
23. D. R. Stull, E. F. Westrum, Jr., and G. C. Sinke,
    Ref. [22].
24. J. T. Edsall and J. Wyman, Biophysical Chemistry,
    Vol. I, Academic, New York, 1958, pp. 232-238.
25. W. P. Jencks, S. Cordes, and J. Carriuolo, J. Biol.
    Chem., $\underline{235}$, 3608 (1960).
26. F. H. Carpenter, J. Am. Chem. Soc., $\underline{82}$, 1111 (1960).
27. Jencks[25] suggests, instead of the nonionized conven-
    tion, the fully ionized version; this also yields a
    pH-independent equilibrium constant, which of course
    has a different value from that for (2.43).
28. J. E. Leffler and E. Grunwald, Rates and Equilibria
    of Organic Reactions, Wiley, New York, 1963, p. 44.

29. P. Love, R. B. Cohen, and R. W. Taft, J. Am. Chem.
     Soc., 90, 2455 (1968).
30. I. M. Klotz, Introduction to Chemical Thermodynamics,
     Benjamin, New York, 1964, p. 25.
31. J. M. Sturtevant, Physical Methods of Organic Chem-
     istry (A. Weissberger, ed.), Vol. I, Part I, Wiley
     (Interscience), New York, 1959, Chap. X; B. D. Kybett,
     T. V. Charlu, A. Chaudhuri, J. Jones, and J. L. Mar-
     grave, Treatise on Analytical Chemistry (I. M. Kolt-
     hoff and P. J. Elving, eds.), Part I, Vol. 8, Wiley
     (Interscience), New York, 1968, Chap. 90.
32. J. Jordan, Treatise on Analytical Chemistry (I. M.
     Kolthoff and P. J. Elving, eds.), Part I, Vol. 8,
     Wiley (Interscience), New York, 1968, Chap. 91.
33. Equilibrium constants can also be estimated by ther-
     mometric titrimetry; see J. J. Christensen, D. P.
     Wrathall, and R. M. Izatt, Anal. Chem., 40, 175 (1968).
34. J. E. Leffler and E. Grunwald, Rates and Equilibria
     of Organic Reactions, Wiley, New York, 1963, pp. 22-
     27.
35. Some authors use the less informative notation ΔΔG,
     etc., to indicate changes in changes.
36. J. E. Leffler and E. Grunwald, Ref. [34], pp. 48-50.
37. E. M. Arnett, W. B. Kover, and J. V. Carter, J. Am.
     Chem. Soc., 91, 4028 (1969).

Chapter 3.   REACTION RATES

The theory, measurement, and interpretation of the rates
of chemical reactions constitute the field of chemical
kinetics. This chapter provides a review of some concepts
basic to kinetic studies, and then treats certain topics
that are of special interest in the kinetics of organic
reactions in solution, for these are the systems that are
most relevant to the analytical chemist. Some subjects
are included because they continue to be loci of confusion
or error for the student or experimentalist (for example,
salt effects and reaction mechanism; the principle of
microscopic reversibility), and others because they are
not treated in detail elsewhere (pH-rate profiles) but are
important to the experimentalist. Jencks's chapter on
practical kinetics [1] should be read by anyone carrying
out solution kinetic studies. More systematic treatments
of kinetics can be found in many texts and reference
works [2].
    Some pertinent topics are omitted from this chapter
because the structure of the book places them elsewhere,
especially in Chap. 4.

3.1   Theory of Reaction Rates

Rates, Rate Constants, and the Activation Energy. An experimental study of the rate of a chemical reaction (constant temperature and pressure conditions are usually
understood) results in estimates of a reactant (or product) concentration c as a function of time t. The tan-
gent to the curve obtained by plotting c against t is the
reaction rate or velocity, at that time, dc/dt, which is
often symbolized v. It is observed that the rate is often
a remarkably simple function of reactant concentrations.
The relationship between reaction rate and reactant concentrations is called the rate equation. For the generalized reaction

$$A + B \rightarrow products \qquad (3.1)$$

the rate equation has the form

$$v = kc_A c_B \qquad (3.2)$$

The proportionality constant k is the <u>rate constant</u> for
the reaction. The power to which the <u>concentration</u> is
raised is the <u>order</u> with respect to that reactant. If
Eq. (3.2) describes (3.1), the reaction is said to be
first order with respect to A, first order with respect
to B, and overall second order. A rate equation may consist of several terms; for example, a very common rate
equation for the hydrolysis of an ester is

$$v = k_H \, c_{ester} \, c_{H^+} + k_w \, c_{ester} + k_{OH} \, c_{ester} \, c_{OH^-}$$

In this equation the concentration of water does not explicitly appear; it has been absorbed into the appropriate
rate constants because, in dilute aqueous solution, c_{H_2O}
is essentially a constant. This is a common practice.[2]
The first term represents acid-catalyzed ester hydrolysis,
and k_H is the second-order rate constant for the acid catalysis. k_w is the first-order rate constant for the uncatalyzed hydrolysis, and k_{OH} is the second-order rate
constant for the hydroxide ion-catalyzed hydrolysis. More
complicated rate equations may be encountered, and their
determination and interpretation are considered subsequently.
        It should be clear that a rate and a rate constant
are quite different quantities [3]. Some reactions proceed via more than one step, and then the rate of the reaction (that is, rate of production of product) is controlled by the rates of the individual steps. If in a
multistep reaction the rate of one of the steps is much
smaller than the rates of the others, the overall reaction
rate will be controlled by this slow <u>rate-determining step</u>.
The reaction rate may be affected by steps preceding the
rate-determining step, but not by steps following it.
Note that the rate-determining step in a multistep reaction
depends upon relative <u>rates</u>, not rate constants; thus
changes in reactant <u>concentrations</u>, which can alter the
relative rates of reaction steps, could change the rate-
determining step.
        The rate constant of a chemical reaction can be em-
pirically related to the absolute temperature by the
Arrhenius equation:

$$k = Ae^{-E_a/RT} \qquad (3.3)$$

The quantity A is the "pre-exponential factor" and E_a is

the experimental activation energy. A theory of reaction
rates is basically an attempt to predict, or at least to
rationalize, observed values of A and E_a. The first
moderately successful theoretical approach to Eq. (3.3)
was the collision theory, which starts with the assumption
that the rate (of a bimolecular reaction) is equal to the
product of the rate Z at which the molecules collide and
the fraction of molecules possessing an energy E at least
equal to the activation energy. According to the Boltzmann
distribution, this fraction is proportional to $e^{-E/RT}$.
The collision frequency Z can be easily calculated, on the
assumption that the molecules are noninteracting incom-
pressible spheres; the theoretical value of Z from simple
collision theory is 2.86×10^{11}. For many reactions in
solution, experimental values of A are very close to this
estimate [4]. Some reactions yield A values very much
smaller or larger than the theoretical Z, and it is customary to write, for Eq. (3.3),

$$k = PZe^{-Ea/RT} \qquad (3.4)$$

Experimental values of the pre-exponential factor may be
quoted in terms of A or PZ (or their logarithms). The
deviation of P from unity is a measure of the discrepancy
between experiment and collision theory.

The collision theory has not been of great use in the
study of reaction mechanisms because it does not readily
lend itself to interpretation in molecular terms [5].

The Transition State Theory. Our understanding of reaction mechanisms has been greatly assisted by the theory
of absolute reaction rates, or, in more common nomenclature, the transition state theory, which is largely associated with Eyring's name [6]. This theory is based on
these postulates: (1) in going from the reactant state
to the product state, molecules must pass through an
intermediate state of higher energy than either the reactants or products; (2) the molecules existing in this
higher energy state are in virtual equilibrium with the
reactant molecules; (3) the rate of the reaction is proportional to the concentration of the molecules in the
higher energy state. The energy levels accessible to a
system in its progress from reactants to products can be
pictured as a three-dimensional surface, with the vertical
(z) dimension representing energy, the y dimension time
(or distance along the process path Reactants → products),
and the x dimension all other degrees of freedom.
Most of the molecules will take the reaction path
corresponding to the minimum possible energy in the x

dimension. Most of the higher energy molecules referred
to in the postulates above will therefore have a configuration characteristic of the minimum possible energy on the
energy barrier separating reactants and products; this
corresponds to a saddle point on the energy surface [7].
The molecule possessing this configuration is called the
activated complex or the transition state for the reaction.
        It is not convenient for our purposes to represent
the course of a reaction as a surface, so a simplified
two-dimensional picture is used. We take a cross section
(not generally a plane section) through the surface along
the path followed by most of the reacting molecules. A
plot of the z value (energy) against the y value (reaction
coordinate) for this section then represents the approximate course of the reaction. This type of reaction coordinate diagram is shown in Fig. 3.1. The transition state
is seen to correspond to the region of highest energy
along the reaction path. Location along the horizontal
axis may be interpreted as a measure of the extent of
conversion of reactants to products.

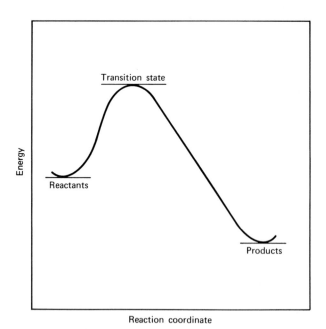

Fig. 3.1. An example of a reaction coordinate diagram.

        The position of the transition state is restricted
by the steep walls of the saddle, but there may be many

alternate paths from reactants to transition state that
are of comparable energy. This means that the path taken
to reach the transition state cannot be precisely defined.
This limitation is a consequence of postulate (2) above,
which invokes an equilibrium between the initial state
and the transition state; the situation is analogous to
that in true thermodynamic equilibrium, which describes
only changes between states [8].

For the quantitative formulation of the transition
state theory, consider the common bimolecular reaction,

$$A + B \rightleftharpoons M^{\ddagger} \rightarrow \text{products} \qquad (3.5)$$

In this equation $M^{\ddagger}$ represents the transition state for
the reaction, and the presumed equilibrium between the
initial and transition states is indicated [9,10]. The
rate of reaction is taken to be equal to the product of
the concentration of activated complexes $c^{\ddagger}$ and the frequency with which activated complexes pass over the energy
maximum to form products. This frequency has been shown
[6] to be a universal quantity equal to kT/h, where k is
the Boltzmann constant and h is Planck's constant. The
rate of reaction is then

$$v = \frac{kT}{h} c^{\ddagger} \qquad (3.6)$$

Applying the equilibrium postulate to Eq. (3.5) allows an
"equilibrium constant" $K^{\ddagger}$ to be defined.

$$K^{\ddagger} = \frac{c^{\ddagger}}{c_A c_B} \qquad (3.7)$$

Combining Eqs. (3.6) and (3.7),

$$v = \frac{kT}{h} K^{\ddagger} c_A c_B \qquad (3.8)$$

Equating this theoretical expression for v with the experimental v as given by Eq. (3.2) gives an equation for the
rate constant k.

$$k = \frac{kT}{h} K^{\ddagger} \qquad (3.9)$$

Equation (3.9) may also be written $k = (RT/Nh)K^{\ddagger}$.

If it is reasonable to write an equilibrium constant for the formation of the transition state, the usual thermodynamic equations can be applied to this process. Thus,

$$\Delta G^{\ddagger} = -RT \ln K^{\ddagger} \tag{3.10}$$

$$\Delta G^{\ddagger} = \Delta H^{\ddagger} - T\Delta S^{\ddagger} \tag{3.11}$$

$$\frac{d \ln K^{\ddagger}}{dT} \quad \frac{\Delta H^{\ddagger}}{RT^2} \tag{3.12}$$

Here $\Delta G^{\ddagger}$, $\Delta H^{\ddagger}$, and $\Delta S^{\ddagger}$ are the free energy, enthalpy, and entropy of activation, respectively; these are standard quantities, but the usual symbol is omitted. These activation quantities can be evaluated from the experimental quantities as follows. From Eq. (3.9) we have $d \ln K^{\ddagger}/dT = d \ln k/dT - 1/T$. This combined with Eq. (3.12) gives

$$\Delta H^{\ddagger} = RT^2 \frac{d \ln k}{dT} - RT \tag{3.13}$$

But from Eq. (3.3), $E_a = RT^2(d \ln k/dT)$. Therefore,

$$\Delta H^{\ddagger} = E_a - RT \tag{3.14}$$

From Eqs. (3.9) and (3.10),

$$\Delta S^{\ddagger} = \frac{\Delta H^{\ddagger}}{T} - R \ln(\frac{T}{k}) - R \ln (\frac{k}{h}) \tag{3.15}$$

The transition state theory represents the rate as directly proportional to concentration rather than to activity [Eq. (3.6)]. As a consequence the observed rate constant is a function of a ratio of activity coefficients. Defining a thermodynamic equilibrium constant for Reaction (3.5),

$$K_0^{\ddagger} = \frac{a^{\ddagger}}{a_A a_B} = K^{\ddagger}\frac{\gamma^{\ddagger}}{\gamma_A\gamma_B} \tag{3.16}$$

which, combined with Eq. (3.9), gives

$$k = \frac{kT}{h} K_0^{\ddagger} \frac{\gamma_A\gamma_B}{\gamma^{\ddagger}} = k_0 \frac{\gamma_A\gamma_B}{\gamma^{\ddagger}} \tag{3.17}$$

where k_0 is the rate constant in the reference state of infinite dilution.

Estimates of the activity coefficients can be made

with the Debye-Hückel equation, yielding an equation pre-
dicting the dependence of rate constant on ionic strength.
Data for numerous reactions are consistent with this formulation, which provides justification for the practice
of writing rates in terms of concentrations (usually
molarities) [11].

3.2 Determination of Rate Parameters

Rate Constants. The basic datum obtained from kinetic
studies is the rate constant k under controlled conditions,
which include temperature, pressure, and composition of
the medium. (For reactions in solution, the pressure is
essentially constant and is seldom explicitly considered
or stated.) Organic reactions can usually be studied
with conditions such that the rate equation is reasonably
simple. The basic necessities for carrying out simple
solution kinetics are presented in this section.
The rate equation for a first-order reaction $A \rightarrow B$ is

$$- \frac{dc_A}{dt} = kc_A \tag{3.18}$$

This can be integrated to give

$$\ln \frac{c_A}{c_A^0} = -kt \tag{3.19}$$

where c_A is the concentration of A at time t and c_A^0 is its
concentration at zero time. Converting to base 10 logarithms,

$$\log \frac{c_A}{c_A^0} = - \frac{kt}{2.303} \tag{3.20}$$

A plot of $\log (c_A/c_A^0)$ or of $\log c_A$ against t will be
linear for a first-order reaction; its slope will be equal
to $-k/2.303$. If the analysis gives c_B rather than c_A as
a function of time, the mass balance relationship $c_A^0 =
c_A + c_B$ shows that a plot of $\log(c_A^0 - c_B)$ versus t will
yield the same results. Another form is $\log (c_B^\infty - c_B)$
versus t, where c_B^∞ is the final (t = ∞) concentration of
product.
The half-life, $t_{1/2}$, is defined to be the time required for the concentration to decrease by a factor of
two. Substituting $c_A = c_A^0/2$ into Eq. (3.19) shows that
$t_{1/2} = 0.693/k$. In a first-order reaction, the half-life

is independent of concentration. This provides a test for
the reaction order; if the reaction is first order, the
experimental rate constant should not change when the
initial reactant concentration is varied. Further evidence
that the reaction is first order is the demonstration of a
linear dependence of log c on time over at least two, and
preferably three or four, half-lives.

If a reaction is extremely slow in terms of the time
available to study it, or if its kinetic behavior becomes
complicated in its later stages, it may be preferable to
employ the "initial rate" method. The concentration is
measured as a function of time for as long as is necessary
to evaluate $-(dc_A/dt)_{t\,=\,0}$, but with the restriction that
a negligible extent of reaction (say 1-2%) has occurred.
Then throughout this period of observation $c_A \approx c_A^0$, and
according to Eq. (3.18), the initial velocity v_0 =
$-(dc_A/dt)_{t\,=\,0}$ is a constant; v_0 is the slope of the plot
of c_A against t. The rate constant is found from the
special form of Eq. (3.18), $k = v_0/c_A^0$.

A first-order rate constant has the dimension $time^{-1}$;
the usual units are sec^{-1}, though min^{-1}, hr^{-1}, and $\overline{day^{-1}}$
are sometimes seen in the literature.

Very few true first-order chemical reactions of the
A   B type are encountered, but this basic rate form occurs
in many more complicated systems, and it accounts for most
of the solution kinetics carried out. One of these is the
reversible first-order reaction,

$$A \underset{k_{-1}}{\overset{k_1}{\rightleftharpoons}} B \qquad\qquad (3.21)$$

for which the rate equation is $-(dc_A/dt) = k_1 c_A - k_{-1} c_B$.
Upon integration [2] Eq. (3.22) is obtained,

$$\ln \frac{c_A - c_A^e}{c_A^0 - c_A^e} = -(k_1 + k_{-1})t = -k_{obs}t \qquad\qquad (3.22)$$

where c_A^e is the equilibrium concentration of A. The plot
of log $(c_A - c_A^e)$ against t yields a straight line with
slope equal to $-k_{obs}/2.303$, where $k_{obs} = k_1 + k_{-1}$; this
interesting result will seem reasonable upon reflection.
When equilibrium is attained, $dc_A/dt = 0$, and $k_1 c_A^e$ =
$k_{-1} c_B^e$; rearranging, $c_B^e/c_A^e = k_1/k_{-1} = K$, where K is the
equilibrium constant. Thus by measuring K and k_{obs}, the
individual first-order constants k_1 and k_{-1} can be evaluated.

A very important reaction scheme is exemplified by

the second-order reaction

$$A + B \rightarrow C + D$$

with the rate equation

$$-\frac{dc_A}{dt} = -\frac{dc_B}{dt} = kc_A c_B$$

If $c_A^0 = c_B^0 = c^0$, integration gives [12]

$$\frac{1}{c} - \frac{1}{c^0} = kt \qquad (3.23)$$

A plot of $1/c$ against t will be linear with slope k. More generally $c_A^0 \neq c_B^0$; then Eq. (3.24) is the integrated form of the rate equation [2].

$$\frac{1}{c_B^0 - c_A^0} \ln \frac{c_A^0 \, c_B}{c_B^0 \, c_A} = kt \qquad (3.24)$$

The linear plot of $\log (c_B/c_A)$ against t will have a slope equal to $k(c_B^0 - c_A^0)/2.303$. Sometimes it is convenient to represent the ratio c_B/c_A in the form $(c_B^0 - x)/(c_A^0 - x)$, where x is the decrease in concentration in time t. The second-order rate constant has the dimensions concentration^{-1}-time^{-1}, and usually is given the units liter/mole-sec, or M^{-1}-sec^{-1}.

This method of carrying out second-order reactions requires that appreciable changes in both c_A^0 and c_B^0 occur during the course of the reaction, which means that these two concentrations should be of the same order of magnitude. A large proportion of the second-order rate constants obtained during the first half of the century were determined under these conditions. The quantity x was evaluated by classical analytical methods, very often by titrimetry. This in turn required that the concentrations be substantial, and led to the widespread use of mixed solvents in order to achieve dissolution of the reactants. Greater flexibility with regard to the optional reaction conditions, as well as simplification in data treatment, has been one consequence of the more sensitive modern analytical techniques. Consider this reaction carried out with $c_B^0 >>> c_A^0$; then the concentration of B will undergo no significant change while c_A goes from c_A^0 to zero. Under these conditions the reaction behaves as if it were a first-order reaction; it is said to be a pseudo-first-order reaction. The rate equation will be

$$- \frac{dc_A}{dt} = k_{obs} c_A \tag{3.25}$$

where

$$k_{obs} = kc_B^0 \tag{3.26}$$

The pseudo-first-order rate constant k_{obs} is evaluated as described for a true first-order rate constant, and the second-order rate constant k is calculated from Eq. (3.26). The linearity of the log c_A versus t plot provides evidence that the reaction is indeed first order with respect to A. Repetitions of the entire experiment at different c_B^0 values should then give essentially the same k value if the reaction is also first order in B. The method can obviously be generalized to reactions involving higher orders and more reactants. It is also useful in simplifying complex schemes; for example, the equilibrium A + B $\rightleftharpoons$ C simplifies to Eq. (3.21) if $c_B^0 >>> c_A^0$. The constancy of k can also be tested graphically by plotting k_{obs} against c_B^0 (or other appropriate functions for different systems); linearity indicates that Eq. (3.26) holds, and the slope is equal to k.

Table 3.I Calculation of Second-Order Rate Constants from Pseudo-First-Order Constants[a,b]

10^4 [Allylamine]/$\underline{M}$	$10^3 k_{obs}/sec^{-1}$	$k/\underline{M}^{-1}sec^{-1}$
1.122	3.75	33.4
2.244	7.56	33.7
4.488	15.3	34.2

[a]For Reaction (3.27) at 25.0° in acetonitrile solution.
[b]W. H. Hong, Ph.D. Dissertation, University of Wisconsin, Madison, 1969.

Table 3.I gives an example of a second-order rate constant calculation from pseudo-first-order rate constants. The reaction is the acylation of allylamine by trans-cinnamic anhydride.

$$(C_6H_5CH=CHCO)_2O + CH_2=CHCH_2NH_2 \xrightarrow{k}$$

$$C_6H_5CH=CHCONHCH_2CH=CH_2 + C_6H_5CH=CHCOOH \tag{3.27}$$

The initial anhydride concentration was about $3 \times 10^{-6}\underline{M}$, and the reaction was observed to be first order with respect to this reactant in the presence of excess allylamine. Table 3.I lists k_{obs} values at several concentrations of allylamine. The constancy of k, obtained as the ratio $k_{obs}/[allylamine]$, indicates that the reaction is first order in amine.

Table 3.II   Acylation Kinetics for n-Butylamine with N-trans-Cinnamoylimidazole[a,b]

[amine]/$\underline{M}$	$10^3 k_{obs}/sec^{-1}$	$10^3 k_{obs}/[amine]$	$10^2 k_{obs}/[amine]^2$
0.1099	0.768	6.99	6.36
0.1507	1.44	9.56	6.34
0.1952	2.37	12.1	6.20
0.2340	3.56	15.2	6.50
0.2751	4.88	17.7	6.43
0.3277	6.74	20.6	6.29

[a]At 25.0° in acetonitrile.
[b]W. H. Hong, Ph.D. Dissertation, University of Wisconsin, Madison, 1969.

More complicated behavior is shown by the acylation of n-butylamine with N-trans-cinnamoylimidazole, $\underline{1}$. The

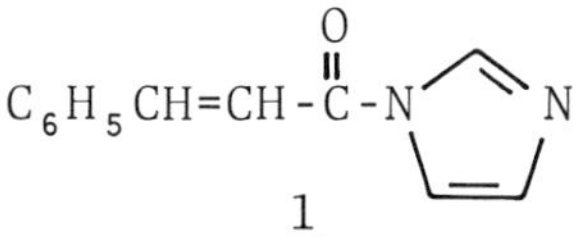

$$C_6H_5 CH{=}CH-\overset{\overset{\textstyle O}{\|}}{C}-N\diagup\diagdown N$$
$$\underline{1}$$

stoichiometric equation has the same form as Eq. (3.27), but the data, given in Table 3.II, show that the reaction (which is first order in cinnamoylimidazole) is not first order in amine. The last column indicates that a reasonable constant is obtained by dividing k_{obs} by the square of the amine concentration. (Another way to show this is by the linearity of a plot of $k_{obs}/[amine]$ against [amine].) The reaction is therefore overall third order, having the rate equation

$$v = k[cinnamoylimidazole][butylamine]^2$$

where $k = 0.0635\ \underline{M}^{-2}\text{-}sec^{-1}$. The significance of this kind of rate equation is considered in a later section.

Throughout this section we have treated equations as if our experimental measurements yield estimates of absolute concentrations (c_A) or relative concentrations (c_A/c_A^0) as a function of time. This is, however, more information than is usually needed to evaluate rate constants. All that is needed is the value of some property, physical or chemical, that changes as a consequence of the chemical reaction; the functional relationship between this property and the concentrations must be known in form, though not quantitatively. Measurement of the course of a reaction by absorption spectroscopy provides an excellent, and perhaps the most important, example; the property is the solution absorbance, and the functional relationship is Beer's law. We apply this to the first-order reaction A → B. The integrated first-order rate equation is log (c_A/c_A^0) = $-kt/2.303$. Thus we seek a relationship between the absorbance and this concentration ratio. Beer's law is assumed to hold for both A and B. The basic requirement is that a wavelength can be found at which the absorbance of the solution undergoes a significant change from its initial value A_0 at $t = 0$ to its final value A_∞ at $t = \infty$. The initial conditions are $c_A = c_A^0$ and $c_B = 0$. Let A_t be the absorbance at time t; ε_A and ε_B are the molar absorptivities of A and B at the experimental wavelength, and b is the path length. Then

$$A_0 = \varepsilon_A b c_A^0$$

$$A_\infty = \varepsilon_B b c_A^0$$

$$A_t = \varepsilon_A b c_A + \varepsilon_B b c_B$$

Combining these equations with the mass balance relationship $c_A^0 = c_A + c_B$ gives the desired equation.

$$\frac{c_A}{c_A^0} = \frac{A_t - A_\infty}{A_0 - A_\infty}$$

Therefore the first-order rate equation can be written

$$\log \frac{A_t - A_\infty}{A_0 - A_\infty} = -\frac{kt}{2.303} \tag{3.28}$$

and a plot of log ($A_t - A_\infty$) against t should have a slope equal to $-k/2.303$. (The quantity $A_t - A_\infty$ is positive if $\varepsilon_A > \varepsilon_B$; for the reverse condition one uses $A_\infty - A_t$[13].)

Table 3.III gives absorbance-time data for the hydrolysis of aspirin anion.

Table 3.III   Rate Study of the Hydrolysis of Aspirin Anion[a]

t/sec	A_t	$\dfrac{A_\infty - A_t}{A_\infty - A_0}$
0	0.007	1.000
180	0.149	0.808
360	0.273	0.640
540	0.383	0.491
720	0.470	0.373
900	0.525	0.298
1200	0.602	0.194
1500	0.650	0.129
1800	0.685	0.081
	0.745	0.000

[a]Conditions: temp. 98°C; pH 5.01; 0.02 M acetate buffer; initial aspirin concentration 2.1×10^{-4} M; followed at 296 nm.

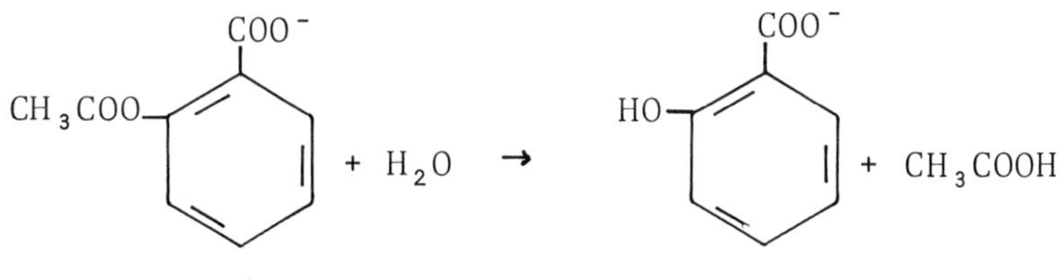

$$CH_3COO\text{-}C_6H_4\text{-}COO^- + H_2O \rightarrow HO\text{-}C_6H_4\text{-}COO^- + CH_3COOH$$

Although water appears in the stoichiometric equation, its concentration remains essentially constant throughout the reaction, and it is not taken into account. The plot according to Eq. (3.28) is shown in Fig. 3.2. The slope is -5.8×10^{-4} sec^{-1}, so the first-order rate constant is 1.34×10^{-3} sec^{-1}. The half-life is 517 sec; the reaction was followed for about $3^{1}/_{2}$ half-lives, corresponding to about 90% completion of the reaction [14,15].

This method of following first-order reactions is valid for any property that is directly proportional to concentration and that is additive. Ultraviolet and visible spectrophotometry are the most widely used techniques, but others may be useful. Infrared spectroscopy, refractometry, optical rotatory dispersion, dilatometry (measurement of volume changes), pressure measurements, nuclear magnetic resonance, and conductimetry are other techniques that may be applied, though the concentration range over which a linear relationship exists may be limited for some of them.

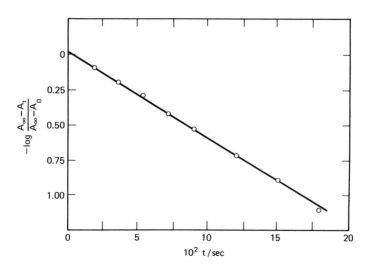

Fig. 3.2. First-order plot of data in Table 3.III for the hydrolysis of aspirin.

The technique requires the value of the property at "infinity" time, that is, when the reaction is essentially complete; in Eq. (3.28) this is the quantity A_∞. This should be measured after the lapse of nine to ten half-lives. Sometimes it is inconvenient or impossible to measure this infinity value, perhaps because the half-life is very long, or because a subsequent reaction of the product occurs, or because it is beyond the capability of the measuring technique. Guggenheim [16] described a method to overcome this limitation. It is necessary that the reaction be first order for this method to be valid. Writing Eq. (3.28) in the exponential form gives

$$(A_t - A_\infty) = (A_0 - A_\infty)e^{-kt}$$

A corresponding equation can be written for a reaction time $(t + \Delta)$, where Δ is a time interval.

$$(A_{t + \Delta} - A_\infty) = (A_0 - A_\infty)e^{-k(t + \Delta)}$$

Subtracting these equations and rearranging,

$$(A_t - A_{t + \Delta}) = (A_0 - A_\infty)e^{-kt}(1 - e^{-k\Delta})$$

or, in logarithmic form,

$$\log(A_t - A_{t+\Delta}) = \frac{-kt}{2.303} + \log(A_0 - A_\infty)(1 - e^{-k\Delta}) \quad (3.29)$$

If Δ is a constant time interval, the final term in Eq. (3.29) is a constant. A plot of $\log(A_t - A_{t+\Delta})$ against t will be linear with slope equal to $-k/2.303$, exactly as with the usual first-order plot. The time interval Δ should be as large as possible; one to two half-lives is usually adequate. With modern instrumental techniques that can yield a continuous tracing of a physical property as a function of time, it is very easy to read the required values of A_t and $A_{t+\Delta}$ from a chart.

Rate constants can be sensitive functions of temperature, ionic strength, and solvent composition (including mixed solvent ratios, buffer components, and pH). Some of these dependencies are treated subsequently. Temperatures should be held constant to within 0.01-0.10°C for work of ordinary accuracy. Often it is desirable to check the dependency of a rate constant on ionic strength or buffer concentration by varying this factor while holding other variables (pH, temperature, solvent) as constant as possible. An appreciable dependence may require evaluation of the apparent rate constant at several values of the variable, and extrapolation to the zero level.

A valuable collection of rate constants from the literature is available in a series of tables prepared by the National Bureau of Standards [17].

Energies and Entropies of Activation. The Arrhenius equation, (3.3), describes the temperature dependence of rate constants. Modifications have been suggested to account for temperature dependence of A and E_a, but these seem unnecessary for most systems. The activation energy E_a is evaluated by means of the logarithmic form of this equation.

$$\log k = \log A - \frac{E_a}{2.303\ RT} \quad (3.30)$$

From the slope of an Arrhenius plot of $\log k$ against $1/T$ the activation energy is calculated by

$$E_a = -(4.576)(slope)\ cal/mole \quad (3.31)$$

The pre-exponential factor A can then be calculated from Eq. (3.30).

The thermodynamic parameters can be calculated with equations given earlier. Thus, from Eq. (3.14),

$\Delta H^{\ddagger} = E_a - RT$; in using this formula T is taken in the middle of the experimental temperature range, and the proper value of R is 1.987 cal/mole-deg. The uncertainty in E_a and $\Delta H^{\ddagger}$ is commonly several hundred calories per mole. $\Delta S^{\ddagger}$ is obtained from Eq. (3.15), which can be written

$$\Delta S^{\ddagger} = \frac{\Delta H^{\ddagger}}{T} - 4.576 \log \frac{T}{k} - 47.22 \qquad (3.32)$$

$\Delta S^{\ddagger}$ is in entropy units. The rate constant k must be expressed in time units of seconds in using this equation. From the relationship $\Delta G^{\ddagger} = \Delta H^{\ddagger} - T\Delta S^{\ddagger}$ the free energy of activation is calculated.

Table 3.IV   Variation of Rate Constant with Temperature in a Vinyl Ether Hydrolysis[a]

$t/°C$	$10^3(°K)/T$	$k/\underline{M}^{-1}\text{-sec}^{-1}$
22	3.39	0.467
30	3.30	0.725
35	3.25	1.033
40	3.19	1.300

[a]For Reaction (3.33) in water, ionic strength 1.0 $\underline{M}$.

Table 3.IV gives rate constant-temperature data for the acid-catalyzed hydrolysis of 4-methoxy-3-buten-2-one, a vinyl ether [18].

$$CH_3OCH{=}CHCOCH_3 + H_2O \xrightarrow{H^+} CH_3OCH_2CHO + CH_3OH \qquad (3.33)$$

The reactions were carried out under pseudo-first-order conditions of constant pH, and second-order constants were obtained as the slopes of plots of k_{obs} against a_{H^+}. Figure 3.3 is the Arrhenius plot, which gives a slope of -2215 deg. E_a is therefore 10.1 kcal/mole and $\Delta H^{\ddagger}$ is 9.5 kcal/mole. From Eq. (3.32), applied at 40°C, $\Delta S^{\ddagger}$ = -27.8 e.u.

The enthalpy of activation is very sensitive to errors in the rate constants. This problem has been quantitatively analyzed by considering the effect of the maximum error in the rate constants on the error in $\Delta H^{\ddagger}$ [19]. For example, it is calculated that if the maximum error in the rate constants is 5%, and the enthalpy of activation is based on rate constant determinations at 0°

and 20°C, the error in $\Delta H^{\ddagger}$ is 0.8 kcal/mole. If two such
$\Delta H^{\ddagger}$ values are being compared, each possesses this uncertainty, so their difference must be substantially greater
than 1.6 kcal/mole before they can be judged significant-
ly different. It is found [19] that the error in $\Delta S^{\ddagger}$ is
linearly related to the error in $\Delta H^{\ddagger}$.

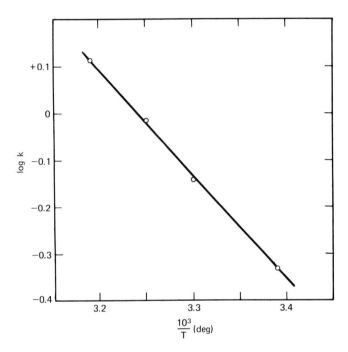

Fig. 3.3. Arrhenius plot for acid-catalyzed hydrolysis
of 4-methoxy-3-buten-2-one; data from Table 3.IV.

The mechanistic interpretation of an experimental
activation energy or enthalpy of activation may be hazar-
dous if the nature of the experimental rate constants is
not well understood. If the observed rate constant is a
function of several rate constants each describing a step
in a complex reaction scheme, the Arrhenius plot with the
observed rate constants may give an activation energy that
is a function of the activation energies of the several
steps (if, indeed, a linear Arrhenius plot is observed at
all). Consider the relatively simple example of the reversible reaction $A \rightleftharpoons B$; as shown earlier, the observed
first-order rate constant is the sum of the rate constants
for the forward and reverse reactions; $k_{obs} = k_f + k_r$, or,

$$k_{obs} = A_f e^{-E_f/RT} + A_r e^{-E_r/RT}$$

In the special case that $E_f = E_r = E$ this takes the form

$$k_{obs} = (A_f + A_r) e^{-E/RT}$$

and the Arrhenius plot will yield the activation energy $E = E_f = E_r$. In general, however, $E_f \neq E_r$, and curvature is possible in the Arrhenius plot [20]. In fact, a curved Arrhenius plot is usually evidence of a complex reaction system.

Another important type of reaction system may lead to misinterpretation of activation energies. If a second-order reaction is cast into a pseudo-first-order reaction by holding one of the reactant concentrations essentially constant, it is often best to prepare the Arrhenius plot with the calculated second-order rate constants rather than with the observed pseudo-first-order rate constants. This is because fast pre-equilibria (usually acid-base equilibria) undergone by the species whose concentration is included in the pseudo-first-order constant may make a contribution to the observed activation energy. Suppose the pseudo-first-order rate constant has the form $k_{obs} = kc_{OH}$, where c_{OH} is the hydroxide ion concentration, and suppose further that k_{obs} is determined at several temperatures with pH being maintained the same at the different temperatures. Then $\ln k_{obs} = \ln k + \ln c_{OH}$, and the terms on the right may be expressed by logarithmic forms of the Arrhenius and van't Hoff equations (remembering that $c_{OH} = K_w/c_H$):

$$\ln k = \ln A - E_a/RT$$

$$\ln c_{OH} = \ln K_w - \ln c_H = \text{constant} - \Delta H_i^0/RT$$

Therefore the observed rate constant is related to temperature by an equation of the Arrhenius form,

$$\ln k_{obs} = \text{constant} - E_{obs}/RT \qquad (3.34)$$

where $E_{obs} = E_a + \Delta H_i^0$. Thus the observed activation energy is different from the correct E_a by the heat of ionization of water. The correct procedure is to calculate the second-order rate constants, using the value c_{OH} obtained from K_w appropriate to the experimental temperatures, and to construct the Arrhenius plot with these rate constants. Similar problems arise when studying catalytic properties of weak acids or bases in pH ranges near their pK_a's; then

the pK_a of the catalyst (which is the species in excess)
must be known at each experimental temperature in order
to calculate the concentration of the reactive species at
that temperature.

$\Delta G^{\ddagger}$ and $\Delta S^{\ddagger}$ values depend upon the choice of standard
state, just as for equilibria discussed in Chap. 2. Nearly
all solution kinetics studies are carried out with the
molarity concentration scale. Unitary free energies and
entropies of activation can be obtained from the conventional quantities as described for equilibria by converting from the molarity basis to the mole fraction scale.
This conversion requires Δq, the change in number of particles between products and reactants as reflected in the
definition of the "equilibrium constant," $K^{\ddagger}$. In this
reaction rate situation, the "product" is the activated
complex, and it is usually true that only one particle
appears in the transition state region. Perhaps the most
frequently encountered class of rate-determining reaction
in organic chemistry is the bimolecular reaction

$$A + B \rightleftharpoons M^{\ddagger} \tag{3.35}$$

for which $\Delta q = -1$. The relationship between the equilibrium constant expressions on the molarity and mole ratio
scales is, by analogy with Eq. (2.22),

$$\frac{K_c^{\ddagger}}{K_y^{\ddagger}} = (M^* \rho_0)^{\Delta q} \tag{3.36}$$

where M^* is the number of moles of solvent per 1000 g and
ρ_0 is the density of the solvent; in very dilute solution
$K_y^{\ddagger}$ approaches $K_x^{\ddagger}$, the constant on the mole fraction scale.
Applying Eq. (3.9) allows Eq. (3.36) to be written
in terms of rate constants.

$$k_y = k_c (M^* \rho_0)^{-\Delta q} \tag{3.37}$$

or $\log k_y = \log k_c - \Delta q \log (M^* \rho_0)$. From Eq. (3.32) the
entropies of activation on the two scales can be written
$\Delta S_c^{\ddagger} = 4.576 \log k_c$ - constant and $\Delta S_y^{\ddagger} = 4.576 \log k_y$ -
constant, since $\Delta H^{\ddagger}$ is the same for both standard states.
These relationships are combined with Eq. (3.37) to give

$$\Delta S_y^{\ddagger} = \Delta S_c^{\ddagger} - 4.576 \Delta q \log (M^* \rho_0) \tag{3.38}$$

In very dilute solution $\Delta S_y^{\ddagger}$ is equal to $\Delta S_x^{\ddagger}$, which is the
unitary entropy of activation. Note that, since for a
rate process Δq is usually zero or negative [21], the

unitary entropy of activation is equal to or more positive
than the conventional entropy of activation. Entropies of
activation have some utility in characterizing the transition state; a greater loss of entropy in the activation
process (more negative $\Delta S^{\ddagger}$) indicates greater constraint
in passing from the initial to the transition state. For
example, a reaction in which two or three molecules must
be brought together to form the transition state will have
a negative entropy of activation. In making comparisons
of entropies of activation for reactions conducted in different solvents the unitary entropies of activation should
be compared, for these do not include the variable mixing
contribution.

Table 3.V lists activation parameters (E_a, A, $\Delta H^{\ddagger}$,
$\Delta S^{\ddagger}$) for several kinds of reactions. [The table is re-
dundant, of course; the pairs of parameters (E_a, A) and
($\Delta H^{\ddagger}$, $\Delta S^{\ddagger}$) contain the same information.] Frequently rate
constant data are recorded in the logarithmic form of the
Arrhenius equation; for example, the kinetic behavior of
the first reaction in Table 3.V can be represented by
log k = 7.532 - 16,200/RT. Notice that most energies of
activation fall in the range 10-25 kcal/mole. The rough
rule of thumb that reaction rates increase by a factor of
two to three for each 10°C rise in temperature is based
on this observation and the integrated form of the
Arrhenius equation,

$$\log \frac{k_2}{k_1} = \frac{E_a(T_2 - T_1)}{2.303 \ RT_1 T_2} \tag{3.39}$$

With Eq. (3.39) it is found, for T_1 = 300°K and T_2 =
310°K, that if E_a = 10 kcal/mole, k_2/k_1 = 1.72; if E_a =
15, k_2/k_1 = 2.24; if E_a = 20, k_2/k_1 = 2.95; and if E_a =
25, k_2/k_1 = 3.86.

3.3   Interpretation of Kinetic Data

Kinetic data may be put to several analytical uses.
One of these is the informed control of reaction rates to
achieve greater reactivity, greater stability, or selectivity. Knowledge of the relevant rate equation and kinetic parameters will then permit prediction of the experimental conditions (temperature, medium, concentrations) required for the desired rate effect. In an intu-
itive or qualitative way this application of kinetics is
widely used, as when a reaction rate is increased by
raising the temperature or adjusting the pH. This chapter
provides the background to place this kind of analytical
practice on a quantitative basis.

Another analytical application of kinetics that has recently been exploited is the direct use of kinetic data to determine concentrations. This is a particularly interesting approach to the analysis of mixtures. Experimental designs for the kinetic analysis of mixtures are based upon the elementary considerations of this chapter; they are not reviewed here [22]. Kinetic methods of enzyme analysis, and the use of enzymes in rate determinations of substrate concentrations, provide further illustration of the direct analytical use of kinetic data [23].

This section is concerned with a third use of kinetic data, their application to the elucidation of reaction mechanisms. The mechanism of a reaction is "understood" when it is possible to specify with reasonable assurance the structures of all reactants and products and of all transition states and intermediates on the reaction path. It is often pointed out that one can never prove that a postulated mechanism is the correct one; all that can be done is to show that it is consistent with available data. There may be other mechanisms, as yet unimagined, that also are consistent with the data, or data yet to be obtained may show the postulated mechanism to be invalid. In this last way one can prove a mechanism wrong, and this is often useful information, because the possible mechanisms are probably few in number.

This section treats some general features of theory and experiment that can be employed in elaborating reaction mechanisms. The following sections, and Chap. 4, deal with some important specifics.

The Rate Equation. The first goal in the kinetic investigation is the determination of the rate equation. This may not be as simple as it appeared in the preceding section. Most organic reactions in solution can be studied under pseudo-first-order conditions, and we treat only this type of situation. The kinetics therefore yield an empirical rate equation of this form:

$$v = k_{obs} c_A \qquad (3.40)$$

where k_{obs} is the pseudo-first-order rate constant, the determination of which has established that the rate equation is first order with respect to A. The next step is to express k_{obs} as a function of concentrations and true rate constants. Presuming that Eq. (3.40) holds, it is known that k_{obs} is not a function of c_A. The detailed analysis of k_{obs} is accomplished by means of systematic investigations of the dependence of k_{obs} on system variables; and the design of these experiments is based upon

Table 3.V Activation Parameters for Some Solution Reactions[a]

Reaction	Medium
$CH_3COOC_2H_5 + H_2O \xrightarrow{H^+} CH_3COOH + C_2H_5OH$	56 wt % acetone
$CH_3COOCH_2C_6H_5 + OH^- \rightarrow CH_3COO^- + C_6H_5CH_2OH$	40% v/v acetone
$C_2H_5Br + OH^- \rightarrow C_2H_5OH + Br^-$	1 liter acetone + 400 g H_2O
$C_6H_5CH_2NO_3 + Br^- \rightarrow C_6H_5CH_2Br + NO_3^-$	acetone
$CH_3Br + I^- \rightarrow CH_3I + Br^-$	water
$(CH_3)_2CHC\ell + OH^- \rightarrow CH_3-CH=CH_2 + C\ell^- + H_2O$	80% ethanol
$CH_2(COOH)COO^- \rightarrow CH_3COO^- + CO_2$	water
$CH_3I + N(C_2H_5)_3 \rightarrow I^- + CH_3N^+(C_2H_5)_3$	benzene
$NaBH_4 + 3H_2O \rightarrow NaH_2BO_3 + 4H_2O$	water

[a]Based on time in seconds, concentrations in moles per liter; Ea and A from Ref. [17].

[b]G. Davies and D. P. Evans, J. Chem. Soc., 1940, 339.

[c]E. Tommila and C. N. Hinshelwood, J. Chem. Soc., 1938, 1801.

[d]R. A. Fairclough and C. N. Hinshelwood, J. Chem. Soc., 1937, 538.

[e]J. W. Baker and W. S. Nathan, J. Chem. Soc., 1936, 236.

[f]E. A. Moelwyn-Hughes, J. Chem. Soc., 1938, 779.

[g]E. D. Hughes and U. G. Shapiro, J. Chem. Soc., 1937, 1277.

[h]G. A. Hall, Jr., J. Am. Chem. Soc., 71, 2691 (1949).

[i]H. C. Raine and C. N. Hinshelwood, J. Chem. Soc., 1939, 1378.

[j]R. L. Pecsok, J. Am. Chem. Soc., 75, 2862 (1953).

E_a/kcal-mole^{-1}	A	$\Delta H^{\ddagger}$/kcal-mole^{-1}	$\Delta S^{\ddagger}$/e.u.	Ref.
16.2	3.4×10^7	15.6	-26	b
12.3	8.1×10^7	11.7	-24	c
18.5	2.5×10^9	17.9	-18	d
18.4	1.7×10^{10}	17.8	-14	e
18.3	1.7×10^{10}	17.7	-14	f
24.8	4.8×10^{10}	24.1	-12	g
28.1	1.48×10^{11}	27.4	-10	h
9.7	2.1×10^4	9.1	-41	i
8.9	8×10^{11}	8.3	-6	j

chemical knowledge and perhaps a few preliminary studies. Some simple examples were shown in Sec. 3.2.

Elucidation of the explicit form of k_{obs} is most challenging for so-called complex reactions, which are reaction systems composed of two or more simple reactions. (If the written reaction scheme contains two or more arrows, it is complex.) Consider the series system,

$$A \xrightarrow{k_1} B \xrightarrow{k_2} C \tag{3.41}$$

The theoretical rate equation for the decrease in A concentration is $v_A = k_1 c_A$. If the analytical measurements permit c_A to be followed, evidently the experimental rate equation will be $v_A = k_{obs}^A c_A$; hence $k_{obs}^A = k_1$. In general, the kinetic behavior with respect to product C will be more complex, because this has the theoretical rate equation $v_C = k_2 c_B$, where c_B is a function of c_A, k_1, and k_2. In a special case, which, however, happened to be widely applicable, the situation becomes very simple. Suppose that over a considerable course of the reaction the change in c_B is negligible relative to the change in concentrations of other species. Then we can set $dc_B/dt = k_1 c_A - k_2 c_B \simeq 0$, whence

$$c_B \simeq \frac{k_1 c_A}{k_2} \tag{3.42}$$

giving, for v_C,

$$v_C = k_1 c_A = v_A$$

The rate of production of C is equal to the rate of loss of A, and $k_{obs} = k_1$ whether A or C is measured. This assumption that the change in c_B is negligible is an example of the steady state approximation. It is equivalent to assuming that c_B is very small relative to c_A and c_C over most of the reaction course. For Reaction (3.41), this requires that $k_2 >> k_1$. With this condition, substance B may be considered as a reactive intermediate in the reaction.

Some rather complicated systems can be rendered manageable by the steady-state assumption. A reasonable description of the nucleophilic reactions of some carboxylic acid derivatives is provided by Eqs. (3.43)-(3.45),

$$S + OH^- \xrightarrow{k_1} P + Q \tag{3.43}$$

$$S + N \overset{k_2}{\underset{k_{-2}}{\rightleftharpoons}} I + Q \tag{3.44}$$

$$I + OH^- \overset{k_3}{\rightarrow} P + N \tag{3.45}$$

where S is the substrate (the compound undergoing reaction), N is an added nucleophile, I is a reactive intermediate (the acylated nucleophile), and P,Q represent products. The rate equation is

$$v = \frac{dc_P}{dt} = k_1 c_S c_{OH} + k_3 c_I c_{OH} \tag{3.46}$$

To the unstable intermediate I the steady-state approximation is (tentatively) applied.

$$\frac{dc_I}{dt} = k_2 c_S c_N - k_{-2} c_I c_Q - k_3 c_I c_{OH} = 0$$

$$c_I = \frac{k_2 c_S c_N}{k_{-2} c_Q + k_3 c_{OH}} \tag{3.47}$$

Combining Eqs. (3.46) and (3.47),

$$v = \left[k_1 c_{OH} + \frac{k_2 k_3 c_N c_{OH}}{k_{-2} c_Q + k_3 c_{OH}} \right] c_S$$

Since the experimental rate equation, under pseudo-first-order conditions of constant pH and excess nucleophile, is $v = k_{obs} c_S$, evidently

$$k_{obs} = k_1 c_{OH} + \frac{k_2 k_3 c_N c_{OH}}{k_{-2} c_Q + k_3 c_{OH}} \tag{3.48}$$

This derivation is characteristic of the application of the steady-state approximation and of the development of relationships between the pseudo-first-order rate constant and the reaction variables. Some of the constants can be determined with the aid of Eq. (3.48), if it is valid; and at least this type of thinking provides a rational and often efficient experimental design. Thus, in the present instance, k_1 is readily determined in the usual manner by setting $c_N = 0$ and measuring k_{obs} at constant pH; $k_1 = k_{obs}/c_{OH}$. In the presence of N and at very high pH the condition $k_3 c_{OH} \gg k_{-2} c_Q$ should be approached, and Eq. (3.48) becomes

$$k_{obs} - k_1 c_{OH} = k_2 c_N$$

A plot of the left-hand side of the equation against c_N, at constant pH, yields k_2. Other limiting cases can be suggested. Agreement of such predicted functional dependencies with the data provides support for the postulated reaction scheme and the associated rate equation.

It is quite possible that the steady-state approximation will fail, or that its validity may depend upon the reaction conditions [24]. Examination of Eqs. (3.44) and (3.45) shows that at very high pH, where the k_3 step is fast, the formation of I will be rate limiting, the concentration of I will be very low, and the steady-state assumption probably will be valid. At lower pH, however, the k_3 step may become very slow and possibly even rate determining, the steady-state assumption will not apply, and the kinetics may become very complicated. Deviations from first-order kinetics under normal pseudo-first-order reaction conditions indicate the presence of an intermediate.

Rate equations often include concentrations of ionizable species, particularly weak acids and bases. It will then be possible to write some of the rate terms in forms that are kinetically equivalent; that is, rate data cannot distinguish between two such representations. Suppose the reactant is a weak acid HA. Then a rate term k[HA] is kinetically indistinguishable from the rate term k'[H+][A-], since the two concentration products are directly related by the acid dissociation constant K_a,

$$K_a[HA] = [H^+][A^-]$$

and it can be seen that $k = k'K_a$. There is no justification for the inclusion of two kinetically equivalent rate terms in a rate equation on the basis of kinetic results alone, and the choice of term to be used is usually guided by mechanistic ideas [25]. A similar situation occurs when two reactants, say HA and HB, are ionizable. Because of the equilibrium

$$HA + B^- \rightleftharpoons HB + A^-$$

the rate terms k[HA][B-] and k'[HB][A-] are kinetically equivalent. Although the charge types of the species are irrelevant in this context, note that kinetically equivalent rate terms always reflect the same net charge and composition; this property, in fact, accounts for their indistinguishability.

The Transition State. The utility of transition state
theory in studying reaction mechanisms is largely manifested by reaction coordinate diagrams, which provide a
medium for qualitative descriptions of reaction features.
The experimentalist will visualize the reaction path in
terms of either enthalpy changes (which have the advantage
of being independent of standard state, and the disadvantage of difficulty of interpretation) or free energy
changes (which are dependent on standard states, but are
immediately relatable to the observable reactivity). The
nature of a free energy reaction profile is illustrated
with the hypothetical reaction

$$R \underset{k_{-1}}{\overset{k_1}{\rightleftharpoons}} I \underset{k_{-2}}{\overset{k_2}{\rightleftharpoons}} P \qquad (3.49)$$

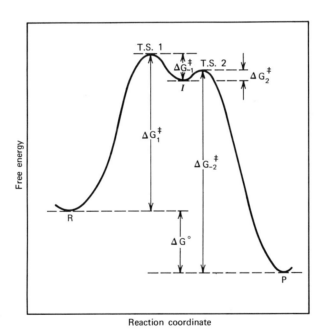

Fig. 3.4. Free energy reaction coordinate diagram for
the reaction $R \rightleftharpoons I \rightleftharpoons P$. (T.S. $\equiv$ transition state.)

Reactant R (which may symbolize more than one reactant
species) is transformed into product(s) P via the intermediate I. All reactions must be reversible for meaning-
ful representation. Figure 3.4 is a reaction diagram of

this reaction with arbitrarily selected energy properties.
The horizontal axis is the reaction coordinate representing progress along the reaction path. Free energy, or
rather the sum of the partial molar free energies, is
plotted vertically. The energy reference level is arbitrary, since only changes in free energy are of concern.
     Evidently the product (final state) is more stable
than the reactant (initial state) by an amount ΔG^0, which
is given by the thermodynamic relationship $\Delta G^0 = -RT \ln K$,
where K is the equilibrium constant for the overall reaction. The reaction path includes an intermediate I that
is unstable with respect to both R and P. The intermediate, however, represents a local energy minimum, which
differentiates it from a transition state; a transition
state corresponds to a local energy maximum. This overall
reaction must have two transition states, one for the re-
action R $\rightleftharpoons$ I and the second for I $\rightleftharpoons$ P. The magnitudes of
the four free energies of activation corresponding to the
four rate constants are given by the vertical distances
from the respective initial state to the appropriate ener-
gy barrier. A larger $\Delta G^{\ddagger}$ is equivalent to a smaller rate
constant; thus from Fig. 3.4 it is seen that $k_2 > k_{-1} > k_1 > k_{-2}$
for this reaction. It is easily shown that the equili-
brium constant can be related to the rate constants by

$$K = \frac{k_1 k_2}{k_{-1} k_{-2}} \qquad\qquad (3.50)$$

thus K is greater than unity, in agreement with the earlier
conclusion.
     For Reaction (3.49) as it proceeds from left to right,
it is obvious from Fig. 3.4 that the first step, formation
of the intermediate, is the rate-determining step of the
overall reaction [26]. The second step, decomposition of
the intermediate to form P, is fast and is not reflected
in the experimental kinetics; it follows that the kinetic
measurements can give no information about this step [27].
     Figure 3.4 shows very clearly that there is no neces-
sary relationship between the rate of a reaction (measured
by $\Delta G^{\ddagger}$, in the figure) and its equilibrium properties (i.
e., ΔG^0). These quantities are often closely related,
however, and such correlations are dealt with in Chap. 4.
     Our concern thus far has been with the vertical coor-
dinates of the free energy reaction diagram and, although
these will usually be employed in a qualitative way to
indicate relative reactivities, they are susceptible of
quantitative statement. The horizontal position, the reaction coordinate, is less clear-cut in its interpretation.
This represents the progress of the reaction along the

multidimensional energy surface, but it is conventional, and useful, to picture it as the fractional progress from reactants to products. The question then occurs: At what point along this reaction coordinate does the transition state appear? A qualitative answer can be given. By definition the transition state represents the maximum energy, that is, the least stable configuration, along the reaction path. It is therefore reasonable to expect the transition state to more closely resemble the less stable of the two other defined states (initial and final states [28]. This is represented graphically by placing the maximum in the reaction coordinate diagram closer to the less stable state. Figure 3.4 incorporates this idea. For the process R $\rightleftharpoons$ I the species I is less stable than R, so the transition state (T.S. 1) is placed closer to I than to R. Similarly for the process I $\rightleftharpoons$ P, the postulate places T.S. 2 closer to I than to P. Since the structures of initial and final states are usually known, this rule provides guidance for describing transition state structures.

The rate equation gives immediate information about the reaction mechanism. Consider first a simple reaction. According to transition state theory, the rate of this reaction is proportional to the concentration of the transition state species (the activated complex). Experimentally, we find that the rate is proportional to some simple function of the reactant concentrations. It is therefore reasonable to postulate that the experimental rate equation gives the composition of the transition state. The data of Tables 3.I and 3.II illustrate the application of this rule. For the acylation of allylamine with cinnamic anhydride the rate equation was found to be v = k[anhydride][amine]; hence we conclude that the transition state for this reaction includes one anhydride molecule and one amine molecule, though of course this merely expresses the net composition of the transition state and says nothing about the mutual disposition of the atoms. In contrast, Table 3.II shows that the rate equation for the acylation of n-butylamine with cinnamoylimidazole is v = k[cinnamoylimidazole][amine]^2; therefore the transition state includes one cinnamoylimidazole molecule and two amine molecules.

In complex reactions the rate equation gives no information about fast steps that occur after the rate-determining step. It may, however, be influenced by fast steps that precede the rate-determining step. Here again the transition state theory provides guidance. As a consequence of the postulate that a pseudo-equilibrium exists between the initial and transition states, it is not possible to obtain detailed information from the rate equation concerning the reaction path between these states. Hence any fast steps (such as proton transfers) preceding the rate-determining step are included as if they were part of the rate-determining step, and the rate term gives an overall stoichiometric composition of the transition state without specifying the order in which the components were brought together.

Earlier the concept of kinetic equivalence of rate terms was introduced. The meaning of this equivalence should now be clearer. Since the rate term merely provides the net stoichiometric composition (including charge) of the transition state, kinetically equivalent rate terms contain identical kinetic information.

The rate equation gives no direct information about the involvement of solvent molecules in the transition state, because it is not possible to obtain, in an unambiguous way, the kinetic dependence upon the solvent concentration. (This is because any substantial alteration in the solvent composition, which would be required to establish the kinetic dependence, will concurrently alter other solvent properties that might affect the rate.) This weakness of kinetics means that the overall order of the reaction may not be the same as the total number of molecules contributing to the composition of the transition state. This latter quantity is called the molecularity of the reaction, and is more precisely defined [29] as the number of molecules necessarily undergoing covalency changes. A bimolecular reaction will usually be overall second order, but if one of the reactants is the solvent it will show first-order kinetics.

Since the entropy of activation is related to the relative constraints on the transition and the initial states, it may tell something about the molecularity of the reaction. A unimolecular reaction should have a $\Delta S^{\ddagger}$ value near zero (value independent of standard state for a first-order process), whereas a bimolecular reaction is expected to have an appreciable negative $\Delta S^{\ddagger}$, reflecting the loss of entropy incurred in bringing two reactant molecules into the single transition state molecule. Within limits such expectations are realized; thus for numerous acid-catalyzed ester hydrolyses believed on other grounds to proceed by a unimolecular mechanism, the $\Delta S^{\ddagger}$ values are positive or small negative numbers; for bimolecular ester hydrolyses $\Delta S^{\ddagger}$ is -15 to -25 e.u. [30].

If the mechanism, that is, the transition state, is known for a reaction in one direction, it is also known for the reverse reaction; it will be the same mechanism run backward. This is a consequence of the principle of microscopic reversibility, which states that, at equilibrium, any reaction and its reverse will have equal rates. This is not a requirement of thermodynamics [31]. The equality of opposing rates is assumed when deriving, for a reversible reaction

$$A \underset{k_{-1}}{\overset{k_1}{\rightleftharpoons}} B$$

the relationship $K = k_1/k_{-1}$ between the equilibrium constant and the rate constants. There is no doubt of the validity of this equation under the conditions (i.e., equilibrium) specified in the principle of microscopic reversibility. It should be observed, however, that rate constants are usually determined far from the equilibrium position, and the equivalence of K and the ratio k_1/k_{-1} for such rate constants is not assured. In fact very few tests (which consist of the independent determination of k_1, k_{-1}, and K) of this equality have been made [32].

The principle of microscopic reversibility is sometimes used as a mechanistic tool to eliminate a mechanism by showing that it is not the exact reverse of the known forward mechanism, and is therefore impossible. The preceding cautionary paragraph is, with amplification, relevant here. It is quite possible that more than one pathway is available for the process. Then the principle says that at equilibrium the favored pathway for the forward reaction is the same as that for the reverse reaction; that is, the rates of forward and reverse reactions are equal for this pathway, and they are likewise equal for each pathway. When the system is not at equilibrium, however, it is possible for the favored reaction in one direction to be different from that in the other direction. Since kinetic studies are not often made with the system at equilibrium, this point may be an important one [33].

3.4 Medium Effects

Types of Effects. By medium effects we mean those effects on reaction rates attributable to properties or constituents, other than reactants, of the reaction medium. More specific effects are treated in Secs. 3.5 (pH effects) and 3.6 (catalysis).

The delta operators [34] used to describe medium and substituent effects on equilibria (Sec. 2.3) are convenient for rate effects. The symbol Δ refers to a change caused by a chemical process (in this context the transformation from initial state to transition state), δ_M signifies a change caused by the medium, and δ_R indicates a substituent effect. These operators can be combined, and they commute. Their use can be illustrated with the data in Table 3.VI, which gives rate constants for the reaction of bromphenol blue and hydroxide ion in water-ethanol mixtures. From these data the medium effect on the rate, expressed as $\delta_M \Delta G^{\ddagger}$, is calculated by means of the transition state theory equation, $k = (kT/\underline{h}) \exp(-\Delta G^{\ddagger}/RT)$. This equation can be written for the reference system (let this be the fully aqueous system), with k_0 and $\Delta G_0^{\ddagger}$ signifying this state, and for a mixed solvent system (k_M, $\Delta G_M^{\ddagger}$). Combining and rearranging gives $\delta_M \Delta G^{\ddagger} = \Delta G_M^{\ddagger} - \Delta G_0^{\ddagger}$.

$$\delta_M \Delta G^{\ddagger} = -2.303 \ RT \ \log \frac{k_M}{k_0}$$

Table 3.VI   Medium Effect on the Bromphenol Blue-Hydroxide Ion Reaction[a]

Wt. % ethanol	$k/\underline{M}^{-1}$-day^{-1}	$\delta_M \Delta G^{\ddagger}$/kcal-mole^{-1}
0.00	25.2	(0.00)
10.2	9.71	0.57
15.4	5.46	0.91
20.6	3.01	1.26
31.5	0.103	3.26

[a]At 25° and zero ionic strength. Data from E. S. Amis and V. K. LaMer, J. Am. Chem. Soc., 61, 905 (1939).

The units in which k is expressed will not affect the value of $\delta_M \Delta G^{\ddagger}$, since the rate constants appear as a ratio. This example clearly shows that the logarithm of a rate constant is related to $\Delta G^{\ddagger}$, and the logarithm of a ratio of rate constants is proportional to $\delta \Delta G^{\ddagger}$. These functions will often be encountered in studies of solvent and substituent effects.

Reaction rates may be affected by added electrolytes [35]. The primary salt effect is a consequence of the dependence of activity coefficients on the ionic strength, as shown in Eq. (3.17). The secondary salt effect is not a true kinetic phenomenon. If the rate equation includes the concentration of a species involved in a fast equilibrium, then this concentration will be dependent upon the

ionic strength because the equilibrium constant is a function of ionic strength. This salt effect can be eliminated, or rather taken into account, by calculating rate
constants on the basis of the actual reactant concentrations, as evaluated by means of equilibrium constants
applicable at the experimental ionic strength.

Rates are also dependent upon the "polarity" of the
reaction solvent. No universally applicable measure of
polarity has yet been devised, though several concepts are
available. The dielectric constant appears in equations
obtained by considering the effect of the solvent on the
electrostatic work required in the activation process.
This quantity, the macroscopic dielectric constant, is not
really a pertinent measure of polarity as it exists in the
vicinity of a solute molecule, but the "microscopic dielectric constant" cannot be determined experimentally.

The organic chemical view of solvent polarity is embodied in its description as the power to solvate charges
in solutes [36]. On this basis, solvents can be listed
in ranked order of polarity without difficulty, as long
as they are sufficiently different. Thus in the series
H_2O, CH_3OH, $CH_3CH_2OCH_2CH_3$, C_6H_5, the solvents are listed
in order of decreasing polarity. Quantitative measures
of solvent polarity related to this concept are considered
in Chap. 4.

Solvent polarity can exert an effect on rates through
a general mechanism easily accounted for by transition
state theory. The reaction rate is determined by the free
energy of activation, which is the difference in free energies of the transition and initial states. If a change
in the medium decreases $\Delta G^{\ddagger}$, a rate increase occurs; a
medium change that increases $\Delta G^{\ddagger}$ will decrease the rate.
According to transition state theory the problem of accounting for the solvent effect on reaction rate can be
replaced by the simpler one of the solvent effect on the
equilibrium $A + B \rightleftharpoons M^{\ddagger}$ (for a bimolecular reaction). The
primary feature of concern is the charge distribution in
the transition state and the reactants. Thus if $M^{\ddagger}$ is
more polar than are the reactants, an increase in solvent
polarity should increase $K^{\ddagger}$ and therefore should increase
the rate. If the transition state is less polar than the
initial state, increase in solvent polarity will decrease
the rate. (In mechanistic investigations this argument
is often turned around; one observes the effect of a solvent change, and then postulates a consistent transition
state structure.)

Interpretations. Mechanistic purposes are usually served
by interpreting solvent effects according to the preceding

idea, namely, that if the change in solvent polarity aids
the transformation of reactants to transition state, the
rate will be increased, and vice versa. A few examples
can illustrate this.

If neutral, nonpolar reactants react to form neutral,
nonpolar products, presumably the transition state will
also be nonpolar, and so little sensitivity to the solvent
polarity is expected. The dimerization of cyclopendadiene

appears to be a good example. The rate constant is similar (within a factor of two) in the gas phase and in a
variety of solvents, including ethanol, acetic acid, benzene, paraffin, and the neat liquid [37]. The entropy of
activation is close to the standard entropy change for the
dimerization equilibrium, indicating that the transition
state resembles closely the dimer. The homolytic decomposition of azo-bis-isobutyronitrile also occurs at a rate
practically independent of the solvent, suggesting that
the transition state is nonpolar [19].

Reactions of neutral molecules to form charged species must pass through a transition state in which a partial charge separation has developed. The Menschutkin
reaction, the formation of a quaternary halide from a
tertiary amine and an alkyl halide, is a good example [38].

$$R_3N + R'CH_2I \rightarrow \left[\begin{matrix} \delta^+ & H \quad H & \delta^- \\ R_3N\text{----} & \underset{\underset{R'}{|}}{C} & \text{----} I \end{matrix} \right] \rightarrow R_3\overset{+}{N}\text{-}CH_2R' \\ I^-$$

We anticipate that an increase in solvent polarity will
"stabilize the transition state relative to the initial
state," therefore decreasing $\Delta G^{\ddagger}$ and increasing the rate
constant. In general the prediction is successful [39].

A second major class of reactions is that between an
ion and a neutral molecule. Ester saponification is an
example:

$$RCOOR' + OH^- \rightarrow RCOO^- + R'OH$$

and the S_N2 reaction with an ionic nucleophile is another:

$$RX + CN^- \rightarrow RCN + X^-$$

Qualitatively, we expect the transition state to be less polar than the initial state, because the transition state will have the same charge, but it will be dispersed over a greater volume. Thus the transition state will be de-stabilized relative to the initial state by highly polar solvents, and it is anticipated that the rate constant will be greater in nonpolar than in polar solvents. This is generally observed, though a reservation must be added about exceptions caused by specific solvent participation.

Reactions between two ions can first be considered qualitatively. If the ions are of opposite charge, the transition state will have no (or less) net charge, and will therefore be less polar than the initial state; an increase in solvent polarity would decrease reaction rate by stabilizing the initial state relative to the transition state [40]. If the ions are of the same charge sign, the charge on the transition state is their sum, which is necessarily 2 or more. Of course the volume of the mole-cule has also increased, but the transition state is still more polar than the initial state [41]. An increase in solvent polarity will therefore stabilize the transition state, decreasing $\Delta G^{\ddagger}$ and increasing the rate constant. The effect is illustrated by the data in Table 3.VI for this reaction:

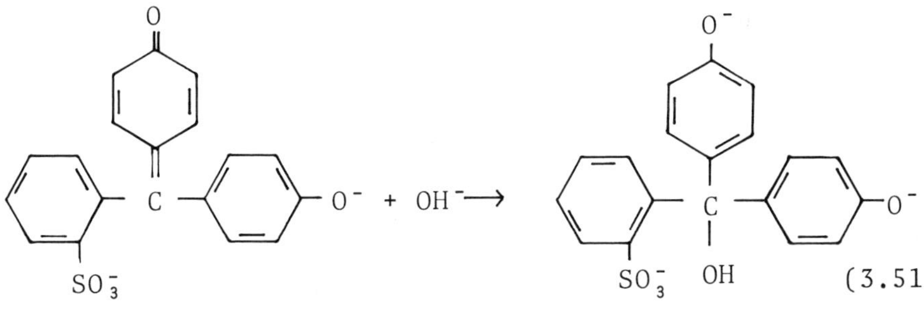

$$\text{(3.51)}$$

The most successful application of quantitative theory to medium effects has been the description of salt effects for ion-ion reactions [42]. Combining Eq. (3.17) with the Debye-Hückel expression for the activity coeffi-cients leads to

$$\ln k = \ln k_0 + \frac{2Z_A Z_B A \sqrt{I}}{1 + aB\sqrt{I}} \tag{3.52}$$

where the identity $Z_{\ddagger} = Z_A + Z_B$ and the approximation $a = a_A = a_B = a_{\ddagger}$ have been used. At low ionic strengths, $\log k$ should therefore be a linear function of the square root of ionic strength, with slope $= 2Z_A Z_B A$. For water

at 25°, A = 0.509; therefore the slope should be about
equal to the product of the ionic charges. The success
of Eq. (3.52) is clear in a remarkable diagram reproduced
by many authors [43].

When the ionic strength becomes appreciable (certain-
ly when it exceeds 0.1 $\underline{M}$), the simple linear relationship
cannot be expected. Note also that if either, or both,
of the reactants is neutral, Eq. (3.52) predicts no primary salt effect. In fact, however, at higher salt concentrations a linear dependence of log k on I is sometimes
observed, and this can be described with the use of an
empirical term proportional to I that is added to the
Debye-Hückel equation for the activity coefficient [44].

Equation (3.52) has been used mechanistically to
learn, from the slope of a plot of log k against √I at low
ionic strengths, the product $Z_A Z_B$ in the reaction [44].
It is important to note, however, that this salt effect
cannot be used to differentiate between kinetically equivalent rate terms. Such terms arise (see Sec. 3.3) when
a fast equilibrium step preceding the rate-determining
step allows the rate equation to be expressed in alternate
forms related by the equilibrium constant; for example,
the rate terms k[HA][B] and k´[A⁻][BH⁺] are kinetically
equivalent. It might seem that the primary salt effect
should distinguish between these, since Eq. (3.52) predicts slopes of 0 and -1, respectively, for these terms.
This view, however, neglects the salt effect on the equilibrium HA + B ⇌ A⁻ + BH⁺, and it can be shown that the
salt effects for the alternate forms of the rate equations
must be identical [45].

3.5   pH Effects

Preliminaries. That the rates of many reactions are
markedly dependent upon the acidity or alkalinity of the
reaction medium has been known for many decades. In this
section the kinetic analysis of reactions in dilute
aqueous solution, in which pH is the accessible measure
of acidity, is presented in sufficient detail to allow
the experimentalist to interpret data for most of the
systems likely to be encountered.

Throughout this section the hydronium ion and hydroxide ion concentrations appear in rate equations. For convenience these are written [H⁺] and [OH⁻]. Usually, of
course, these quantities will have been estimated from a
pH, so they are conventional activities rather than concentrations. However, our present concern is with the
formal analysis of rate equations, and we can conveniently

assume that activity coefficients are unity or are at
least constant. The basic experimental information is k,
the pseudo-first-order rate constant, as a function of
pH. Within a series of such measurements the ionic
strength should be held constant. If the pH is maintained
constant with a buffer, k should be measured at more than
one buffer concentration (but at constant pH) to see if
the buffer affects the rate. If such a dependence is observed, the rate constant should be measured at several
buffer concentrations and extrapolated to zero buffer to
give the correct k for that pH.

Except for those reactions whose characteristic rate
constants vary linearly with the hydronium or hydroxide
ion concentration, the most effective presentation of pH-
rate data is a graphical one. Two kinds of plot ("pH-rate
profiles") are commonly seen: (i) The observed rate constant is plotted on the vertical axis against pH on the
horizontal axis. This plot is particularly suited for the
estimation of dissociation constants; (ii) the logarithm
of the rate constant is plotted against pH. This plot has
two advantages--the order with respect to H^+ or OH^- is
readily apparent from linear portions of the curve, and a
great range of values of the rate constant can be exhibited
on one graph. The initial goal of the kinetic analysis
is to express k as a function of $[H^+]$, pH-independent rate
constants, and appropriate acid-base dissociation constants. Then numerical estimates of these constants are
obtained. The theoretical pH-rate profile can now be
calculated and compared with the experimental curve. A
quantitative agreement indicates that the proposed rate
equation is consistent with experiment. It is advisable
to use other information (such as independently measured
dissociation constants) to support the kinetic analysis.

The manner of the analysis is to locate characteristic features of the pH-rate curve--linear segments, maxima, minima, and inflection points--and to relate the value
of pH at which they occur to the parameters of the
(assumed) rate equation. Since many rate equations are
quite complicated, the only feasible differentiation is
with respect to $[H^+]$, yielding the derivative $dk/d[H^+]$.
This, however, does not correspond to the slope of either
of the plots employed in kinetic graphing, and must be
converted. The appropriate conversion equations, which
are obtained from the basic relations d ℓn u = du/u and
ℓn u = 2.303 log u, are

$$\frac{dk}{d\ pH} = -2.3[H^+]\frac{dk}{d[H^+]} \qquad (3.53)$$

$$\frac{d \log k}{d \ pH} = -\frac{[H^+]}{k} \frac{dk}{d[H^+]} \qquad (3.54)$$

$$\frac{d^2 k}{d \ pH^2} = -2.3[H^+] \frac{d^2 k}{d \ pH \ d[H^+]} \qquad (3.55)$$

$$\frac{d^2 \log k}{d \ pH^2} = -2.3[H^+] \frac{d^2 \log k}{d \ pH \ d[H^+]} \qquad (3.56)$$

One useful fact demonstrated by these formulas is that the location of maxima and minima is unaffected by the type of plot, so the derivative $dk/d[H^+]$ can be used in locating these features.

Throughout this section attention is restricted to rate equations that include concentrations of only the substrate, H^+, and OH^-. The observed first-order rate constant, therefore, contains concentrations of only H^+ and OH^- (the quantity $[OH^-]$ is often replaced by $K_w/[H^+]$). The substrate may be ionizable. Rate equations containing the concentration of additional solutes (especially catalytic additives) are considered in Sec. 3.6.

Curves without Inflection Points. If a pH-rate curve does not exhibit an inflection, then very probably the substrate does not undergo an ionization in this pH range. The kinds of substrates that are often of interest that lead to such simple curves are nonionizable compounds subject to hydrolysis, such as esters and amides. Reactions other than hydrolysis may be characterized by similar behavior if catalyzed by H^+ or OH^-. The general rate equation is

$$v = k_1 [H^+]^n [S] + k_2 [S] + k_3 [OH^-]^m [S] \qquad (3.57)$$

although sometimes a rate equation may contain terms in (say) both the first order and second order in some catalytic species. S represents the substrate. The experimental first-order rate constant, determined at essentially constant pH, is defined by $v = k[S]$, therefore

$$k = k_1 [H^+]^n + k_2 + k_3 [OH^-]^m \qquad (3.58)$$

Plots of k or $\log k$ against pH for this equation yield a so-called U-graph or a V-graph. The analysis of this curve will provide estimates of k_1, k_2, k_3, n, and m.

We introduce an approximation that is subsequently used many times, and that is indispensable. This is to consider only a portion of the curve and to neglect those terms describing the rest of the curve. It is necessary

to exercise some chemical discretion in applying such
approximations. The relative values of the rate constants
and concentrations determine the approximations that can
be safely made, and the level of uncertainty that one may
be willing to introduce in this way is gauged by a consideration of the experimental error in the raw data. Consider, in the present case, the very acid region ($[H^+]$ is
large, pH is low). Then in most cases Eq. (3.58) reduces
to (3.59) since $[OH^-] = K_w/[H^+]$.

$$k = k_1[H^+]^n \qquad (3.59)$$

Taking logarithms,

$$\log k = \log k_1 - npH \qquad (3.60)$$

Thus in the low pH region a plot of log k against pH
should give a straight line with slope equal to -n, yielding the order with respect to hydronium ion. Usually
$n = 1$.
        Similar reasoning shows that in regions of high basi-
city a plot of log k against pH will be linear with slope
+m. Usually in a log k-pH plot a positive constant slope
indicates dependence upon hydroxide ion, and a negative
constant slope shows dependence upon hydronium ion. This
is one of the conveniences of the log k-pH plot. If k_2
is not negligible relative to the other terms, evidently
a straight line will not be obtained; however, at very
low or very high pH this would be a most unusual circumstance. An example of the kind of graph described is
shown in Fig. 3.5, in which log k is plotted against pH
for the alkaline hydrolysis of ethyl p-nitrobenzoate [46].
The slope of the line is +1.00, hence the reaction is
first order in hydroxide ion.
        The pH at which the minimum occurs in a plot of Eq.
(3.58) is calculated for the most common case, $n = m = 1$.
The derivative $dk/d[H^+] = k_1 - k_3K_w/[H^+]^2$ is set equal to
zero, giving

$$pH_{min} = \frac{1}{2} pK_w + \frac{1}{2} \log \frac{k_1}{k_3} \qquad (3.61)$$

If $k_1 = k_3$, $pH_{min} = 1/2\ pK_w$. This is an unusual condition; it has been observed [47] in the hydrolysis of acetamide at 100°. Since $pK_w = 12.32$ at 100°, the minimum
rate in this reaction occurs at pH 6.16. For ester hydrolyses, k_3 is usually greater than k_1 and the minimum
is observed near pH 5-6 (at room temperature). Equation
(3.61) is used in the construction of a calculated pH-rate
profile, when it allows the position of the minimum to be

easily located. This equation evidently yields the pH at
which the reaction rate is minimal, and so it is useful in
selecting the pH at which the substrate is most stable.

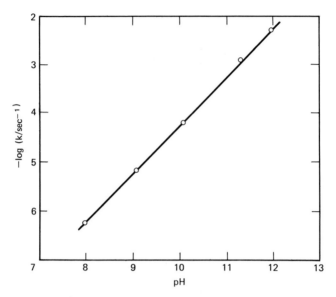

Fig. 3.5. Log k-pH plot for the alkaline hydrolysis of
ethyl p-nitrobenzoate.

The constant k_1 is readily calculated from data at
low pH, where Eq. (3.59) holds, since n is now known.
Similarly k_3 can be calculated from data at high pH.

If k_2 is sufficiently large that it cannot be neglected (which will be revealed by the inconstancy of the
"constant" calculated as above), then Eq. (3.58) may be
put into the forms

$$k = k_2 + k_1[H^+]^n \quad \text{(at low pH)} \qquad (3.62)$$

$$k = k_2 + k_3[OH^-]^m \quad \text{(at high pH)} \qquad (3.63)$$

A plot of k against $[H^+]^n$ at low pH should give a straight
line with intercept k_2 and slope k_1; at high pH the plot
of k against $[OH^-]^m$ similarly permits the evaluation of
k_2 and k_3.

Figure 3.6 is a pH-rate profile for the hydrolysis of
p-nitrophenyl acetate. The slopes of the straight-line
portions are -1, 0, and +1, reading in the acid to base
direction, and this system can be described by

$$k = k_1[H^+] + k_2 + k_3[OH^-] \qquad (3.64)$$

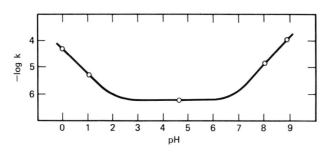

Fig. 3.6.   pH-rate profile for the hydrolysis of p-nitrophenyl acetate at 25° in aqueous solution containing 1%
acetonitrile; ionic strength = 0.1 $\underline{M}$ except at pH 0.

The constants k_1 and k_2 were evaluated from the measurements at low and high pH as described earlier, and k_2 was
calculated from the measurement at pH 4.65 after taking
into account any significant contribution from the k_1 and
k_3 terms.   The smooth curve in Fig. 3.6 was calculated
with Eq. (3.64) and the estimated rate constants, which
have these values [48]: $k_1 = 5.34 \times 10^{-5}$ $\underline{M}^{-1}$-sec^{-1}; $k_2 =$
5.5×10^{-7} sec^{-1}; $k_3 = 12.6$ $\underline{M}^{-1}$-sec^{-1}.
        If a substrate contains an ionizable group that dis-
sociates in the pH range being kinetically studied, one
may expect a change in direction in the pH-rate curve in
that pH region corresponding to the ionization of the
substrate; this is because the conjugate acid and base
forms are unlikely to undergo reaction at the same velocity.   Such an identity of rates has apparently been observed, however, in the hydrolysis of nicotinamide [49].
The pH-rate curve is essentially V-shaped with no inflection corresponding to the ionization of the substrate.

Sigmoid Curves.   An inflection point in a pH-rate profile
suggests a change in the nature of the reaction caused by
a change in the pH of the medium.   The usual reason for
this behavior is an acid-base equilibrium of a reactant.
Here we consider the simplest such system, in which the
substrate is a monobasic acid (or monoacidic base).   It
is pertinent to consider the mathematical nature of the
acid-base equilibrium.   Let HS represent a weak acid.
(The charge type is irrelevant.)   The acid dissociation
constant, $K_a = [H^+][S^-]/[HS]$, is taken to be appropriate
to the conditions (temperature, ionic strength, solvent)
of the kinetic experiments.   The fractions of solute in

the conjugate acid and base forms are given by

$$F_{HS} = \frac{[HS]}{S_t}$$

$$F_S = \frac{[S^-]}{S_t}$$

where S_t, the total molar concentration of solute, is

$$S_t = [HS] + [S^-]$$

Combining these equations,

$$F_{HS} = \frac{[H^+]}{[H^+] + K_a} \tag{3.65}$$

$$F_S = \frac{K_a}{[H^+] + K_a} \tag{3.66}$$

Figure 3.7 shows F_{HS} and F_S plotted against pH, according to Eqs. (3.65) and (3.66), for a weak acid of $pK_a = 4.0$. Because of their appearance such curves are called S-shaped or sigmoid curves.

These curves have some interesting properties. At any given pH, evidently $F_{HS} + F_S = 1$. At the point where the two curves cross, $F_{HS} = F_S = 0.5$, and from Eqs. (3.65) and (3.66), at this point $[H^+] = K_a$, or $pH = pK_a$. That this point corresponds to the inflection point can be shown by taking the second derivative d^2F/dpH^2 and setting this equal to zero; one finds $pH_{infl} = pK_a$. In the limit as $[H^+]$ becomes much greater than K_a, F_{HS} approaches unity, and F_S approaches zero (though theoretically they never attain these limiting values). Similarly, as $[H^+]$ becomes much less than K_a, F_{HS} approaches zero and F_S approaches unity.

Now suppose the substrate of a chemical reaction is a weak acid, with both the conjugate acid HS and conjugate base S^- being capable of undergoing reaction. Usually these two species will react at different rates because of the considerable difference in their electronic configurations. The rate equation for this system is

$$v = k'[HS] + k''[S^-] \tag{3.67}$$

Combining this with the preceding equations,

$$v = \left[\frac{k'[H^+] + k''K_a}{[H^+] + K_a} \right] S_t \qquad (3.68)$$

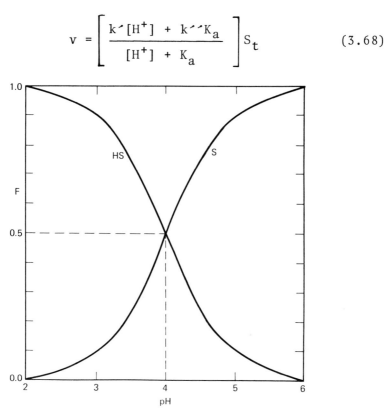

Fig. 3.7. Variation with pH of the fractions F_{HS} (conjugate acid) and F_S (conjugate base) for an acid with $pK_a = 4.0$.

Since the experimental rate equation is $v = kS_t$, the first-order rate constant k becomes,

$$k = \frac{k'[H^+] + k''K_a}{[H^+] + K_a} \qquad (3.69)$$

which relates k to the substrate dissociation constant, the hydronium ion concentration, and the pH-independent rate constants characteristic of the reactions of the two forms of the substrate.

If k' is much larger than k'', Eq. (3.69) takes the form of Eq. (3.65) for the fraction F_{HS}; thus we may expect the experimental rate constant to be a sigmoid function of pH. If k'' is larger than k', the k-pH plot should resemble the F_S-pH plot. Equation (3.69) is a very important relationship for the description of pH effects on reaction rates. Most sigmoid pH-rate profiles can be quantitatively accounted for with its use. Relatively minor modifications [such as the addition of rate terms first order in H^+ or OH^- to Eq. (3.67)] can often extend the description over the entire pH range.

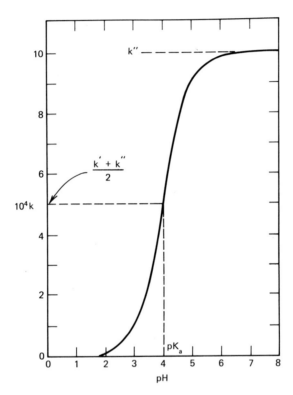

Fig. 3.8. Plot of Eq. (3.69) with pK_a = 4.0, k' = 1×10^{-6} sec^{-1}, k'' = 1×10^{-3} sec^{-1}.

     We consider first the k-pH plot corresponding to Eq. (3.69). Figure 3.8 is such a plot calculated with Eq. (3.69) and the typical values pK_a = 4.0, k' = 1×10^{-6} sec^{-1}, and k'' = 1×10^{-3} sec^{-1}. Note that this plot is similar to the F_S-pH plot in Fig. 3.7. Differentiating Eq. (3.69) to find $dk/d[H^+]$ and using the identity equation (3.53) gives the first derivative of the k-pH plot. The second derivative is found with the aid of Eq. (3.55). Imposing the condition $k' \neq k''$ and setting the second derivative equal to zero leads to a simple expression for the pH at the inflection point.

$$pH_{infl} = pK_a$$

This property of the sigmoid curve permits K_a to be easily estimated. This is an advantage of the k-pH plot. If the inflection point cannot be accurately located, the dissociation constant may still be estimated. Let $[H^+] = K_a$ in Eq. (3.69); then Eq. (3.70) results.

$$k = \frac{k´ + k´´}{2} \qquad (3.70)$$

That is, pK_a = pH at the point where Eq. (3.70) holds. Since the larger of the two constants is usually much greater than the smaller one, this often may be interpreted that pK_a = pH when $k = k_{max}/2$ (see Fig. 3.8). Graphical methods for estimating K_a by using all of the kinetic data are considered later.

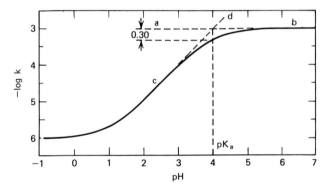

Fig. 3.9. Log k-pH plot of Eq. (3.69) with $pK_a = 4.0$, $k´ = 1 \times 10^{-6}$ sec^{-1}, $k´´ = 1 \times 10^{-3}$ sec^{-1}.

The log k-pH plot can display a large range in values on the vertical axis. The properties of this curve are different ·from those of the k-pH plot. Figure 3.9 is a plot of log k against pH for the same system graphed in Fig. 3.8. The first derivative is found with the help of Eq. (3.54) and the second derivative is obtained from this and Eq. (3.56). If the special case $k´ = k´´$ is excluded, this leads to

$$pH_{infl} = pK_a - \frac{1}{2} \log \frac{k´´}{k´} \qquad (3.71)$$

This dependence of pH_{infl} on the rate constants as well as the dissociation constant has sometimes been overlooked by authors evaluating log k-pH curves, and the literature contains examples of such plots that have been erroneously used to estimate pK_a.

It was shown above that, for the k-pH plot, $pK_a = pH$ when $k = k_{max}/2$. In terms of the log k-pH plot, this means that $pK_a = pH$ when $\log k = \log k_{max} - 0.30$. Figure 3.9 shows the relationship expressed by this statement. An alternate method is developed from Eq. (3.69); in the special case that $k'' >> k'$, this becomes $k = k'' K_a/([H^+] + K_a)$. Taking logarithms and imposing the condition $[H^+] >> K_a$,

$$\log k = pH + \log k'' - pK_a \qquad (3.72)$$

Thus, under these conditions, the log k-pH plot is linear with slope +1. The extension of this segment of the plot is shown as c-d in Fig. 3.9. When $\log k = \log k''$, $pH = pK_a$; that is, the pH at which c-d and a-b intersect is numerically equal to the pK_a. A similar treatment can be given for the case $k' >> k''$.

The dissociation constant is most accurately estimated from kinetic data when all of the data points are used in the evaluation. There are several ways to do this. The Henderson-Hasselbalch equation

$$pK_a = pH - \log \frac{[S^-]}{[HS]}$$

can be combined with the equations $k = k' F_{HS} + k'' F_S$ and $F_{HS} + F_S = 1$ to give

$$pK_a = pH - \log \frac{k - k'}{k'' - k} \qquad (3.73)$$

pK_a can be calculated with Eq. (3.73) at each of the experimental pH's in the rising portion of the sigmoid curve; values of k' and k'' can be estimated from the extreme low and high pH regions. Alternatively the third term can be plotted against pH; $pK_a = pH$ at the point where the logarithmic term equals zero [50]. The slope of this plot should be unity.

In suitable circumstances the parameters K_a, k', and k'' of Eq. (3.69) may be evaluated by means of linear extrapolation methods. Consider the case in which $k'[H^+] >> k'' K_a$. Then Eq. (3.69) becomes

$$k = \frac{k'[H^+]}{[H^+] + K_a} \qquad (3.74)$$

Equation (3.74) is of the form $x = y(a + by)$, with $x = k$, $y = [H^+]$, $a = K_a/k´$, and $b = 1/k´$. Three corresponding linear equation are:

$$\frac{1}{k} = \frac{K_a}{k´[H^+]} + \frac{1}{k´} \qquad (3.75)$$

A plot of $1/k$ against $1/[H^+]$ should be linear; from the slope and intercept K_a and $k´$ can be evaluated.

$$\frac{k}{[H^+]} = -\frac{k}{K_a} + \frac{k´}{K_a} \qquad (3.76)$$

$k/[H^+]$ is plotted against k.

$$\frac{[H^+]}{k} = \frac{[H^+]}{k´} + \frac{K_a}{k´} \qquad (3.77)$$

$[H^+]/k$ is plotted against $[H^+]$.

When $k´´K_a \gg k´[H^+]$, three linear equations, suitable for evaluating $k´´$ and K_a, can be written as shown for the first case.

The kinetic analysis of the sigmoid pH-rate profile will yield numerical estimates of the pH-independent parameters K_a, $k´$, and $k´´$. With these estimates the apparent constant k is calculated using the theoretical equation over the pH range that was explored experimentally. Quantitative agreement between the calculated line and the experimental points indicates that the model is a good one. A further easy, and very pertinent, test is a comparison of the kinetically determined K_a value with the value obtained by conventional methods under the same conditions.

Many sigmoid rate curves have been reported. A typical example is provided by the hydrolysis of phthalamic acid [51].

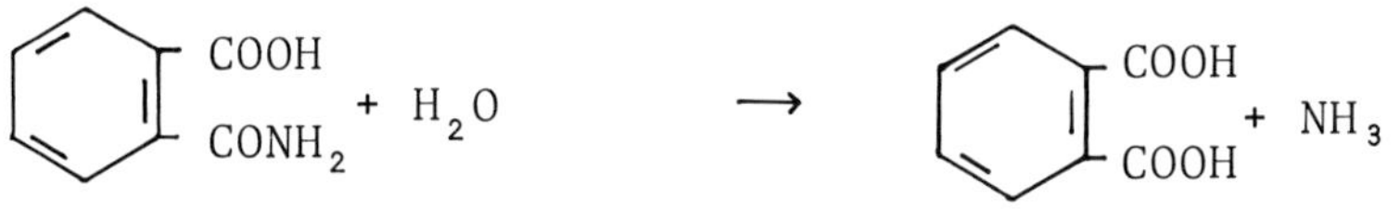

$$(3.78)$$

This compound undergoes hydrolysis of the amide group intramolecularly catalyzed by the neighboring carboxylic acid group. The rate equation, in the pH range 1-5, is given by Eq. (3.67) with $k´´$ essentially equal to zero; thus Eq. (3.74) describes the pH-rate curve. The kinetic results are shown as the k-pH plot [Fig. 3.10(a)] and as the log k-pH plot [Fig. 3.10(b)]. Application of the simple methods described earlier gives $pK_a = 3.6$ and

$k' = 2.35 \times 10^{-4}$ sec^{-1}. The original paper [51] also pre-
sents a plot of these data according to Eq. (3.75).

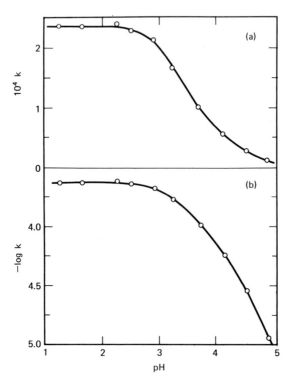

Fig. 3.10. (a) pH-rate profiles plotted as k-pH and (b)
log k-pH for the hydrolysis of phthalamic acid [51].

     The hydrolysis of aspirin [Eq. (3.79), R = CH$_3$] is a
classic example demonstrating a sigmoid pH-rate

(3.79)

effect [52]. Figure 3.11 shows this curve for trimethyl-
acetylsalicylic acid [Eq. (3.79), R = C(CH$_3$)$_3$] [53]. The
extreme left-hand segment of the curve has a slope of -1
and the extreme right-hand slope is +1, suggesting rate
terms first order in [H$^+$] and [OH$^-$], respectively. The
intermediate portion is sigmoid. Writing the substrate as
HS, the rate equation immediately suggested by Fig. 3.11
is therefore

$$v = k^0 [HS][H^+] + k^{\prime}[HS] + k^{\prime\prime}[S^-] + k^{\prime\prime\prime}[S^-][OH^-] \qquad (3.80)$$

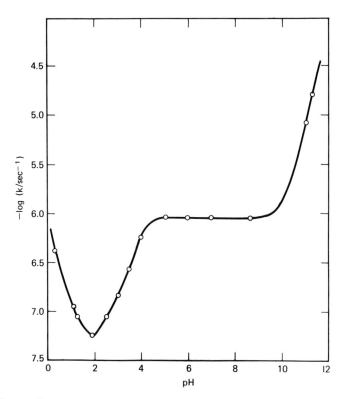

Fig. 11. pH-rate profile for the hydrolysis of trimethyl-acetylsalicylic acid at 25 (aqueous solution containing 0.5% ethanol) [53].

The pH-independent "plateau" from about pH 5 to 9 represents reaction of the acylsalicylate anion [54]. It is obvious from the pH-rate profile that $k^{\prime\prime}$ is much larger than $k^{\prime}$. The theoretical equation for k, the observed first-order rate constant, is derived in the usual way from Eq. (3.80).

$$k = \frac{k^0[H^+]^2 + k^{\prime}[H^+] + k^{\prime\prime}K_a + k^{\prime\prime\prime}K_a[OH^-]}{[H^+] + K_a} \qquad (3.81)$$

k^0 is evaluated from k values at very low pH, when the other terms become negligible, and $k^{\prime\prime\prime}$ from the data at very high pH. From the plateau region, $k^{\prime\prime}$ can be estima-

ted, because in this range only the $k^{\prime\prime}$ term is important.
The value of $k^{\prime}$ can then be estimated from data in the pH
2-3 range, where the k^0, $k^{\prime}$, and $k^{\prime\prime}$ terms may all contribute to k, by making appropriate corrections with the
known constants k^0, $k^{\prime\prime}$, and K_a.

Bell-Shaped Curves. A frequently encountered pH-rate profile exhibits a bell-like shape or "hump," with two inflection points. This graphical feature is essentially
two sigmoid curves back-to-back. By analogy with the earlier analysis of the sigmoid pH-rate curve, where the shape
was ascribed to an acid-base equilibrium of the substrate,
we find that the bell-shaped curve can usually be accounted
for in terms of two acid-base dissociations of the substrate. The substrate can be regarded, for this analysis,
as a dibasic acid H_2S, where the charge type is irrelevant;
we take the neutral molecule as an example. The acid
dissociation constants are

$$K_1 = \frac{[H^+][HS^-]}{[H_2S]} \qquad (3.82)$$

$$K_2 = \frac{[H^+][S^{2-}]}{[HS^-]} \qquad (3.83)$$

The fractions of solute in each form are given by $F_{H_2S} = [H_2S]/S_t$, $F_{HS} = [HS^-]/S_t$, and $F_S = [S^{2-}]/S_t$, where S_t, the
total molar concentration of substrate, is $S_t = [H_2S] + [HS^-] + [S^{2-}]$. Combining these leads to expressions for
the fractions of solute as functions of the dissociation
constants and the hydronium ion concentration.

$$F_{H_2S} = \frac{[H^+]^2}{[H^+]^2 + K_1[H^+] + K_1K_2} \qquad (3.84)$$

$$F_{HS} = \frac{K_1[H^+]}{[H^+]^2 + K_1[H^+] + K_1K_2} \qquad (3.85)$$

$$F_S = \frac{K_1K_2}{[H^+]^2 + K_1[H^+] + K_1K_2} \qquad (3.86)$$

Figure 3.12 shows F_{H_2S}, F_{HS}, and F_S plotted against
pH for an acid with $pK_1 = 5.0$ and $pK_2 = 10.0$. Evidently
with such widely spaced dissociation constants the solution contains, at any one pH, significant fractions of
only two species. The fraction of monoanion rises essentially to unity at one point. The pH at which the mono-

anion fraction achieves its maximum value is calculated
by differentiating Eq. (3.85) and setting the result equal
to zero; this gives

$$pH_{max} = \frac{1}{2}(pK_1 + pK_2) \qquad (3.87)$$

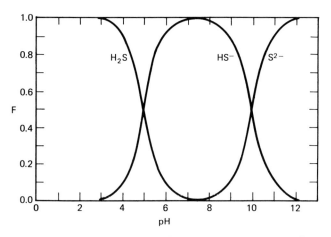

Fig. 3.12.   Distribution of acid-base species for an acid
with $pK_1 = 5.0$ and $pK_2 = 10.0$.

The intersection of the curves F_{H_2S} and F_{HS} occurs at
pH = pK_1, as can be found by setting Eqs. (3.84) and
(3.85) equal.   Also, when $F_{HS} = F_S$, pH = pK_2.   These are
general relationships.   We note, at this point, that the
function F_{HS} has the previously mentioned bell shape, and
it is this function that will be of later kinetic interest.
    In order to find the inflection points in a plot of
F_{HS} against pH, the second derivative $d^2\ F_{HS}/d\ pH^2$ is set
to zero.   The result is a quartic in $[H^+]$, which is not
reproduced.   When K_1 is much larger than K_2 (by at least
three orders of magnitude), the location of the inflection
points becomes particularly simple.   Then at low pH, in
the region of the left inflection point, $[H^+]$ is much
greater than the quantity $(K_1K_2)^{\frac{1}{2}}$, and

$$pH_{infl}^{left} = pK_1 \qquad (3.88)$$

Similarly, when $(K_1K_2)^{\frac{1}{2}} >> [H^+]$, the location of the right
inflection point is given by

$$pH_{infl}^{right} = pK_2 \qquad (3.89)$$

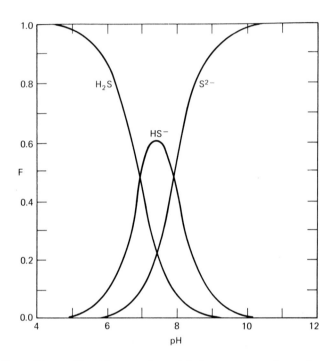

Fig. 3.13. Distribution of acid-base species for an acid
with pK_1 = 7.0 and pK_2 = 8.0.

These simple results will not be applicable if pK_1 and pK_2
are fairly close. Figure 3.13 is a plot of the species
distribution for a hypothetical dibasic acid with pK_1 =
7.0 and pK_2 = 8.0. The inflection points on the F_{HS} curve
do not coincide with the pK's of the acid. It is also important to notice that F_{HS} never reaches unity, and that
within the approximate pH range 6 to 8, appreciable fractions of all three forms coexist [55].
     Now suppose that only the monoanionic form of the di-
basic acid H_2S undergoes reaction, and that neither the
hydronium nor the hydroxide ion is directly involved. The
kinetic scheme is therefore

$$HS^- \longrightarrow products \qquad (3.90)$$

and the rate equation is $v = k'[HS^-]$. The concentration
of monoanion is given by $F_{HS}S_t$, and the experimental rate
equation, at constant pH, is $v = kS_t$. These lead to

$$k = \frac{k'K_1[H^+]}{[H^+]^2 + K_1[H^+] + K_1K_2} \qquad (3.91)$$

Equation (3.91) has the same form as Eq. (3.85) for F_{HS}, so the simple scheme embodied in Eq. (3.90) can account for a bell-shaped curve when k is plotted against pH.

The observed kinetics are seldom as simple as this. Usually one or both of the other forms of the substrate also undergo reaction, with the usual reactions being hydronium ion catalysis of H_2S and hydroxide ion catalysis of S^{2-}. The kinetic scheme is then

$$H_2S + H^+ \xrightarrow{k^0} \text{products}$$

$$HS^- \xrightarrow{k'} \text{products}$$

$$S^{2-} + OH^- \xrightarrow{k''} \text{products}$$

and the rate equation is

$$v = k^0[H_2S][H^+] + k'[HS^-] + k''[S^{2-}][OH^-] \quad (3.92)$$

This leads, after the usual development, to Eq. (3.93) for the pseudo-first-order rate constant k.

$$k = \frac{k^0[H^+]^3 + k'K_1[H^+] + k''K_1K_2K_w/[H^+]}{[H^+]^2 + K_1[H^+] + K_1K_2} \quad (3.93)$$

The analysis of a k-pH curve in terms of Eq. (3.93) is treated by making approximations that are equivalent to ignoring some of the rate terms in certain pH regions. A common type of system is analyzed as an example; the approach can be modified to suit a particular demand. The evaluation of k^0 and k'' can nearly always be accomplished from rate data at very low and high pH, respectively. We are concerned with k', K_1, and K_2. Figures 3.14 and 3.15 are k-pH and log k-pH plots for a hypothetical system described by Eq. (3.93). The analysis assumes this equation and these types of parameters.

Often k'' is smaller than k^0 because attack on the dianion by hydroxide is disfavored. Moreover, the maximum often occurs on the acid side of neutrality. In the pH region near the maximum it will therefore often be permissible to set $k'' = 0$ in Eq. (3.93). The result is differentiated and the derivative is set equal to zero, giving, for the hydronium ion concentration at the maximum in the bell,

$$[H^+]^2_{max} \simeq \frac{k'K_1K_2}{k' - 3k^0K_2}$$

where quartic and cubic terms in $[H^+]$ have been neglected.

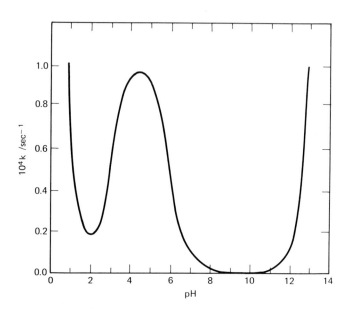

Fig. 3.14.   Theoretical k-pH plot according to Eq. (3.93) for a substrate with $pK_1 = 3.0$, $pK_2 = 6.0$, $k^0 = 1 \times 10^{-3}$ $\underline{M}^{-1}\text{-sec}^{-1}$, $k^{\prime} = 1 \times 10^{-4}$ sec^{-1}, $k^{\prime\prime} = 1 \times 10^{-3}$ $\underline{M}^{-1}\text{-sec}^{-1}$.

If $k^{\prime} >> 3k^0 K_2$, this becomes

$$pH_{max} \simeq \frac{1}{2}(pK_1 + pK_2) \tag{3.94}$$

The value of the observed rate constant at the maximum is found by substituting $[H^+]_{max} \simeq (K_1 K_2)^{\frac{1}{2}}$ into Eq. (3.91), giving the approximate result [56]

$$k_{max} \simeq \frac{k^{\prime}}{1 + 2(K_2/K_1)^{\frac{1}{2}}} \tag{3.95}$$

which may be useful in estimating $k^{\prime}$.

The placement of the minimum that occurs to the left of the maximum will not be affected by the $k^{\prime\prime}$ term, and it will usually be permissible to neglect the second dissociation of the substrate. Thus Eq. (3.91) simplifies to

$$k = \frac{k^0 [H^+]^2 + k^{\prime} K_1}{[H^+] + K_1}$$

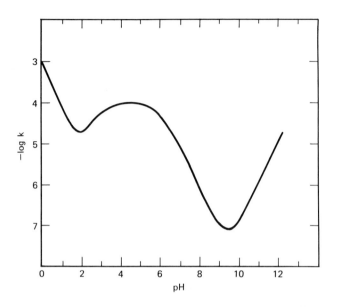

Fig. 3.15. Theoretical log k-pH plot for the substrate described in Fig. 3.14.

Setting the first derivative to zero gives Eq. (3.96) for the hydronium ion concentration at the left-hand minimum.

$$[H^+]_{min}^{left} = K_1 [(1 + \frac{k'}{k^0 K_1})^{\frac{1}{2}} - 1] \qquad (3.96)$$

The left-hand inflection point is obtained by neglecting K_2 terms. The derivative d^2k/dpH^2 is set equal to zero. As a very approximate solution, higher-order terms are neglected,

$$[H^+]_{infl}^{left} \simeq \frac{k' K_1}{4k^0 K_1 + k'} \qquad (3.97)$$

or, in even more approximate form, $pH_{infl}^{left} \simeq pK_1$. In many cases, however, the location of the left-hand inflection point in the k-pH plot will not be given by these relationships.

In the region of the right-hand inflection point both the k^0 and k'' terms can often be neglected. The second derivative d^2k/dpH^2 is then set to zero. As a first approximation all terms higher than the linear one are neglected:

$$pH_{infl}^{right} \simeq pK_2 \qquad (3.98)$$

Very often this will give a good estimate of K_2; K_1 can then be found with Eq. (3.94).

When estimates of k^0, k', k'', K_1, and K_2 have been obtained, a calculated pH-rate curve is developed with Eq. (3.93). If the experimental points follow closely the calculated curve, it may be concluded that the data are consistent with the assumed rate equation. The constants may be considered adjustable parameters that are modified to achieve the best possible fit. The dissociation constants K_1 and K_2 thus derived from kinetic data should be in reasonable agreement with the dissociation constants obtained (under the same experimental conditions) by other means.

The rate equation (3.92), upon which this analysis has been based, may be found inadequate to account in full for the data. Obvious alterations are the addition of the rate terms $k_1[H_2S]$ and $k_2[S^{2-}]$. These terms added to those of Eq. (3.92) include all of the possibilities (other than terms containing concentrations to orders other than unity); any other term involving these species would be kinetically equivalent to one of these five terms.

Kinetic schemes other than that embodied in Eq. (3.92) can give rise to a bell-shaped curve. Several of these include a reactant other than the substrate, and are not considered here; briefly, the rate term responsible for the bell-shaped curve will include two ionizable species or (as in the case treated in the preceding pages) a species whose concentration depends upon two ionization steps [57]. A less obvious scheme that can lead to a bell-shaped curve has been recognized by Zerner and Bender [58]. The substrate is a weak acid or base, but possesses only one ionizable group, and no other ionizable reactant is involved (other than water). The second inflection in the curve is ascribed to an ionizable group created in an intermediate. An example has been discovered in the hydrolysis of o-carboxyphthalimide [58].

$$(3.99)$$

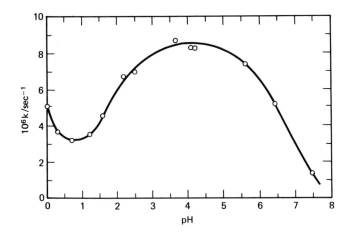

Fig. 3.16. pH-rate profile for the hydrolysis of mono-
ethyl dihydrogen phosphate in aqueous solution at 100.1°.

Figure 3.16 shows a bell-shaped pH-rate profile for
the hydrolysis of monomethyl dihydrogen phosphate [59].
Other examples are the hydrolysis of o-carboxyphenyl hydrogen succinate [60] and the hydration of fumaric acid
[61].

3.6 Catalysis

Types of Catalysis. A definition is useful in concentrating attention upon related phenomena, though it can be
made so restrictive that the resulting attention is unproductive. Definitions of catalysis take several forms,
not all of them equivalent. The classical definition is:
a catalyst is a substance that increases the rate of a
chemical reaction without affecting the position of equilibrium. Another way to express this is that the catalyst
increases the rate but is not itself consumed in the overall reaction [62]. Bell [63] has generalized the concept,
defining a catalyst to be a substance whose concentration
appears in the rate equation to a higher power than it
does in the stoichiometric equation. It is usually argued
that, since a catalyst increases the rate in the forward
direction and it does not alter the equilibrium constant,
it must increase the rate in the rearward direction by
exactly the same factor. The earlier discussion (Sec. 3.3)
on the principle of microscopic reversibility is pertinent
here in showing that this conclusion is not necessarily

correct in general, because the rates are usually studied
far from equilibrium.

It is probably more useful to regard as a catalyst
any substance that increases the rate constant of a chemical reaction, other than by a medium effect, regardless of
the ultimate fate of this substance [64, 65]. In the terms
of the transition state theory, a catalyst makes accessible
a reaction path with a lower free energy of activation than
is available in its absence [66].

The manner in which catalysis occurs can be illustrated in a general way by considering the catalysis of the
reaction A + B → M + N by catalyst X. Here is a possible
kinetic scheme:

$$A + X \rightleftharpoons C$$

$$C + B \longrightarrow M + N + X$$

And here is a different scheme:

$$A + X \rightleftharpoons C' + M$$

$$C' + B \longrightarrow N + X$$

(Consumption of X by a subsequent fast reaction with M or
N may be of experimental importance, but is irrelevant to
a consideration of the catalysis.) Catalysis by X means
that the rate constant for reaction between the intermediate (C or C') and B must be larger than that for the reaction of A and B. The catalyst may accomplish this by
bringing the reactant molecules close together [67], thus
increasing their probability of reaction by increasing
their collision probability and perhaps by assisting them
to attain the relative orientation required for reaction.
Some form of noncovalent complex formation may be involved
in this process. Catalysis may also proceed via covalent
bond changes.

Catalysis in aqueous solution by hydronium ion and
hydroxide ion is called specific acid and specific base
catalysis, respectively. These terms could be generalized
to include catalysis by the lyonium and lyate ions of any
dissociating solvent. If the substrate is ionizable (say
HS), a rate term $k[HS][OH^-]$ is kinetically indistinguishable from the term $k'[S^-]$. Thus the mechanistic involvement of a catalytic hydroxide ion is not proved, in such
a case, by the rate equation.

Catalysis by a base other than hydroxide ion (that
is, by a Brønsted base) is general base catalysis, and
general acid catalysis involves a proton acid other than
hydronium ion. This is not yet a complete definition,

however. It must be added that general base and general
acid catalysts are those that exert their catalytic influence by accepting or donating a proton. Catalysts are
known that operate by mechanisms other than proton transfer. A nucleophilic catalyst is a general base but it
acts by forming a bond (donating an electron pair) to an
atom other than hydrogen--usually carbon. An electrophilic catalyst is one that acts as an electron pair acceptor
(a Lewis acid); metal ions are examples. Much of the
later discussion, especially in Sec. 13.1, is concerned
with examples of catalysis.

     According to the principle of microscopic reversibi-
lity, the transition state is the same for the forward
and rearward steps of a reaction at equilibrium. This
means (keeping in mind the reservation that rate studies
are usually made far from the equilibrium position) that
the type of catalysis in the rearward step is derivable
from knowledge about the catalysis in the forward step;
it is simply the catalysis run in reverse. Thus if the
forward step is general base catalyzed (donation of an
electron pair to a proton), the reverse reaction is general acid catalyzed (donation of a proton to an electron
pair).

Experimental Study. The detection and quantitative evaluation of catalytic phenomena is accomplished by determining rate equations and rate constants. Much of this chapter has dealt with these problems. Section 3.5 is, in
this context, a quantitative description of specific acid
and specific base catalysis. The more general types of
catalysis lead to more complicated rate equations. These
equations will include concentrations of the substrate
(which may be ionizable), of the catalyst (also ionizable),
and possibly of H^+ and OH^-, if the solvent is water. Two
kinds of experimental manipulation are used: (1) The
pseudo-first-order rate constant is studied as a function
of total catalyst concentration at constant pH; it is often convenient to use the catalyst as the buffer. Let B
and BH^+ represent the conjugate base and acid forms of
this buffer, with the formal concentration being B_t; thus
B_t = [B] + [BH^+]. Then in this experiment k is measured
as a function of B_t while the ratio [B]/[BH^+] is held
constant. If k increases with B_t, evidently one, or both,
of the species catalyzes the reaction. (2) The pseudo-
first-order rate constant is measured as a function of pH,
with one of the concentrations (B_t, [B], or [BH^+]) held
constant.

     Since there are many possible rate equations, a gen-
eral treatment is not feasible. Actually the rate equations observed are usually fairly simple. Some typical

experimental designs are described. Suppose the substrate
is nonionizable and is subject to the usual first-order
specific acid and specific base catalysis, and is also
subject to first-order general acid and general base catalysis. The experimental first-order rate constant k is
then given by

$$k = k_1[H^+] + k_2 + k_3[OH^-] + k_b[B] + k_a[BH^+] \quad (3.100)$$

At a given pH the quantity $k_0 = k_1[H^+] + k_2 + k_3[OH^-]$ is
a constant. Equation (3.100) can be written

$$k = k_0 + (F_B k_b + F_{BH} k_a) B_t \quad (3.101)$$

where $F_B = [B]/B_t$ and $F_{BH} = [BH^+]/B_t$. Since F_B and F_{BH}
depend only on the pH and the pK_a of the acid BH^+, at constant pH, k should be a linear function of B_t. Now suppose
the experiment is repeated at some higher pH value. If
the experimental slope is larger in this second experiment,
evidently the general base term is more important than the
general acid term, and vice versa if the slope is smaller.
The rate constants can be evaluated by rearranging Eq.
(3.101) to

$$\frac{k - k_0}{B_t} = k_a + (k_b - k_a) F_B \quad (3.102)$$

A plot of $(k - k_0)/B_t$, which is the apparent second-order
catalytic rate constant, against F_B should be linear if
Eq. (3.100) describes the system. The extrapolated intercept at $F_B = 0$ is equal to k_a, and the intercept at $F_B = 1$
is equal to k_b [68]. This plot can make use of data taken
at any value of pH and B_t.
     Some authors prefer to plot k against [B] rather than
B_t; in the case (a common one) that k_a is negligible, the
slope is then equal to k_b. Curvature in these plots indicates that Eq. (3.100) is not fully applicable. Positive
curvature suggests a rate term of higher order in the
catalyst; a simple example was treated in Table 3.II.
Negative curvature can be caused by complex formation of
the reactants or by a change in rate-determining step [69].
     A rate constant describing the catalytic effect of a
general base is not, by itself, sufficient evidence that
the rate enhancement is caused by general base catalysis.
It may be a consequence of a nucleophilic catalysis. The
distinction has been carefully explored for reactions of
carboxylic acid derivatives. Several forms of evidence
can be used to decide if a particular catalytic process
is general base or nucleophilic catalysis. This problem
is discussed in Sec. 13.1; see especially Table 13.III.

Catalytic data may be presented and analyzed in the form of pH-rate profiles. The analysis follows the lines presented in Sec. 3.5 for systems containing only substrate and solvent species. Usually an inflection point in the k-pH plot is a manifestation of an ionization process. Sigmoid and bell-shaped curves are observed. Thus a system composed of a nonionizable substrate S and a monoprotic base catalyst B will give a sigmoid curve if k is plotted against pH (at constant B_t). Bell-shaped curves can arise from a reaction between ionizable substrate and catalyst, as in the scheme

$$HS + B \longrightarrow products$$

which is kinetically equivalent to

$$S^- + BH^+ \longrightarrow products$$

Suppose, for example, that a carboxylic acid and an amine react to give a bell curve; it is tempting to postulate that the catalysis proceeds with the free carboxyl acting as an acid and the amine functioning as a base. The data, however, can equally well support the mechanism in which the acid is $R-NH_3^+$ and the base is $R'-COO^-$.

It is not necessary that the substrate be ionizable to obtain a bell-shaped curve. These kinetically equivalent schemes will give such a profile:

$$S + HA + B \longrightarrow products$$

$$S + A^- + BH^+ \longrightarrow products$$

The acid and base groups may be in the same molecule. Bell-shaped pH-rate profiles observed in enzyme-catalyzed reactions are usually interpreted according to this scheme, with the catalytic acid and base being assigned to the active site of the enzyme [70]. The apparent pK_a values estimated from such curves may differ appreciably from the actual dissociation constants if the mechanism includes pre-equilibria (other than the acid dissociation equilibria) prior to the rate-determining step [71].

Problems

1. Calculate the percent completion of a first-order reaction for the elapse of 0, 1, 2, ..., 10 half-lives. Plot the percent completion against the number of half-lives.

2. Show that the slope of a plot of log (k/T) against $1/T$ is equal to $-2.303\Delta H^{\ddagger}/R$.

3. Derive an equation relating $\Delta S^{\ddagger}$ to A at all temperatures.

    Answer: $\Delta S^{\ddagger} = 4.576 \log A - 4.576 \log T - 49.21$.

4. For the kinetically equivalent rate terms $k[HA][B^-]$ and $k'[HB][A^-]$, find the relationship between k and k'.

    Answer: $k = k' K_a^{HA}/K_a^{HB}$

5. Ester saponification proceeds through a tetrahedral intermediate as shown:

    $$RCOOR' + OH^- \underset{k_{-1}}{\overset{k_1}{\rightleftharpoons}} \underset{\underset{O^-}{|}}{\overset{\overset{OH}{|}}{R-C-OR}} \overset{k_2}{\longrightarrow} RCOO^- + R'OH$$

    Applying the steady-state approximation to the intermediate, derive an equation relating the second-order hydrolytic rate constant to the constants k_1, k_{-1}, and k_2.

    Answer: $k_{OH} = \dfrac{k_1 k_2}{k_{-1} + k_2}$

6. For the parallel first-order reaction

    $$A \overset{k_B}{\longrightarrow} B$$

    $$A \overset{k_C}{\longrightarrow} C$$

    relate the observed first-order rate constant k_{obs} to the constants k_B and k_C for measurements based on analyses of A, B, and C, respectively. Hint: use the equality $[B]/[C] = k_B/k_C$.

    Answer: $k_{obs} = k_B + k_C$ in each case.

7. Show that a plot of $\log v_0$ against $\log c_0$ gives the order of the reaction with respect to reactant C, where v_0 and c_0 are initial rate and concentration, respectively, and concentrations of all other reactants are held constant.

8. For a reaction that is followed spectrophotometrically,
    let $v_0' = dA/dt$ be the initial rate in absorbance
    units per second. Relate v_0' to v_0, the initial rate
    in moles per liter per second.

    Answer: $v_0 = v_0'/b(\epsilon_s - \epsilon_p)$, where b is path length,
    and ϵ_s and ϵ_p are molar absorptivities of substrate
    and product.

9. These absorbance-time data were obtained for the hy-
    drolysis of N-trans-cinnamoylimidazole at 25° in pH
    10.00 aqueous solution at ionic strength 0.3 $\underline{M}$.

t/sec	A
15	0.767
30	0.608
45	0.482
60	0.383
75	0.303
90	0.240
105	0.192
∞	0.000

    Calculate the pseudo-first-order rate constant and the
    second-order rate constant.

10. These rate constants were measured for the hydrolysis
    of methyl ethylene phosphate at 25° [R. Kluger, F.
    Covitz, E. Dennis, L. D. Williams, and F. H. West-
    heimer, J. Am. Chem. Soc., $\underline{91}$, 6066 (1969)].

pH	k/min^{-1}	pH	k/min^{-1}
2.01	0.21	7.00	0.0023
3.01	0.019	8.00	0.0090
4.00	0.0028	8.50	0.021
5.00	0.0016	9.00	0.050
6.00	0.0014	9.50	0.18
		10.00	0.61

    (a) Convert these to rate constants with units of
        sec^{-1}. Plot the log k-pH rate profile.
    (b) Determine the rate equation and evaluate the rate
        constants.

11. These rate constants are for the acetolysis of exo-2-
     tosyloxybicyclo[2.2.1]-heptan-7-one ethylene glycol
     ketal [P. G. Gassman and J. G. MacMillan, J. Am. Chem.
     Soc., 91, 5527 (1969)].

$t/^\circ C$	$10^4 k/sec^{-1}$
80.0	2.71
90.0	8.05
100.0	25.6

Make the Arrhenius plot and estimate $\Delta H^{\ddagger}$ and $\Delta S^{\ddagger}$.

12. Enzyme catalysis proceeds according to this scheme:

$$E + S \underset{k_{-1}}{\overset{k_1}{\rightleftharpoons}} ES \overset{k_2}{\longrightarrow} E + P$$

where E is enzyme, S substrate, P products, and ES is
an enzyme-substrate complex. Let $E_t = [E] + [ES]$,
and $S_t >>> E_t$. Apply the steady state assumption to
ES and derive the Michaelis-Menten equation,

$$v = \frac{V_{max}[S]}{K_m + [S]}$$

where $V_{max} = k_2 E_t$ and $K_m = (k_{-1} + k_2)/k_1$.

13. Show that a plot of $1/v$ against $1/[S]$ for an enzyme
     catalyzed reaction (a Lineweaver-Burk plot) is
     linear, and show how V_{max} and K_m can be obtained.

14. Suppose the rate equation for the aminolysis of an
     ester E is [61]

$$v = k_1 [E][OH^-] + k_2 [E][RNH_2] + k_3 [E][RNH_2]^2$$

$$+ k_4 [E][RNH_2][OH^-] + k_5 [E][RNH_2][RNH_3^+]$$

Develop an experimental design that will permit all
of the rate constants to be determined.

15. Consider the second-order reaction $A + B \rightarrow C$ carried
     out in this flow system:

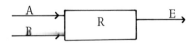

Solutions of A and B are fed into the reaction vessel R, where they are mixed. Solution flows out through exit E. The flow rates are adjusted so that a steady-state is reached in R, with the concentrations a, b, and c of substances A, B, and C being invariant with time. Let the volume of R be V liters, the rate of outflow be u liters/sec. Derive an equation giving k as a function of a, b, c, u, and V.

Answer: $k = uc/abV$

16. These rate constants are for the degradation of the antibiotic clindamycin at pH 4.0 in 0.2 $\underline{M}$ citrate buffers [T. O. Oesterling, J. Pharm. Sci., $\underline{59}$, 63 (1970)].

$t/^\circ C$	$10^6 k/sec^{-1}$
47	0.0124
53	0.0279
59	0.0831
70	0.249
80	0.969
92	3.62

Calculate the apparent activation energy.

References

1. W. P. Jencks, Catalysis in Chemistry and Enzymology, McGraw-Hill, New York, 1969, Chap. 11.
2. See, for example, S. W. Benson, The Foundations of Chemical Kinetics, McGraw-Hill, New York, 1960; A. A. Frost and R. G. Pearson, Kinetics and Mechanisms, 2nd ed., Wiley, New York, 1961; S. L. Friess, E. S. Lewis, and A. Weissberger (eds.), Investigation of Rates and Mechanisms of Reactions, Part I, Vol. VIII of Technique of Organic Chemistry, 2nd ed., Wiley (Interscience), New York, 1961; K. J. Laidler, Chemical Kinetics, 2nd ed., McGraw-Hill, New York, 1965.
3. E. D. Hughes, C. Ingold, and A. J. Parker, J. Chem. Soc., 1960, p. 4400. The rate constant is numerically equal to the rate when the concentrations are unity.
4. Standard states are on the molar scale.
5. For a survey of the collision theory, and a comparison of the theory with experiment, see E. A. Moelwyn-

Hughes, The Kinetics of Reactions in Solution, 2nd
ed., Clarendon, Oxford, 1947, Chap. III.

6.   S. Glasstone, K. J. Laidler, and H. Eyring, The Theory
of Rate Processes, McGraw-Hill, New York, 1941; H.
Eyring and E. M. Eyring, Modern Chemical Kinetics,
Reinhold, New York, 1963.

7.   This is not strictly accurate. The barrier is the
energy required to reach the saddle point, as described. But most molecules will take a path somewhat higher on the potential energy surface than the
saddle; this is because the reaction trajectory on
the surface is curved and centrifugal potential will
carry the trajectory along a path higher than the
saddle. The actual energy required is the threshold.
For reactions in which the threshold is much larger
than the thermal energy, the experimental activation
energy is a good estimate of the threshold. Cf. M.
Menziger and R. L. Wolfgang, Angew. Chem. Intern. Ed.,
8, 438 (1969).

8.   J. E. Leffler and E. Grunwald, Rates and Equilibria
of Organic Reactions, Wiley, New York, 1963, p. 64.

9.   The "double dagger" superscript always refers to the
transition state or to the process Initial state $\rightleftharpoons$
transition state.

10. This is not a true equilibrium in the thermodynamic
sense. When A and B react to form the transition
state, this species continues along the reaction path
to form products; its probability of reverting to
reactants is very small. But the transition state
can also be formed from the reverse direction, with
production of A and B; and the assumption of equilibrium appears to be justified and useful.

11. V. K. LaMer, Chem. Revs., 10, 179 (1932); Glasstone,
Laidler, and Eyring, Ref. [6], pp. 427-430. See also
K. Denbigh, The Principles of Chemical Equilibrium,
2nd ed., Cambridge Univ. Press, Cambridge, 1966, pp.
445-448. This theory is developed more fully in Sec.
3.4; Eq. (3.52) is the relationship referred to here.

12. This includes the special case that A and B are identical molecules, i.e., the reaction $2A \rightarrow$ products.

13. For ease in comparing plots superimposed on the same
graph, and for formal presentations, it is preferable
to plot the logarithm of the ratio $(A_t - A_\infty)/(A_0 - A_\infty)$,
because this "normalizes" all plots to the same
intercept.

14. This calculation is made by combining the definition
$n = t/t_{1/2}$ (n is number of half-lives) with $\ln$
$(c_A/c_A^0)^{1/2} = -kt$ to give $\log (c_A/c_A^0) = -0.30n$. The
percent reaction is $100(1 - c_A/c_A^0)$.

15. The plot can also be made by plotting on semi-logarithmic paper; this has the advantages of speed in plotting and ease in reading $t_{1/2}$ directly from the plot as the time corresponding to 50% reaction. Remember to take logarithms of the ordinates when calculating the slope from semi-log paper.
16. E. A. Guggenheim, Phil. Mag., 2, 538 (1926).
17. Tables of Chemical Kinetics (Homogeneous Reactions), NBS Circular 510, Supplement 1, Supplement 2, and Monograph 34, National Bureau of Standards, U. S. Dept. Commerce, Washington, D. C. (1951 through 1961).
18. L. R. Fedor and J. McLaughlin, J. Am. Chem. Soc., 91, 3594 (1969).
19. R. C. Petersen, J. H. Markgraf, and S. D. Ross, J. Am. Chem. Soc., 83, 3819 (1961); K. B. Wiberg, Physical Organic Chemistry, Wiley, New York, 1964, p. 378. See also Sec. 4.1.
20. S. P. Eriksen and H. Stelmach, J. Pharm. Sci., 54, 1029 (1965).
21. There can be exceptions; see S. Glasstone, K. J. Laidler, and H. Eyring, The Theory of Rate Processes, McGraw-Hill, New York, 1941, pp. 190, 447.
22. See S. Siggia, Quantitative Organic Analysis via Functional Groups, 3rd ed., Wiley, New York, 1963, Chap. 25; Kinetics in Analytical Chemistry (H. B. Mark, Jr. and G. A. Rechnitz, eds.), Wiley (Interscience), New York, 1968.
23. H. U. Bergmeyer (ed.), Methods of Enzymatic Analysis, Academic, New York, 1963.
24. Oxime formation illustrates this latter possibility; see W. P. Jencks, Catalysis in Chemistry and Enzymology, McGraw-Hill, New York, 1969, p. 590.
25. For some examples see Sects. 12.1 and 13.1.
26. This is correct as far as it goes; but the figure is in terms of rate constants, that is, of rates at unit concentrations. Since rates are concentration dependent, it is conceivable that the rate-determining step could be altered by a sufficiently drastic change in concentrations.
27. There are several ways to obtain this information, however. The equilibrium constant contains some information about the product, as demonstrated by Eq. (3.50). The reaction could be studied in the reverse direction, that is, starting with P. Finally, it may be possible to prepare I, or to generate it in situ, and to study its subsequent conversion to P.
28. This guideline is sometimes called the Hammond postulate; see G. S. Hammond, J. Am. Chem. Soc., 77, 334 (1955) and J. R. Murdoch, ibid., 94, 4410 (1972).

29. C. K. Ingold, Structure and Mechanism in Organic
     Chemistry, Cornell Univ. Press, Ithaca, New York,
     1953, p. 315.
30. L. L. Schaleger and F. A. Long, Advan. Phys. Org.
     Chem., $\underline{1}$, 1 (1963).
31. K. Denbigh, The Principles of Chemical Equilibrium,
     2nd ed., Cambridge Univ. Press, Cambridge, 1966,
     p. 448.
32. J. H. Sullivan, J. Chem. Phys., $\underline{51}$, 2288 (1969). For
     the reaction $2I + H_2 \rightleftharpoons 2HI$ the equality holds, but a
     discrepancy was found for the process $C\ell + H_2 \rightleftharpoons HC\ell$
     + H.
33. R. L. Burwell and R. G. Pearson, J. Phys. Chem., $\underline{70}$,
     300 (1966); R. M. Krupka, H. Kaplan, and K. J. Laid-
     ler, Trans. Faraday Soc., $\underline{62}$, 2754 (1966).
34. J. E. Leffler and E. Grunwald, Rates and Equilibria
     of Organic Reactions, Wiley, New York, 1963, p. 22.
35. R. P. Bell, Acid-Base Catalysis, Oxford, London,
     1941, Chap. II.
36. C. K. Ingold, Structure and Mechanism in Organic
     Chemistry, Cornell Univ. Press, Ithaca, New York,
     1953, p. 347.
37. A. Wasserman, Monatsh., $\underline{83}$, 543 (1952); K. B. Wiberg,
     Physical Organic Chemistry, Wiley, New York, 1964,
     p. 376.
38. It is conventional to enclose postulated transition
     state structures in brackets.
39. K. B. Wiberg, Physical Organic Chemistry, Wiley, New
     York, 1964, pp. 379-388.
40. Since we can only measure differences between the
     transition and initial states, this specification of
     initial state stabilization is arbitrary, and is used
     because it is conceptually convenient. The only non-
     arbitrary description of this prediction is that an
     increase in solvent polarity will increase $\Delta G^{\ddagger}$.
41. The free energy is related to the square of the charge,
     and this dependence is the justification for this
     statement.
42. V. K. LaMer, Chem. Revs., $\underline{10}$, 179 (1932); C. W.
     Davies, Progr. Reaction Kinetics, $\underline{1}$, 161 (1961); E. S.
     Amis, Solvent Effects on Reaction Rates and Mechanisms,
     Academic, New York, 1966.
43. V. K. LaMer Chem. Revs., $\underline{10}$, 179 (1932); S. Glasstone,
     K. J. Laidler, and H. Eyring, The Theory of Rate Pro-
     cesses, McGraw-Hill, New York, 1941, p. 429; K. B.
     Wiberg, Ref. [39], p. 394; K. J. Laidler, Chemical
     Kinetics, 2nd ed., McGraw-Hill, New York, 1965, p.
     221.
44. C. W. Davies, Progr. Reaction Kinetics, $\underline{1}$, 161 (1961).

45. A. A. Frost and R. G. Pearson, Kinetics and Mechanism, 2nd ed., Wiley, New York, 1961, p. 307.

46. K. A. Connors and M. L. Bender, J. Org. Chem., $\underline{26}$, 2498 (1961).

47. T. C. Bruice and F.-H. Marquardt, J. Am. Chem. Soc., $\underline{84}$, 365 (1962).

48. The reaction described by the k term is variously called the "water," "uncatalyzed," or "spontaneous" reaction, and some authors write $k_2 = k_W[H_2O]$, expressing $[H_2O]$ as the molar concentration in calculating k_W.

49. P. Finholt and T. Higuchi, J. Pharm. Sci., $\underline{51}$, 655 (1962).

50. W. P. Jencks, Catalysis in Chemistry and Enzymology, McGraw-Hill, New York, 1969, p. 583.

51. M. L. Bender, Y.-L. Chow, and F. Chloupek, J. Am. Chem. Soc., $\underline{80}$, 5380 (1958).

52. L. J. Edwards, Trans. Faraday Soc., $\underline{46}$, 723 (1950); $\underline{48}$, 696 (1952).

53. E. R. Garrett, J. Am. Chem. Soc., $\underline{79}$, 3401 (1957).

54. This plateau region accounts for the absence from the pharmaceutical market of neutral solutions of aspirin, which are too unstable to be a feasible dosage form.

55. Dixon and Webb have described the properties of these functions (or rather their reciprocals) in greater detail; cf. M. Dixon and E. C. Webb, Enzymes, 2nd ed., Academic, New York, 1964, pp. 116-128.

56. R. A. Alberty and V. Massey, Biochim. Biophys. Acta, $\underline{13}$, 347 (1954).

57. A change in rate-determining step can also produce a bell-shaped curve, as in oxime formation; W. P. Jencks and M. Gilchrist, J. Am. Chem. Soc., $\underline{86}$, 5616 (1964).

58. B. Zerner and M. L. Bender, J. Am. Chem. Soc., 83, 2267 (1961).

59. C. A. Bunton, D. R. Llewellyn, K. G. Oldham, and C. A. Vernon, J. Chem. Soc., 1958, p. 3574.

60. H. Morawetz and I. Oreskes, J. Am. Chem. Soc., $\underline{80}$, 2591 (1958).

61. M. L. Bender and K. A. Connors, J. Am. Chem. Soc., $\underline{83}$, 4099 (1961); $\underline{84}$, 1980 (1962); J. L. Bada and S. L. Miller, ibid., $\underline{91}$, 3948 (1969).

62. I. Amdur and G. G. Hammes, Chemical Kinetics, McGraw-Hill, New York, 1966, p. 152.

63. R. P. Bell, Acid-Base Catalysis, Oxford, London, 1941, p. 3.

64. A. A. Frost and R. G. Pearson, Kinetics and Mechanism, 2nd ed., Wiley, New York, 1961.

65. Only homogeneous systems are being considered. The

important field of heterogeneous catalysis (which has
probably not received sufficient attention as to its
analytical possibilities) will not be treated; cf. E.
K. Rideal, Concepts in Catalysis, Academic, New York,
1968.

66. S. Glasstone, K. J. Laidler, and H. Eyring, The Theory
of Rate Processes, McGraw-Hill, New York, 1941, p.
197.

67. Jencks calls this catalysis by approximation; W. P.
Jencks, Catalysis in Chemistry and Enzymology, McGraw-
Hill, New York, 1969, Chap. 1. It is called the pro-
pinquity effect by T. C. Bruice and S. J. Benkovic,
Bioorganic Mechanisms, Vol. I, Benjamin, New York,
1966, p. 119.

68. W. P. Jencks, Ref. [67], pp. 163-166.

69. W. P. Jencks, Ref. [67], pp. 571-576.

70. M. Dixon and E. C. Webb, Enzymes, 2nd ed., Academic
New York, 1964, pp. 128-145.

71. T. C. Bruice and G. L. Schmir, J. Am. Chem. Soc., 81,
4552 (1959).

Chapter 4.   EXTRATHERMODYNAMIC RELATIONSHIPS

The magnitude of a change in a thermodynamic property of
a system is determined only by the values of the property
in the initial and final states of the system. An infinite number of routes may exist between these two states,
and each route necessarily results in the same net change
in the property. Thus knowledge of the overall change
cannot distinguish between alternate paths. The concepts
and methods of chemical kinetics provide considerable insight into the actual paths (mechanisms) accessible to the
process, because kinetics is concerned with changes in
properties between an initial state and some intermediate
state, namely the transition state. Chapters 2 and 3
dealt with the thermodynamic and kinetic approaches to
chemical reactions.
      The great generality of thermodynamics is a conse-
quence of its minimal use of specific and detailed models;
on the other hand, it is the absence of such models that
prevents thermodynamics from providing insight into molecular mechanisms. The combination of detailed models with
the concepts of thermodynamics is called the extrathermodynamic approach. Since it involves model building, the
technique lacks the rigor of thermodynamics, but it can
provide information not otherwise accessible. Extrathermodynamic relationships often take the form of correlations among rates and equilibria, and the models used to
account for these include hypotheses of chemical bonding,
transmission of electronic effects, and interactions among
electronic effects.
      The literature in this area is extensive and some of
the concepts and symbolism may be transitory. This chapter reviews the field at a level and with a coverage adequate for the experimentalist to use the standard relationships and to follow their use in the mechanistic literature. Research on the meaning of the extrathermodynamic
relationships themselves is beyond our present needs; the
interested reader can explore these ideas further in the
leading references given here.

4.1 Principles

Linear Free Energy Relationships. The most common manifestation of extrathermodynamic relationships is a linear
correlation between the logarithms of rate or equilibrium
constants for one reaction series and the logarithms of
rate or equilibrium constants of a second reaction series,
both sets being subjected to the same variation, usually
of structure. For illustration, suppose the logarithm of
the rate constants for a reaction series B is linearly
correlated with the logarithm of the equilibrium constants
for a reaction series A, with substituent changes being
made in both series. The empirical correlation is

$$\log k_B = m \log K_A + b \qquad (4.1)$$

As shown earlier, these logarithmic terms are linearly related to free energy changes:

$$\log k_B = \frac{-\Delta G_B^{\ddagger}}{2.3RT} + \log \frac{kT}{h} \qquad (4.2)$$

$$\log K_A = \frac{-\Delta G_A^0}{2.3RT} \qquad (4.3)$$

Combining Eqs. (4.1)-(4.3),

$$\Delta G_B^{\ddagger} = m \Delta G_A^0 + b' \qquad (4.4)$$

Such correlations are therefore called linear free energy
relationships. Often it is convenient to express the
correlation in terms of ratios of constants by referring
all members of the series to a reference member of the
series; thus the correlation in Eq. (4.1) can be expressed

$$\log \frac{k_B}{k_B^0} = m \log \frac{K_A}{K_A^0} \qquad (4.5)$$

where k_B^0 and K_A^0 are the constants for the reference substituent. With the symbolism adopted in Secs. 2.3 and
3.4, Eq. (4.5) can be written

$$\delta_R \Delta G_B^{\ddagger} = m \delta_R \Delta G_A^0 \qquad (4.6)$$

Though linear free energy relationships are not a
necessary consequence of thermodynamics, their occurrence
suggests the presence of a real connection between the

correlated quantities, and the nature of this connection
can be explored. This treatment follows Leffler and Grunwald [1]. Standard free energy changes ΔG^0 will pertain
to either equilibrium or rate processes. Consider the
reaction of a molecule that, for convenience, we divide
into two zones, R and X. Zone R includes the variable
substituent and X is the reaction site. The standard free
energy of the substance is considered to be an additive
function of the free energies of these two zones plus a
term describing the interaction between the two zones.

$$G_{RX}^0 = G_R^0 + G_X^0 + I_{R,X} \tag{4.7}$$

Equation (4.7) represents the model, which will be applied
to a reaction series A, in which a set of compounds varying only in the substituent R undergoes the common reaction

$$RA \rightleftharpoons RA' \tag{4.8}$$

Writing $\Delta G_{RA}^0 = G_{RA'}^0 - G_{RA}^0$ and applying Eq. (4.7),

$$\Delta G_{RA}^0 = (G_{A'}^0 - G_A^0) + (I_{R,A'} - I_{R,A}) \tag{4.9}$$

in which the G_R^0 term has disappeared because it does not
undergo any change in the chemical reaction. A second
reaction series B, with the same variations in substituents but a different reaction site, is similarly described:

$$RB \rightleftharpoons RB' \tag{4.10}$$

$$\Delta G_{RB}^0 = (G_{B'}^0 - G_B^0) + (I_{R,B'} - I_{R,B}) \tag{4.11}$$

Variations in the substituent R within the reaction series
are described by the terms $\delta_R \Delta G_A^0$ and $\delta_R \Delta G_B^0$.

$$\delta_R \Delta G_A^0 = \Delta G_{RA}^0 - \Delta G_{R_0 A}^0 \tag{4.12}$$

$$\delta_R \Delta G_B^0 = \Delta G_{RB}^0 - \Delta G_{R_0 B}^0 \tag{4.13}$$

where R_0 is the reference substituent. Equations (4.12)
and (4.13) are combined with (4.9) and (4.11):

$$\delta_R \Delta G_A^0 = (I_{R,A'} - I_{R,A}) - (I_{R_0,A'} - I_{R_0,A}) \tag{4.14}$$

$$\delta_R \Delta G_B^0 = (I_{R,B'} - I_{R,B}) - (I_{R_0,B'} - I_{R_0,B}) \tag{4.15}$$

It is necessary to introduce one further assumption to
make this model lead to a linear free energy relationship;
this is a separability postulate,

$$I_{R,X} = I_R \cdot I_X \qquad (4.16)$$

Then, from Eqs. (4.14) and (4.15),

$$\delta_R \Delta G_A^0 = (I_R - I_{R_0})(I_{A'} - I_A) \qquad (4.17)$$

$$\delta_R \Delta G_B^0 = (I_R - I_{R_0})(I_{B'} - I_B) \qquad (4.18)$$

Combining these equations,

$$\delta_R \Delta G_B^0 = \left(\frac{I_{B'} - I_B}{I_{A'} - I_A}\right) \delta_R \Delta G_A^0 \qquad (4.19)$$

Equation (4.19) has the form of a linear free energy relationship [compare with Eq. (4.6)]. The quantity in
parentheses is independent of the nature of the substituent, depending only upon the reaction types; it is called
the reaction parameter. Now suppose that reaction series
A is selected as a standard reaction; then $\delta_R \Delta G_A^0$ becomes
dependent only on the substituent, and is called the substituent parameter. (For the standard reaction, the
reaction parameter is arbitrarily set equal to unity.)
Wells [2] has given an equivalent treatment.
      Later sections describe several empirical relation-
ships with the form of Eq. (4.19). Some free energy relationships require expression as four-parameter equations,
of the form

$$\delta_R \Delta G_B^0 = m \delta_R \Delta G_1^0 + n \delta_R \Delta G_2^0 \qquad (4.20)$$

Such an equation implies two interaction mechanisms in the
model [3].
      The preceding discussion has been concerned with
linear free energy relationships between rates or equilibria of two different reactions. We consider next the
possible relationship between the rate constant and the
equilibrium constant of the same reaction. Since ΔG^0 is
independent of the reaction path, there is no rigorous
theoretical requirement that ΔG^0 be simply related to $\Delta G^{\ddagger}$.
It is nevertheless common to find that rates parallel
equilibria. A linear free energy relationship between the
rates and equilibria of the same reaction series, subjected
to variations in substituents or the medium, can be simply

accounted for [4]. We consider the effect of the change
(substituent or medium) on the free energy of the transition state. Presumably the transition state resembles the
reactants and products in many ways, being intermediate to
them in structure and electronic distribution. Then it is
supposed that changes in $G^{\ddagger}$ can be represented as a linear
combination of changes in the free energies of reactants
and products,

$$\delta G^{\ddagger} = a\delta G_P^0 + b\delta G_R^0 \qquad (4.21)$$

where a and b are constants of the system. The values of
a and b are fixed by the further postulate that, although
$G^{\ddagger}$ is greater than either G_R^0 or G_P^0, changes in $G^{\ddagger}$ will be
intermediate in magnitude to the corresponding changes in
G_R^0 or G_P^0. Equation (4.21) then can be written,

$$\delta G^{\ddagger} = \alpha\delta G_P^0 + (1-\alpha)\delta G_R^0 \qquad (4.22)$$

where α must be between 0 and 1. Rearranging Eq. (4.22)
and invoking the commutation property that $\Delta\delta G = \delta\Delta G$,

$$\delta\Delta G^{\ddagger} = \alpha\delta\Delta G^0 \qquad (4.23)$$

Equation (4.23) is a linear free energy relationship between rates and equilibria, with the proportionality constant being susceptible to a simple physical interpretation; α may be considered the fractional distance along
the reaction coordinate at which the transition state is
located. That is, α is the degree to which the transition
state resembles the product. Sometimes the value of α can
be roughly predicted, as with the Hammond postulate [5],
which says that the transition state will more closely
resemble the less stable of the two states (initial or
final) in the reaction.
    The constancy of α within a reaction series, as im-
plied by Eq. (4.23), can only be expected if the changes
in substituents or the medium are moderate. Cohen and
Marcus [6] give a semi-theoretical relationship between
rates and equilibria,

$$\Delta G^{\ddagger} = \frac{\lambda}{4}(1 + \frac{\Delta G^0}{\lambda}) \qquad (4.24)$$

where λ is a parameter having the dimensions of ΔG. Since
α is the slope of the plot of $\Delta G^{\ddagger}$ against ΔG^0, then

$$\alpha = \frac{1}{2}(1 + \frac{\Delta G^0}{\lambda}) \qquad (4.25)$$

The dependence of α upon ΔG^0 is slight but clearly evident for numerous reactions when a wide range in ΔG^0 is accessible [6].

Enthalpy-Entropy Relationships. It can be shown [7] that if a linear free energy relationship, Eq. (4.6), is observed over a range of temperatures, and if the enthalpy and entropy changes are temperature independent, then the enthalpy changes must be directly proportional to the entropy changes for the reaction series. Let us start with the proposition that a real effect of this type has been demonstrated for a given reaction series; we write this as

$$\delta\Delta H^{\ddagger} = \beta\delta\Delta S^{\ddagger} \tag{4.26}$$

This can be combined with the general relationship

$$\delta\Delta G^{\ddagger} = \delta\Delta H^{\ddagger} - T\delta\Delta S^{\ddagger} \tag{4.27}$$

to give

$$\delta\Delta G^{\ddagger} = (\beta-T)\delta\Delta S^{\ddagger} \tag{4.28}$$

$$\delta\Delta G^{\ddagger} = (1 - \frac{T}{\beta})\delta\Delta H^{\ddagger} \tag{4.29}$$

The proportionality constant β has the dimension of absolute temperature, and it is called the isokinetic temperature [8]. It has the significance that when $T = \beta$, $\delta\Delta G^{\ddagger} = 0$; that is, all substituent (or medium) effects on the free energy change vanish at the isokinetic temperature. At this temperature the ΔH and $T\Delta S$ terms exactly offset each other, giving rise to the term compensation effect for isokinetic behavior.

Two extreme situations should be noted. If $\beta = 0$, then $\delta\Delta G^{\ddagger} = -T\delta\Delta S^{\ddagger}$, and the reaction series is entirely entropy controlled; it is said to be isoenthalpic. If $1/\beta = 0$, then $\delta\Delta G^{\ddagger} = \delta\Delta H^{\ddagger}$, and the series is enthalpy controlled, or isoentropic.

The demonstration of a real isokinetic relationship, Eq. (4.26), has two practical consequences: first, it suggests caution in the interpretation of substituent effects or medium effects when the experimental temperature is near the isokinetic temperature. As we have seen, when $T = \beta$ the reaction appears to be totally insensitive to variation in structure (or medium). As T is varied from below β to above β the reaction parameter will change sign, which must be taken into account in an interpretation of the meaning of the reaction parameter. A literature survey of apparent isokinetic relationships reveals that a large fraction of β values falls in the range 300-400°, which is a common experimental range [9]; the possibility that the experimental T is close to β is therefore very real.

The second consequence of an isokinetic relationship is that it can tell something about the mechanism of the process. The usual interpretation follows these lines: ΔH^0 is related to the strength of the interactions involved in the process. In a series of closely related compounds, arrangement in order of increasingly negative ΔH^0 will therefore correspond to progressively stronger bonding. The stronger the bonding, the greater the loss of translational and rotational freedom, hence the more negative the ΔS^0 for the process. Thus a compensatory correlation between ΔH^0 and ΔS^0 seems reasonable, though this qualitative argument does not account for the form of the relationship.

If more than one interaction mechanism is contributing to the overall ΔH and ΔS values, these experimental quantities may be difficult to interpret unless they are resolved into the separate contributions. Hepler has analyzed the ionization of some weak acids in aqueous solution in terms of two interaction mechanisms [10]. The basic assumption is that the experimental enthalpy and entropy changes can be expressed as the sums of internal (within the molecule) and external (solute-solvent interaction) contributions. For the proton exchange reaction

$$HB + A^- \rightleftharpoons B^- + HA \tag{4.30}$$

where HB is a substituted acid and HA is the unsubstituted reference acid, this assumption gives

$$\delta\Delta H^0 = \delta\Delta H_{int} + \delta\Delta H_{ext} \tag{4.31}$$

$$\delta\Delta S^0 = \delta\Delta S_{int} + \delta\Delta S_{ext} \tag{4.32}$$

From the symmetry of Reaction (4.30), it is reasonably assumed that $\delta\Delta S_{int}$ is negligible. An isokinetic relationship is assumed for the external contributions: $\delta\Delta H_{ext} = \beta\delta\Delta S_{ext}$. Combining these gives

$$\delta\Delta H^0 = \delta\Delta H_{int} + \beta\delta\Delta S^0 \tag{4.33}$$

Several kinds of evidence suggested that β is about 280°. With this value and the experimental $\delta\Delta H^0$ and $\delta\Delta S^0$ values, $\delta\Delta H_{int}$ could be calculated. Substituting Eq. (4.33) into $\delta\Delta G^0 = \delta\Delta H^0 - T\delta\Delta S^0$ yields

$$\delta \Delta G^0 = \Delta \delta H_{int} + (\beta - T) \delta \Delta S^0 \qquad (4.34)$$

When $(\beta - T)$ is small this becomes

$$\delta \Delta G^0 \simeq \delta \Delta H_{int} \qquad (4.35)$$

The $\delta \Delta H_{int}$ values calculated with Eq. (4.33) do in fact agree closely with experimental $\delta \Delta G^0$ values. This agreement provides justification for the practice of rationalizing acid strengths in terms of electron displacement models (inductive and resonance effects), for these are internal energy effects. Hepler's approach to resolving multiple interactions may be pertinent to other types of processes [11].

The existence, or at any rate the widespread occurrence, of isokinetic relationships has been debated in the recent literature. It is acknowledged that many linear enthalpy-entropy relationships are known [12]; the controversial point is that many of these may be artifacts rather than facts. Petersen et al. [13] analyzed the errors in $\Delta H^{\ddagger}$ and $\Delta S^{\ddagger}$, showing that the error in $\Delta S^{\ddagger}$ is directly proportional to the error in $\Delta H^{\ddagger}$. A reaction series can generate an apparent isokinetic relationship solely through the operation of this error effect, which predicts that a plot of $\Delta H^{\ddagger}$ against $\Delta S^{\ddagger}$ will be linear with slope about equal to the experimental temperature. The observation that many apparent isokinetic temperatures are in the experimental range may be related to this error effect.

The error contour in the ΔH-ΔS plane is a narrow elipse [14] with the slope of the major axis being T. It has been suggested that to distinguish between a true isokinetic relationship and a spurious error correlation the spread of enthalpy and entropy values should be several times the maximum error. Another criterion is that the points on the line fall in a rational, rather than a random order, as established by correlation with other properties. Petersen [15] believes that a linear enthalpy-entropy relationship can never be, by itself, a sufficient demonstration of an isokinetic relationship; he concludes that the only unambiguous evidence for an isokinetic relationship is the intersection of all log (k/T) - $1/T$ plots for the series at a common point (which will correspond to $1/\beta$).

Exner [16] has made an important contribution to this problem. Let k_1 and k_2 be the rate constants for a member of a reaction series at temperatures T_1 and T_2, with $T_2 > T_1$, and let k_1^0 and k_2^0 be the corresponding values for the reference member of the series. Then Eqs. (4.36) and (4.37) are easily derived for the reaction series.

$$\delta_R \Delta H^{\ddagger} = R \left(\frac{T_1 T_2}{T_2 - T_1}\right)\left(\ell n \frac{k_2}{k_2^0} - \ell n \frac{k_1}{k_1^0}\right) \qquad (4.36)$$

$$\delta_R \Delta S^{\ddagger} = R \left(\frac{1}{T_2 - T_1}\right)\left(T_2 \, \ell n \frac{k_2}{k_2^0} - T_1 \, \ell n \frac{k_1}{k_1^0}\right) \qquad (4.37)$$

Equation (4.36) shows that, in general, a plot of log k_2 against log k_1 is not expected to be linear. Linearity in such a plot can be assured, however, if an isokinetic relationship, $\delta_R \Delta H^{\ddagger} = \beta \delta_R \Delta S^{\ddagger}$, is followed by the system. By incorporating this relationship into Eqs. (4.36) and (4.37), we obtain:

$$\ell n \frac{k_2}{k_2^0} = \frac{T_1 (T_2 - \beta)}{T_2 (T_1 - \beta)} \, \ell n \frac{k_1}{k_1^0} \qquad (4.38)$$

Thus a linear plot of log k_2 against log k_1 for a reaction series implies an isokinetic relationship for the series. The reason that this plot is a reliable test for such a relationship is that the errors in k_1 and k_2 are independent (unlike the errors in $\Delta H^{\ddagger}$ and $\Delta S^{\ddagger}$). From the slope q of the straight line the isokinetic temperature β can be found:

$$\beta = \frac{T_1 (1 - q)}{(T_1/T_2) - q} \qquad (4.39)$$

The ratio T_1/T_2 should be appreciably smaller than unity to achieve reliable estimates of β; unfortunately most reported studies are not suited to accurate analysis by this method because the temperature studies were designed for conventional Arrhenius plotting.

Figure 4.1 is an Exner-type plot of log k_2 against log k_1 for the alkaline hydrolysis of benzoate esters C_6H_5COOR, where R may be aliphatic or aromatic [17,18]. The satisfactorily straight line is evidence that a single isokinetic relationship fits all of the data, and therefore that these compounds belong to a single reaction series subject to a common mechanism.

Several types of enthalpy-entropy behavior can be distinguished for series of related reactions. The classification in Table 4.I is given by Exner [16b] who feels that β should be interpreted as a constant of extrapolation with little physical meaning; β is believed always to lie well outside the experimental range. Case 3c, which includes a large number of the reported apparent isokinetic relationships, has not yet been verified by the log k_2-log k_1 criterion. Isoenthalpic behavior, case 2, should be diagnosed with care, because in the limit as the temperature interval T_2-T_1 becomes small all systems will approach the condition q = 1. Lumry and Rajender [8], in an extended discussion, conclude that the case 3c pattern is real, and relate its occurrence to the solvent (water) structure.

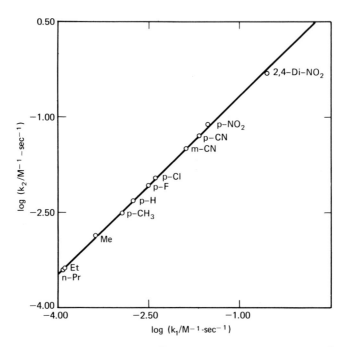

Fig. 1.   Plot of log k_2 (35°) against log k_1 (20°) for the alkaline hydrolysis of benzoate esters.

It has been convenient to assume that enthalpy changes are temperature independent, that is, that $\Delta C_p = 0$. In fact, however, this first approximation may not always be valid [19], and then small $\delta \Delta G$ or $\delta \Delta H$ values can readily be misinterpreted. Even the relative positions of the members of a series may then vary with the temperature.

4.2   Substituent Effects in the Substrate

The designation of one reactant as the substrate and another as the reagent is arbitrary but useful in discussing chemical reactivity. The substrate always underdoes some change in the reaction. A catalyst is always considered to be a reagent. When a reaction is studied under

the usual pseudo-first-order conditions, the concentration
of the substrate does not appear in the apparent first-
order rate constant, whereas the reagent concentration
does. In the reaction between an acid anhydride and an
amine, the physical organic chemist might regard the amine
as the reagent, whereas the analytical chemist may consider the anhydride as the reagent. This dual use of the
term should not cause any confusion here. The present
section concerns linear free energy relationships describing substituent effects in substrates; very generally, a
series of substrates R-X varying in the R substituent is
subjected to a common reaction at site X, and the quanti-
ties $\delta_R\Delta G$ or $\delta_R\Delta G^{\ddagger}$ are correlated with a standard reaction series.

Table 4.I   Enthalpy-Entropy Relationships in Reaction
Series

Case	Slope q[a]	β	Relationship	Occurrence[b]
1	T_1/T_2	∞	Isoentropic	Uncertain
2	1	0	Isoenthalpic	Occasional
3a	$<T_1/T_2$	$>T_2$	⎫	Frequent
b	>1	$<T_1$	⎬ Compensation	Occasional
c	<0	$<T_2$ and $>T_1$	⎭	Unknown
4	$>T_1/T_2$ and <1	>0	Augmentation	Occasional

[a]Slope of log k_2 - log k_1 plot; $T_2>T_1$.
[b]According to Exner [16b].

Aromatic Substrates. The classic example, and still the
most useful one, of a linear free energy relationship is
the Hammett equation [20], which correlates rates and
equilibria of many side-chain reactions of meta- and para-
substituted aromatic compounds. The standard reaction is
the aqueous ionization equilibrium at 25° of meta- and
para-substituted benzoic acids.

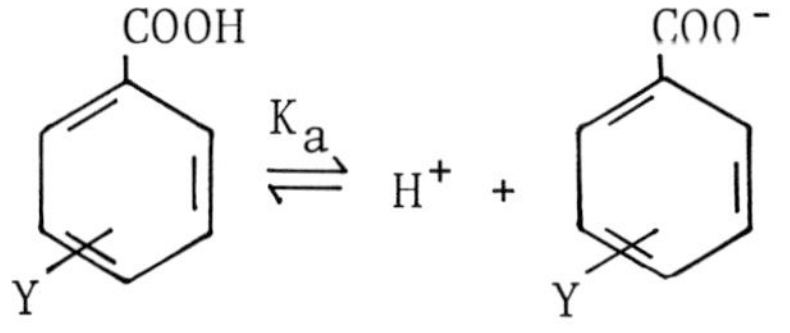

$$(4.40)$$

The reference compound is benzoic acid. The substituent parameter, which is called the Hammett substituent constant σ, is defined by

$$\sigma = \log \frac{K_a}{K_a^0} = pK_a^0 - pK_a \qquad (4.41)$$

σ is related to $\delta_R \Delta G^0$ by

$$\delta_R \Delta G^0 = -2.3\ RT\ \sigma \qquad (4.42)$$

The Hammett equation is written as

$$\log \frac{k}{k^0} = \rho\sigma \qquad (4.43)$$

where k is a rate or equilibrium constant for m- or p-substituted substrates and ρ is the Hammett reaction constant. Equation (4.43) obviously has the same form as Eqs. (4.5) and (4.6).

The form of Eq. (4.43) suggests, by comparison with Eq. (4.19), that a single interaction mechanism is operative between the substituent R and the reaction site X. That this is not necessarily true is shown as follows. Suppose two interaction mechanisms control the substituent effect. Then a Hammett relationship may be written as in

$$\log \frac{k}{k^0} = \rho_1\sigma_1 + \rho_2\sigma_2 \qquad (4.44)$$

which can be rearranged to

$$\log \frac{k}{k^0} = \rho_1\left(\sigma_1 + \frac{\rho_2}{\rho_1}\sigma_2\right) \qquad (4.45)$$

If the two interaction mechanisms maintain the same relative importance throughout the reaction series, that is, if the ratio ρ_2/ρ_1 is constant, Eq. (4.45) becomes identical in form with Eq. (4.43). This in fact appears to be the usual situation in Hammett correlations [21], and the two interaction mechanisms are commonly discussed in terms of the inductive and resonance effects of electronic displacement.

Two points of view have developed concerning the "best" values of the Hammett substituent constants. One of these, exemplified by the remarks of Jaffé [22] and Ehrenson [23], treats σ as an adjustable parameter whose value should be chosen to best fit the entire body of experimental data. The assignment of σ values in this context becomes a statistical problem of curve fitting, best handled by computer methods. The advantage of the method

is that it provides substituent constants capable of generating ρ values, and of regenerating rate and equilibrium
constants, with reasonable accuracy over the entire range
of reactions used in establishing the values. One of its
disadvantages is that these σ values are subject to periodic change as new data become available.

The other point of view [24,25] is that the σ values
should be based on the original defining relation of
Hammett, Eq. (4.41). Besides the advantage of providing
essentially permanent σ values, this approach takes a
fundamentally different, and more optimistic, view of deviations from Hammett plots. In this context such deviations represent different mechanistic effects in the correlated reaction series and the standard reaction series,
and the deviations can guide mechanistic interpretations.
The statistical method of evaluating substituent constants
tends to minimize deviations. The dispute really concerns
the utility of linear free energy relationships. If their
principal use is for data storage and retrieval, then the
statistically evaluated σ values will, on the average,
provide more accurate estimates of rate and equilibrium
constants. If they are helpful primarily as tools for
mechanistic interpretation, the definition of σ in terms
of a standard reaction is to be preferred, because it
yields substituent constants of definite value and unambiguous meaning (even though the meaning may not be understood). The practice of estimating "secondary" σ values
by measuring ρ for a given reaction series and then calculating σ for a group subjected to this reaction, without
having available its corresponding benzoic acid dissociation constant, has contributed to the range of published
σ values for a single substituent.

We take the view of McDaniel and Brown [24] that the
Hammett substituent constants should be defined by Eq.
(4.41) [26]. Table 4.II summarizes these constants based
on the ionization of meta- and para-substituted benzoic
acids. It is recommended that ρ values be based only on
the underlined σ values, for these are calculated from
thermodynamic constants.

The Hammett equation is said to be followed when a
plot of log k against σ is linear; this is taken to mean
that the correlation coefficient is at least 0.95, and
preferably above 0.98. Figure 4.2 is a Hammett plot for
the hydrolysis of substituted ethyl p-biphenylcarboxylates
[27], 1. The slope of the line, and therefore ρ, is
+0.583⁻[28]. This is an exceptionally good example, for
the correlation coefficient is 0.999 [22]. Table 4.III
lists 25 reactions that are correlated by the Hammett
equation [22]. Besides the reaction constant ρ, the table

Table 4.II. Hammett Substituent Constants Based on Ionization of Benzoic Acids[a,b]

Group	Meta position		Para position	
	σ_m	Uncertainty	σ_p	Uncertainty
$-CH_3$	-0.069	0.02	-0.170	0.02
$-CH_2CH_3$	-0.07	0.1	-0.151	0.02
$-CH(CH_3)_2$	--		-0.151	0.02
$-C(CH_3)_3$	-0.10	0.03	-0.197	0.02
$-C_6H_5$	0.06	0.05	-0.01	0.05
$-CF_3$	0.43	0.1	0.54	0.1
$-CN$	0.56	0.05	0.660	0.02
$-COCH_3$	0.376	0.02	0.502	0.02
$-CO_2C_2H_5$	0.37	0.1	0.45	0.1
$-CO_2H$	0.37[c]	0.1	0.45[c]	0.1
$-CO_2^-$	-0.1	0.1	0.0	0.1
$-CH_2Si(CH_3)_3$	-0.16	>0.1	-0.21	>0.1
$-Si(CH_3)_3$	-0.04	0.1	-0.07	0.1
$-Si(C_2H_5)_3$	--		0.0	0.1
$-N_2^{+}$[d]	1.76	0.2	1.91	0.2
$-NH_2$	-0.16	0.1	-0.66	0.1
$-NHCH_3$	--		-0.84	0.1
$-N(CH_3)_2$	--		-0.83	0.1
$-NHCOCH_3$	0.21	0.1	0.00	0.1
$-N(CH_3)_3^{+}$	>0.88	>0.2	0.82	>0.2
$-NO_2$	0.710	0.2	0.778	0.02
$-PO_3H^-$	0.2	>0.1	0.26	>0.1
$-AsO_3H^-$	--		-0.02	>0.1
$-OCH_3$	0.115	0.02	-0.268	0.02
$-OC_2H_5$	0.1	0.1	-0.24	0.1
$-O(CH_2)_2CH_3$	0.1	0.1	-0.25	0.1
$-OCH(CH_3)_2$	0.1	0.1	-0.45	0.1
$-O(CH_2)_3CH_3$	0.1	0.1	-0.32	0.1
$-O(CH_2)_4CH_3$	0.1	0.1	-0.34	0.1
$-OC_6H_5$	0.252	0.02	-0.320	0.02
$-OH$	0.121	0.02	-0.37	0.04
$-OCOCH_3$	0.39	0.1	0.31	0.1
$-SCH_3$	0.15	0.1	0.00	0.1
$-SC_2H_5$	--		0.03	0.1
$-SCH(CH_3)_2$	--		0.07	0.1
$-SH$	0.25	0.1	0.15	0.1
$-SCOCH_3$	0.39	0.1	0.44	0.1
$-SCN$	--		0.52	0.1
$-S(O)CH_3$	0.52	0.1	0.49	0.1
$-SO_2CH_3$	0.60	0.1	0.72	0.1

Table 4.II   (continued)

Group	Meta position		Para position	
	σ_m	Uncertainty	σ_p	Uncertainty
$-SO_2NH_2$	0.46	0.1	0.57	0.1
$-S(CH_3)_2{}^+$	1.00	>0.1	0.90	>0.1
$-SO_3{}^-$	0.05	>0.1	0.09	>0.1
$-SeCH_3$	0.1	0.1	0.0	0.1
$-F$	0.337	0.02	0.062	0.02
$-C\ell$	0.373	0.02	0.227	0.02
$-Br$	0.391	0.02	0.232	0.02
$-I$	0.352	0.02	0.18	0.1
$-\beta$-naphthyl	--		0.042	0.02

[a] From McDaniel and Brown, Ref. [24].   Underlined values
are based on thermodynamic pK_a values.
[b] $pK_a^0 = 4.203$ at 25.0° for benzoic acid.
[c] Assumed identical with that for $-CO_2C_2H_5$.
[d] From E. S. Lewis and M. D. Johnson, J. Am. Chem. Soc.,
81, 2070 (1959).

gives a value for k^0 (from the smooth line), which provides all the information needed to estimate k for any
member of the series, if the corresponding σ is available,
by means of Eq. (4.43).

The reaction constant ρ is a quantitative measure of
the susceptibility of the reaction to the influence of
substituents.   Three factors may contribute to the value
of ρ [22]:   (a) the transmission of electronic effects to
the reaction site; (b) the susceptibility of the reaction
to electronic effects; (c) the effects of reaction conditions.   The first of these effects is illustrated by ρ
for the ionization of ArCOOH, $ArCH_2COOH$, and $ArCH_2CH_2COOH$
(ρ values 1.000, 0.489, and 0.212, respectively).   The
insulating effect of the methylene group interposed between the aromatic ring and the reaction site produces a
progressive decrease in ρ, amounting to a factor of about
2.3 for a $-CH_2-$ group; this is the ratio ρ_2/ρ_3 for reac-

Table 4.III Hammett ρ Values for Some Equilibrium and Rate Processes[a]

No.	Reaction	Solvent	t/°C	ρ	-log k^0 [b]
	Equilibria				
1a	Ionization of ArCOOH	H_2O	25	(1.000)[c]	4.203
1b	Ionization of ArCOOH	CH_3OH	25	1.537	6.514
1c	Ionization of ArCOOH	C_2H_5OH	25	1.957	7.206
2	Ionization of $ArCH_2COOH$	H_2O	25	0.489	4.297
3	Ionization of $ArCH_2CH_2COOH$	H_2O	25	0.212	4.551
4	Ionization of ArCH=CHCOOH	H_2O	25	0.466	4.447
5	Ionization of ArCH=CHCOCOOH	H_2O	25	-0.054	1.971
6	Ionization of ArOH	H_2O	25	2.113	9.847
7	Ionization of $ArNH_3^+$	H_2O	25	2.767	4.557
8	$2ArNH_2 + Ag^+ = Ag(ArNH_2)_2^+$	59% C_2H_5OH	25	-1.452	-3.18
9	ArCHO + HCN = ArCHOHCN	95% C_2H_5OH	20	-1.492	2.138
10	$ArCOCH_3 + H_2 = ArCHOHCH_3$	H_2O	25	1.630	3.913
	Rates				
11	$ArCOOH + CH_3OH + H^+ → ArCOOCH_3$	CH_3OH	25	-0.229	3.841
12	$ArCOOH + HN_3 → ArNH_2 + CO_2$	CCl_2=CHCl	40	-1.415	5.289
13	$ArCOOCH_3 + OH^- → ArCOO^-$	60% Me_2CO	25	2.229	2.075
14	$ArCOOC_2H_5 + OH^- → ArCOO^-$	60% Me_2CO	25	2.265	2.557
15	$ArCOOC_2H_5 + H^+ → ArCOOH$	60% Me_2CO	100	0.106	4.086
16	p-$ArC_6H_4COOC_2H_5 + OH^- →$ p-$ArC_6H_4COO^-$	88% C_2H_5OH	40	0.583	2.636
17	$ArCH_2COOC_2H_5 + OH^- → ArCH_2COO^-$	87.83% C_2H_5OH	30	0.824	1.813
18	$ArCH_2CH_2COOC_2H_5 + OH^- →$ $ArCH_2CH_2COO^-$	87.83% C_2H_5OH	30	0.489	2.198

No.	Reaction	Solvent	Temp.		
19	$ArCH=CHCOOC_2H_5 + OH^- \rightarrow$ $ArCH=CHCOO^-$	87.83% C_2H_5OH	30	1.329	2.752
20	$ArCOCl + H_2O \rightarrow ArCOOH$	95% Me_2CO	25	1.782	4.200
21	$ArCONH_2 + OH^- \rightarrow ArCOO^-$	60% C_2H_5OH	100.1	1.100	3.523
22	$ArCONH_2 + H^+ \rightarrow ArCOOH$	60% C_2H_5OH	99.6	-0.222	3.806
23	$(ArCO)_2O + H_2O \rightarrow ArCOOH$	75% dioxane	58.25	1.568	5.113
24a	$ArOCOC_6H_5 + OH^- \rightarrow ArOH$	60% Me_2CO	-20	1.051	1.801
24b	$ArOCOC_6H_5 + OH^- \rightarrow ArOH$	60% Me_2CO	-10	1.034	1.408
24c	$ArOCOC_6H_5 + OH^- \rightarrow ArOH$	60% Me_2CO	0	0.976	1.015
24d	$ArOCOC_6H_5 + OH^- \rightarrow ArOH$	60% Me_2CO	15	0.930	0.490
25	$ArNH_2 + C_6H_5COCl \rightarrow ArNHCOC_6H_5$	C_6H_6	25	-2.781	2.888

aAbstracted from the compilation by Jaffé (Ref. [22]), where original references may be found.

bValue of log k on the least-squares regression line where σ = 0; the time unit is seconds.

cBy definition.

tions 2 and 3, Table 4.III. Similarly, the ratio ρ_{17}/ρ_{18} is about 1.7. That this effect of distance between substituent and reactive site is not the only controlling factor can be seen by comparing the ionization of benzoic and cinnamic acids (ρ_1/ρ_4 = 2.1) and the saponification of ethyl benzoates and ethyl cinnamates (ρ_{16}/ρ_{19} = 0.44). These different ratios reflect a difference in susceptibility of the two processes to substituent changes.

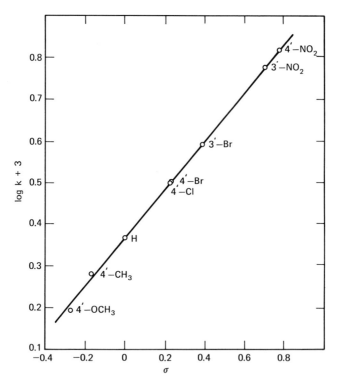

Fig. 4.2. Hammett plot for the alkaline hydrolysis of substituted ethyl p-biphenylcarboxylates at 40° in 88.7% (w/w) ethanol [27].

It is apparent from Table 4.III that a high positive value of ρ suggests that reaction is facilitated by low electron density at the reaction site [see, for example, Reaction (13)], and a negative ρ indicates facilitation by high electron density [see Reaction (22)]. It is better to consider relative rho values, however, because the preceding statement is an oversimplification, as can be seen by comparing Reactions (14) and (15). Both reactions have

positive ρ values, but ρ_{15} (acid catalysis) is much smaller than ρ_{14} (base catalysis). Much (though not all) of
the difference can be ascribed to this factor of their
different susceptibilities to substituent electronic effects at the reaction site. Relative ρ values therefore
provide information about the reaction mechanism. ρ values
near zero should be interpreted with caution for two reasons: first, since ρ is controlled by more than one factor, a small ρ does not necessarily mean that the reaction
is insensitive to electronic effects; second, if the concept of an isokinetic temperature within the experimental
range should turn out to be a real possibility (see Sec.
4.1), a small ρ might simply mean that the experimental
and isokinetic temperatures are very similar.

   The effect of reaction conditions on ρ can be very
pronounced. Reaction (1) in Table 4.III shows how the
solvent can influence ρ. This effect may be related to
the ease of transmission of electronic effects, not only
through the molecule but also by a field effect through
the solvent. It may therefore be related to the dielectric constant of the medium. The data for Reaction (24)
show the effect of temperature on ρ. The expected effect
can be found by writing the Hammett equation in this form:

$$\delta_R \Delta G^{\ddagger} = -2.3 \ RT \ \log \frac{k}{k^0} = -2.3 \ RT \ \rho\sigma = \delta_R \Delta H^{\ddagger} - T\delta_R \Delta S^{\ddagger}$$

$$(4.46)$$

$$\rho = -\frac{\delta_R \Delta H^{\ddagger}}{2.3RT\sigma} + \frac{\delta_R \Delta S^{\ddagger}}{2.3R\sigma} \qquad (4.47)$$

The substituent constant is defined at a given temperature
($25°$). If $\delta_R \Delta H^{\ddagger}$ and $\delta_R \Delta S^{\ddagger}$ are independent of temperature,
ρ should be linearly related to $1/T$ if all other condi-
tions are held constant [29]. If an isokinetic relation-
ship applies for the system, $\delta_R \Delta H^{\ddagger} = \beta\delta_R \Delta S^{\ddagger}$, and Eq.
(4.47) becomes

$$\rho = \frac{\delta_R \Delta H^{\ddagger}}{-2.3RT\sigma} + \frac{\delta_R \Delta H^{\ddagger}}{2.3R\beta\sigma} \qquad (4.48)$$

This equation also predicts that ρ varies with $1/T$ [30].
Though available experimental data are not adequate for a
careful test of this prediction, they are at least consistent with theory [19,22,31].

   That the Hammett equation should be so widely successful [32] has stimulated considerable discussion, because this success is difficult to explain. The treatment
in Sec. 4.1 leading to Eq. (4.19) offers a rationalization, but it fails to account for the constancy of the

reaction parameter. Since the free energy is determined by the enthalpy and the entropy, as $\delta_R \Delta G = \delta_R \Delta H - T\delta_R \Delta S$, with $\delta_R \Delta H$ and $\delta_R \Delta S$ being, in general, capable of independent variation, it is surprising that $\delta_R \Delta G$ seems so often to vary linearly with σ. Hepler's treatment [10,11], discussed in Sec. 4.1, accounts for the observed relationships very satisfactorily, provided that it is acceptable to invoke an isokinetic temperature close to the experimental temperature. This is precisely the case that Exner [16b] feels is most dubious. Ritchie and Sager [19a] have proposed that the isokinetic temperature is not a constant for a reaction series, and that β/T is temperature dependent, leading to appreciable compensation in $\delta_R \Delta H - T\delta_R \Delta S$ at all temperatures.

The Hammett equation is generally applicable to side-chain reactions of meta- and para-substituted benzoic acid derivatives. It fails dramatically when applied to ortho-substituted benzoic acid derivatives [33]. This departure from the behavior expected on the basis of the meta and para representatives is called the ortho effect or proximity effect. Hammett [33] noted that the proximity effect tends to parallel entropy changes, and he concluded that simple steric effects could not entirely account for the proximity behavior. Taft [34] has attempted to separate the polar and steric contributions to the ortho effect. Recently Charton [35] has concluded that the ortho effect is purely electronic in nature, with no important steric contribution.

Occasionally a Hammett plot shows a nonlinear but smooth correlation. This may consist of a smooth curve or of discrete straight-line segments. Two causes have been identified.

1. Change in rate-determining step. If the reaction is complex, each step (rate or equilibrium) will have a characteristic susceptibility to substituent effects, and therefore its own ρ value. The observed ρ will be primarily determined by the ρ value of the rate-determining step. If this step changes at some position along the σ axis, the observed ρ will change, leading to curvature in the Hammett plot. Since any step can a priori have any ρ value, the net result could be curvature either concave upward or concave downward. Jencks [36] has described the analysis of such plots.

2. Change in mechanism. Within a reaction series any change in mechanism must be such as to provide a smaller energy of activation for the reaction (otherwise the mechanism wouldn't change). If a substituent effect can produce a change in mechanism, the result must be curvature that is concave upward in the Hammett plot [37].

Most of the research and speculation on linear free
energy relationships has concerned the nature of substituent constants. In general terms, these are measures of
the electronic effects of substituents. Since the electronic effect of a substituent can be the result of more
than one mechanism of electronic displacement, it is not
surprising that a single σ value is inadequate to describe
the effect of a given substituent in all types of reactions. Hammett [38] observed, for example, that σ_p for
the nitro group should be 1.27 for the reactions of aniline and phenol derivatives, and 0.78 for other substrates.
Such enhanced sigma values are attributed to direct resonance (conjugative) interaction between the substituent
and the reaction site. These constants, symbolized σ⁻,
apply primarily to electron-withdrawing groups in reactions aided by low electron density at the reaction site
[39], that is, reactions in which a negative charge is
developed at the reaction site. For groups in the meta
position, which cannot undergo extensive resonance delocalization with the reactive site, the Hammett σ_m values
are used for correlating these reactions; likewise for
para-substituted groups incapable of extensive resonance
interaction, the σ_p constants are adequate. Table 4.IV
gives σ⁻ constants for those para-substituted groups ex-
hibiting these enhanced effects. σ⁻ has yielded improved
correlations for aromatic nucleophilic substitutions [40].

Reactions that occur with the development of an electron deficiency, such as aromatic electrophilic substitutions, are best correlated by substituent constants based
on a more appropriate defining reaction than the ionization of benzoic acids. Brown and Okamoto [41] adopted the
rates of solvolysis of substituted phenyldimethylcarbinyl
chlorides (t-cumyl chlorides) in 90% aqueous acetone at
25° to define electrophilic substituent constants symbolized σ⁺. Their procedure was to establish a conventional
Hammett plot of log (k/k^0) against σ_m for 16 meta-substituted t-cumyl chlorides, since meta substituents cannot
undergo significant direct resonance interaction with the
reaction site. The resulting ρ value of -4.54 was then
used in a modified Hammett equation,

$$\log \frac{k}{k^0} = \rho\sigma^+ \tag{4.49}$$

to calculate σ⁺ constants for a total of 41 substituents.
This procedure has the result of placing the σ⁺ and σ constants on essentially the same scale. The σ⁺ constants
are given in Table 4.V. Some of the reactions correlated
by Eq. (4.49) are listed in Table 4.VI.

Table 4.IV   Enhanced Para-Substituent Constants,[a] σ^-

Substituent	σ^-	Substituent	σ^-
$-CO_2H$	0.728	$-CN$	1.000[b]
$-CO_2CH_3$	0.636	$-NO_2$	1.27[c]
$-CO_2C_2H_5$	0.678	$-SO_2CH_3$	1.049
$-CO_2C_4H_9$	0.674	$-C_6H_4N=NC_6H_5$	1.088
$-CO_2CH_2C_6H_5$	0.667	$-CH=CH-C_6H_5$	0.619
$-CONH_2$	0.627	$-N_2^+$	3.2[d]
$-CHO$	1.126		
$-COCH_3$			

[a]From Jaffé's collection [22] except as indicated.
[b]A. I. Biggs and R. A. Robinson, J. Chem. Soc., 1961,
p. 388, give 0.983 and 0.874 based on ionization of
anilines and phenols, respectively.
[c]Biggs and Robinson give 1.24.
[d]E. S. Lewis and M. D. Johnson, J. Am. Chem. Soc., 81,
2070 (1959).

It appears, on the basis of the preceding discussion,
that a substituent effect can give rise to a range of sigma values depending upon the electronic requirements of
the particular reaction and upon the capability of the
substituent to affect the electron density at the reaction
site.   Direct resonance interaction seems to be the principal mechanism contributing to variability in sigma values.
For this reason a set of sigma values has been proposed in
which direct resonance interaction is absent; these substituent constants will therefore be quantitative measures
of the inductive transmission of electronic effects.
These "normal" or "primary" sigma values were taken to be
the Hammett σ_m values for "well-behaved" substituents that
are expected to undergo no resonance delocalization interactions.   van Bekkum et al. [42], symbolized these by σ^n,
and Taft [43] by σ^0.   Taft also developed σ^0_{para} values by
selecting reaction series (numbers 2, 3, 17 in Table
4.III) with methylene groups between the aromatic ring
and the reaction site.   The σ^n and σ^0 scales are nearly
identical with each other and with the Hammett σ_m values
except for para substituents with resonance capabilities.
Taft and Lewis [44] have attempted to evaluate the
inductive and resonance contributions to Hammett σ values
by making the assumptions that inductive and resonance
effects are additive, that the inductive effects are equal

in the meta and para positions, and that the resonance effect in the meta position is smaller than in the para position by a constant fraction. The results show that, in general, there is no scale of resonance effects that is independent of reaction type and conditions, unlike the inductive effects, which are reasonably independent of reaction type. Ritchie and Sager [19a] have discussed the conditions required for the separability of polar, resonance, steric, and solvation effects. It appears that the assumption of additivity is questionable in these systems.

Table 4.V Electrophilic Substitution Constants [41], σ^+

Substituent	σ_m^+	σ_p^+
-OCH$_3$	0.047	-0.778
-SCH$_3$	0.158	-0.604
-CH$_3$	-0.066	-0.311
-C$_2$H$_5$	-0.064	-0.295
-CH(CH$_3$)$_2$	-0.060	-0.280
-C(CH$_3$)$_3$	-0.059	-0.256
-C$_6$H$_5$	0.109	-0.179
-(2-naphthyl)		-0.135
-H	0	0
-Si(CH$_3$)$_3$	0.011	0.021
-F	0.352	-0.073
-Cl	0.399	0.114
-Br	0.405	0.150
-I	0.359	0.135
-COOH	0.322	0.421
-COOCH$_3$	0.368	0.489
-COOC$_2$H$_5$	0.366	0.482
-CF$_3$	0.520	0.612
-CN	0.562	0.659
-NO$_2$	0.674	0.790
-COO$^-$(K$^+$)	-0.028	-0.023
-N(CH$_3$)$_3^+$(Cl$^-$)	0.359	0.408

At the present time correlations of aromatic reactions are best attempted with σ, σ^-, or σ^+ values, as appropriate to the reaction type. The best correlations for most aromatic side chain reactions will be observed for meta-substituted substrates, for which the Hammett σ_m values are the proper substituent constants.

Table 4.VI   Reaction Constants for σ^+ Correlations [41]

No.	Reaction	t/°C	ρ
	Electrophilic nuclear substitutions		
1	Bromination of monosubstd. benzenes by Br_2 in HOAc	25	-12.14
2	Chlorination of monosubstd. benzenes by Cl_2 in HOAc	25	-8.06
3	Nitration of monosubstd. benzenes by HNO_3 in CH_3NO_2 or Ac_2O	0 or 25	-6.22
4	Bromination of monosubstd. benzenes by HOBr and $HClO_4$ in 50% dioxane	25	-5.78
5	Protonolysis of aryltrimethyl-silanes by H_2SO_4 in HOAc		-4.60
6	Brominolysis of aryltrimethyl-silanes in HOAc	25	-6.04
7	Brominolysis of benzeneboronic acids in 20% HOAc and 0.40 $\underline{M}$ NaBr	25	-4.44
8	Bromination of polymethylbenzenes by Br_2 in CH_3NO_2	25	-8.10
	Electrophilic side chain reactions		
9	Solvolysis of t-cumyl chlorides in 90% acetone[a]	25	-4.54
10	Solvolysis of t-cumyl chlorides in ethanol	25	-4.67
11	Acid-catalyzed rearrangement of α-phenylethyl chlorocarbonates in dioxane	80	-3.01
12	Equil. Consts. for carbonium ion formation from triphenylcarbinols in aq. H_2SO_4	25	-3.64
13	Equil. Consts. for ionization of triphenylcarbinyl chlorides in SO_2	0	-3.73

[a]Model reaction.

Aliphatic Substrates. The Hammett type of correlation
fails with aliphatic substrates [45] just as it does with
ortho-benzoates. Taft [34] provided the first successful
correlations of aliphatic reactions by developing quantitatively an early suggestion of Ingold's [46] that polar
and steric effects might be separated by measuring the
relative rates of alkaline and acid hydrolysis of esters.
Taft's treatment requires these assumptions: (1) $\delta_R \Delta G^{\ddagger}$
terms are additive functions of polar, resonance, and
steric effects; (2) in the acidic and alkaline hydrolysis
rates of the same ester, the steric and resonance effects
are the same; (3) the polar effects are much greater in
alkaline hydrolysis than in acid hydrolysis. The differ-
ence $\delta_R \Delta G^{\ddagger}_{alkaline}$ - $\delta_R \Delta G^{\ddagger}_{acid}$ for an ester hydrolysis
should, if these assumptions are valid, be a measure of
the polar effect in the substituent. Taft defined a polar substituent constant σ^* based on these ideas:

$$\sigma^* = \frac{1}{2.48}\left[\log\left(\frac{k}{k^0}\right)_B - \log\left(\frac{k}{k^0}\right)_A\right] \tag{4.50}$$

$(k/k^0)_B$ is the relative rate of alkaline hydrolysis of
ester RCOOR´, with k^0 being the rate constant for R = CH_3;
$(k/k^0)_A$ is the relative rate of acid hydrolysis of the
same ester with the same solvent and temperature; and 2.48
is a normalizing factor that places σ^* on about the same
scale as the Hammett σ values (see typical ρ values for
ester hydrolyses in Table 4.III). Table 4.VII gives these
polar substituent constants. Taft [34] also tabulates σ^*
for ortho-substituents.
    An equation can now be written by analogy with the
Hammett equation.

$$\log\frac{k}{k^0} = \rho^*\sigma^* \tag{4.51}$$

Equation (4.51) provides good correlations for many aliphatic reactions. The scope of this relationship is illustrated by Table 4.VIII. The reaction constant ρ^* has
been interpreted along the lines described for the Hammett
ρ values.
    Correlations with σ^* in carboxylic acid derivative
reactions have been most successful for variations in the
acyl portion, R in RCOX. Variation in the alkyl portion
of esters, R´ in RCOOR´, has not led to many good correlations, though use of relative rates of alkaline and acidic
reactions, as in the defining relation, can generate
linear correlations. The failure to achieve satisfactory
correlations with σ^* for such substrates may be a conse-

quence of the different steric effects of substituents in the acyl and alkyl locations [47]. It has been shown [48] that solvolysis rates of some acetates are related to the pK_a of the "leaving group," that is, of the parent alcohol.

Table 4.VII Polar Substituent Constants [34], σ^*, for Aliphatic Substrates X-R[a]

R	σ^*	R	σ^*
$-CCl_3$	2.65	$-CH_2OH$	0.56
$-CHF_2$	2.05	$-CH_2OCH_3$	0.52
$-CO_2CH_3$	2.00	$-(CH_2)_2NO_2$	0.50
$-CHCl_2$	1.94	$-H$	0.49
$-CH_2N(CH_3)_3^+$	1.90	$-CH=CHC_6H_5$	0.41
$-COCH_3$	1.65	$-CH(C_6H_5)_2$	0.41
$-C\equiv CC_6H_5$	1.35	$-(CH_2)_2Cl$	0.39
$-CH_2SO_2CH_3$	1.32	$-CH=CHCH_3$	0.36
$-CH_2CN$	1.30	$-(CH_2)_2CF_3$	0.32
$-CH_2F$	1.10	$-CH_2C_6H_5$	0.22
$-CH_2CO_2H$	1.05	$-CH_2CH=CHCH_3$	0.13
$-CH_2Cl$	1.05	$-(CH_2)_3CF_3$	0.12
$-CH_2Br$	1.00	$-CH(CH_3)C_6H_5$	0.11
$-CH_2I$	0.85	$-(CH_2)_2C_6H_5$	0.08
$-CH_2CF_3$	0.92	$-CH(C_2H_5)C_6H_5$	0.04
$-CH_2OC_6H_5$	0.85	$-(CH_2)_3C_6H_5$	0.02
$-CH(OH)C_6H_5$	0.77	$-CH_3$	(0.00)
$-CH_2COCH_3$	0.60	$-CH_2C_6H_{11}(cyclo)$	-0.06
$-C_6H_5$	0.60	$-C_2H_5$	-0.10
$-n-C_3H_7$	-0.12	$-C_5H_9(cyclo)$	-0.20
$-i-C_4H_9$	-0.13	$-s-C_4H_9$	-0.21
$-n-C_4H_9$	-0.13	$-CH(C_2H_5)_2$	-0.23
$-C_6H_{11}(cyclo)$	-0.15	$-CH_2Si(CH_3)_3$	-0.26
$-CH_2C_4H_9(tert)$	-0.17	$-CH(Me)(CMe_3)$	-0.28
$-i-C_3H_7$	-0.19	$-t-C_4H_9$	-0.30

[a]Some values have been rounded off.

The pK_a of alcohols has been correlated with σ^* [34,49], but this relationship requires σ^* to refer to R in RCH_2OH. When pK_a is plotted against σ^* for R in ROH, the linear correlation is lost, presumably because of a steric effect. This steric influence is paralleled in ester reactions, and Robinson and Matheson [50] have demonstrated good linear free energy relationships between many ester solvolysis rates and pK_a of the parent alcohols.

Table 4.VIII   Some Reactions Correlated by σ^{*}[a]

No.	Reaction	Solvent	$t/^{\circ}C$	ρ^{*}
1	Hydrolysis of $RCO_2C_2H_5$			(2.48)[b]
2	Ionization of RCO_2H	water	25	1.72
3	Acid hydrolysis of $RCH(OC_2H_5)_2$	50% dioxane	25	-3.65
4	Ionization of RCH_2OH	isopropyl alcohol	27	1.36
5	Sulfation of alcohols, ROH	H_2SO_4/H_2O	25	4.60
6	Acetone iodination catalyzed by RCOOH	water	25	1.14
7	Nitramide decompn. catalyzed by $RCOO^-$	water	15	-1.43
8	Acid hydrolysis of $H_2C(OR)_2$	water	25	-4.17
9	$\log (k_B/k_A)$ for hydrolysis of CH_3COSR	43% acetone	30	1.49
10	Ionization of RNH_3^+[c]	water	25	3.14
11	Ionization of $RC(NO_2)_2H$[d]	water	25	3.60

[a] From Taft, Ref. [34], except as noted.
[b] Defining reaction.
[c] H. K. Hall, Jr., J. Am. Chem. Soc., 79, 5441 (1957); pK_a plotted against $\Sigma\sigma^{*}$ for R plus the H atoms.
[d] M. E. Sitzmann, H. G. Adolph, and M. J. Kamlet, J. Am. Chem. Soc., 90, 2815 (1968).

Because of the dearth of accurate pK_a values for alcohols, they then used this correlation to justify adoption of a secondary standard, the rate of alkaline hydrolysis of esters of 3,5-dinitrobenzoic acid at 25° in 40% acetonitrile. $\delta_R\Delta G^{\ddagger}$ for this reaction, with the methyl ester as the reference compound, was used as the substituent parameter to give linear free energy relationships for many ester reactions. Interestingly, some electrophilic aromatic substitution reactions of substrates with aliphatic substituents also gave linear free energy relationships with this substituent parameter. For later use let us define $\sigma'' = \log (k/k^0)$, where k/k^0 is the relative rate of hydrolysis of the 3,5-dinitrobenzoates reported by Robinson [51].
    Since the effects of acyl and alkyl group variations on rates can be individually correlated for ester substrates, concurrent variations in both components may be

susceptible to correlation. Table 4.IX gives relative rates of alkaline hydrolysis of aliphatic esters $R^i-CO_2-R^j$; the rates are relative to that for methyl acetate [52]. Let k_{ij} be the rate constant for the ester with acyl group R^i and alkyl group R^j. Then j Taft correlations can be written for the columns:

$$\log \frac{k_{ij}}{k_{rj}} = \rho^*{}_j \sigma^*{}_i \quad (j \text{ constant}) \qquad (4.52)$$

where r is the reference acyl group. Similarly, i correlations of the alkyl variations (rows) can be written with analogous equations, using the alkyl substituent parameters σ'' defined earlier:

$$\log \frac{k_{ij}}{k_{is}} = \rho''{}_i \sigma''{}_j \quad (i \text{ constant}) \qquad (4.53)$$

where s is the reference alkyl group. Now Eq. (4.53) is written for k_{rj},

$$\log \frac{k_{rj}}{k_{rs}} = \rho''{}_r \sigma''{}_j \qquad (4.54)$$

where k_{rs} is the rate constant for a single "master" reference compound. Eliminating k_{rj},

$$\log \frac{k_{ij}}{k_{rs}} = \rho^*{}_j \sigma^*{}_i + \rho''{}_r \sigma''{}_j \qquad (4.55)$$

Equation (4.55) expresses all k_{ij} in terms of j + 2 quantities; these are the $\rho^*{}_j$ values, $\rho''{}_r$, and k_{rs}. If the several $\rho^*{}_j$ are nearly identical, they can be replaced by the mean value $\rho^*{}_j$, and then Eq. (4.55) gives all k_{ij} with just three quantities, $\rho^*{}_j$, k_{rs}, and $\rho''{}_r$.

Table 4.IX gives k_{ij}/k_{rs} values, where k_{rs} is for methyl acetate. Taft plots for the columns (R^j = Me, Et, n-Pr, i-Pr, and n-Bu only) gave $\rho^*{}_j$ = +4.26 ± 0.18. The correlation of log (k_{rj}/k_{rs}) with $\sigma''{}_j$ gave $\rho''{}_r$ = 0.99. Equation (4.55) then generated the 27 k_{ij} values with a mean error of 0.145 log units in log (k_{ij}/k_{rs}). Though not precise enough for mechanistic investigations, this correlation is satisfactory for most analytical applications, in which an estimate of saponification rate is used to adjust the solution alkalinity and to select the reaction time.

Taft's [34] separation of polar and steric factors

Table 4.IX Relative Rates of Alkaline Hydrolysis of Aliphatic Esters [52],[a] R^1-CO_2-R^j

Acyl Substituent R^i	Alkyl Substituent, R^j									
	Me	Et	n-Pr	i-Pr	n-Bu	i-Bu	s-Bu	n-Am	i-Am	n-Hex
H	111.9	64.6	51.0	23.3	46.2	42.8	13.4	36.8	35.0	--
Me	1.000	0.438	0.311	0.093	0.282	0.223	0.042	0.236	0.221	0.209
Et	0.717	0.308	0.196	0.052	0.157	--				
n-Pr	0.337	0.133	--							
i-Pr	0.257	--								

[a]At 25° in 37.27% acetone. The second-order rate constant for the reference compound, methyl acetate, is 0.151 $\underline{M}^{-1}$-sec^{-1}.

[53] is based upon the additivity postulate. This assumption has been criticized, and the interpretation of σ^* as a quantitative measure of polar effects is open to serious doubt [19a,35]. However, as an empirical tool for the description of aliphatic rates and equilibria it is still very useful.

4.3   Substituent Effects in the Reagent

Acid-Base Catalysis. The rate constant for a general acid or general base catalyzed reaction increases as the acid or base strength of the catalyst is increased. For many such systems satisfactory linear free energy relationships are observed. These are commonly written as Eq. (4.56) for general acid catalysis and Eq. (4.57) for general base catalysis,

$$k_A = G_A K_a{}^{\alpha} \tag{4.56}$$

$$k_B = G_B \left(\frac{1}{K_a}\right)^{\beta} \tag{4.57}$$

where K_a is the acid dissociation constant of the conjugate acid. These relationships are called Brønsted equations [54]. Their applicability is established by the linearity of plots of $\log k_A$ (or $\log k_B$) against $\log K_a$ (or $\log 1/K_a$), the slopes being α and β, respectively. A Brønsted relationship holds well only for a series of acids or bases that are structurally very similar. It is commonly said that the Brønsted coefficients α and β must lie within the limits 0 and 1. In qualitative terms these limits have this meaning: Taking acid catalysis and Eq. (4.56) for illustration, a negative value of α would mean that increasing acid strength results in decreased catalytic activity, and this is inconsistent with the notion of acid catalysis as a proton transfer phenomenon. An α value greater than unity, on the other hand, would mean that $\log k_A$ is increasing faster than is $\log K_a$ within the reaction series, and this too is not consistent with the normal view of acid catalysis. All known Brønsted plots for nitrogen and oxygen acids and bases appear to yield α and β values between 0 and 1; examples are given by Bell [55].

When Brønsted plots apply to the rates and equilibria of the same reaction, a further restriction can be written for the coefficients. Let the generalized reaction scheme be

$$A \; \underset{k_B}{\overset{k_A}{\rightleftharpoons}} \; B \tag{4.58}$$

and assume that Brønsted relationships apply to both the forward (acid-catalyzed) and reverse (base-catalyzed) steps [56], the rate constant k_A correlating with the equilibrium constant $K_A = k_A/k_B$, and k_B with $1/K_A$. Then, applying Eqs. (4.56) and (4.57) in the equivalent form common to all linear free energy relationships:

$$\log \frac{k_A}{k_A{}^0} = \alpha \log \frac{K_A}{K_A{}^0} \tag{4.59}$$

$$\log \frac{k_B}{k_B{}^0} = \beta \log \frac{(1/K_A)}{(1/K_A{}^0)} \tag{4.60}$$

where the superscripted quantities refer, as usual, to a reference compound. Equation (4.61) follows.

$$\frac{(k_A/k)_B}{(k_A{}^0/k_B{}^0)} = \frac{K_A{}^{(\alpha \, + \, \beta)}}{K_A{}^{0 \, (\alpha \, + \, \beta)}} \tag{4.61}$$

Therefore $\alpha + \beta = 1$.

Recently Bordwell et al. [57] have found Brønsted slopes larger than 1 and smaller than 0 for proton removals from carbon acids. For this reaction series

$$ArCHMeNO_2 + H_2O \; \underset{k_B}{\overset{k_A}{\rightleftharpoons}} \; (ArCMeNO_2)^- + H_3O^+$$

the Brønsted coefficient α is 1.31; therefore β for the reverse reaction is -0.31. This unusual finding has been interpreted as a consequence of the relatively drastic effect of ionization upon the reaction site in these compounds [57-59].

Notice that this type of Brønsted relationship, between rates and equilibria of the same reaction series, has been discussed in Sec. 4.1, where the quantity α was roughly interpreted as the fractional displacement of the transition state along the reaction coordinate. Brønsted coefficients outside the range 0-1 clearly cannot be interpreted in this way [57]. Leffler and Grunwald [60] point out that the more common type of Brønsted relationship between a catalytic rate constant and the K_a of the

catalyst often reflects a rate-equilibrium relationship
within the detailed reaction mechanisms. It has been
customary to interpret Brønsted coefficients mechanistically, but caution is appropriate because of the possibility that these quantities are not as simple as they seem.

The existence of Brønsted relationships affects the
experimental problem of detecting general acid or base
catalysis. This is clearly shown by an example given by
Bell [55a]. Consider the reaction under study as carried
out in an aqueous solution containing 0.10 M acetic acid
and 0.10 M sodium acetate, and suppose that the Brønsted
equation applies. Three catalytic species are present;
these are H_3O^+, with pK_a = -1.74, H_2O, pK_a 15.74, and
HOAc, pK_a 4.76 [61]. The concentrations of these acids
are 1.76×10^{-5} M, 55.5 M, and 0.10 M, respectively.
Then the fraction of the total catalytic effect produced
by any one of the acids is given by

$$\text{Fraction of catalysis by acid } i = \frac{(K_a^i)^\alpha [i]}{\sum\limits_i (K_a^i)^\alpha [i]} \quad (4.62)$$

Table 4.X gives these fractions for the three acids assuming hypothetical Brønsted coefficients of 0.0, 0.5, and
1.0. α clearly measures the susceptibility of the reaction to catalysis by acids of corresponding strength.
When α = 0, the reaction is very sensitive to the weakest
acid, which is the solvent; if α = 1, the lyonium ion,
which is the strongest possible acid, is the most effective catalyst. It therefore appears that general acid
(or general base) catalysis can only be detected if the
Brønsted coefficient is in the intermediate range. This
is correct if the Brønsted equation is accurately followed.
If, however, the slope is unity but the point corresponding to H_3O^+ should show a negative deviation from the line,
then general acid catalysis can be detected. This fortuitous circumstance occurs in the general acid catalysis of
the hemiacetal formed by the addition of hydrogen peroxide
to p-chlorobenzaldehyde [62].

When a Brønsted plot includes acids or bases with
different numbers of acidic or basic sites, "statistical"
corrections are sometimes applied; in effect, the rate
and equilibrium constants are corrected to a "per functional group" basis. If an acid has p equivalent dissociable protons and its conjugate base has q equivalent
sites for proton addition, the statistically corrected
forms of the Brønsted relationships are

$$\frac{k_A}{p} = G_A\left(\frac{qK_a}{p}\right)^\alpha$$

$$\frac{k_B}{q} = G_B\left(\frac{p}{qK_a}\right)^\beta$$

The equivalency of sites required for the application of these equations is seldom found in practice, though many authors apply these corrections.   For most purposes they seem to be unnecessary [55b].

Table 4.X   Acid Catalysis in 0.1 $\underline{M}$ Acetic Acid-0.1 $\underline{M}$ Sodium Acetate Buffer

Brønsted α	Percentage of catalysis by		
	H_3O^+	HOAc	H_2O
0.0	0.0003	0.2	99.8
0.5	23.7	76.1	0.14
1.0	99.8	0.2	1.0×10^{-11}

Nucleophilic Reactivity. If an acidic or basic reagent acts other than as a proton transfer agent, that is, by attacking an atom other than hydrogen, the validity of acid ionization as a model process is less likely than it is in general acid-base catalysis.   It is not unusual, however, to find limited parallels between reactivity and acid-base strength.   Correlations are sought between the logarithm of the rate constant and the pK_a of the reagent; though formally equivalent to Brønsted plots, such correlations are less susceptible to interpretation because the model process and the reaction under study are so different.   The usual type of reaction studied in this way involves a substrate subjected to attack by a series of nucleophiles (bases).   The reaction of p-nitrophenyl acetate with nucleophiles is a typical example [63].   These Brønsted-type plots often seem to be scatter diagrams until the points are collated into groups related by specific structural features.   Thus p-nitrophenyl acetate gives four separate, but parallel, lines for reactions with pyridines, anilines, imidazoles, and oxygen nucleophiles [64].   Figure 4.3 shows such a plot for the reaction of

trans-cinnamic anhydride with primary and secondary aliphatic amines to give substituted cinnamamides [65]. All
of the primary amines without substituents on the α-carbon
(R-CH₂-NH₂) fall on a good line of slope 0.62; cyclopentyl
amine also lies on this line. If this line is characteristic of "normal" behavior, most of the deviations become
qualitatively explicable. The line drawn through the
secondary amines (slope 1.98) connects amines with the
structure R-CH₂-NH-CH₂-R. The different steric requirements in the acylation reaction and in the model process
(protonation) can account for the negative deviations,
nearly all of which occur with amines carrying α-substi-
tuents. The nucleophilicity [66] of the alicyclic amines
is greater than their basicity would suggest; this enhanced reactivity might be a consequence of C-N-C bond
angle contraints [67]. Some of the behavior in Fig. 4.3
may follow from the use of a model process in water to
correlate a reaction in acetonitrile.

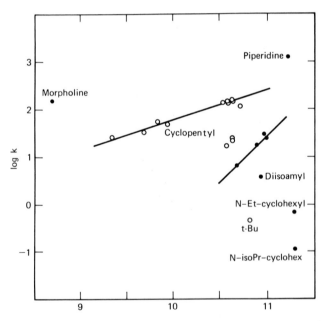

Fig. 4.3. Brønsted-type plot for reaction of aliphatic
amines with cinnamic anhydride at 25° in acetonitrile
[64]; the pK$_a$ values are for the conjugate acids in water.
Open circles: primary amines; closed circles: secondary
amines.

Acid-base strength is used as a model process for correlating nucleophilic reactions not because it is a very good model, but mainly because it is a convenient one. It is, of course, not wholly inappropriate, as we have seen. But it is not to be expected that the chemical reactivity of a series of nucleophiles toward a common substrate (attack usually being on carbon) should accurately reflect the equilibrium affinity of these nucleophiles for the solvated proton. The differing steric requirements have already been noted. A second difference is in the polarizability of the substrate, and a third limitation is in comparing rates with equilibria. Since affinity for hydrogen is not an adequate model, alternatives have been sought. Some measurements have been made of the carbon basicity of nucleophiles, for example, equilibrium constants for the reaction

$$R_3C-X + Y^- \rightleftharpoons R_3C-Y + X^-$$

where Y^- is the nucleophile. It is found [68] that carbon basicity parallels hydrogen basicity for the same nucleophiles.

Another approach is to adopt a model process that is very similar to the reactions of interest. Swain and Scott [69] selected as a standard reaction the nucleophilic substitution reaction of methyl bromide in water at $25°$.

$$CH_3Br + Y^- \rightarrow CH_3Y + Br^- \tag{4.63}$$

A nucleophilic constant n is defined

$$n = \log \frac{k_{MeBr}}{k^0_{MeBr}} \tag{4.64}$$

where k_{MeBr} is the rate constant of Reaction (4.63) with the nucleophile Y^-, and k^0_{MeBr} is the rate constant with the standard nucleophile, H_2O. Then a linear free energy relationship is written in the usual form,

$$\log \frac{k}{k^0} = sn \tag{4.65}$$

where s is a substrate constant, which for the standard reaction is defined to be unity. Values of nucleophilic constants then provided reasonable correlations of many substitution reactions.

Edwards has retained hydrogen basicity as a model process for nucleophilic reactions, and has added, in

linear combination, a measure of polarizability [70].
Edwards' equation is

$$\log \frac{k}{k^0} = aE_N + bH_N \qquad (4.66)$$

where

$$H_N = pK_a + 1.74 \qquad (4.67)$$

$$E_N = E^0 + 2.60 \qquad (4.68)$$

E^0 is the standard potential of the reaction $2Y^- = Y_2 + 2e^-$. The additive constants set H_N and E_N equal to zero
for H_2O. The quantities a and b are parameters measuring
susceptibility to the corresponding nucleophilic parameters. Though not as simple to use as linear free energy
relationships [71] based solely on readily accessible rate
or equilibrium constants, and therefore less suitable for
purely empirical correlations, Edwards' approach makes
use of an interesting model process. The basis of the
oxidation reaction as a model for nucleophilicity is the
analogy between oxidation (loss of electrons) and nucleophilic displacement (donation of an electron pair).
Equation (4.66) has been called the oxibase scale of nucleophilicities [72].

4.4   Medium Effects [73]

Measures of Solvent Polarity. Correlations of $\delta_M\Delta G^0$ and
$\delta_M\Delta G^{\pm}$ have not been as successful as have substituent
effect correlations. In part this is a consequence of
our relative ignorance of the liquid state on the molecular level. In this section the influence of solvent polarity on rates and equilibria is briefly considered.
    Grunwald and Winstein [74] defined a solvent "ioniz-
ing power" parameter $\underline{Y}$ by

$$\underline{Y} = \log \frac{k_M^{t-BuC\ell}}{k_0^{t-BuC\ell}} \qquad (4.69)$$

where $k_0^{t-BuC\ell}$ is the rate constant for the solvolysis of
t-butyl chloride in 80 vol. % [75] ethanol and $k_M^{t-BuC\ell}$ is
the corresponding rate constant in another solvent. This
reaction was selected as the model process because it was
believed to occur by an essentially pure S_N1 process, with
ionization $(RX \rightarrow R^+ + X^-)$ being rate determining. A linear

free energy equation is then written in the familiar form.

$$\log \frac{k}{k_0} = m\underline{Y} \qquad (4.70)$$

Ideally the parameter m should be characteristic of only the substrate, while $\underline{Y}$ should be a function of the solvent; the equation is expected to apply to reactions very similar to the standard reaction, that is, S_N1 substitutions. Table 4.XI gives $\underline{Y}$ values measured by Fainberg and Winstein [76]. Notice that the $\underline{Y}$ of the standard solvent lies about midway between the extremes. $\underline{Y}$ is not simply related to solvent composition over wide ranges, although over limited composition ranges $\underline{Y}$ can be described as a linear function of composition. The great differences in $\underline{Y}$ for pure alcohols (methanol, -1.09; ethanol, -2.03; isopropyl alcohol, -2.73; tert-butyl alcohol, -3.26) are surprising, but are removed by placing the $\underline{Y}$'s on a molar basis [73b].

It is found that m is solvent dependent [77]. The R part of substrate RX cannot be made drastically different from that in the model substrate without causing dispersion into separate lines for different binary solvents. The leaving group X introduces another type of specificity [77].

Swain has proposed three-parameter [78] and four-parameter [79] equations that correlate solvolysis data about as effectively as the Grunwald-Winstein equation [77].

Kosower has taken a different kind of model process as a measure of solvent polarity. The process is the electronic absorption transition of 1-ethyl-4-carbomethoxy-pyridinium iodide:

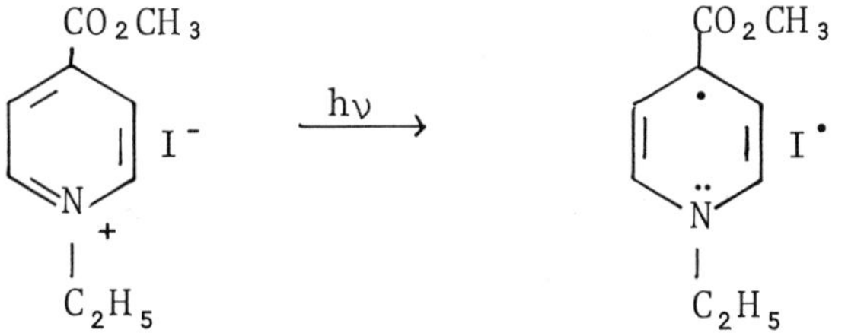

and the solvent polarity parameter is the transition energy (at the longest wavelength band maximum) in kilocalories per mole [73c,80]. This parameter, symbolized Z, is calculated with the relation $E_T = h\nu$, where ν is the frequency of the band maximum. Converting frequency to wavelength, and energy from ergs per photon to kilocalories per einstein, gives

$$\underline{Z} = 2.859 \times 10^5/\lambda_{max} \qquad (4.72)$$

where λ_{max} is in angstroms.

Table 4.XI   The Solvent Polarity Parameter $\underline{Y}$ for Aqueous
Mixed Solvents [76] at 25°

Vol. % of organic solvent[a]	Organic Solvent					
	Ethanol	Methanol	Dioxane	Acetone	Acetic acid[b]	Formic acid[b]
0	3.49	3.49	3.49	3.49	3.49	3.49
10	3.31	3.28	3.22	3.23	--	--
20	3.05	3.03	2.88	2.91	--	--
25	2.91	--	--	2.69	2.84	3.10
30	2.72	2.75	2.46	2.48	--	--
40	2.20	2.39	1.95	1.98	2.31	--
50	1.66	1.97	1.36	1.40	1.94	2.64
60	1.12	1.49	0.72	0.80	1.52	--
70	0.60	0.96	0.01	0.13	--	--
75	--	--	--	--	--	--
80	(0.00)	0.38	-0.83	-0.67	--	2.32
90	-0.75	-0.30	-2.03	-1.86	--	2.22
100	-2.03	-1.09	--	--	-1.64	2.05

[a] x vol. % solution prepared by mixing x volumes of
   organic solvent with 100-x volumes of water.
[b] Containing about 0.07 $\underline{M}$ lithium salts.

The basis for the view that $\underline{Z}$ is a measure of polarity can be seen in the simplified model shown in Fig. 4.4,
which is given by Kosower [80]. The ground state of the
alkylpyridinium iodide complex is essentially the ion
pair shown as (a) in the figure. This species has a large
dipole moment with its negative pole oriented toward the
iodide. The highly polar ground state will induce a more-
or-less ordered solvent structure in its immediate vicinity; the more polar the solvent, the more this "cybotac-
tic region" will stabilize the ground state.

After the transition to the excited state, depicted
as (b) in Fig. 4.4, the associated dipole moment will be
much reduced in magnitude and will be oriented normal to
the ground state dipole, that is, it will now be in the

plane of the ring. The excited state is less polar than
the ground state. According to the Franck-Condon principle, the solvent molecules organized about the ground
state dipole cannot readjust during the electronic transition, so the excited state is essentially unsolvated.
Thus the more polar the solvent the greater the loss in
solvation stabilization upon excitation. The Z value is
therefore larger for more polar solvents.

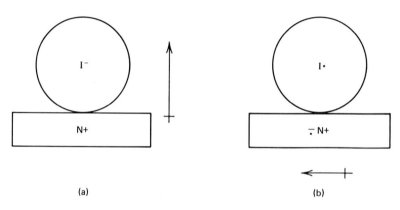

(a)   (b)

Fig. 4.4. Model of the alkylpyridinium iodide complex:
(a) ground state; (b) excited state (after Kosower [80]).

    Z values are easily obtained from Eq. (4.72) for
solvents having Z in the approximate range 63-86. In more
polar solvents the charge-transfer band is obscured by
the pyridinium ion ring absorption, and in nonpolar solvents 1-ethyl-4-carbomethoxy-pyridinium iodide is insoluble. By using the more soluble pyridine-1-oxide as a
secondary standard and obtaining an empirical equation
between Z and the transition energy for pyridine-1-oxide,
it is possible to measure the Z values of nonpolar solvents. Table 4.XII gives Z values for many pure and mixed
solvents.
    One of the interesting features of Z values is their
correlation with Y [80]. Straight line plots of Y against
Z are obtained for binary solvents containing at least 5%
water, and the lines for several binary solvents meet at
the independently determined Y value for pure water. Z
also correlates linearly with other proposed measures of
solvent polarity, in particular with the transition energies E_T of a pyridinium phenol betaine [81] and Ω, the
logarithm of the endo/exo ratio in the cyclopentadiene-
methyl acrylate Diels-Alder addition [82].

Table 4.XII $\underline{Z}$ Values[a]

A. Pure Solvents

Solvent	$\underline{Z}$	Solvent	$\underline{Z}$
H_2O	94.6	$CHCl_3$[b]	63.2
CH_3OH	83.6	CH_3COCH_3	65.7
CH_3CH_2OH	79.6(79.3)[c]		
$CH_3CH_2CH_2OH$	78.3	$HCON(CH_3)_2$	68.5
$CH_3(CH_2)_3OH$	77.7	CH_3CN	71.3
$(CH_3)_2CHOH$	76.3	C_5H_5N	64.0
$(CH_3)_3COH$	71.3	$(CH_3)_3CCH_2CH(CH_3)_2$	60.1
$HOCH_2CH_2OH$	85.1	CH_3SOCH_3	71.1(71.2)[c]
$HCF_2CF_2CH_2OH$	86.3	$HCONH_2$	83.3
$HCF_2(CF_2)_3CH_2OH$	84.8	$cyclo\text{-}C_3H_5COCH_3$	65.4
CH_2Cl_2	64.2	CH_3COOH	79.2

B. Binary Mixtures

Vol. % of organic compound	CH_3OH-H_2O	$(CH_3)_2CHOH-H_2O$	$(CH_3)_3COH-H_2O$	Dioxane$-H_2O$	$CH_3COCH_3-H_2O$
99					68.1
97.5	84.1				
95	84.5	79.3	76.5		72.9
93					74.8
92.5	84.9				
90	85.5	81.5	80.4	76.7	76.6
87.5	85.8				

Wt. % of H_2O	$H_2O-C_2H_5OH$[c]	$H_2O-CH_3SOCH_3$[c]
0	79.3	71.2
5	80.4	73.6
10	81.2	75.8
15	82.5	--
20	82.6	79.1
25	83.9	--
30	84.7	82.8
40	86.6	84.4
50	87.9	87.5
60	89.2	89.3
70	90.5	90.6
75	91.1	91.6
80	91.8	92.0
85	92.5	92.4
90	93.1	93.0
95	93.8	93.7
100	94.5	94.4

Additional values:

83.9   83.3

85	79.2	78.7
80	80.2	80.7
75	81.7	82.1
70	82.8	83.2
65		84.3
60		85.5

aFrom Kosower [80] except as noted.
bContaining 0.13 $\underline{M}$ ethanol.
cFrom J. A. Gowland and G. H. Schmid, Can. J. Chem., 47, 2953 (1969).

Ion pairs can exist as equilibrium mixtures of "intimate" or "solvent-separated" ion pairs. The charge-transfer absorption by alkylpyridinium ion pairs appears to be due to the intimate ion-pair fraction, and the dependence of absorption intensity on solvent composition can be related to the intimate ion-pair equilibrium [83]. The study of such equilibria constitutes a potential application of the medium effect on spectral properties of alkylpyridinium complexes. Recently Mukerjee and Desai [84] have utilized measurements with 1-ethyl-4-carbomethoxypyridinium iodide (they measured a band-shape factor, the ratio of absorbances at 330 nm and 370 nm, instead of $\underline{Z}$) to show that the "effective" dielectric constants of aqueous salt solutions are significantly different from the macroscopic dielectric constants.

The dependence of acid-base strength on solvent composition has been extensively studied, with progress but no definitive results being achieved. An acid dissociation constant is influenced by both the solvent's "polarity" and by its acid-base properties. For the practical purpose of interpolating pK_a values at intermediate compositions of a mixed solvent, it is easy to construct smooth curves from plots of pK_a against some measure of solvent composition. For extrapolation to compositions not studied, or for theoretical interpretation, the mathematical form of the relationship between pK_a and solvent composition should be known; unfortunately, only limited correlations are available. For example, pK_a values for several amines give linear relationships with $\underline{Z}$ and with weight percent water in ethanol-water and dimethyl-sulfoxide-water mixtures. However, the correlations cover the range 30-100% H_2O (in EtOH-H_2O) and 50-100% H_2O (in DMSO-H_2O) [85]; when this is extended to the entire range of 0-100% water, the linear correlation is lost, the behavior being worst for neutral acids.

Grunwald and his co-workers [86] have correlated relative acid strengths in aqueous organic solvents by defining solvent parameters analogous to $\underline{Y}$ values. Glover has taken into consideration the role of the solvent in the ionization process [87]. He writes, for the ionization of HA in solvent S,

$$nS + HA \rightleftharpoons xSH^+ + ySA^- \qquad (4.73)$$

with the solvent being explicitly introduced in the equilibrium constant expression.

$$K_a{}^0 = \frac{(xSH^+)(ySA^-)}{(S)^n(HA)} \qquad (4.74)$$

The parentheses represent activities. The usual expression for the thermodynamic acid ionization constant is

$$K_a = \frac{(xSH^+)(ySA^-)}{(HA)} \qquad (4.75)$$

Combining these gives

$$pK_a = pK_a{}^0 - n \log (S) \qquad (4.76)$$

From plots of pK_a against $\log (S)$, where (S) is taken as the molality of the solvent species involved in the ionization process, values of $pK_a{}^0$ and n were determined. In order to obtain a constant $pK_a{}^0$ over the entire solvent composition range, it was necessary to assume that a 2:1 dioxane-to-water complex forms, and that at dioxane concentrations smaller than the stoichiometric ratio only "free" water enters the ionization reaction, whereas at dioxane concentrations above this ratio only "free" dioxane is involved. The resulting pK_a^0 values then gave a fair account, in Eq. (4.76), of experimental pK_a values in other solvents.

This approach is very similar to Gurney's treatment [88]. Equation (2.20) from Sec. 2.2 is given in logarithmic form as

$$pK = pK_x - \Delta q \log (M^*) \qquad (4.77)$$

where pK is the conventional thermodynamic constant on the molality scale, pK_x is the corresponding constant on the mole fraction scale, Δq is the change in number of particles in the ionization process, and M* is the molality of the solvent. Equations (4.76) and (4.77) are equivalent, with $\Delta q = 1 - n$ and M* = (S) (though specific systems may require different interpretations).

Strongly Acid Solutions. Most organic compounds are bases, that is, they are capable of accepting a proton. The best studied organic bases are the moderately strong ones, which will receive a proton in dilute aqueous solution; amines are the most important examples. The pK_a value of the protonated base, referred to the infinitely dilute aqueous solution, is the usual measure of base strength, and the pH of the solution is a quantitative measure of solvent acidity, or ability to transfer a proton.

Many functional groups are too weakly basic to be appreciably protonated [89] in dilute aqueous solution. Nearly all of the oxygen functional groups belong to this

category, for example. In order to measure the basicity
of such substances in terms of a pK_a value, it is necessary to devise a measure of solution acidity. In strongly
acid solutions the pH scale is inapplicable; the problem
is that any operationally significant change in acidity
can only be accomplished with a concomitant change in the
medium.

Consider a base B of such strength that it can be
protonated in dilute aqueous solution in the acidic range,
say pH 1-2. In the conventional manner the acid dissociation constant K_{BH^+} is defined,

$$K_{BH^+} = \frac{a_{H^+}\, a_B}{a_{BH^+}} = \frac{a_{H^+}\, f_B c_B}{f_{BH^+} c_{BH^+}} \tag{4.79}$$

where the activity coefficients become unity in the infinitely dilute aqueous solution. If B is an indicator
base, the concentration ratio c_B/c_{BH^+} can be measured
spectrophotometrically. Thus K_{BH^+} can be evaluated.

Select now a second indicator base C that is weaker
than B by roughly an order of magnitude; thus a solvent
can be found of such acidity that a significant fraction
of both B and C will be protonated, but this will no long-
er be a dilute aqueous solution, so the individual activity coefficients will in general deviate from unity. For
this solution containing low concentrations of both B and
C,

$$pK_{CH^+} - pK_{BH^+} = -\log \frac{c_C}{c_{CH^+}} + \log \frac{c_B}{c_{BH^+}} - \log \frac{f_C f_{BH^+}}{f_{CH^+} f_B} \tag{4.80}$$

It is reasonable to expect that the ratio $f_C f_{BH^+}/f_{CH^+} f_B$
will not be markedly different from unity, and this has
been checked by establishing the reasonable constancy of
the right-hand side of Eq. (4.80) as the acid concentration is altered. Thus Eq. (4.81) can be written.

$$pK_{CH^+} - pK_{BH^+} = -\log \frac{c_C}{c_{CH^+}} + \log \frac{c_B}{c_{BH^+}} \tag{4.81}$$

Since the concentration ratios can be measured spectrophotometrically (in separate solutions usually) and pK_{BH^+}
is known, the unknown pK_{CH^+} is obtained. This procedure
can now be repeated with a base D that is slightly weaker
than C, using C as the reference. In this stepwise manner
a series of pK_a determinations can be made over the acidity
range from dilute aqueous solution to highly concentrated
mineral acids [90].

Writing Eq. (4.79) in logarithmic form,

$$pK_{BH^+} = -\log \frac{c_B}{c_{BH^+}} - \log \frac{a_{H^+}f_B}{f_{BH^+}} \qquad (4.82)$$

The last term is conveniently designated according to

$$h_0 = \frac{a_{H^+}f_B}{f_{BH^+}} \qquad (4.83)$$

$$H_0 = -\log h_0 \qquad (4.84)$$

Thus Eq. (4.82) becomes

$$H_0 = pK_{BH^+} + \log \frac{c_B}{c_{BH^+}} \qquad (4.85)$$

H_0, the <u>acidity function</u> introduced by Hammett, is a measure of the ability of the solvent to transfer a proton to a base of neutral charge [91]. In dilute aqueous solution h_0 becomes equal to a_{H^+} and H_0 is equal to pH, but in strongly acid solutions H_0 will differ from both pH and $-\log c_{H^+}$. The determination of H_0 is accomplished with the aid of Eq. (4.85) and a series of neutral indicator bases whose pK_a values have been established as outlined above. Table 4.XIII lists H_0 values for some aqueous solutions of common mineral acids [92].

Analogous acidity functions have been defined for bases of other charge types, notably H_{-2} for negatively charged bases and H_R for bases that ionize with the production of a carbonium ion:

$$ROH + H^+ \rightleftharpoons R^+ + H_2O$$

A summary of the several acidity functions has recently appeared [93].

An important use of acidity functions has been for the measurement of the strengths of very weak bases. The procedure utilizes spectrophotometric measurements of the concentration ratio (measurements may be made in the ultraviolet range) in solutions of known acidity function, and application of Eq. (4.85). A major problem is the estimation of the spectra of the extreme forms (protonated and unprotonated) of the base, for the spectra are subject to a medium effect. Corrections must be applied. Alternate treatments of the data are available [94].

The validity of Eq. (4.81) requires that the activity coefficient ratio term vanish in Eq. (4.80), which is

equivalent to the condition $f_B/f_{BH^+} = f_C/f_{CH^+}$. It is this
requirement that leads to the definition of different
acidity functions for indicators of different charge type.
As a first approximation this equality is a very good one,
but it is not universal. In particular, it is most closely met in solvents of high dielectric constant; in low
dielectric constant solvents, such as acetic acid, ion
pairs and higher aggregates are the predominant species
[95].

Table 4.XIII H_0 Values at 25°[a],[b]

Acid concn./$\underline{M}$	HCℓ	HCℓO$_4$	H$_2$SO$_4$
0.1	+0.98	--	+0.83
0.25	+0.55	--	+0.44
0.5	+0.20	--	+0.13
0.75	-0.03	-0.04	-0.07
1.0	-0.20	-0.22	-0.26
1.5	-0.47	-0.53	-0.56
2.0	-0.69	-0.78	-0.84
2.5	-0.87	-1.01	-1.12
3.0	-1.05	-1.23	-1.38
3.5	-1.23	-1.47	-1.62
4.0	-1.40	-1.72	-1.85
4.5	-1.58	-1.97	-2.06
5.0	-1.76	-2.23	-2.28
5.5	-1.93	-2.52	-2.51
6.0	-2.12	-2.84	-2.76
7.0	-2.56	-3.61	-3.32
8.0	-3.00	-4.33	-3.87
9.0	-3.39	-5.05	-4.40
10.0	-3.68	-5.79	-4.89

[a]From M. A. Paul and F. A. Long, Chem. Revs., <u>57</u>, 1
  (1957).
[b]In aqueous solutions.

     Arnett et al. [96] have summarized the limitations
of the acidity function concept, and have proposed that
the enthalpy of solution of the weak base in a strong
acid be used as a quantitative measure of basicity. Heats
of protonation were determined calorimetrically for the
dissolution of bases in concentrated sulfuric acid and in
fluorosulfuric acid, FSO$_3$H. A good linear correlation

was obtained between ΔH and pK_a, covering 40 kcal/mole in the enthalpies and 22 units in pK_a.

Bell [97] has calculated H_0 values with fair accuracy by assuming that the increase in acidity in strongly acid solutions is due to hydration of hydrogen ions, and that the hydration number is four. The addition of "neutral" salts to acid solutions produces a marked increase in acidity [92], and this too is probably a hydration effect in the main. Critchfield and Johnson [98] have made use of this salt effect to titrate very weak bases in concentrated aqueous salt solutions.

The principal use of H_0 has been in the study of reaction mechanisms in acid-catalyzed reactions [99]. For many years the interpretation of kinetic data was based on the Zucker-Hammett hypothesis [99,100], which in an extreme form says that acid-catalyzed reactions belong to one of these two categories:

1. The rate-determining step is the unimolecular transformation of the protonated substrate. According to this scheme it can be shown that a plot of log k_{obs} against $-H_0$ will be linear with unit slope. (k_{obs} is the pseudo-first-order rate constant.)
2. The rate-determining step is bimolecular, with a molecule of water reacting with the protonated substrate. This scheme leads to the prediction that a plot of log k_{obs} against log c_{H^+} will be linear with unit slope.

The key difference is that category 2 includes a molecule of water in the transition state. The criterion of linearity with $-H_0$ or with log c_{H^+} has been used to place a reaction in one of these categories. During the late 1950's, however, it became clear that these categories represent extremes, and that not all acid-catalyzed reactions must necessarily belong to one of them. Two kinds of kinetic evidence led to this view: many reactions did not give really satisfactory linear plots with either variable, and the slopes of linear plots often were significantly different from unity.

Bunnett [101] has found that plots of (log k_{obs} + H_0) against log a_{H_2O}, where a_{H_2O} is the water activity, are often linear. The slope is called w. Sometimes a plot of (log k_{obs} - log c_{H^+}) against log a_{H_2O} is linear, with a slope labeled w*. If the substrate undergoes protonation through the acidity range covered by the kinetic data, a correction must be applied, because the quantity H_0 is related to the fraction protonated.

A range of w values is observed, and by utilizing independent mechanistic evidence, Bunnett proposed the empirical mechanistic criteria in Table 4.XIV to replace the Zucker-Hammett criteria.

Table 4.XIV Mechanistic Criteria Based on w and w*

Substrate type	w	w*	Role of H_2O in r.d.s.
Protonated on O or N	-2.5 to 0.0		Is not involved
Protonated on O or N	+1.2 to +3.3	<-2	Acts as nucleophile
Protonated on O or N	>+3.3	>-2	Acts as general acid-base
Hydrocarbon-like bases	about 0		Acts as general acid-base

The w and w* values have been interpreted as functions of the difference in number of water molecules between the initial and transition states. This interpretation is certainly not fully correct, for other effects seem to contribute to these parameters, but the hydration hypothesis probably is qualitatively valid. Thus, this hypothesis predicts that $\Delta S^{\ddagger}$ will decrease as w increases, as a consequence of the increasing hydration of transition states, and a rough correlation is observed.

Problems

1. Show how the constants a and b in Eq. (4.66) can be evaluated.

   Answer: Plot $E_N^{-1} \log (k/k_0)$ against H_N/E_N.

2. What condition is required for the equation $\rho_1/\rho_2 = T_2/T_1$ to be correct?

3. Estimate pK_a values for p-ethoxybenzoic acid and m-iodobenzoic acid.

4. Find the value of b in Eq. (4.1) in terms of reference quantities.

   Answer: $b = \log (k_B^0/K_A^{0\,m})$

5. These rate constants are for the reaction of ozone
    with $R-C_6H_4CH=CH_2$ in CCl_4 at 25° [A. J. Whitworth,
    R. Ayoub, Y. Rousseau, and S. Fliszár, J. Am. Chem.
    Soc., 91, 7128 (1969)]. Determine the ρ value.

R	$k/M^{-1}\text{-sec}^{-1}$
$p-CH_3$	5.29
H	3.64
$p-Cl$	2.25
$m-Cl$	1.70
$m-NO_2$	0.84

6. P. D. Bolton, C. Johnson, A. R. Katritzky, and S. A.
    Shapiro, J. Am. Chem. Soc., 92, 1567 (1970), report
    a linear enthalpy-entropy relationship for the ioniza-
    tion of very weak bases studied by the indicator over-
    lap method in strong acid solutions. Estimate the
    "isokinetic" temperature β.

7. The slope of the Exner plot in Fig. 4.1 is 0.933.
    Calculate the isokinetic temperature.

    Answer: β = 1091°.

    References

1. J. E. Leffler and E. Grunwald, Rates and Equilibria
    of Organic Reactions, Wiley, New York, 1963, pp. 140-
    141.
2. P. R. Wells, Chem. Revs., 63, 171 (1963); Linear Free
    Energy Relationships, Academic, New York, 1968.
3. J. E. Leffler and E. Grunwald, Ref. [1], p. 143.
4. J. E. Leffler and E. Grunwald, Ref. [1], pp. 156-161.
5. G. S. Hammond, J. Am. Chem. Soc., 77, 334 (1955); see
    also Section 3.3.
6. A. O. Cohen and R. A. Marcus, J. Phys. Chem., 72,
    4249 (1968).
7. J. E. Leffler and E. Grunwald, Ref. [1], pp. 155-156.
8. J. E. Leffler, J. Org. Chem., 20, 1202 (1955); the
    same relations can apply to equilibrium processes, in
    which case β would be the isoequilibrium temperature.
    Schowen has suggested the term isergonic temperature
    to include both situations; R. L. Schowen, J. Pharm.
    Sci., 56, 931 (1967). R. Lumry and S. Rajender, Bio-
    polymers, 9, 1125 (1970) prefer compensation tempera-
    ture.

9.   J. E. Leffler and E. Grunwald, Ref. [1], pp. 324-342.
10. L. G. Hepler, J. Am. Chem. Soc., 85, 3089 (1963).
11. R. L. Schowen, J. Pharm. Sci., 56, 931 (1967).
12. J. E. Leffler, J. Org. Chem., 20, 1202 (1955); J. E. Leffler and E. Grunwald, Ref. [1], pp. 324-342.
13. R. C. Petersen, J. H. Markgraf, and S. D. Ross, J. Am. Chem. Soc., 83, 3819 (1961).
14. J. E. Leffler and E. Grunwald, Ref. [1], pp. 323-324; J. E. Leffler, J. Org. Chem., 31, 533 (1966).
15. R. C. Petersen, J. Org. Chem., 29, 3133 (1964). For extensions of this argument see E. R. Plante and R. C. Paule, J. Chem. Phys., 53, 3770 (1970) and J. R. McCreary and R. J. Thorne, ibid., 53, 3771 (1970).
16. (a) O. Exner, Nature, 201, 488B (1964); R. McGregor and B. Milicevic, ibid., 211, 523B (1966); (b) O. Exner, Coll. Czech. Chem. Commun., 29, 1094 (1964); (c) O. Exner, Nature, 227, 366 (1970).
17. R. J. Washkuhn, V. K. Patel, and J. R. Robinson, J. Pharm. Sci., 60, 736 (1971).
18. The 25° data in Fig. 4.1 are included in Table 13.X.
19. (a) C. D. Ritchie and W. F. Sager, Progr. Phys. Org. Chem., 2, 323 (1964); (b) W. B. Kover and J. V. Carter, J. Am. Chem. Soc., 91, 4028 (1969).
20. L. P. Hammett, Physical Organic Chemistry, McGraw-Hill, New York, 1940, Chap. VII.
21. J. E. Leffler and E. Grunwald, Ref. [1], pp. 192-194.
22. H. H. Jaffé, Chem. Revs., 53, 191 (1953).
23. S. Ehrenson, Progr. Phys. Org. Chem., 2, 195 (1964).
24. D. H. McDaniel and H. C. Brown, J. Org. Chem., 23, 420 (1958).
25. K. B. Wiberg, Physical Organic Chemistry, Wiley, New York, 1964, pp. 408-409.
26. Perhaps the major role for linear free energy relationships in analytical chemistry is to provide estimates of rate and equilibrium constants that will guide the selection of reaction conditions. Such estimates need not be highly accurate to be useful for this purpose, so it doesn't make much difference which definition of σ is used. For mechanistic uses, the standard reaction definition seems superior.
27. E. Berliner and L. H. Liu, J. Am. Chem. Soc., 75, 2417 (1953).
28. Sigma values can be used to decide which members of a reaction series will be studied. Inclusion of both the 4´-chloro- and 4´-bromo-substituted esters in Fig. 4.2 did not significantly strengthen this study.
29. This is not strictly possible, however, because the temperature and dielectric constant cannot be independently varied.

30. Ritchie and Sager [19a], however, conclude that an isokinetic relationship removes the 1/T dependence of ρ. A direct proportionality between ρ and 1/T requires that the final term in (4.47) or (4.48) be negligible.

31. R. W. Taft, Jr. and I. C. Lewis, J. Am. Chem. Soc., 81, 5343 (1959); T. E. Bitterwolf, R. E. Linder, and A. C. Ling, J. Chem. Soc., 1970B, p. 1673.

32. In 1953 Jaffé [22] gave 204 reactions correlated by the Hammett equation, and new examples appear frequently.

33. L. P. Hammett, Ref. [20], pp. 204-207.

34. R. W. Taft, Jr., Steric Effects in Organic Chemistry (M. S. Newman, ed.), Wiley, New York, 1956, Chap. 13.

35. M. Charton, J. Am. Chem. Soc., 91, 619, 624 (1969).

36. W. P. Jencks, Catalysis in Chemistry and Enzymology, McGraw-Hill, New York, 1969, pp. 480-483.

37. J. E. Leffler and E. Grunwald, Ref. [1], pp. 187-191; J. O. Schreck, J. Chem. Educ., 48, 103 (1971).

38. L. P. Hammett, Ref. [20], p. 188.

39. Jaffé [22] denoted these by σ*, but this symbol has since found wider use for aliphatic substituent constants [34].

40. J. F. Bunnett, F. Draper, Jr., P. R. Ryason, P. Noble, Jr., R. G. Tonkyn, and R. E. Zahler, J. Am. Chem. Soc., 75, 642 (1953).

41. H. C. Brown and Y. Okamoto, J. Am. Chem. Soc., 80, 4979 (1958).

42. H. van Bekkum, P. E. Verkade, and B. M. Wepster, Rec. Trav. Chim. Pays-Bas, 78, 815 (1959).

43. R. W. Taft, Jr., J. Phys. Chem., 64, 1805 (1960).

44. R. W. Taft, Jr. and I. C. Lewis, J. Am. Chem. Soc., 80, 2436 (1958); 81, 5343 (1959).

45. To a certain extent the adjective aliphatic is defined, in this context, by the criterion of non-adherence to the Hammett equation.

46. C. K. Ingold, J. Chem. Soc., 1930, 1032.

47. R. W. A. Jones and J. D. R. Thomas, J. Chem. Soc., 1966B, p. 661.

48. T. C. Bruice, T. H. Fife, J. T. Bruno, and N. E. Brandon, Biochemistry, 1, 7 (1962); J. F. Kirsch and W. P. Jencks, J. Am. Chem. Soc., 86, 837 (1964); W. P. Jencks and M. Gilchrist, ibid., 90, 2622 (1968).

49. P. Ballinger and F. A. Long, J. Am. Chem. Soc., 82, 795 (1960).

50. J. R. Robinson and L. E. Matheson, J. Org. Chem., 34, 3630 (1969).

51. J. R. Robinson, Anal. Chem., 39, 1178 (1967). See also, however, the discussion of Fig. 13.5.

52. S. Sun and K. A. Connors, J. Pharm. Sci., $\underline{58}$, 1150
     (1969).
53. The full Taft equation is $\log(k/k^0) = \rho^*\sigma^* + sE_s$,
     where E_s is a steric parameter defined by E_s =
     $\log(k/k_0)_A$ [see Eq. (4.50)] and s is a reaction con-
     stant. For a recent critical review of the Taft
     treatment, see J. Shorter, Quart. Rev., Chem. Soc.,
     $\underline{24}$, 433 (1970).
54. J. N. Brønsted and K. J. Pedersen, Z. Physik. Chem.,
     $\underline{108}$, 185 (1924).
55. (a) R. P. Bell, Acid-Base Catalysis, Oxford Univ.
     Press, London, 1941, Chap. V; (b) The Proton in
     Chemistry, Cornell Univ. Press, Ithaca, N. Y., 1959,
     Chap. X.
56. This is an application of the principle of microsco-
     pic reversibility.
57. F. G. Bordwell, W. J. Boyle, Jr., J. A. Hautala, and
     K. C. Yee, J. Am. Chem. Soc., $\underline{91}$, 4002 (1969).
58. R. A. Marcus, J. Am. Chem. Soc., $\underline{91}$, 7224 (1969).
     See also A. J. Kresge, ibid., $\underline{92}$, 3210 (1970), who
     suggests that these deviant Brønsted slopes are caused
     by electronic interactions within the transition
     state that are not possible in the initial or final
     states.
59. Satchell and Satchell have observed Brønsted slopes
     greater than unity in rate-equilibrium correlations
     of Lewis-acid reactions: R. S. Satchell and D. P.
     N. Satchell, Proc. Chem. Soc., $\underline{1964}$, 362.
60. J. E. Leffler and E. Grunwald, $\overline{Ref.}$ [1], pp. 238-241.
61. Ionic strength effects are neglected. The pK_a values
     for H_3O^+ and H_2O are based on the conventional defi-
     nitions, with $\overline{K}_w = 1.0 \times 10^{-14}$ and the concentration
     of water taken as 55.5 M. Though formally correct,
     and universally used, these pK_a values may not be
     the proper ones because of the special character of
     these two acids in aqueous solutions. This problem
     has been discussed by Bell (Ref. [55a], pp. 91-95)
     and Jencks (Ref. [36], pp. 171-173).
62. E. G. Sander and W. P. Jencks, J. Am. Chem. Soc., $\underline{90}$,
     4377 (1968).
63. W. P. Jencks and J. Carriuolo, J. Am. Chem. Soc., $\underline{82}$,
     1778 (1960).
64. T. C. Bruice and R. Lapinski, J. Am. Chem. Soc., $\underline{80}$,
     2265 (1958).
65. W.-H. Hong and K. A. Connors, J. Pharm. Sci., $\underline{57}$,
     1789 (1968).
66. <u>Nucleophilicity</u> simply means reactivity of a nucleo-
     philic reagent. The concept is discussed in the
     following paragraphs.

67. L. R. Fedor, T. C. Bruice, K. L. Kirk, and J. Mein-
     wald, J. Am. Chem. Soc., 88, 108 (1966).
68. J. F. Bunnett, C. F. Hauser, and K. V. Nahabedian,
     Proc. Chem. Soc., 1961, p. 305; A. J. Parker, ibid.,
     1961, p. 371.
69. C. G. Swain and C. B. Scott, J. Am. Chem. Soc., 75,
     141 (1953).
70. J. O. Edwards, J. Am. Chem. Soc., 76, 1540 (1954);
     78, 1819 (1956).
71. Note that standard potentials are proportional to
     standard free energy changes.
72. R. E. Davis, R. Nehring, W. J. Blume, and C. R.
     Chuang, J. Am. Chem. Soc., 91, 91 (1969); R. E.
     Davis, L. Suba, P. Klimishin, and J. Carter, ibid.,
     91, 104 (1969).
73. For reviews see (a) J. E. Leffler and E. Grunwald,
     Ref. [1], Chap. 8; (b) P. R. Wells, Linear Free
     Energy Relationships, Academic, New York, 1968, Chap.
     4; (c) E. M. Kosower, An Introduction to Physical
     Organic Chemistry, Wiley, New York, 1968, Part Two.
74. E. Grunwald and S. Winstein, J. Am. Chem. Soc., 70,
     846 (1968).
75. Prepared by mixing 80 volumes of ethanol with 20
     volumes of water.
76. A. H. Fainberg and S. Winstein, J. Am. Chem. Soc.,
     78, 2770 (1956).
77. S. Winstein, A. H. Fainberg, and E. Grunwald, J. Am.
     Chem. Soc., 79, 4146 (1957).
78. C. G. Swain and D. C. Dittmer, J. Am. Chem. Soc.,
     75, 4627 (1953); C. G. Swain, D. C. Dittmer, and L.
     E. Kaiser, ibid., 77, 3737 (1955).
79. C. G. Swain, R. B. Mosely, and D. E. Bown, J. Am.
     Chem. Soc., 77, 3731 (1955).
80. E. M. Kosower, J. Am. Chem. Soc., 80, 3253, 3261,
     3267 (1958).
81. K. Dimroth, C. Reichardt, T. Siepmann, and F. Bohl-
     mann, Ann., 661, 1 (1963); this compound permits
     direct E_T measurements in polar solvents.
82. J. A. Berson, Z. Hamlet, and W. A. Mueller, J. Am.
     Chem. Soc., 84, 297 (1962).
83. A. Ray and P. Mukerjee, J. Phys. Chem., 70, 2138
     (1968).
84. P. Mukerjee and N. R. Desai, Nature, 223, 1056 (1969).
85. J. A. Gowland and G. H. Schmid, Can. J. Chem., 47,
     2953 (1969).
86. E. Grunwald and B. J. Berkowitz, J. Am. Chem. Soc.,
     73, 4939 (1951); B. Gutbezahl and E. Grunwald, ibid.,
     75, 559, 565 (1953); H. P. Marshall and E. Grunwald,
     ibid., 76, 2000 (1954).

87. D. J. Glover, J. Am. Chem. Soc., _87_, 5275, 5279 (1965).

88. R. W. Gurney, Ionic Processes in Solution, Dover Publications, New York, 1962, Chap. 6.

89. This common phrase is inaccurate, since any given molecule either is protonated or it is not. (Even this statement could be argued about.) The meaning to be taken is that the fraction of molecules in the protonated form is not detectable or accurately determinable.

90. L. P. Hammett, Physical Organic Chemistry, McGraw-Hill, New York, 1940, pp. 262-267.

91. L. P. Hammett and A. J. Deyrup, J. Am. Chem. Soc., _54_, 2721 (1932); L. P. Hammett, Ref. [20], pp. 267-271.

92. M. A. Paul and F. A. Long, Chem. Revs., _57_, 1 (1957).

93. C. J. O'Connor, J. Chem. Educ., _46_, 686 (1969); H_0 is tabulated as a function of acid molarity, molality, and % w/w composition, and the water activity is given for these solutions. See also C. H. Rochester, Acidity Functions, Academic, New York, 1970.

94. J. F. Bunnett and F. P. Olsen, Chem. Commun., 1965, p. 601; Can. J. Chem., _44_, 1899 (1966). These authors develop a linear free energy relationship using as a model process the ionization of a hypothetical base with $pK_{BH^+} = 0$. Examples of its application are given by C. C. Greig and C. D. Johnson, J. Am. Chem. Soc., _90_, 6453 (1968).

95. For low mineral acid concentrations in glacial acetic acid, such as are encountered in titrations, a pH scale has been defined; see I. M. Kolthoff and S. Bruckenstein, J. Am. Chem. Soc., _79_, 1 (1957).

96. E. M. Arnett, R. P. Quirk, and J. J. Burke, J. Am. Chem. Soc., _92_, 1260 (1970).

97. R. P. Bell, Ref. [55b], pp. 81-84.

98. F. E. Critchfield and J. B. Johnson, Anal. Chem., _30_, 1247 (1958); _31_, 570 (1959).

99. F. A. Long and M. A. Paul, Chem. Revs., _57_, 935 (1957).

100. L. Zucker and L. P. Hammett, J. Am. Chem. Soc., _61_, 2791 (1939).

101. J. F. Bunnett, J. Am. Chem. Soc., _83_, 4956, 4968, 4973, 4978 (1961).

Chapter 5. SELECTIVITY

By selectivity is meant, in general terms, the relative response to a common process of a series of substrates or reagents [1]. The analytical chemist usually seeks the identity, amount, or concentration of a substance in a sample that must be presumed to be a mixture. The types of mixtures can vary greatly: besides a component of primary interest the sample may contain products of its decomposition; a synthetic reaction product may contain isomers or closely related by-products, as well as a solvent, catalysts, and excess reagents; pharmaceutical dosage forms may include several active ingredients, their decomposition products, and pharmaceutical adjuncts such as flavoring and coloring agents, surface-active agents, binders and lubricants, etc.; and biological samples, which may be the most complex of all, can contain dozens of components, some of them very similar.

There are four approaches to the achievement of analytical selectivity; or rather there are four convenient ways to think about this problem, with considerable overlap among them. These are: (1) differentiation by means of equilibrium differences; (2) differentiation by means of kinetic differences; (3) differentiation in the final measurement; and (4) separation of the components. The last of these, though perhaps the single most powerful approach to mixture analysis, is not our present concern, except to note that separation techniques are based on the preceding phenomena. Thus liquid-liquid extraction and countercurrent distribution achieve separations because of differentiating equilibria. The several forms of chromatography take advantage of differential migration rates that result from both equilibrium and rate phenomena [2]. Differential dialysis [3] is based on rate differences.

In our assertion (Chap. 1) that chemical analytical methods can be partitioned into two components, namely, a chemical reaction or reactions followed by a terminal observation, we note that selectivity can be introduced (or lost!) in each of these components. The next two sections treat rate and equilibrium control of selectivity in the reaction step, and the third section discusses selectivity achieved in the final measurement.

Although high selectivity is commonly considered to be a major goal in analytical development, and this chapter is devoted to this cause, it should be remembered that sometimes a nonselective assay will be the method of choice. The nature of the sample, the information required, and the practical matters of time and facilities may determine this choice; but even more to the point are analytical situations in which a general method is preferable to a selective one because selectivity is provided by another component of the procedure. This is very commonly the case in chromatographic separations, when the selectivity is accomplished by chromatography, and the detection method can then be (in fact, usually should be) very general in its response.

5.1   Kinetic Control

Selectivity-Reactivity Relationships. We consider a bimolecular reaction between a substrate and a reagent. Upon each encounter of these two species there is a probability P that reaction will occur. If the solution contains two substrates S_1 and S_2, each characterized by a probability of reaction P_1 and P_2 with the common reagent, evidently the ratio P_2/P_1 is a measure of the selectivity of this process for S_2 relative to S_1. If the two substrates are not markedly dissimilar, the ratio P_2/P_1 will be similar to the ratio of rate constants, k_2/k_1. Leffler and Grunwald [4] define the selectivity as

$$\begin{array}{c}\text{Selectivity of the reagent for } S_2 \\ \text{relative to } S_1 = RT \; \ell n \; (k_2/k_1)\end{array} \qquad (5.1)$$

The reactivity of a reagent can be defined as its rate constant with a standard substrate. Equation (5.1), therefore, relates selectivity to reactivity. The significance of this statement can be developed by imagining that a substrate is progressively changed in such a way that its probability of reaction increases toward the limit of unity. Comparing two substrates, each of whose reactivity is made successively higher in this way, the ratio P_2/P_1 will tend toward unity, as will its estimator k_2/k_1. Thus in the limit of "infinite" reactivity, the selectivity becomes zero [5].

Figure 5.1 shows a moderate example of a selectivity-reactivity relationship. The data, which are from Table 4.IX, are rate constants (relative to methyl acetate) for the alkaline hydrolysis of a series of aliphatic acetates (k_{Me}) and formates (k_H). The rate constant for the acetate is a measure of reactivity, and the ratio k_H/k_{Me} is a measure of the selectivity of OH^- for the substrate HCOOR relative to CH_3COOR. Evidently the selectivity decreases as the reactivity increases. If the formates and acetates can be separately correlated with alkyl substituent constants $\sigma^{\prime\prime}$ [see Eq. (4.53)] to give

$$\log \frac{k_H}{k_H{}^0} = \rho_H{}^{\prime\prime}\sigma^{\prime\prime} \qquad (5.2)$$

$$\log \frac{k_{Me}}{k^0{}_{Me}} = \rho_{Me}{}^{\prime\prime}\sigma^{\prime\prime} \qquad (5.3)$$

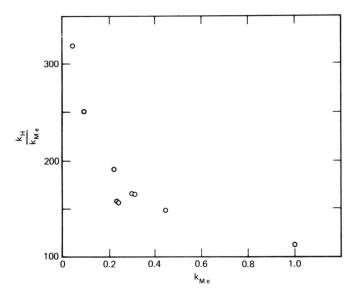

Fig. 5.1. Selectivity-reactivity relationship for the alkaline hydrolysis of a series of acetate (k_{Me}) and formate (k_H) esters (data from Table 4.IX).

then it is readily shown that

$$\log \frac{k_H}{k_{Me}} = \left(\frac{\rho^{\prime\prime}{}_H}{\rho^{\prime\prime}{}_{Me}} - 1 \right) \log k_{Me} + \text{constant} \qquad (5.4)$$

A log-log plot of the data in Fig. 5.1 is linear, having
the equation

$$\log \frac{k_H}{k_{Me}} = -0.25 \log k_{Me} + 2.06 \qquad (5.5)$$

with which it is calculated that the selectivity of this
process would vanish if the acetate member of a pair of
substrates had a reactivity corresponding to $k_{Me} = 6.58$.
      Another way to consider a selectivity-reactivity
relationship is to compare the relative effects of a series of substituents on a pair of reactions. This is what
is done when Hammett plots are made for a pair of reactions and their ρ values are compared. The slope of a
linear free energy relationship is a function of the sensitivity of the process being correlated to structural or
solvent changes. Thus in a family of closely related
linear free energy correlations, the one with the steepest
slope is the most selective, and the one with the smallest
slope is the least selective [6]. Moreover, the intercept
(or some arbitrarily selected abscissa value, usually log
k_0 for the reference substituent) should be a measure of
reactivity in each reaction series. Thus a correlation
should exist between the slopes (selectivity) and intercepts (reactivity) of a family of related linear free
energy relationships. It has been suggested that the
slopes and intercepts should be linearly related [7], but
the conditions required for linearity are seldom met, and
it is instead common to find only a rough correlation,
indicative of normal selectivity-reactivity behavior [6].
The Brønsted slopes, β, for the halogenation of a series
of carbonyl compounds catalyzed by carboxylate ions show
a smooth but nonlinear correlation with log k_0 [4].
Jencks [8] uses the selectivity-reactivity relationship
between Brønsted slopes and nucleophilic reactivity to
distinguish between general acid catalysis and specific
acid-general base catalysis.
      In contrast to the comparison of a pair of reactions,
with variations in substituents, Brown and his co-workers
compared a pair of substituents subjected to a large number of reactions [9]. These were all electrophilic aromatic substitutions, the most extensively studied substrates being benzene and toluene. When a monosubstituted
benzene derivative undergoes substitution of hydrogen,
there are three possible sites for the entering group:
ortho (two equivalent positions), meta (two), and para
(one). Thus the observed rate constant for the substitution reaction does not give a complete picture of the
course of the reaction, because it does not distinguish
among the positional isomers. If $k_{C_6H_5Y}$ is the observed

rate constant for reaction of the monosubstituted derivative:

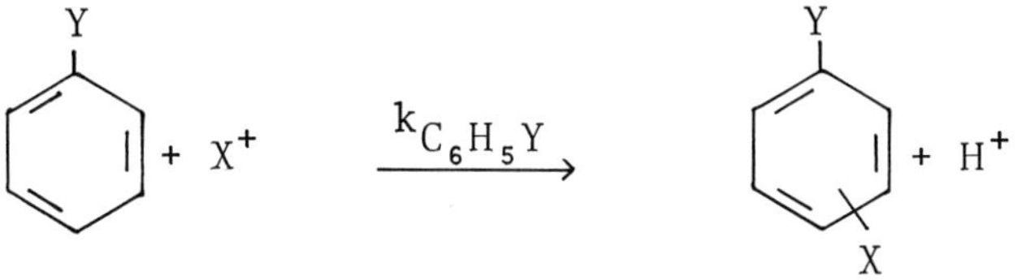

$$+ X^+ \quad \xrightarrow{k_{C_6H_5Y}} \quad + H^+ \qquad (5.6)$$

then $k_{C_6H_5Y} = 2k_{o-Y} + 2k_{m-Y} + k_{p-Y}.$
The corresponding reaction for benzene,

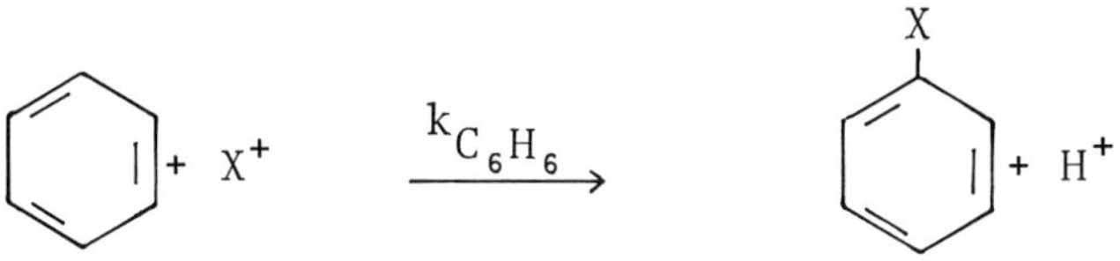

$$+ X^+ \quad \xrightarrow{k_{C_6H_6}} \quad + H^+ \qquad (5.7)$$

leads to $k_{C_6H_6} = 6k_H$, because the reagent has an equal
probability of attacking each of the six carbon atoms.
The isomer distributions can be related to the individual
rate constants, for Reaction (5.6), by

$$\frac{(\% \ ortho)}{100} = \frac{2k_{o-Y}}{k_{C_6H_5Y}} \qquad (5.8)$$

$$\frac{(\% \ meta)}{100} = \frac{2k_{m-Y}}{k_{C_6H_5Y}} \qquad (5.9)$$

$$\frac{(\% \ para)}{100} = \frac{k_{p-Y}}{k_{C_6H_5Y}} \qquad (5.10)$$

Partial rate factors are defined as rates of substitution
at each position relative to benzene, on a per site basis,
thus

$$o_f^Y = \frac{k_{o-Y}}{k_H} \qquad (5.11)$$

$$m_f^Y = \frac{k_{m-Y}}{k_H} \qquad (5.12)$$

$$p_f^Y = \frac{k_{p-Y}}{k_H} \qquad (5.13)$$

Combining these relationships gives the following equations for calculation of partial rate factors [9-11]:

$$o_f^Y = \frac{3k_{C_6H_5Y}}{k_{C_6H_6}} \times \frac{(\% \text{ ortho})}{100} \qquad (5.14)$$

$$m_f^Y = \frac{3k_{C_6H_5Y}}{k_{C_6H_6}} \times \frac{(\% \text{ meta})}{100} \qquad (5.15)$$

$$p_f^Y = \frac{6k_{C_6H_5Y}}{k_{C_6H_6}} \times \frac{(\% \text{ para})}{100} \qquad (5.16)$$

Partial rate factors have the significance that a factor greater than unity represents activation of that site by the substituent Y relative to hydrogen, and a factor less than unity represents deactivation. [For a fuller discussion see Sec. 7.1.]

Brown [9,12] defined a selectivity factor S_f by

$$S_f = \log \frac{p_f^Y}{m_f^Y} \qquad (5.17)$$

Evidently S_f is a measure of intramolecular selectivity; since it involves a ratio, the contribution of the benzene substitution rate disappears, and the selectivity factor expresses the selectivity of the reagent X^+ in Eq. (5.6) for the para position relative to the meta position. Each individual partial rate factor, on the other hand, is expressive of an intermolecular selectivity; thus p_f^Y is a measure of the selectivity of the reagent for the para position in C_6H_5Y relative to benzene. It was observed that Eq. (5.18), where c_{Me} is a constant [13], is satisfied for a large number of electrophilic substitutions of toluene [9].

$$\log p_f^{Me} = c_{Me} S_f \qquad (5.18)$$

Equation (5.18) is a selectivity-reactivity relationship, with lower values of S_f denoting lower selectivity. Lower values of p_f correspond to greater reactivity,

with the limit being a partial rate factor of unity for
an infinitely reactive electrophile. This selectivity-
reactivity relationship is followed for the electrophilic
substitution reactions of many substituted benzenes,
though toluene is the best studied of these.
     Brown [9] uses a related test for selectivity by
plotting the logarithm of the partial rate factor for a
single substituent against ρ for a variety of reactions;
the appropriate ρ values are obtained from conventional
linear free energy plots for a series of monosubstituted
benzenes subjected to a common reagent against the electrophilic substituent constant σ^+, as in Eq. (4.49).

Interference. Most organic analytical methods are "total
change" methods; that is, the sample compound is subjected
to a chemical reaction and, after reaction is "complete,"
some measure of reagent consumption or product generation
is obtained. In order for such a method to be specific
for one compound (the sample, S) in the presence of another
similar compound (arbitrarily designated the interference,
I), S must react essentially completely while the other,
in the same time period, remains essentially unreacted.
The accuracy required in the analysis will determine the
necessary selectivity. A quantitative treatment is simple
if pseudo-first-order conditions are assumed for both S
and I. Then Eqs. (5.19) and (5.20) give the concentrations of S and I as functions of time.

$$[S] = [S]_0 e^{-k_S t} \qquad (5.19)$$

$$[I] = [I]_0 e^{-k_I t} \qquad (5.20)$$

The fraction of S unreacted is symbolized x, so x =
$[S]/[S]_0$; therefore, x = exp $(-k_S t)$, or,

$$t = \frac{-\ln x}{k_S} \qquad (5.21)$$

Equation (5.21) gives the time required for 100 (1-x)%
of sample compound S to undergo reaction. The fraction
of I remaining in this length of time is obtained by substituting Eq. (5.21) into (5.20).

$$[I] = [I]_0 e^{(k_I/k_S)\ln x} \qquad (5.22)$$

An indicator of the extent of interference is the apparent
percent recovery of S, which is the total fraction of

reaction relative to the initial concentration of S, mul-
tiplied by 100.

$$\frac{\text{Apparent \% recovery of S}}{100} = \frac{[S]_0 - [S]}{[S]_0} + \frac{[I]_0 - [I]}{[S]_0}$$

(5.23)

Combining Eqs. (5.19), (5.22), and (5.23),

$$\frac{\text{Apparent \% recovery}}{100} = (1-x) + \frac{[I]_0}{[S]_0} (1-e^{(k_I/k_S)\ln x})$$

(5.24)

The apparent recovery is therefore a function of the time
(through x), of the selectivity (through k_I/k_S), and of
the ratio of initial concentrations of the interference
and the sample. A simpler expression can be obtained
directly from (5.19) and (5.20); thus,

$$\frac{\log([S]/[S]_0)}{\log([I]/[I]_0)} = \frac{k_S}{k_I}$$

(5.25)

with which the rate constant is related to fractions of
unreacted S and I [14]. In this form, however, the role
of the concentration ratio $[I]_0/[S]_0$ in determining the
analytical error is obscured.

Table 5.1 gives apparent recoveries calculated with
Eq. (5.24) for three values of x, four of $[I]_0/[S]_0$, and
five of k_I/k_S. (Values of k_I/k_S greater than unity are
relatively uninteresting; the recovery is then the sum of
the reacted S and essentially all of the I.) This table
illustrates some potential dangers in testing analytical
methods on potential samples. For example, the apparent
recovery of 99.0% for a system containing 9.1 mole percent
of an interference with k_I/k_S = 1 occurs if x = 0.10.
Thus it is important to vary the time of reaction if the
purity of the test substance is in doubt. If a method is
used to obtain an analytical result for the total (sample
plus interference) compounds in a mixture, the time of
reaction must be based on the slowest reacting component.

The minimization of error caused by an interfering
component in an analytical sample can be achieved only
through the quantities x and k_I/k_S, because $[I]_0/[S]_0$ is
a characteristic of the sample. As Table 5.I shows, if
either $[I]_0/[S]_0$ or k_I/k_S is small, good results can
usually be achieved with a reaction time corresponding to

$x = 0.001$. Limits on the ratios of initial concentrations and rate constants can be assigned by adopting a permissible error limit and using Eq. (5.24). Suppose an error of 0.5% is allowable and the sample contains 25 mole percent of an interfering substance. That is, let percent recovery = 100.0, $x = 0.005$, and $[I]_0/[S]_0 = 1/3$. Then Eq. (5.24) gives $k_I/k_S = 0.00285$ as the maximum relative rate consistent with the desired accuracy.

Table 5.I Apparent Percent Recovery of Sample S in Presence of Inteference I[a]

			$[I]_0/[S]_0$		
x	k_I/k_S	0.01	0.1	1.0	10
	1.0	90.9	99.0	180	990
	0.1	90.2	92.1	110.6	296
0.10	0.01	90.0	90.2	92.3	113
	0.001	90.0	90.0	90.2	92.0
	0.0001	90.0	90.0	90.0	90.0
	1.0	100.0	108.9	198	1089
	0.1	99.4	102.7	136	468
0.01	0.01	99.1	99.5	103.5	144
	0.001	99.0	99.1	99.5	103.5
	0.0001	99.0	99.0	99.1	100.0
	1.0	100.9	109.9	200	1090
	0.1	100.4	104.9	150	600
0.001	0.01	100.0	100.6	106.7	168
	0.001	99.9	99.9	100.0	100.5
	0.001	99.9	99.9	99.9	99.9

[a]Calculated with Eq. (5.24).

As a means of minimizing error by interference, variation in x is limited by the desired accuracy; in a typical organic analysis x must be 0.01 or smaller. Manipulation of the selectivity, k_I/k_S, offers the best possibility for minimizing interference. This is treated in the following paragraphs.

<u>Modification of Selectivity and Reactivity</u>. The analyst may find it convenient to influence the <u>absolute</u> rate

constant of a reaction, that is, the reactivity, usually
to decrease the reaction time; or he may need to alter
the relative rates of reactions, thus altering the selectivity in order to minimize interference. As we saw earlier, these properties are often related in an inverse
sense; an increased reactivity is usually accompanied by
a decreased selectivity. This observation may be helpful
in modifying or choosing an analytical method for a particular sample. For example, thiol esters (RCOSR´) and
oxygen esters (RCOOR´) have about the same reactivity
toward hydroxide ion, but with amine nucleophiles, thiol
esters are much more reactive [15].

$$RCOSR´ + R´´NH \rightarrow RCONHR´´ + R´SH$$

This selectivity is controlled in part by a mechanistic
difference in the two cases, and a better illustration of
selectivity-reactivity behavior may be provided by hydroxamic acid formation to determine a thiol ester in the
presence of an oxygen ester.

$$RCOSR´ + H_2NOH \rightarrow RCONHOH + R´SH$$

At high pH the reaction is catalyzed by hydroxide ion,
and both the thiol ester and the oxygen ester form the
hydroxamic acid. Under neutral conditions the thiol ester forms the hydroxamic acid whereas the oxygen ester
does not react [16].

Interesting and valuable selectivity patterns have
been revealed in systematic studies of numerous reducing
agents, especially lithium aluminum hydride, aluminum hydride, lithium trimethoxyaluminohydride, lithium tri-t-
butoxyaluminohydride [17], and diborane [18]. Within a
series of related agents, such as the nucleophilic aluminum hydride series, rough selectivity-reactivity correlation occurs, but diborane, which is a mild electrophilic
reagent, has a different pattern of selectivity toward
functional groups. Thus acyl halides are easily reduced
by all of the aluminum hydrides, but they react very slowly with diborane. However, diborane reduces carboxylic
acids as readily as the powerful LiAlH$_4$, whereas lithium
tri-t-butoxyaluminohydride is not an effective reducing
agent for this functional group. Selectivity behavior of
this type, which has been exploited for synthetic purposes,
should find valuable analytical applications [19].

The sample and reagent having been fixed, only four
variables remain to the analyst for the control of reaction rates. These are pressure, temperature, concentration, and reaction medium. Pressure is not a convenient
experimental variable, so most reactions in the liquid

state are carried out at atmospheric pressure. As the
results in Table 5.I demonstrate, really useful improvements in selectivity require relative rate changes of at
least a factor of ten. Changes in absolute rate constants
of this order are also sometimes required, although even
a factor of two may be a welcome alteration.

Analytical reaction rates are often increased by
raising the temperature. It was earlier pointed out
(Chap. 3) that a rise of 10°C commonly increases the rate
constant by a factor of 2 to 3. Obviously the relative
rate of a pair of reactions will undergo only a moderate
change with temperature. A plot of log k_1 against log k_1
for a reaction series, k_2 and k_1 being the rate constants
for a member of the series at T_2 and T_1 ($T_2>T_1$), usually
has a slope smaller than unity, meaning that the selectivity of the reagent decreases with increasing temperature
[20]. This is consistent with normal selectivity-reactivity behavior. Decreased temperature does not appear
capable of increasing selectivity to an analytically useful extent [21].

The effect of concentration on rate is described by
the rate equation for the reaction. Absolute rates can
be altered over many orders of magnitudes by changing the
reactant concentrations. In a bimolecular reaction with
rate equation $v = k[R][S]$, the rate v is concentration-
dependent and the rate constant k is independent of concentration. Often pseudo-first-order conditions are
arranged, with $[R] \gg [S]$; then $v = k_{obs}[S]$, where $k_{obs} =$
$k[R]$. The time required for "completion" of reaction is
a function of the half-life, $t_{1/2} = 0.693/k_{obs}$. Obviously,
then, the half-life depends upon $[R]$ but not upon $[S]$. If
S is an analytical sample and R is a reagent, the time of
reaction can be decreased by increasing the reagent concentration. An obvious example occurs in any reaction
involving hydroxide ion, with k_{obs} increasing tenfold for
each unit increase in pH. Even if such a system deviates
from pseudo-first-order conditions during the reaction,
as in the saponification of esters using a titrimetric
finish [23], the time of reaction will be shortened by increasing the reagent concentration though quantitative
treatment will require the second-order equation.

The response of the relative rate for two substrates
reacting with a common reagent (as in the sample-interfer-
ence mixture discussed earlier) clearly depends upon the
forms of the individual rate equations. If the reagent
appears to the same order in the two rate equations, as
is often likely, changes in reagent concentration will alter the two rates identically, thus having no effect on
the selectivity. If, however, the reagent appears to a
different order in the two equations, very dramatic changes

in selectivity can be achieved by varying the reagent concentration. Two substrates may undergo substitution reactions with a common nucleophile but with different rate
equations, as when one of them reacts in a bimolecular
mechanism and the other unimolecularly. Then the selectivity will vary with the nucleophile concentration. Among
the best examples of the dependence of relative rates on
concentration are those provided by reactions involving
H^+ or OH^- in aqueous solution. The formal treatment of
Sec. 3.5 provides a means for selecting the pH optimal for
the assay. Figure 5.2 shows the pH-rate profiles [24] for
the hydrolysis of the acetoxy group in aspirin, 1, and its
methyl ester, 2.

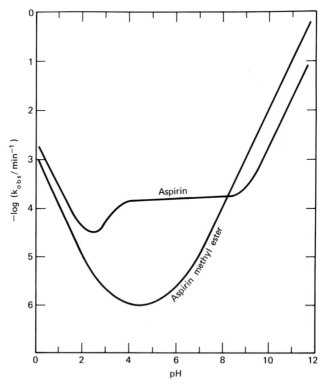

Fig. 5.2. pH-rate profiles for aspirin and aspirin methyl
ester at $25°$ [24].

The relative rate ratio k_1/k_2 is about 160 at pH 4.5 and
0.15 above pH 9. Control of pH can therefore bring about

a reversal in the rate ratio, which can be varied by a factor of 10^3 in this way.

COOH

—OCOCH₃

1

COOCH₃

—OCOCH₃

2

Even when two substrates react with identical rate equations, so that selectivity cannot be altered by changing the reagent concentration, an advantage may be realized from a concentration decrease if the rate constants are very large. It may happen that the selectivity ratio is satisfactory for discrimination between the two substrates, but that the absolute rates of both are so large that it is experimentally difficult to stop the reaction at an appropriate time. Reduction of the reagent concentration will reduce both rates, possibly to useful levels. The lowest possible reagent concentration can be achieved by the generation of reagent in a slow preliminary reaction; the reagent becomes a reactive intermediate, for which the two substrates compete. The result is that the rates of the two substrates remain in the same ratio as if the reagent had been added batch-wise, but both become smaller.

Complex formation has interesting potentialities for modifying selectivity. When a substrate interacts with another molecule (a ligand) through noncovalent bonding interactions, the product, called a complex, may have a reactivity different from the uncomplexed substrate. The complex formation is an equilibrium process. Such a system offers two routes to selectivity control: different equilibrium responses of substrates to the common ligand, and different kinetic responses of the complexed substrates to the common reagent. Usually the reactivity of the substrate is decreased upon complexation; it is not unusual to find reaction essentially completely inhibited. However, the observed decrease in reactivity also depends upon the fraction of substrate in the complexed form. A twofold decrease is readily achieved with many complexes between planar molecules in aqueous solution [25], and order-of-magnitude effects have been observed with inclusion complexes of cycloamyloses [26].

Salt effects on reactivity and selectivity will normally be moderate. Changes in the solvent may, however, lead to major rate effects. One way to consider these, for reactions correlated by the Grunwald-Winstein $\underline{Y}$ solvent parameter, is by comparison of substrate parameters, m. This is analogous to comparing ρ values for sensitivity to substituent effects; that is, m measures the sensitivity of the reaction rate to changes in solvent polarity. If two substrates have different m values in a common solvent series, their selectivity (relative reactivities) will be a function of solvent composition.

5.2   Equilibrium Control

Leveling and Differentiating Effects. The selectivity of an equilibrium process for a substrate S_2 relative to another substrate S_1 can be discussed in terms of the function $\log (K_2/K_1)$, where K_1 and K_2 are the appropriate equilibrium constants. This type of selectivity function occurred in several situations in the preceding discussion of rate control.

　　To concentrate attention let us compare the strengths of a pair of acids as measured by their equilibrium constants for reaction with a common solvent. The possible outcomes (keeping in mind the experimental realities) are two [27]:

　　1. Each acid reacts essentially completely with the solvent. Suppose the strongly acidic solute HX is dissolved in the basic solvent S. If the reaction

$$HX + S \rightleftharpoons SH^+ + X^- \qquad (5.26)$$

"goes to completion"--that is, essentially all HX is transformed into SH^+--the solvent is said to be a leveling solvent for HX. It is evidently impossible to compare the strengths of the two acids HX and HY in a leveling solvent for these acids, for they will be identical in strength. In fact, these acids will be leveled to the strength of the lyonium ion, which is equivalent to saying that the lyonium ion is the strongest acid that can exist in the solvent. A similar statement can be made about the lyate ion as the strongest possible base. It is in this manner that the mineral acids appear to be of identical strength in water; though they are actually of quite different strengths, they are all strong enough to convert water quantitatively to the hydronium ion. Solute bases can behave analogously to acids; thus glacial acetic acid, being fairly acidic, should be a leveling solvent for many bases, and it is observed that all bases that are (in water) stronger than aniline are leveled by acetic acid. They are thus transformed into the strongest base, acetate ion, in this solvent.

    2. The second possibility is that the solute does not react completely with the solvent. Let the acid HA be dissolved in solvent S,

$$HA + S \rightleftharpoons SH^+ + A^- \qquad (5.27)$$

If the reaction does not go completely to the right, the solvent S is called a differentiating solvent for HA. The value of such a system will be appreciated by considering the second acid HB for which S is also a differentiating solvent. The extent of reactions between HA and the solvent and between HB and the solvent is characterized by the equilibrium constants K_{HA} and K_{HB}, respectively. These constants may be taken as measures of the acid strengths of HA and HB relative to the reference base S. This is the procedure adopted in measuring acid strengths in water, where $S \equiv H_2O$ and the constants are the conventional acid dissociation constants. Mineral acids are differentiated by glacial acetic acid because of its low basicity [28].

    These ideas form the basis for acid-base titrimetry in nonaqueous solvents [29]. Acidity and basicity can be greatly enhanced with leveling solvents, which permit the determination of total acidity or basicity in a sample. In a leveling solvent, however, the selectivity is zero, so mixtures cannot be differentially titrated. For this purpose a differentiating solvent must be chosen. For acids the solvent should be nonbasic, and for bases it should be nonacidic. Aprotic solvents are good differentiating solvents.

    These effects of solvents are readily seen in potentiometric titration curves. Figure 5.3 shows such curves for the titration of the weak base antipyrine, 3 (pK_a 1.5 in water), in water and in glacial acetic acid [30]. Clearly the analysis is impossible in water and is feasible

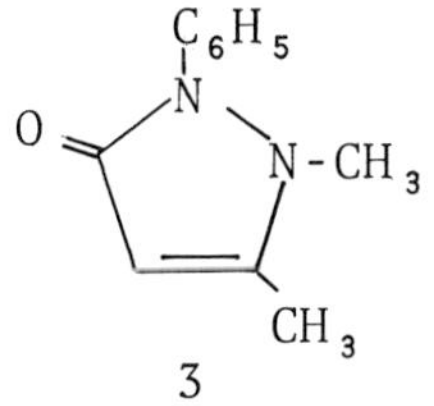

$$\underset{\underline{3}}{}$$

in acetic acid. Figure 5.4 shows a similar experiment with the weaker base caffeine, 4 (pK_a 1 in water), in

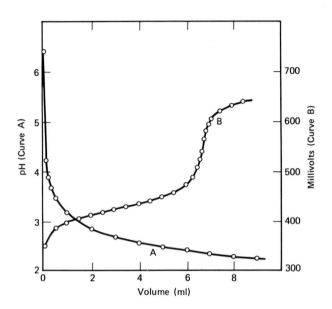

Fig. 5.3. Potentiometric titration of identical samples of antipyrine. Curve A: titration in water with 0.1 $\underline{N}$ hydrochloric acid; curve B: titration in glacial acetic acid with 0.1 $\underline{N}$ acetous perchloric acid.

4

the solvents acetic acid and acetic anhydride [30]. A differentiating titration of picric acid, 2,4-dinitro-phenol, 2-nitrophenol, and phenol (aqueous pK_a's 0.6, 4.1, 7.2, and 10.0, respectively), in tert-butyl alcohol solvent, is given in Fig. 5.5 [31].

Crabb and Critchfield [32 developed two measures of solvent differentiating power for potentiometric titra-tions: (1) the slope of a plot of ΔHNP against pK_a (H$_2$O) for a series of acids in the solvent (ΔHNP is the half-

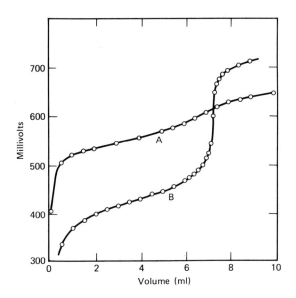

Fig. 5.4. Potentiometric titration of identical samples of caffeine with 0.1 N acetous perchloric acid. Curve A: titration in glacial acetic acid; Curve B: titration in acetic anhydride.

neutralization potential relative to a reference acid). A larger slope is associated with a greater differentiating power; (2) the change in potential, ΔE, through the first end point in the titration of a mixture of p-nitrophenol, pK_a (H_2O) = 7.14, and m-chlorophenol, pK_a (H_2O) = 8.85. A larger value of ΔE is interpreted as indicative of a greater power of differentiation. Table 5.II lists these measures of selectivity [32]. The two measures correlate fairly well. A rough correlation of differentiating power with dielectric constant is apparent. The selectivity of pyridine, which is much more basic than water, is surprising; this may be partly a consequence of its low dielectric constant, because the extent of the titration reaction (salt formation) is controlled not only by the strengths of the acid and base, but also by the dissociation constant of the salt formed. Small differences in this quantity might be revealed in a low dielectric constant solvent. Pyridine might also have a specific solvation effect through its ability to form complexes with the sample acids.

The theory of differentiating acid-base titrations

has been most thoroughly treated for aqueous solution;
many descriptions are available [33].

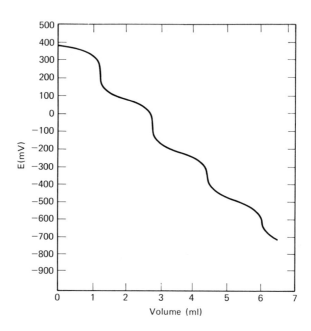

Fig. 5.5. Titration of picric acid, 2,4-dinitrophenol,
2-nitrophenol, and phenol in tert-butyl alcohol with
tetrabutylammonium hydroxide [31].

The leveling-differentiating concept provides an
equilibrium analogy to the selectivity-reactivity relationships among rate phenomena. In very rough terms, the
larger the equilibrium constants for a pair of substrates
in a common reaction, the smaller the differentiating
power of the process for the two substrates. In the same
way that Hammett ρ values measure sensitivity to structural
changes for rate processes, so they may be interpreted for
equilibria. An inverse relationship is expected between
ρ and log k_0 for closely related series. Table 4.III contains examples, as for the ionization of substituted benzoic acids in water (ρ = 1.000, log k_0 = -4.203), methanol
(ρ = 1.537, log k_0 = -6.514), and ethanol (ρ = 1.957, log
k_0 = -7.206); and for the ionization of phenols in water,
ρ = 2.113, log k_0 = -9.847.

Equilibrium limitations manifest themselves very
clearly in reactions that involve the making or breaking

of covalent bonds [34]. The equilibrium constants of closely related substrates undergoing a common reaction seldom differ by orders of magnitude (unlike the more favorable acid-base reactions). Interference is therefore a common phenomenon.

Table 5.II Differentiating Power of Solvents for Potentiometric Acid Titrations [32]

Solvent	Slope[a]/mV-pK$_a$$^{-1}$	ΔE[b]/mV	Dielectric constant
tert-Butyl alcohol	157	95	11.5
Pyridine	137	72	12.3
Tetrahydrofuran	137	64	7.4
Isopropyl alcohol	130	64	18.3
Dimethylsulfoxide	125	56	46.7
Acetone	100	36	20.7
Acetonitrile	87	22	37.5
Ethylene glycol	62	--	37.0
Water	57	0	79.5

[a]Slope of the ΔHNP versus pK$_a$ (H$_2$O) plot.
[b]Potential break at the first end point in the titration of p-nitrophenol and m-chlorophenol.

If a substrate can undergo parallel reactions to give more than one product, the product distribution is evidence of another kind of selectivity. If the reactions are not significantly reversible, the product ratio will be kinetically controlled (see Sec. 5.1), but if the reactions are reversible it will be thermodynamically (i.e., equilibrium) controlled [35]. Yet another type of selectivity is that in which the products may differ in their stereochemistry, as when a reaction may produce optical isomers, or cis-trans, syn-anti, axial-equatorial, etc., pairs. The product distribution is a function of the stereoselectivity of the process. If only one of the possible isomers is formed, the process is said to be stereospecific [36].

Attainment of Selectivity. One way to achieve analytical selectivity for samples containing two or more components is somehow to alter the equilibrium constants for the components so as to achieve a dispersion of values. A very effective way to do this for acid-base titrations is

through choice of solvent, as discussed above. Temperature of course affects equilibrium constants, which (unlike rate constants) may either increase or decrease, thus
affording possibilities to alter selectivity in special
circumstances.

If the components of the sample do not respond to an
analytical process with the requisite selectivity, possibly some derivative will. This approach can be utilized
in combinations of chemical reactions with separation
techniques. Thus a cation exchange column in the iron
(III) form will retain antipyrine by complex formation,
whereas caffeine passes through the column [37]. Cinnamic
acid, 4-methoxycinnamic acid, and 3,4-dimethoxycinnamic
acid are separated by liquid-liquid partition chromatography on a column in which 7-(2,3-dihydroxypropyl)-
theophylline is incorporated in the internal phase; the
three acids form complexes with this xanthine to different extents, and the greater the complex stability constant the more the acid is retarded on the column [38].

Conventional functional group methods may incorporate
derivatization to alter equilibrium constants. Often
multiple analyses, with one analytical method per component, are used to analyze mixtures. Such approaches take
advantage of equilibrium differences. A mixture of an
alcohol and an amine can be titrated with an acid to give
the amount of amine. An acetylation reaction on a separate portion of the sample will yield the total composition, and the alcohol is found by difference.

A mixture of an aliphatic primary and an aliphatic
secondary amine can be titrated with acid to give the sum
of the two. A separate portion can be treated with an
aldehyde to form the Schiff base of the primary amine.

$$RNH_2 + R'CHO \rightleftharpoons R'CH = NHR + H_2O$$

The secondary amine will not form a Schiff base, and can
be titrated without interference. In effect the equilibrium constant for the titration reaction of the primary
amine has been drastically altered by its conversion to
the Schiff base, which is a very weak base.

The effective pK_a of boric acid can be varied 5 units
simply by adding mannitol in varying concentrations [39].
This is a very specific effect.

An interesting analogy can be made between conventional equilibrium-controlled functional group analytical
methods (for example, the determination of carbonyl compounds by oxime formation), Eq. (5.28), and separation
techniques based on phase distribution, Eq. (5.29).

$$Sample + reagent \rightleftharpoons products \qquad (5.28)$$

Solute in phase A $\rightleftharpoons$ solute in phase B   (5.29)

For Eq. (5.29) the equilibrium constant is called the
partition coefficient or distribution coefficient. The
usual analytical reaction is carried out as a batch process, in exactly the way that liquid-liquid extraction in
a separatory funnel is a batch process. The two opera-
tions are equilibrium limited in the same way. Separation
techniques have achieved high resolution through small
differences in selectivity (ratios of partition coefficients) by means of cascade processes, such as countercurrent distribution and partition chromatography. If the
concept of the cascade process could be applied to chemical equilibria, great selectivity should be possible. The
procedure requires physical separation of products and
reactants (if the analogy is extended), and this is generally difficult. An application of the principle has
been made by Kemp et al, [40] to analyze the racemic
content of enantiomers. Successive isotopic dilution,
in a cascade manner, based on separation of enantiomers,
achieves a sensitivity in the range 0.001-1% of racemate.

5.3   The Final Measurement

It may happen that adequate selectivity has been achieved
in the reaction component of the analysis, but the resolution of the final observation may degrade the overall
selectivity. For example, hydrochloric acid and sulfuric
acid are differentiated by glacial acetic acid, but when
they are titrated (with an alkali acetate titrant) potentiometrically, the mixture is not resolved. A conducto-
metric titration, however, yields three end points, one
corresponding to hydrochloric acid and two of them to the
two protons of sulfuric acid [41].
   Each measurement technique has several variables
under the analyst's control, and these are potential
sources of selectivity. Titrimetry offers several types
of end-point detection, each with its characteristic
capabilities. Spectrophotometry's variables are solvent
composition, pH, and wavelength. The 5,5-disubstituted
barbituric acids, 5, are dibasic acids, with pK_1 ~ 8 and
pK_2 ~ 12. Upon ionization their ultraviolet spectra
undergo marked changes, as shown in Fig. 5.6 [42]. Suitable combination of pH and wavelength can avoid some
interferences.
   Control of pH to produce a conjugate acid or base is
a simple example of a major analytical technique, namely,
derivatization. One of the goals of forming derivatives
is to overcome interferences. Spectrophotometry offers

many examples, most of them fitting the pattern of a
colorless compound (that is, with no absorption in the
visible range) that is converted to a derivative absorbing
visible light.

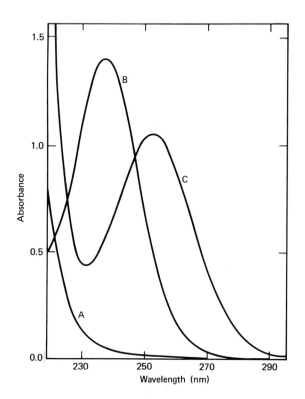

Fig. 5.6.   Ultraviolet absorption spectra of barbital.
A: in 0.1 N sulfuric acid (un-ionized form); B: in pH
9.9 buffer (singly ionized form); C: in 1 N sodium hydroxide (doubly ionized form).   The concentration is
2.50 mg per 100 ml.

The power of gas chromatography as a tool for identification can be increased simply by subjecting the column
effluent to simple functional group colored derivative
tests.   In this way individual peaks can be assigned to
functional classes, and it is even possible to show that
some peaks correspond to unresolved mixtures of different
classes of compounds [43].
Instead of generating a functional group with stronger absorption properties, as in the common derivatization

procedures of spectrophotometry, it may be profitable to eliminate the absorption of a chromophore. This will usually involve the saturation of a double or triple bond, and therefore addition or reduction reactions will be necessary. An excellent example is afforded by a method to identify and determine aldehydes through acid-catalyzed acetal formation [44].

$$R-\overset{O}{\overset{\|}{C}}H + 2CH_3OH \overset{H^+}{\rightleftharpoons} R-\overset{OCH_3}{\overset{|}{\underset{|}{C}}}H + H_2O \qquad (5.30)$$

Destruction of the carbonyl chromophore leads to profound spectral changes. The decrease in absorption upon acetal formation is proportional to the aldehyde concentration. By measuring the change in absorption, we eliminate most interference. Figure 12.2 illustrates this method.

Problems

1. Express S_f [Eq. (5.17)] in terms of isomer distributions.

   Answer: $S_f = \log \dfrac{2 \times (\% \text{ para})}{(\% \text{ meta})}$

2. Derive Eq. (5.4) and obtain an explicit expression for the constant term.

3. The rate constants for iodination of acetone and methyl ethyl ketone are 1.68×10^{-2} sec^{-1} and 9.0×10^{-3} sec^{-1}, respectively, at 25°. For a determination of acetone based on this reaction, estimate the apparent percent recovery expected if the acetone sample contains 10% (w/w) methyl ethyl ketone and the reaction time is 5 min.

   Answer: 107.7%.

4. What is the maximum percentage concentration of o-methyl ethyl benzoate permissible in a sample of m-methyl ethyl benzoate whose purity is to be determined by saponification, if the maximum allowable error is 1%? The relative rate of alkaline hydrolysis is 0.18 (ortho/meta).

   Answer: 1.74% (i.e., $[I]_0/[S]_0 = 0.0178$ when $x = 0.01$).

References

1. For our purposes selectivity and specificity are
    synonymous. Some authors speak of absolute specifi-
    city, meaning (usually in connection with enzyme cata-
    lysis) that a process occurs with a single substrate
    only; and relative specificity, implying a range of
    responses with different substrates.
2. J. C. Giddings, Dynamics of Chromatography, Part I.
    Principles and Theory, Dekker, New York, 1965.
3. L. C. Craig, Advan. Anal. Chem. Instr., 4, 35 (1965).
4. J. E. Leffler and E. Grunwald, Rates and Equilibria
    of Organic Reactions, Wiley, New York, pp. 162-168.
5. Since the designation of one reactant as the substrate
    and the other as the reagent is arbitrary, a similar
    statement could be made about a pair of reagents re-
    acting with a common substrate.
6. M. L. Bender, Chem. Revs., 60, 53 (1960).
7. S. I. Miller, J. Am. Chem. Soc., 81, 101 (1959).
8. W. P. Jencks, Catalysis in Chemistry and Enzymology,
    McGraw-Hill, New York, 1969, pp. 195-198.
9. L. M. Stock and H. C. Brown, Advan. Phys. Org. Chem.,
    1, 35 (1963), and references therein.
10. L. M. Stock, Aromatic Substitution Reactions, Prentice-
    Hall, Englewood Cliffs, N. J., 1968, p. 45.
11. C. K. Ingold, Structure and Mechanism in Organic
    Chemistry, Cornell Univ. Press, Ithaca, N. Y., 1953,
    pp. 245-247.
12. H. C. Brown and C. R. Smoot, J. Am. Chem. Soc., 78,
    6255 (1956).
13. If Hammett equations applied to the individual varia-
    bles, these would be: $\log p_f^{Me} = \rho\sigma_{p-Me}$ and $\log$
    $(p_f^{Me}/m_f^{Me}) = \rho(\sigma_{p-Me} - \sigma_{m-Me})$, giving

    $$c_{Me} = \frac{\sigma_{p-Me}}{\sigma_{p-Me} - \sigma_{m-Me}}$$

    The experimental value of c_{Me}, from the plot of
    Eq. (5.18), is 1.325; the value calculated with Ham-
    mett sigma constants and the above equation is 1.68.
    It was this discrepancy that led to the introduction
    of σ^+ values for electrophilic reactions (see Sec.
    4.2).
14. J. S. Fritz and G. S. Hammond, Quantitative Organic
    Analysis, Wiley, New York, 1957, p. 150; H. B. Mark,
    Jr. and G. A. Rechnitz, Kinetics in Analytical Chem-
    istry, Wiley (Interscience), New York, 1968, pp. 72-
    74.
15. T. C. Bruice and S. J. Benkovic, Bioorganic Mechanisms

Vol. I, Benjamin, New York, 1966, pp. 271-275.

16. W. P. Jencks, S. Cordes, and J. Carriuolo, J. Biol. Chem., 235, 3608 (1960).

17. H. C. Brown and N. M. Yoon, J. Am. Chem. Soc., 88, 1464 (1966).

18. H. C. Brown, P. Heim, and N. M. Yoon, J. Am. Chem. Soc., 1637 (1970).

19. The best entry into the literature of organic synthetic reagents is through M. Fieser and L. Fieser, Reagents for Organic Synthesis, Vol. (1967), Vol. 2 (1969), Wiley, New York.

20. O. Exner, Nature, 201, 488B (1964). These plots are discussed in Sec. 4.1.

21. Some reactions undergo large rate accelerations in ice, relative to the rate at the same temperature in water [22]. This phenomenon appears to be a concentration effect, the reactants being concentrated in a liquid phase. These observations suggest some analytical possibilities.

22. R. E. Pincock and T. E. Kiovsky, J. Am. Chem. Soc., 88, 4455 (1966); T. E. Kiovsky and R. E. Pincock, ibid., 88, 4704 (1966).

23. The alkali is determined before and after reaction with the sample, so an accurate determination requires that a substantial change occur in the alkali concentration; this violates the pseudo-first-order condition.

24. T. St. Pierre and W. P. Jencks, J. Am. Chem. Soc., 90, 3817 (1968).

25. P. A. Kramer and K. A. Connors, J. Am. Chem. Soc., 91, 2600 (1969).

26. M. L. Bender, Mechanisms of Homogeneous Catalysis from Protons to Proteins, Wiley (Interscience), New York, 1971, pp. 373-382.

27. Throughout this section it is assumed that rate limitations are absent or are allowed for.

28. I. M. Kolthoff and A. William, J. Am. Chem. Soc., 56, 1007 (1934).

29. W. Huber, Titrations in Nonaqueous Solvents, Academic, New York, 1967.

30. K. A. Connors, A Textbook of Pharmaceutical Analysis, Wiley, New York, 1967, Chap. 2.

31. J. S. Fritz and L. W. Marple, Anal. Chem., 34, 921 (1962).

32. N. T. Crabb and F. E. Critchfield, Talanta, 10, 271 (1963).

33. S. Bruckenstein and I. M. Kolthoff, Treatise on Analytical Chemistry (I. M. Kolthoff and P. J. Elving, eds.), Part I, Vol. 1, Wiley (Interscience), New York, 1959, Chap. 12; J. E. Ricci, Hydrogen Ion

Concentration, Princeton Univ. Press, New Jersey, 1952; H. Freiser and Q. Fernando, Ionic Equilibria in Analytical Chemistry, Wiley, New York, 1963; J. N. Butler, Ionic Equilibrium: A Mathematical Approach, Addison-Wesley, Reading, Massachusetts, 1964; E. Bishop, Comprehensive Analytical Chemistry (C. L. Wilson and D. W. Wilson, eds.), Vol. 1B, Elsevier, Amsterdam, 1960, Chap. VII.2.

34. These are often relatively slow reactions, so the reagent is used in excess to decrease analysis time. Thus even when equilibrium constants are large, direct titration is not always feasible.

35. E. S. Gould, Mechanism and Structure in Organic Chemistry, Holt, Rinehart and Winston, New York, 1959, p. 172.

36. For a review see S. I. Miller, Advan. Phys. Org. Chem., $\underline{6}$, 185 (1968).

37. E. Sjöström and L. Nykänen, J. Am. Pharm. Assoc., $\underline{47}$, 248 (1958).

38. E. Sondheimer and I. E. Pollak, J. Chromatogr., $\underline{8}$, 413 (1962).

39. G. Marinenko and C. E. Champion, Anal. Chem., $\underline{41}$, 1208 (1969).

40. D. S. Kemp, S. W. Wang, G. Busby III, and G. Hugel, J. Am. Chem. Soc., $\underline{92}$, 1043 (1970).

41. T. Higuchi and C. R. Rehm, Anal. Chem., $\underline{27}$, 408 (1955).

42. K. A. Connors, Pharmaceutical Analysis (T. Higuchi and E. Brochmann-Hanssen, eds.), Wiley (Interscience), New York, 1961.

43. C. Merritt, Jr., Ancillary Techniques of Gas Chromatography (L. S. Ettre and W. H. McFadden, eds.), Wiley (Interscience), New York, 1969.

44. E. P. Crowell, W. A. Powell, and C. J. Varsel, Anal. Chem., $\underline{35}$, 184 (1963).

Chapter 6. SENSITIVITY

In the most general formulation all analytical methods
involve the measurement of some property x and its relationship to a concentration c; the equation relating
these is

$$x = f(c) \tag{6.1}$$

For all methods the sensitivity is defined by

$$\text{sensitivity} = \frac{dx}{dc} \tag{6.2}$$

The magnitude and interpretation of sensitivity depend
upon the nature of x, the units of c, and the uses to be
made of the information.

6.1 Sensitivity Limitations

Limits of Detection. The application of statistical methods in interpreting analytical data, evaluating errors,
and designing experiments is now well accepted [1]. Sta-
tistics enters any consideration of sensitivity limits,
though in a relatively simple way. For convenience the
required formulas are given here. In a series of n re-
plicate determinations of a property x, several statistical parameters are useful. The average or mean, $\bar{x}$, is
the first moment, defined by

$$\bar{x} = \frac{1}{n} \sum_{i=1}^{n} x_i \tag{6.3}$$

The second moment about the mean is the variance, s^2.

$$s^2 = \frac{1}{n-1} \sum_{i=1}^{n} (x_i - \bar{x})^2 \tag{6.4}$$

The standard deviation, s, is the square root of the

191

variance; s is an experimental estimate of a hypothetical quantity σ, which is a measure of the spread or scatter of all members constituting the population of which the n analytical determinations constitute a sample [2]. The standard deviation of the mean, $s_{\overline{x}}$, is given by Eq. (6.5); evidently the distribution function characterizing the means has less scatter than the function describing the individual observations.

$$s_{\overline{x}} = \frac{s}{\sqrt{n}} \qquad (6.5)$$

Relative standard deviations can be expressed on a fractional basis, $s/\overline{x}$; as percent, $10^2 \, s/\overline{x}$; parts per thousand, $10^3 \, s/\overline{x}$; or parts per million, $10^6 \, s/\overline{x}$.

If the variances of the separate components of a complete analytical method are independent, the overall variance of the method is given by the sum of the contributing variances. Suppose, for example, that a spectrophotometric analysis is characterized by independent variances in weighing, dilution of solutions, and the spectrophotometric measurement. Then the variance of the analysis is given by

$$s^2 = s_{weighing}^2 + s_{dilution}^2 + s_{measuring}^2 \qquad (6.6)$$

It follows that the standard deviation s is not an additive function of the contributing standard deviations. Equation (6.6) applies also to fractional standard deviations. It is clear that a significant decrease in s^2 can only be achieved by reducing the value of its principal component.

Specification of a minimum level of detection for an analytical method is a statistical problem relating to the signal-to-noise ratio of the method. The observation x may have a contribution from the substance to be determined as well as a contribution--the analytical noise-- from random perturbations of many sources. Such noise may arise from vibrations of the balance table, light source intensity fluctuations, reagent variability, uncertainties in time, volume, or other readings, etc. The magnitude of the analytical noise should be evaluated for each analytical method by measuring the mean $\overline{x}_b$ and standard deviation of the mean $s_{\overline{b}}$ of a series of blank analyses. Kaiser [3] recommends that at least 20 blank analyses be made for this purpose.

Now in order for a measured value to be accepted as a real indicator of the presence of the sample substance, x must be some value larger than $\overline{x}_b$. Since $\overline{x}_b$ has associated with it a range of fluctuations described by $s_{\bar{b}}$, clearly x must be larger than $\bar{x}_b$ by some multiple of $s_{\bar{b}}$ before x can be accepted as significantly different from $\bar{x}_b$. Writing x_{min} as this minimum value of x,

$$x_{min} = \bar{x}_b + ks_{\bar{b}} \qquad (6.7)$$

The corresponding concentration is, from Eq. (6.1),

$$x_{min} - \bar{x}_b = f(c_{min}) \qquad (6.8)$$

A suitable value having been given to k, Eqs. (6.7) and (6.8) fix the limiting concentration c_{min} for the analytical method [3,4].

Assuming a particular frequency distribution for x (usually the normal distribution or student's t distribution), and adopting an arbitrary "confidence limit," the value of k can be selected. This very common approach is not sufficiently conservative, however, because the distribution may diverge from the assumed one, and because the estimates of $\bar{x}_b$ and $s_{\bar{b}}$ are themselves approximations. It appears that k = 3 is generally suitable for use in Eq. (6.7); this is an empirical choice but it is reasonably conservative [3]. In the absence of specific knowledge of the frequency distribution, this choice of k should not be interpreted in terms of sharply defined confidence limits.

This development has led to a limit of detection that is the minimum concentration that can be detected. Often the minimum detectable amount is specified. Though simply related, these different detection limits should be kept in mind when evaluating analytical methods. A remarkable number of definitions of detection limits have appeared [5]. To some of these the term sensitivity has been applied, but the use of this word should be consistent with its definition in Eq. (6.2).

A few examples will illustrate the range of detection limits that have been devised. Feigl [6] gives a "limit of identification" in micrograms (µg) of sample. Sandell's [7] sensitivity is the number of micrograms of element, converted to the colored product, that in a column of cross section 1 cm^2 shows an absorbance of 0.001. This is equivalent to the number of micrograms per milliliter required to give an absorbance of 0.001 when the path length is 1 cm, because $µg/cm^2 = µg\text{-}cm/cm^3$. Sawicki et al. [8] define a concentration limit for the colorimetric determination of azulene as the concentration required to give an absorbance of 0.1 in a 1-cm cell; this concentration was expressed in parts per million and was also

converted to minimum amount determinable based on the
volume of solution in the test procedure. Korenman [9]
uses a "limiting concentration" that is a ratio of parts
of solvent per gram of sample; thus the limiting concentration in the colorimetric determination of dichromate
with diphenylcarbazide is 1:5 × 10^7. For gravimetric
analysis, a quantity has been defined as pD = -log (c_{min}/
mg-ml^{-1}), where c_{min} is the limiting concentration that
forms a precipitate [10].

The ultimate sensitivity of the analytical method is
determined by the final measurement technique, which is
obvious, and it can also be a function of the preliminary
treatment, in particular of the chemical reactions that
precede the quantitative measurement. Sometimes the primary purpose of these preliminary steps is to increase the
sensitivity.

The rate and equilibrium laws of chemistry are statistical laws, which means that they apply with high precision only when large numbers of molecules take part.
For ordinary experimental reproducibility to be the limiting factor in observed precision, a sample probably should
contain 10^5 to 10^6 molecules, that is, about 10^{-18} mole.
Below this limit, detection methods based on laws of large
numbers will exhibit erratic response, though qualitative
tests may still be applicable to 10^{-20} mole [11]. Such
limits are not far beyond some widely used methods. For
example, the measurement of pH = 12 in 1 mℓ of solution
represents the determination of 10^{-15} mole of hydronium
ion.

Capabilities of Measurement Techniques. Spectrophotometry
is an important analytical tool and a convenient one with
which to examine sensitivity and detection limits. The
functional relationship written as Eq. (6.1) is Beer's
law,

$$A = \varepsilon bc \qquad\qquad (6.9)$$

where A is absorbance, ε molar absorptivity, b path length
in centimeters, and c concentration in moles per liter.
Then, according to Eq. (6.2), the sensitivity of spectrophotometric analysis is (in units of $\underline{M}^{-1}$),

$$\text{sensitivity} = \frac{dA}{dc} = \varepsilon b \qquad\qquad (6.10)$$

Note that the sensitivity is independent of the spectro-
photometer used (except for the trivial restriction that
the instrument accept the usual range of b values, that
is, cuvets).

The limit of detection in spectrophotometry does

depend upon the particular instrument, as well as upon
the sample and assay method. These dependencies enter
through Eqs. (6.7) and (6.8). The blank analyses required
for the evaluation of $\overline{x}_b$ ($\overline{A}_b$ in the present instance) and
$s_{\overline{b}}$ must be carefully designed and executed to duplicate
as closely as possible the conditions used in the sample
measurement. These conditions include the instrument and
its settings, the solvent, temperature, time, reagents,
and the analytical procedure, only the sample substance
being omitted. Now suppose that in a particular spectrophotometric analysis these values were established for
the blank: $\overline{A}_b$ = 0.021 and $s_{\overline{b}}$ = 0.003. Then by Eq. (6.7)
with k = 3 we find A_{min} = 0.030 as the minimum sample absorbance that can be interpreted as evidence for the
presence of the test substance. The corresponding concentration is obtained with Beer's law if the absorptivity
is known; this quantity c_{min} is clearly a function of the
sensitivity, εb, as well as of the procedural limits
(through $s_{\overline{b}}$). In this example we have $A_{min} - \overline{A}_b$ = 0.009
= $\varepsilon b c_{min}$.
    In qualitative terms, the intensity of electronic
absorption can be classified as follows, where ε_{max} is the
molar absorptivity at the band maximum:

$$\log \varepsilon_{max} =$$

0 - 1	very weak
1 - 2	weak
2 - 3	moderate
3 - 4	strong
4 - 5	very strong

    Braude [12] gave the following theoretical estimate
of the maximum possible value of ε. In order that a photon may be absorbed, it must strike a molecule roughly
within the area defined by the molecular dimensions. The
transition probability P that an electronic transition
will occur is the proportion of photon "hits" that lead
to absorption; P = (No. of photons absorbed)/(No. of
photons striking). Let light of intensity I fall on a
layer of solution of thickness db and unit cross section,
containing c moles per liter of absorbing solute. The
solution is taken to be very dilute so there is no super-
position of molecules in the direction of light propagation. If the average cross-sectional area per molecule
is a square centimeters, the absorbing area in the solution is given by

Absorbing area = area/molecule × total no. of molecules

$$= a \times \text{no. molecules/liter} \times \text{no. liters}$$

$$= a \times cN \times \frac{(db/cm)(1/cm^2)}{1000/cm^3 - liter^{-1}}$$

$$= \frac{acNdb}{1000} \qquad\qquad (6.11)$$

The relative loss in intensity, $-dI/I$, is equal to the ratio (absorbing area)/(total area) times the transition probability. Upon integration over the path length b,

$$\log \frac{I_0}{I} = \frac{PacNb}{2.303} \qquad\qquad (6.12)$$

Comparison with Beer's law, $\log (I_0/I) = \varepsilon bc$, gives

$$\varepsilon = \frac{PaN}{2.303} \qquad\qquad (6.13)$$

The maximum value of P is unity, and a is roughly 10 Å^2 ($10^{-15}cm^2$). The maximum value of ε is therefore about 3×10^5 liter/mole-cm. Braude introduces a statistical factor of one-third to account for absorption being maximal only for a particular orientation.

This calculation is roughly consistent with the largest recorded ε values, examples of which are shown in Table 6.I. The preceding treatment of detection limits, together with these maximum absorptivities, allows an estimate of the minimum concentration determinable by spectrophotometry. Cell compartments routinely accept cuvets with path lengths of 0.1 to 10 cm, providing 100-fold adjustment in the sensitivity. Suppose the earlier numerical example concerned a solute for which $\varepsilon = 1 \times 10^5$, which is very large but accessible. The minimum significant absorbance, corrected for the blank, was 0.009. Then in a 10-cm cell the limiting concentration is $c_{min} = 0.009/(10)(10^5) = 9 \times 10^{-9}M$. The minimum amount determinable depends upon the volume required; this may be 30 ml in a 10-cm cell, giving 2.7×10^{-4} μmole as the limit of detection. These limits can seldom be attained in practice because ε is usually in the approximate range 10^3 to 10^4.

The product εb has been used here as the measure of spectrophotometric sensitivity. Sandell's [7] sensitivity measure, described earlier, is closely related. ε is measured in units of liters/mole-centimeter, or 1000 cm^2/mole; if a is the absorptivity in units of cm^2/mole, then $a = 1000\varepsilon$. The Sandell sensitivity is the concentration

in microgram per milliliter required to give an absorbance
of 0.001 in a 1-cm cell. Converting a to the units square
centimeters per microgram gives a = $\varepsilon/10^3 MW$, where MW is
the solute molecular weight. Applying Beer's law with
A = 0.001, b = 1 cm, and c_s = Sandell sensitivity, then
c_s = 0.001/ab, or

$$\text{Sandell sensitivity } c_s = \frac{MW}{\varepsilon} \qquad (6.14)$$

This relationship is convenient for the interconversion
of sensitivities on weight and molar bases.

Table 6.I   Examples of High-Intensity Electronic Absorption

Compound	λ_{max}/nm	$\log \varepsilon_{max}$	Ref.
Naphthalene	220	5.05	[12]
Pentacene	310	5.45	[12]
Dilithium α,β,-γ,δ- tetraphenylporphine	440[a]	5.79[a]	b
Diethyl-tetraacetylene	226, 234	5.30, 5.45	c
Copper (II) diphenyl- carbohydrazide	495	5.20	d
Azulene coupling product with 4-azobenzene diazonium ion	620	5.04	[8]

[a]Read from graph.
[b]G. D. Dorough, J. R. Miller, and F. M. Huennekins,
J. Am. Chem. Soc., 73, 4315 (1951).
[c]J. B. Armitage, E. R. H. Jones, and M. C. Whiting,
J. Chem. Soc., 1952, p. 2014.
[d]R. W. Turkington and F. M. Tracy, Anal. Chem., 30,
1699 (1958).

Other physical techniques can be described in much
the same way as in this treatment of spectrophotometry.
Sensitivity can always be defined by Eq. (6.2), though
the result may not be as simple as for spectrophotometry.
If the analytical signal is not linear in concentration,
the sensitivity will be a function of concentration.
Polarimetry is an example of such a method; see Problems
2 and 4 for other examples. The expression for sensitivity may include an instrumental variable, as in fluorimetry. At very low concentrations the fluorescence

intensity is given by $F = 2.303\ I_0\ \Phi bc$, where I_0 is the incident intensity and Φ is the quantum yield. The sensitivity is therefore $2.303\ I_0\Phi\varepsilon b$, which depends upon the incident intensity [13]. This is a major difference between fluorimetry and absorption spectroscopy. Fluorimetry provides maximal sensitivities approximately 2 to 3 orders of magnitude greater than in absorption spectroscopy.

Gas chromatographic detector sensitivity can be expressed in terms of output potential per unit concentration of solute in the effluent [14]. This parameter is defined as the output in millivolts per microgram of solute per cubic centimeter of carrier gas; its units are written $(mV - cm^3)/\mu g$. Typical sensitivities of thermal conductivity cells are $1 - 10(mV - cm^3)/\mu g$ and for flame ionization detectors $1000(mV - cm^3)/\mu g$ [15]. Many special-purpose detectors have been combined with gas chromatographs, and the range in sensitivities is enormous [16]. Odor seems to be one of the most sensitive properties, though its quantitative description is difficult [16,17].

Titrimetric analysis can be discussed in the same terms. Let the titrant normality and volume be N_t and V_t, respectively, and let the corresponding sample quantities be N_s and V_s. Then the equation relating the analytical signal V_t to the unknown concentration N_s is $V_t = (V_s/N_t)N_s$. The sensitivity is therefore given by V_s/N_t. Of course there are experimental difficulties in the way of enhancing the sensitivity of an assay on this basis; thus reduction of N_t ultimately is limited by uncertainties in the endpoint detection. The physical significance of V_s/N_t as the sensitivity of titrimetry is that an increase in this ratio results in a larger signal (titrant volume) per unit change in sample concentration.

Titrimetry can be remarkably sensitive. Belcher has developed titration procedures for many standard analyses on the microgram level [18]. Typical sample weights are in the range 30-50 μg, which may correspond to a few tenths of a micromole. Titrations can be performed with syringe burets of total capacity 0.5 ml or less [19], readings being made to 0.1 μliter. Robinson et al. [20] have carried out microtitrations by a serial dilution technique, using a transfer loop and a spot plate instead of a buret and titration flask.

6.2   Chemical Amplification

Concentrating Solutions.   Whether an analytical method requires a larger amount of solute or a larger concentra-

tion than is provided by a solution sample, a simple method for amplifying the analytical signal is available:
increase the solute concentration by decreasing the quantity of solvent. There are several ways in which this
can be accomplished [21]. All of them are essentially
separation techniques.

Evaporation of some of the solvent will increase the
solute concentration. Obviously the solute must be essentially nonvolatile and stable under the conditions of the
evaporation. This process is simple a distillation, and
the principles are well known. The experimental variables
are pressure, temperature, and time. It is sometimes
helpful to recall that the vapor pressure of a liquid is
related to the temperature by

$$\log p = -\frac{\Delta H_V}{2.3RT} + \text{constant} \qquad (6.15)$$

where ΔH_V is the heat of vaporization. For approximate
purposes ΔH_V may be considered independent of temperature.
If the boiling points of a liquid are known at two or more
pressures, a plot of log p against 1/T will allow an estimate to be made of the boiling point at other pressures.
This information can be useful in selecting the distillation conditions, for it may be necessary to use a reduced
pressure, and therefore a low temperature, to avoid solute
decomposition.

Simple evaporation of solvent will concentrate all
solute species, including interfering substances. Sometimes the solute can be distilled from a solution; if all
of the solute but not all of the solvent distills, a concentration has been achieved, and removal from interferences also may be possible. The removal of ammonia by
steam distillation in the Kjeldahl procedure is an example.
The microdiffusion analysis method is related [22].

Liquid-liquid extraction can be an effective concentrating method; the solute is quantitatively partitioned
into a smaller volume of a second solvent immiscible with
the first. Let c_u and c_ℓ be equilibrium concentrations
of solute in the upper and lower phases, and let the
volumes of these phases be V_u and V_ℓ. Then the partition
coefficient is $K = c_u/c_\ell$ and the volume ratio is $U = V_u/V_\ell$. Defining p as the fraction of solute in the upper
phase at equilibrium, it follows [23] that $p = KU/(1 + KU)$.
The fraction q in the lower phase must then be $q = 1/(1 + KU)$. Rearranging the expression for p,

$$KU = \frac{p}{1 - p} \qquad (6.16)$$

Suppose the concentration is to be accomplished in a single extraction. The limitations are easily obtained with Eq. (6.16). We suppose, for the sake of the example, that the initial dilute solution serves as the lower phase, and we require that 99.9% of the solute be transferred to the upper phase in a single extraction; thus p = 0.999 and KU = 999. If we wished to effect a 100-fold concentration (U = 0.01), K would have to be about 10^5. In this way the feasibility of the partitioning can be estimated [24]. If the solute cannot be quantitatively extracted in a single stage, the number of extractions, n, required to achieve any desired degree of extraction can be estimated with [23]

$$\text{Total fraction extracted} = 1 - q^n \qquad (6.17)$$

Chromatography has been applied to concentrate solutions. The ion exchange method allows the ionic solutes in a large volume of dilute solution to be retained on a column. The solute species are then displaced with a smaller volume of a powerful eluting agent. This concentrating process is achieved in a different way for macromolecules by molecular sieves or gels whose pores admit small molecules but exclude macromolecules. The solvent can be selectively absorbed by such gels, leaving a more concentrated solution of the macromolecular solute.

Precipitation and coprecipitation are concentration techniques widely practiced in inorganic analysis but seldom used for organic analysis. Simple precipitation may be ineffective in very dilute solutions because of the appreciable solubility of the precipitate. Coprecipitation procedures utilize an added component that is precipitated along with the sample solute; the sample precipitate, being contained with the carrier precipitate through mixed crystals or adsorption, is then readily handled [21].

An unusual method for intensifying electronic absorption spectra achieves a five-fold increase in absorption by cytochromes simply by quick freezing in aqueous solution [25]. Greater amplification was obtained by forcing the aqueous solution from a syringe into liquid nitrogen, forming small ice pellets. These were homogenized to give a fine snow, which was tamped into a cuvet for spectral observation. A direct 20-fold intensification was observed which, corrected to the melted volume, was equivalent to a 45-fold amplification. This increase is more profound than a mere concentration effect.

Derivative Formation. A major means for signal amplification is by chemical modification, the formation of derivatives. This is a frequent basis for spectrophotometric analysis in which the spectral measurement is preceded by a chemical reaction. The production of a light-absorbing derivative is not always motivated by a need for sensitivity (selectivity and ease of measurement are other goals), but increased sensitivity is often a result. There are many examples [26].

When a functional group is derivatized prior to spectrophotometric observation, its fate in the reaction has a bearing on the practical analysis. Two extreme possibilities arise: either (1) the functional group is sufficiently modified in the reaction that it becomes part of a new chromophore or (2) a chromophore from the reagent is simply attached to the function group, which does not itself constitute an important part of the chromophore. In the first instance it is seldom necessary to remove the unreacted (excess) reagent, because a new chromophore is created. The second route involves the addition of a spectral "tag" to the molecule, which may then have to be separated from the excess reagent, which is spectroscopically very similar. Consider, as examples, some amine determinations. The first class includes the well-known diazotization and coupling method for primary aromatic amines, which is discussed in detail in Chap. 7; this reaction sequence converts the amino group into part of the strongly absorbing azo group.

$$\text{Ar-NH}_2 + \text{HNO}_2 \xrightarrow{\text{HC}\ell} \text{Ar-N}_2^+\text{C}\ell^- + 2\text{H}_2\text{O} \qquad (6.18)$$

$$\text{Ar-N}_2^+ + \text{C}_6\text{H}_5\text{-R} \rightarrow \text{Ar-N=N-C}_6\text{H}_4\text{-R} + \text{H}^+ \qquad (6.19)$$

The excess of coupling reagent, C_6H_5R in Eq. (6.19), does not interfere in the subsequent measurement of the diazo compound.

The second class of derivatization is exemplified by the acylation of aliphatic amines with trans-cinnamic anhydride [27].

$$\text{R-NH}_2 + (\text{C}_6\text{H}_5\text{CH=CHCO})_2\text{O} \rightarrow \text{C}_6\text{H}_5\text{CH=CHCONHR} + \text{C}_6\text{H}_5\text{CH=CHCOOH}$$

$$(6.20)$$

Absorption by the excess anhydride and by the cinnamic acid interferes with the cinnamamide absorption, so a separation is necessary before the spectrophotometric measurement [28].

Simple ionization processes may be considered derivatizations, and the spectral consequences may be dramatic. Figure 5.6 shows an example. A less clear-cut, though also profound, change can be observed in the N-ethylmaleimide (NEM) method for thiols, which proceeds in neutral aqueous solution.

$$
\text{R-SH} + \begin{array}{c} \text{CH} - \text{C} \overset{\displaystyle O}{\diagup} \\ \| \qquad\qquad \diagdown \\ \qquad\qquad\qquad \text{N-C}_2\text{H}_5 \\ \diagup \\ \text{CH} - \text{C} \underset{\displaystyle O}{\diagdown} \end{array} \rightarrow \begin{array}{c} \text{R-S-CH} - \text{C} \overset{\displaystyle O}{\diagup} \\ | \qquad\qquad \diagdown \\ \qquad\qquad\qquad \text{N-C}_2\text{H}_5 \\ \diagup \\ \text{CH}_2 - \text{C} \underset{\displaystyle O}{\diagdown} \end{array} \qquad (6.21)
$$

Each mole of thiol results in the loss of one mole of NEM, with a corresponding decrease in absorbance; λ_{max} for NEM is 300 nm, with $\varepsilon_{max} = 6.20 \times 10^2$. The method is therefore not highly sensitive, being applicable to minimum concentrations of about 10^{-4}M [29]. By carrying out the reaction in isopropyl alcohol and then making the solution alkaline, an intense red color is developed, presumably from an ionized form of the addition product. This modification extends the limiting concentration to about 10^{-7}M [30].

Noncovalent interactions (molecular complex formation) can lead to spectral changes of analytical utility, as in the determination of tertiary aromatic amines by complex formation with tetracyanoethylene, 1 (TCNE) [31].

$$
\begin{array}{c} \text{N} \equiv \text{C} \diagdown \qquad\qquad \diagup \text{C} \equiv \text{N} \\ \qquad\qquad \text{C} = \text{C} \\ \text{N} \equiv \text{C} \diagup \qquad\qquad \diagdown \text{C} \equiv \text{N} \end{array}
$$

1

Fluorescent derivatives can provide great sensitivity. As with absorption spectroscopy, fluorimetry can be carried out on covalently bonded and noncovalently bonded derivatives. An important example of covalent derivatization is provided by the reaction of amino acids with dansyl chloride (5-dimethylaminonaphthalene-1-sulfonyl chloride) [32], 2, to give strongly fluorescent dansyl amino acids, 3. The lower limit of fluorometric detection on thin-layer plates is 10 pmoles [33] of dansyl amino acid [34].

Noncovalent interactions between macromolecules and so called fluorescent probes result in changes in the

fluorescent properties of the probe molecule [35]. These
interactions have been primarily useful in studying the
regions of the macromolecule involved in the noncovalent
bonding.

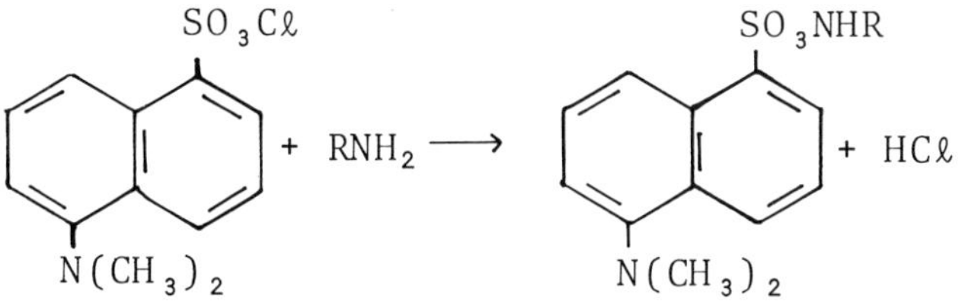

$$SO_3Cl \qquad\qquad SO_3NHR$$

2 + RNH$_2$ $\longrightarrow$ 3 + HCl $\qquad$ (6.22)

N(CH$_3$)$_2$ $\qquad\qquad$ N(CH$_3$)$_2$

2 $\qquad\qquad\qquad$ 3

Halogenated derivatives enable the sensitivity of the
electron capture detector to be applied in gas chromatographic analysis of many nonhalogen compounds. Thus phenols are converted to their chloroacetates by acylation
with chloroacetic anhydride; linear reponse was observed
for 0.1-3 ng of the chloroacetates [36]. Similar sensitivity in steroid determination by electron capture gas
chromatography follows conversion of the steroid to highly
fluorinated derivatives [37].

Derivatives labeled with radioisotopic atoms, introduced through the reagent, provide a capability for high
sensitivity. The technique can be illustrated by the
determination of hydroxy groups, which are acetylated with
^{14}C-labeled acetic anhydride [38]. The specific activity
of the acetylated product is measured. The sensitivity is
a function of the specific activity of the reagent.

Derivative formation can increase sensitivity by
stoichiometric multiplication, as in inorganic methods
proposed by Emich [39] and reviewed by Weisz [40]. In
this approach the sample compound is subjected to a reaction whose product can be cycled to multiply its yield.
The scheme shown below illustrates this multiplication
effect.

$$S + O_2 \rightarrow SO_2$$

$$SO_2 + 2H_2S \rightarrow 3S + 2H_2O \text{ (multiplication factor 3)}$$

$$3S + 3O_2 \rightarrow 3SO_2$$

$$3SO_2 + 6H_2S \rightarrow 9S + 6H_2O \text{ (multiplication factor } 3^2)$$

.
.
.

etc.

Catalysis. Figure 3.5 in Sec. 3.5 is a linear plot of the logarithm of a first-order rate constant against pH, over the pH range 8-12. This graph constitutes a standard curve for the determination of pH by measuring the rate constant of this reaction under the same conditions. This is a simple example of the use of kinetics to measure concentrations. In fact, kinetic methods were at one time used to measure pH.

Kinetic measurements can provide great sensitivity. If the system is arranged so that the substance to be determined is a catalyst, then the rate equation will take the form of Eq. (6.23), or a similar form:

$$v = k(\text{substrate})(\text{reagent})(\text{catalyst}) \qquad (6.23)$$

If the catalyst is regenerated, its concentration remains constant. Equation (6.23) is used analytically by measuring (1) the initial rate, (2) the pseudo-first-order constant, (3) the time for a given extent of reaction to occur, or (4) the extent of reaction occurring in a given time. All of these quantities can be related to the catalyst concentration to establish a working curve [41]. The sensitivity of the method is related to the rate constant. The method has been applied to the determination of numerous metal ions, which function as catalysts in redox reactions [41].

Amplification of the signal can reach impressive levels by taking advantage of the cycling of the catalyst. Let us first treat this schematically. Assume the kinetic scheme

$$S_1 + C \xrightarrow{k_1} P_1 + C' \qquad (6.24)$$

$$S_2 + C' \xrightarrow{k_2} P_2 + C \qquad (6.25)$$

where S_1 and S_2 are substrates, P_1 and P_2 are products, and C and C' are different forms (e.g., oxidized and reduced, or protonated and unprotonated) of the catalyst. Suppose the catalyst is the substance whose concentration is to be determined. By using large concentrations (relative to the catalyst concentration) of the substrates, pseudo-first-order conditions apply. (The reactions could involve additional reactants, also at high concentration;

these could be absorbed into the rate constants k_1 and k_2.) By mixing all reactants and allowing the reaction to proceed, each molecule of C initially present will result in the production of many molecules of P_2. The ratio $[P_2]/[C]_0$, where $[P_2]$ is the concentration of P_2 at time t and $[C]_0$ is the catalyst concentration at zero time, may be called the amplification factor. The way in which this depends on the system parameters and variables is easily developed. The relevant rate equations are

$$\frac{d[P_2]}{dt} = k_2 [S_2][C'] \qquad (6.26)$$

$$\frac{d[C]}{dt} = - \frac{d[C']}{dt} = k_2 [S_2][C'] - k_1 [S_1][C] \qquad (6.27)$$

If the steady-state approximation can be individually applied to C and C', Eq. (6.28) results [42].

$$\frac{[C']}{[C]} = \frac{k_1 [S_1]}{k_2 [S_2]} \qquad (6.28)$$

Applying the mass balance equation,

$$[C]_0 = [C] + [C'] \qquad (6.29)$$

leads, upon combination with (6.26) and (6.28), to

$$\frac{d[P_2]}{dt} = \frac{k_1 k_2 [S_1][S_2][C]_0}{k_1 [S_1] + k_2 [S_2]} = k_c \qquad (6.30)$$

The reaction follows pseudo-zero-order kinetics. Integrating gives $[P_2] = k_c t$, so

$$\text{amplification factor} = \frac{[P_2]}{[C]_0} = \frac{k_c}{[C]_0} t \qquad (6.31)$$

Suppose as an example that $k_1 = 10$ M^{-1}-sec^{-1}, $k_2 = 5$ M^{-1}-sec^{-1}, $S_1 = 0.1$ M, $S_2 = 0.1$ M; then according to Eqs. (6.30) and (6.31), the amplification factor is t/3. That is, for each mole of C initially present, 1/3 mole of P_2 will be produced per second. Thus in 30 min, 600 moles of P_2 will be produced per mole of C. A real example drawn from biochemical analysis is used for the analysis of the coenzymes nicotinamide-adenine dinucleotide, NAD (formerly called diphosphopyridine nucleotide, DPN) or nicotinamide-adenine dinucleotide phosphate, NADP

(formerly triphosphopyridine nucleotide, TPN). The oxidized forms of these coenzymes are shown as structure 4.

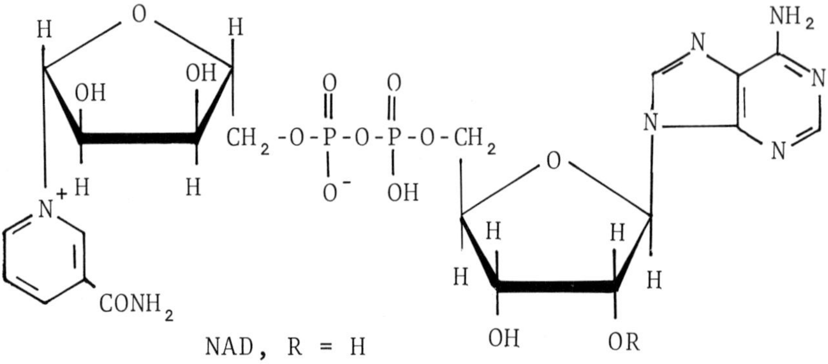

NAD, R = H
NADP, R = PO(OH)$_2$

4

Structures 5 and 6 show how the oxidized (5) and reduced
(6) forms are related [43]. NAD and NADP are hydrogen
carriers, being essential reactants in very many enzyme-
catalyzed redox reactions.

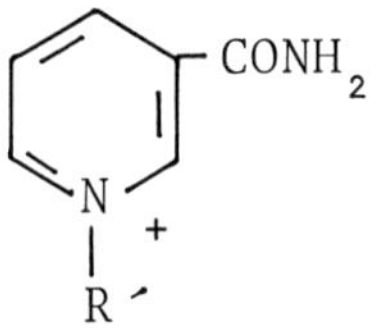

5   6

(NAD$^+$; NADP$^+$)   (NADH; NADPH)

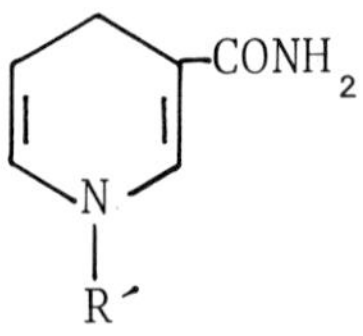

Lowry et al. [44] utilize Reactions (6.32) and (6.33),
which may be compared with (6.24) and (6.25).

$$\alpha\text{-ketoglutarate} + NH_4^+ + NADPH \xrightarrow{\text{glutamate} \atop \text{dehydrogenase}}$$

$$\text{glutamate} + NADP^+ \tag{6.32}$$

$$\text{glucose-6-phosphate} + \text{NADP}^+ \xrightarrow{\begin{array}{c}\text{glucose-6-phosphate}\\\text{dehydrogenase}\end{array}}$$

$$\text{6-phosphogluconate} + \text{NADPH} \qquad (6.33)$$

With the recommended conditions, each molecule of NADP results in the formation of $5\text{-}10 \times 10^3$ molecules of 6-phosphogluconate in 30 min. The 6-phosphogluconate is then determined by adding 6-phosphogluconate dehydrogenase and excess NADP^+.

$$\text{6-phosphogluconate} + \text{NADP}^+ \longrightarrow$$

$$\text{6-phospho-2-oxogluconate} + \text{NADPH} \quad (6.34)$$

Finally the NADPH produced in Reaction (6.34) is measured fluorometrically.

The method can be used to determine either NADP^+ or NADPH, since it is irrelevant which of the reactions, (6.32) or (6.33), is written first on paper. Notice that in these reactions the coenzyme takes the role of a catalyst; in the measurement reaction, Eq. (6.34), the coenzyme acts as a substrate in excess. The procedure described here can be used to determine NADP in the concentration range 1×10^{-9} to 5×10^{-8} $\underline{M}$, and amounts as small as 10^{-15} mole.

A second "cycle" can be operated by running the analysis as described, destroying the excess NADP^+ from the 6-phosphogluconate reaction, taking an aliquot of this reaction mixture (which now contains $5\text{-}10 \times 10^3$ molecules of NADPH for each molecule of NADPH in the original sample), and subjecting this aliquot to a second treatment by the entire amplification procedure. In this way an amplification factor of 4.5×10^7 was achieved [44].

Such an analytical method has applications far beyond the simple determination of the coenzyme. A large number of enzyme systems require these coenzymes [45], and suitable arrangement of coupled reactions and reactant concentrations will permit the indirect analysis of substrates that can be connected to the NAD and NADP system. In this way as little as 1 pmole of prostaglandin E_1 can be determined [46].

The mechanisms of amplification in processes that can be actuated by a single particle have not yet been elucidated. A single photon can stimulate a response in the dark-adapted eye. Similarly, one photon (or very few) can lead to an entire grain becoming developable on a photographic film. Blood clotting probably occurs as a

result of very few molecules of a clotting factor being
released. In each of these instances the amplification
factor must be enormous. The blood clotting mechanism
appears to develop amplification through a cascade process
in which each clotting factor in the chain is a proenzyme
that, upon activation, activates the next in the chain.
Thus each factor becomes an enzyme (catalyst) capable of
magnifying its numbers by its action on the next factor
in line. Wald [47] has suggested by analogy that the
vision process might develop amplification in a similar
sequence of proenzyme-enzyme steps, resulting in a "biochemical photomultiplier." Perhaps the amplification of
the vision or clotting processes could be connected to a
chemical system for analytical application. The simplest
way to do this might be to use the biological system in
situ to amplify the initial analytical signal. Something
of this sort is done when compounds are detected by their
odors [16,17]. A further example is the observation of
responses (food capture, tentacle writhing, and mouth
opening) of Hydra pseudoligactis after exposure to thiol-
oxidizing agents; the nature and degree of response is
related to the reagent concentration [48].
     Cascade processes can be applied as noncatalytic cy-
clic processes to achieve sensitivity, as in the multiple
separation of enantiomers by which Kemp et al. were able
to measure 0.001-1% of racemate [49].

Problems

1. Calculate the Sandell sensitivity for a substrate with
    molecular weight 300 and $\varepsilon_{max} = 2 \times 10^4$.

    Answer: $c_S = 0.015$ µg/cm^2.

2. From Problem 12, Chapter 3, the Michaelis-Menten
    equation for enzyme-catalyzed reactions is

    $$v = \frac{V_{max}[S]}{K_m + [S]}$$

    What is the sensitivity of rate measurements for
    determining substrate concentration?

    Answer: sensitivity $= \dfrac{dv}{d[S]} = \dfrac{K_m V_{max}}{(K_m + [S])^2}$

3. When concentration is expressed in g/100 ml (i.e., as

% w/v), the corresponding spectrophotometric absorptivity is symbolized $A_{1 \, cm}^{1\%}$. Derive an equation relating $A_{1 \, cm}^{1\%}$ and ε for the same solute.

Answer: $\varepsilon = A_{1 \, cm}^{1\%} \left(\dfrac{MW}{10} \right)$

4. Obtain a formula for sensitivity in a spectrophotometric analysis when Beer's law is not obeyed.

Answer: sensitivity $= \varepsilon b + bc(d\varepsilon/dc)$

References

1. See, for example, W. L. Gore, Statistical Methods for Chemical Experimentation, Wiley (Interscience), New York, 1952; F. J. Linning, J. Mandel, and J. M. Peterson, Anal. Chem., 26, 1102 (1954); C. A. Bennett and N. L. Franklin, Statistical Analysis in Chemistry and the Chemical Industry, Wiley, New York, 1954; E. C. Wood, Comprehensive Analytical Chemistry (C. L. Wilson and D. W. Wilson, eds.), Vol. IA, Elsevier, Amsterdam, 1959, Chap. II.4; K. Doerffel, Z. Anal. Chem., 185, 1 (1962).

2. It is not necessary that $\bar{x}$ and s be calculated from the same set of n observations; good estimates of $\bar{x}$ can often be obtained from fewer observations than are required for a good estimate of s. The estimators $\bar{x}$ and s are unrelated.

3. H. Kaiser, Anal. Chem., 42, 26A, No. 4 (1970); Z. Anal. Chem., 209, 1 (1965).

4. R. Püschel, Microchim. Acta, 1968, p. 82.

5. L. A. Currie, Anal. Chem., 40, 586 (1968); A. Hubaux and G. Vos, ibid., 42, 849 (1970).

6. F. Feigl, Spot Tests in Organic Analysis, 6th ed., Elsevier, Amsterdam, 1960.

7. E. B. Sandell, Colorimetric Determination of Traces of Metals, 3rd ed., Wiley (Interscience), New York, 1959, p. 83.

8. E. Sawicki, T. W. Stanley, and W. Elbert, Proceedings of International Symposium on Microchemical Techniques, University Park, Pa., 1961 (N. D. Cheronis, ed.), Wiley (Interscience), New York, 1962, p. 633.

9. I. M. Korenman, Analytical Chemistry of Low Concentrations, Israel Program for Scientific Translations, Jerusalem, Daniel Davey and Co., Hartford, Conn., 1968, p. 19.

10. C. L. Wilson and D. W. Wilson (eds.), Comprehensive Analytical Chemistry, Vol. IA, Elseview, Amsterdam,

1959, p. 232.
11. N. D. Cheronis (ed.), Proceedings of the Symposium on Submicrogram Experimentation, Arlington, Va., 1960, Wiley (Interscience), New York, 1961, pp. 10-15.
12. E. A. Braude, Determination of Organic Structures by Physical Methods (E. A. Braude and F. C. Nachod, eds.), Academic, New York, 1955, Chap. 4.
13. See, for alternative measures of fluorescence sensitivity, D. W. Ellis, Fluorescence and Phosphorescence Analysis (D. M. Hercules, ed.), Wiley (Interscience), New York, 1966, Chap. 2.
14. M. Dimbat, P. E. Porter, and F. H. Stross, Anal. Chem., 28, 290 (1956).
15. A. B. Littlewood, Gas Chromatography, Academic, New York, 1962, p. 243.
16. L. S. Ettre and W. H. McFadden, Ancillary Techniques of Gas Chromatography, Wiley (Interscience), New York, 1969.
17. F. Feigl, Ref. [6], p. 69.
18. R. Belcher, Submicro Methods of Organic Analysis, Elsevier, Amsterdam, 1966.
19. T. F. Kelley, Anal. Chem., 37, 1078 (1965).
20. J. R. Robinson, H. Stelmach, and S. P. Eriksen, Anal. Chem., 42, 495 (1970).
21. Korenman, Ref [9], Chapter IV.
22. E. J. Conway, Microdiffusion Analysis and Volumetric Error, 5th ed., Crosby Lockwood and Son, London, 1962.
23. See, for example, K. A. Connors, A Textbook of Pharmaceutical Analysis, Wiley, New York, 1967, Chap. 14.
24. A very rough estimate of K may often be made by taking the ratios of the solubilities of the solute in the two solvents. For empirical methods of estimation, and for a valuable collection of experimental partition coefficients, see A. Leo, C. Hansch, and D. Elkins, Chem. Revs., 71, 525 (1971).
25. W. B. Elliot and G. F. Doebbler, Nature, 198, 690 (1963).
26. F. D. Snell and C. T. Snell, Colorimetric Methods of Analysis, 3rd ed., Van Nostrand, New York, 1948-1954; N. L. Allport and J. W. Keyser, Colorimetric Analysis, 2nd ed., Vol., Chapman and Hall, London, 1957; N. L. Allport and J. E. Brocksopp, ibid., Vol. 2, 1963.
27. W. H. Hong and K. A. Connors, Anal. Chem., 40, 1273 (1968).
28. This analysis is an example of one of the few spectrophotometric derivatization procedures that does not rely on a working curve prepared by carrying "known" samples through the same procedure as the unknown. Instead, since both the reaction and separation steps are quantitative, it is possible to base

the calculation on the amide product, whose absorptivity can be determined independently of the analytical method.

29. E. Roberts and G. Rouser, Anal. Chem., 30, 1291
     (1958); N. M. Alexander, ibid., 30, 1292 (1958).
30. J. Broekhuysen, Anal. Chim. Acta, 19, 542 (1958).
31. G. H. Schenk, P. Warner, and W. Bazzelle, Anal. Chem.,
     38, 907 (1966).
32. W. R. Gray and B. S. Hartley, Biochem. J., 89, 59P
     (1963).
33. The prefixes for submultiple units are, with their
     symbols: deci (d), 10^{-1}; centi (c), 10^{-2}; milli
     (m), 10^{-3}; micro (μ), 10^{-6}; nano (n), 10^{-9}; pico
     (p), 10^{-12}.
34. M. S. Arnott and D. N. Ward, Anal. Biochem., 21, 50
     (1967). DANS acid has been recommended as a standard
     for measuring quantum yields: C. M. Himel and R. T.
     Mayer, Anal. Chem., 42, 130 (1970).
35. G. M. Edelman and W. O. McClure, Accounts Chem. Res.,
     1, 65 (1968).
36. R. J. Argauer, Anal. Chem., 40, 122 (1968).
37. M. A. Kirschner and J. P. Taylor, Anal. Biochem., 30,
     346 (1969).
38. R. H. Benson and R. B. Turner, Anal. Chem., 32, 1464
     (1960); 33, 344 (1961). For a review see G. Ayrey,
     D. Barnard, and T. H. Houseman, Chem. Revs., 71, 371
     (1971).
39. F. Emich, Mikrochemie, 13, 283 (1933).
40. H. Weisz, Mikrochim. Acta, 1970, p. 1057.
41. H. B. Mark, Jr., and G. A. Rechnitz, Kinetics in
     Analytical Chemistry, Wiley (Interscience), New York,
     1968, Chap. 3.
42. It is obvious that $d([C] + [C']) /dt = 0$, but it does
     not necessarily follow that $d[C]/dt = 0$ and $d[C']/dt = 0$.
43. The reduced forms have a broad absorption band with
     λ_{max} = 340 nm and ε_{max} = 6.22 $\times$ 10^3; the oxidized
     forms are transparent at this wavelength.
44. O. H. Lowry, J. V. Passonneau, D. W. Schulz, and M.
     K. Rock, J. Biol. Chem., 236, 2746 (1961).
45. See M. Dixon and E. C. Webb, Enzymes, 2nd ed., Aca-
     demic, New York, 1964, pp. 361-370, 672-679.
46. E. Änggård, F. M. Matschinsky, and B. Samuelsson,
     Science, 163, 479 (1969).
47. G. Wald, Science, 150, 1028 (1965).
48. H. M. Lenhoff, E. M. Kosower, and N. S. Kosower,
     Nature, 224, 171 (1969).
49. D. S. Kemp, S. W. Wang, G. Busby III, and G. Hugel,
     J. Am. Chem. Soc., 92, 1043 (1970). See also Sec. 5.2.

Chapter 7.   ELECTROPHILIC AROMATIC SUBSTITUTION

7.1   Nature of the Reaction

Survey of the Reaction.   The title of this chapter is
based on the nomenclatural conventions described in Chap.
1.   Thus all of these reactions involve substitution by
an electrophilic reagent upon an aromatic substrate, the
overall reaction having the form of

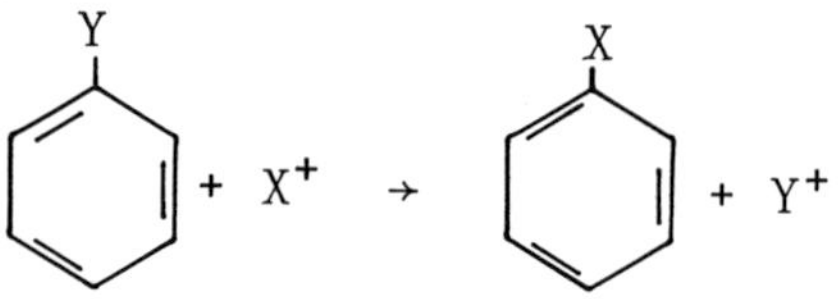

$$\text{(7.1)}$$

This section briefly treats the scope of this class of
reactions and provides an introduction to the later more
detailed description.   Many reviews are available [1].
     The type of substitution is specified by a name com-
posed of the electrophile [X in (7.1)], the syllable "de,"
the name of the leaving group Y, and the suffix "ation."
The substituent ^{1}H is named "proto," though its participation is usually understood and is not specifically denoted.   Table 7.I lists a few aromatic substitution reactions to illustrate this system.   The actual electrophilic
species may be different from the formal species given in
the table.
     The study of aromatic substitutions must be concerned
with the following questions, which are considered in later
sections.

     1. In a monosubstituted benzene derivative subjected
to attack by an electrophile, how is the orientation of
the substitution (i.e., ortho, meta, or para to the existing substituent) influenced by the nature of the substituent?
     2. How does the nature of the substituent in a sub-
stituted aromatic substrate affect the reactivity of the
substrate relative to the unsubstituted compound, and how

does it affect the reactivity of one site relative to
another within the same substrate?
    3. How does the nature of the electrophile affect
orientation and reactivity?
    4. What is the mechanism (or mechanisms) of electro-
philic aromatic substitution?

Table 7.I   Examples of Electrophilic Aromatic Substitutions

Electrophile $[X^+$ in Eq. (7.1)]	Leaving group $[Y^+$ in Eq. (7.1)]	Name of reaction
$C\ell^+$	H^+	chlorination
NO_2^+	H^+	nitration
SO_3	H^+	sulfonation
Me_3C^+	H^+	alkylation
CH_3CO^+	H^+	acetylation
$Hg(C\ell O_4)^+$	H^+	mercuration
H^+	D^+	dedeuteration
H^+	R^+	protodealkylation
H^+	SiR_3^+	protodesilylation
SiR_3^+	Li^+	silyldelithiation
I^+	SnR_3^+	iodo destannylation

Investigations into questions (1) and (2) have had a great
influence on the development of electronic theories of organic chemistry. It is not necessary to follow the early
ideas, and we use the present views directly. Thus electron releasing and withdrawing properties of substituents
by the inductive and resonance (mesomeric) effects are
recognized and invoked both qualitatively and quantitatively to account for substituent effects on orientation
and reactivity.
    In order to provide a background for discussion,
later sections are anticipated by pointing out that electrophilic aromatic substitution is believed to take place
via an intermediate, underline{1} in Eq. (7.2), in which both the
attacking electrophile and the departing group are bonded
to the same carbon.

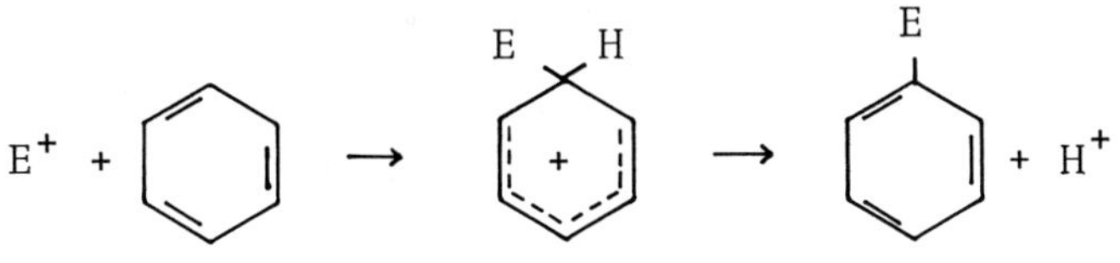

$$E^+ + \begin{array}{c}\bigcirc\end{array} \longrightarrow \begin{array}{c}\overset{E\quad H}{\bigcirc}\end{array} \longrightarrow \begin{array}{c}\overset{E}{\bigcirc}\end{array} + H^+ \qquad (7.2)$$

underline{1}

Selectivity and Reactivity. When a monosubstituted benzene undergoes monosubstitution, three possible disubsti-
tuted benzenes may be produced. In the absence of any
effect by the substituent in the substrate, the purely
statistical product distribution of 40% ortho, 40% meta,
and 20% para substitution is expected.

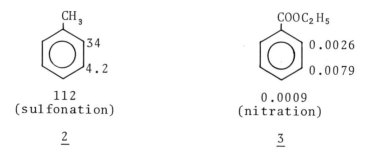

The observed product distribution is found to depend upon
the natures of R and of E^+. We first consider the effect
of the substituent R upon the site of substitution (its
directive effect) and upon the rate of substitution.
        The description of substituent effects is accomplished
with concepts and quantities introduced in Chaps. 4 and 5.
Equations (5.14), (5.15), and (5.16) define partial rate
factors o_f, m_f, and p_f, which are rates of substitution at
the indicated position, relative to a single position in
benzene [2]. A partial rate factor greater than unity
signifies that the site is more reactive than a site in
benzene; the group R is then said to be activating. Deactivating groups lead to partial rate factors less than
unity. In structures 2 and 3 partial rate factors are
shown for the sulfonation of toluene [3] and the nitration
of ethyl benzoate [2]. Evidently the methyl group is

activating and the carbethoxy group is deactivating for
the reactions shown. Table 7.II gives many more partial
rate factors from the extensive summary of Stock and Brown.
        Provided that the product distribution is under kine

Table 7.II Partial Rate Factors for Electrophilic
Aromatic Substitution Reactions on Monosubstituted
Benzenes [3]

Substituent	Partial rate factor		
	o_f	m_f	p_f

Nitration ($AcONO_2$, Ac_2O, 25°)

Substituent	o_f	m_f	p_f
-H	(1.00)	(1.00)	(1.00)
-CH_3	46.5	2.1	48.5
-C_2H_5 (0°)	31.4	2.3	69.5
-$CH(CH_3)_2$ (0°)	14.8	2.4	71.6
-$C(CH_3)_3$	3.8	3.8	57.7
-C_6H_5	41	--	38
-OC_6H_5	117	--	230
-F	0.037	--	0.77

Bromination (Br_2, $HOAc$-$MeNO_2$, 25°)

Substituent	o_f	m_f	p_f
-CH_3	600	5.5	2420
-C_2H_5	465	--	1800
-$CH(CH_3)_2$	180	--	1200
-$C(CH_3)_3$	4.97	6.09	806
-C_6H_5	37.5	0.3	2920
-OC_6H_5	--	--	1.0×10^8
-OCH_3	8.7×10^7	2.0	1.1×10^{10}
-OH	--	--	3.7×10^{12}
-$NHCOCH_3$	--	--	1.2×10^9
-Br	--	5.3×10^{-4}	0.0618
-$C\ell$ (30°)	--	5.6×10^{-4}	0.145
-I (30°)	--	0.0022	0.0802
-F (30°)	--	0.0010	4.62
-CN (30°)	--	8.7×10^{-7}	3.1×10^{-6}

Acetylation ($MeCOC\ell$, $A\ell C\ell_3$, $C_2H_4C\ell_2$, 25°)

Substituent	o_f	m_f	p_f
-CH_3	4.5	4.8	749
-C_2H_5	1.0	10.4	753
-$CH(CH_3)_2$	0.0	11.5	745
-$C(CH_3)_3$	0.0	13.1	658
-C_6H_5	--	0.3	248
-OCH_3	--	--	1.8×10^6
-F	--	--	1.51

215

Table 7.II   continued

Substituent	Partial rate factor		
	o_f	m_f	p_f

Acetylation ($MeCOC\ell$, $A\ell C\ell_3$, $C_2H_4C\ell_2$, 25°)

$-C\ell$	0.0	3×10^{-4}	0.125
$-Br$	--	--	0.084

Mercuration ($Hg(OAc)_2$, HOAc, 25°)

$-CH_3$ [a]	4.98	2.25	32.9
$-C(CH_3)_3$	0.0	3.41	17.2
$-C_6H_5$	0.0813	0.773	6.42
$-NHCOCH_3$	--	--	277
$-OC_6H_5$	--	--	194
$-OCH_3$	186	1.2	2310
$-F$	0.63	0.040	2.98
$-C\ell$	0.075	0.060	0.36
$-Br$	0.070	0.054	0.27

[a]Medium contained perchloric acid.

tic control, the partial rate factors reflect the fraction
of substitution at each site [see Eqs. (5.15), (5.15),
(5.16)]. For example, the gallium bromide catalyzed ethyl-
ation of toluene gives 55.7% ortho, 9.9% meta, and 34.4%
para substitution, with the relative rate toluene:benzene
being 5.70. These figures give partial rate factors o_f =
9.51, m_f = 1.7, p_f = 11.8. Again the activating influence
of the methyl group is seen. Moreover, examination of
such data, as in 2 and Table 7.II, reveals regular patterns
in substituent directive effects. Figure 7.1 is a plot of
the logarithms of partial rate factors for substitutions
on toluene (data from Table 7.II). Clearly the methyl
group is activating for all positions and with all four
electrophiles. Moreover, it appears that the ortho and
para positions are activated more profoundly than is the
meta position. This phenomenon is widely observed, namely,
that activating substituents are ortho, para directing [4].
Methoxybenzene (anisole) provides the classic example of
o,p-orientation by an activating substituent.
     A second type of behavior is illustrated by 3, which

shows meta-orientation prevailing over o,p-orientation with a deactivating substituent. Relatively few examples of the meta directive effect by deactivating groups are available, in part because the slowness of reactions of deactivated substrates makes their study difficult. The third type of effect is shown by halobenzenes, which are seen from Table 7.II to be deactivated but o,p-directing.

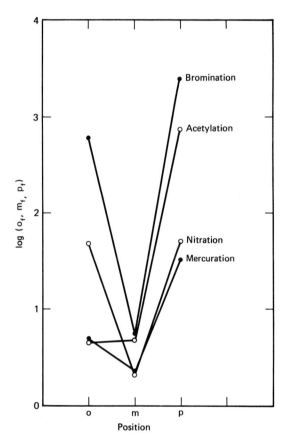

Fig. 7.1. Logarithms of partial rate factors for some electrophilic substitutions of toluene, showing the ortho, para-directing effect of the activating methyl group.

Rationalizations of substituent directive effects in electrophilic aromatic substitution apply the ideas of inductive and resonance electron displacement effects to the substitution reaction (7.3), where para substitution is shown as an example. It is reasonable to expect that the

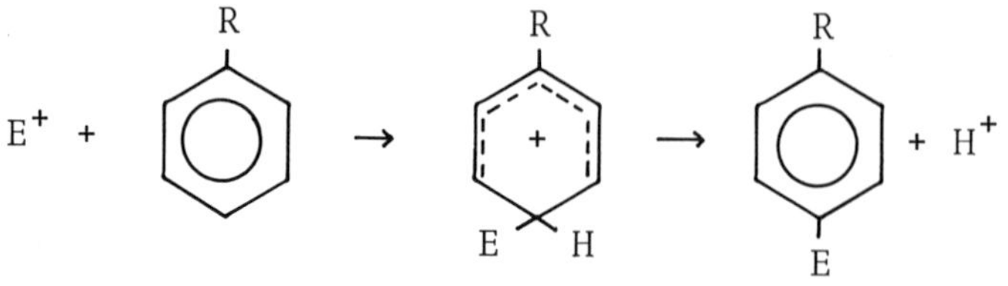

$$E^+ \; + \quad \text{(4)} \quad \rightarrow \quad \text{(5)} \quad \rightarrow \quad + \; H^+ \qquad (7.3)$$

electrophilic reagent will preferentially attack sites with the highest electron density. One approach is therefore to attempt to correlate the directive effect with the substituent's influence on electron density at the ortho, meta, and para positions of the substrate, 4. The weakness of this method is its implicit assumption that differences in rates of substitution at the several sites of 4 are consequences of initial state energy differences, whereas the rates are really determined by free energy changes between the initial and transition states. Nevertheless, the method is reasonably successful in accounting for the general pattern of directive effects. As an example of an electron-donating substituent consider the methoxy group in anisole. The canonical valence-bond structures 6a-d show the essential features of the resonance effect upon electron distribution.

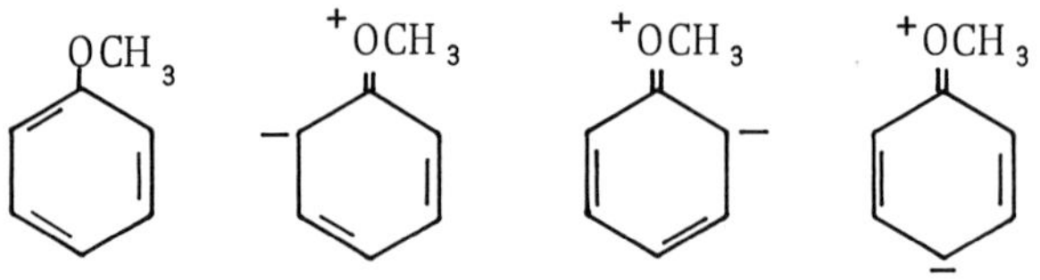

6

Structure 7 is another representation of the same effect.

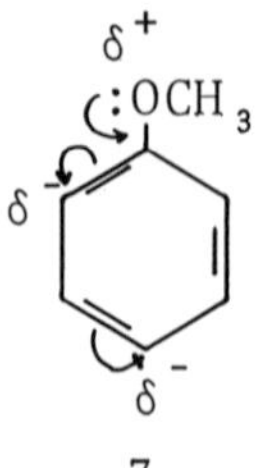

7

The electron-donating methoxy group increases the electron density of the ring, relative to benzene, accounting for the observed activation. Furthermore, the increase in electron density is maximal at the ortho and para positions, which is consistent with the experimental product distribution.

Structures 8 and 9 illustrate this initial state description for benzaldehyde, which is deactivated and meta-directed.

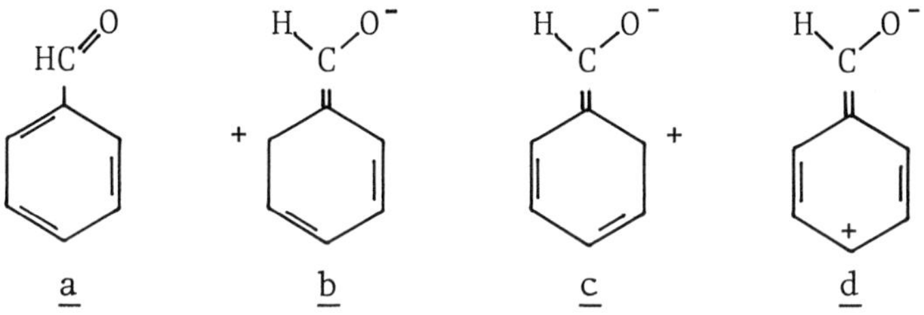

HC=O and related resonance structures.

a   b   c   d

8

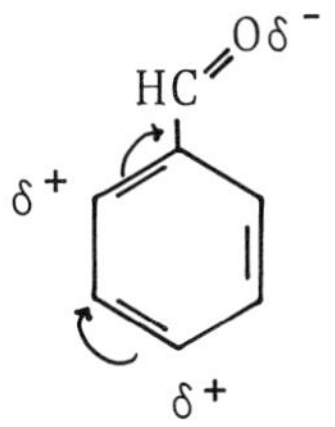

9

A drift of electron density out of the ring accounts for the deactivation, and the ortho and para positions are deactivated more than the meta position.

The halogens are electronegative relative to hydrogen, so in the halobenzenes they withdraw electrons from the ring via the inductive effect. However, since they possess an unshared electron pair capable of resonance with the ring, they are electron donating by this effect. The net result is that they are usually deactivating o,p-directing substituents.

It was indicated above that more confidence might be

placed in an explanation that takes into account the transition state of the reaction.   If the rate-determining step
is the formation of the benzenonium ion intermediate 5
(which is often referred to as a σ-complex, or as a Wheland
intermediate) [5], the rate constant is determined by the
free energy change on passing from the initial state to
the transition state for the formation of the intermediate.
The benzenonium ion is much less stable than is the substrate, 4, so, according to the Hammond postulate [6], the
transition state probably closely resembles the intermediate, which therefore should be a good model for the transition state.   In considering the directive effect of a
substituent, the relative rates for o, m, and p substitution are determined solely by the relative stabilities of
the respective transition states, and, by assumption, of
the Wheland intermediates, as shown in Eq. (7.4).   The
analysis of substituent effects on the substitution process therefore is accomplished by considering the effect
of the substituent R on the stabilities of the three
intermediates.

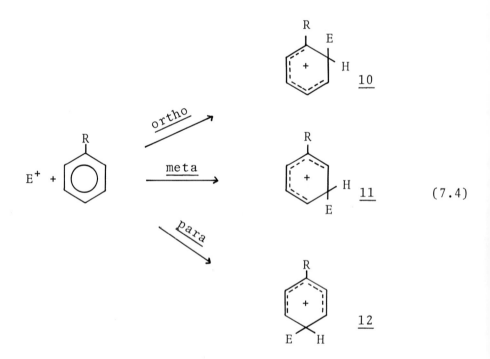

$$(7.4)$$

A substituent that donates electron density to the ring, thus reducing the positive charge on the ring, will stabilize the transition state relative to that for benzene, and therefore will be activating. If the substrate withdraws electrons from the ring, the transition state will be destabilized and the rate will be decreased; such a group is deactivating [7]. Provided that R cannot enter into direct resonance interaction with the ring, the canonical valence-bond structures contributing to 10, 11, and 12 are shown below.

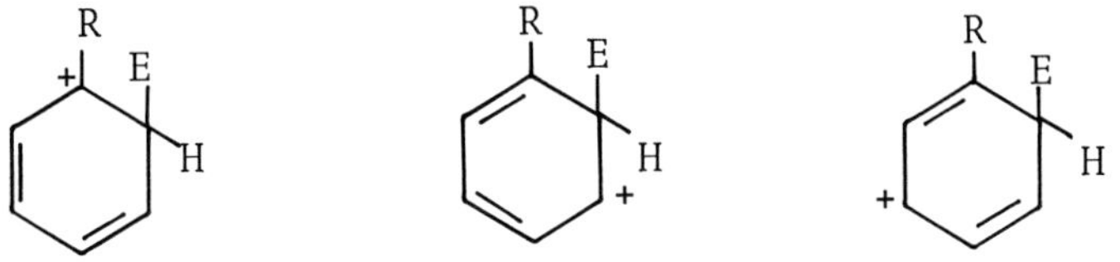

R
+ E
H
a

R
E
H
+
b

R
E
H
+
c

10

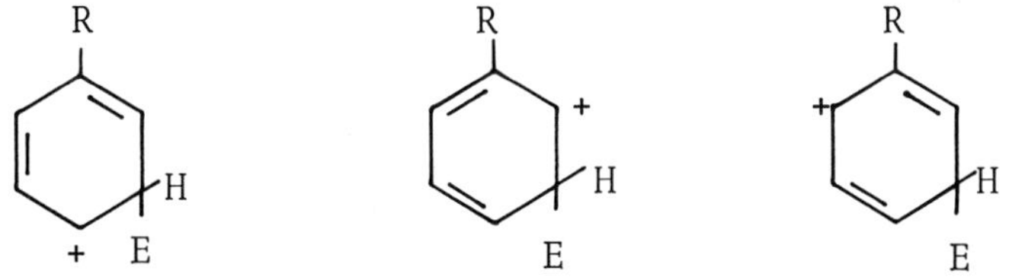

R
H
+ E
a

R
+
H
E
b

R
+
H
E
c

11

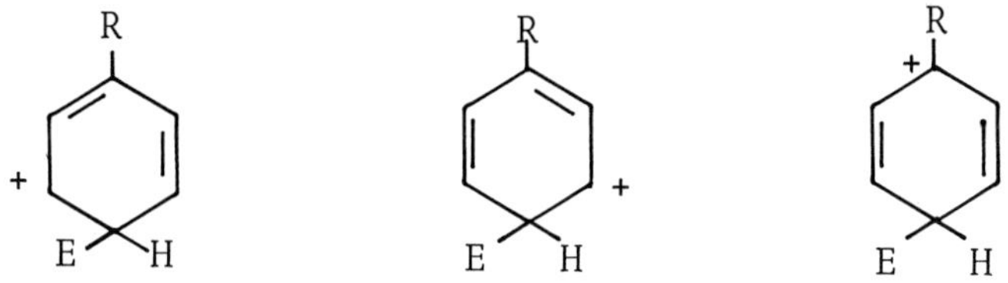

R
+
E H
a

R
+
E H
b

R
+
E H
c

12

For each intermediate, part of the positive charge is localized in the positions ortho and para to the incoming electrophile. If the substitution is ortho or para to R, this places a partial positive charge on the carbon bonded to R (10a, 12c). Therefore if R is electron donating, the ortho and para substitution intermediates will be stabilized relative to the meta, whereas if R is electron withdrawing, o,p will be destabilized.

If resonance is possible between R and the ring, a further valence-bond structure, illustrated for a para substitution on chlorobenzene, contributes to the resonance hybrid for o, p-substitution (but not for m-substitution).

The resonance effect might oppose or reinforce the inductive effect, as described earlier.

A further directive effect has been implicated in reactions of some positively charged monosubstituted benzenes. These strongly deactivated substrates would be expected to be substituted exclusively in the meta position, but appreciable para substitution is observed, as the data in Table 7.III show [8]. In each of these substrates, reaction occurs via the cationic species. Since the effect of monosubstitution on the free energy of activation of benzene, $\delta_R\Delta G^{\ddagger}$, is logarithmically related to the partial rate factors, these data show that $\delta_R\Delta G_p^{\ddagger}/\delta_R\Delta G_m^{\ddagger}$ is about unity. This result, as Ridd points out [8], suggests a ring-deactivating effect by the charged substituent that does not strongly discriminate between the meta and para positions. A direct electrostatic (field) effect could account for the observed orientation. Such electrostatic deactivation would be exerted to about the same extent at each of these positions, with the precise balance being determined by the detailed charge distribution in the transition state.

The reasoning applied to monosubstituted benzenes can be extended to the reactions of more complex substrates. Structures 13 and 14 show resonance structures contributing to the intermediates formed during 1-substitution and 2-substitution, respectively, of naphthalene [9].

Table 7.III   Isomer Distribution and Partial Rate Factors
for the Nitration of Anilinium Ions [8]

Substrate	Percent isomer[a]		Partial rate factor	
	m	p	$10^8 m_f$	$10^8 p_f$
$PhNH_3^+$	62	38	162	195
$PhNH_2Me^+$	70	30	57	49
$PhNHMe_2^+$	78	22	12.3	7.1
$PhNMe_3^+$	89	11	4.2	1.0

[a]In 98% sulfuric acid at 25°.

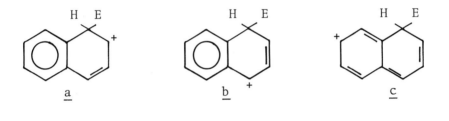

The intermediate formed from 1-substitution has contributions from two benzenoid structures, 13a and 13b, each of
which is itself stabilized by delocalization; 2-substitu-
tion leads to one such structure, 14a.   Accordingly it is
expected that the 1-position will be more reactive to electrophilic substitution than will the 2-position.   The partial rate factors for nitration at these positions are 470

and 50, respectively, in agreement with prediction [10].
The observed activation is accounted for by the extensive
delocalization of positive charge.

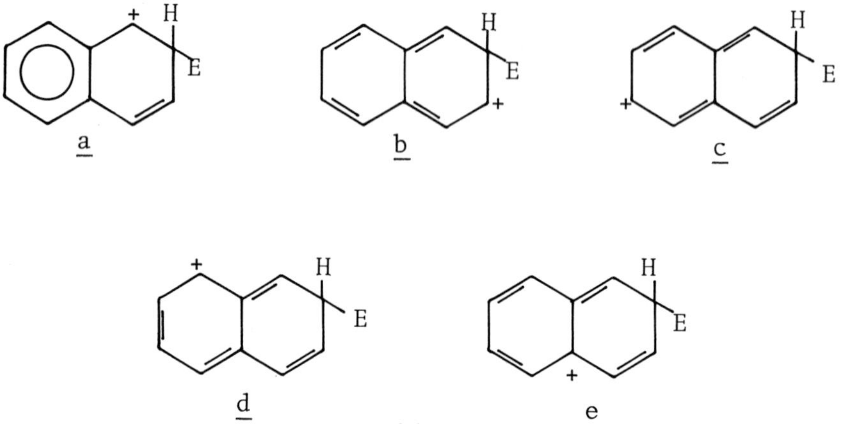

14

An electron-withdrawing substituent on the napthalene
system deactivates and directs into the unsubstituted ring.
An electron-donating group activates the molecule and
directs into the substituted ring.
    Anthracene, according to similar thinking, should be
attacked at the 9-position, because this leads to an intermediate whose principle resonance contribution comes from
the highly delocalized structure 15. In fact, this struc-

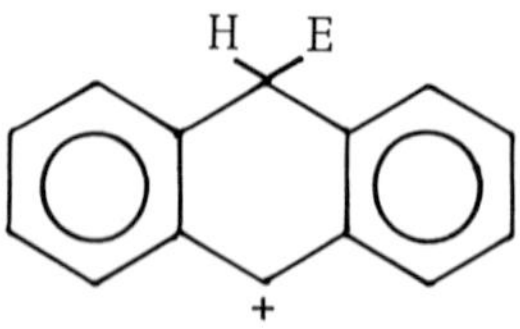

15

ture is sufficiently stable, relative to the anthracene
nucleus, that 9, 10 addition, instead of substitution,
often occurs:

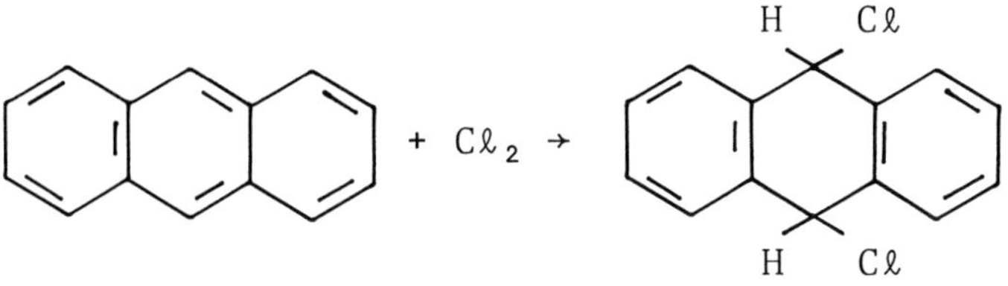

Heteroatoms in aromatic rings can be treated as substituents. Thus nitrogen is electronegative relative to carbon, and is deactivating in the pyridine ring. Pyridine is very resistant to electrophilic substitution [11] The presence of an electron-releasing substituent on the pyridine ring opposes the deactivating influence of the nitrogen, and allows substitution to take place.

The ideas presented thus far do not account for the variability in observed ratios of ortho/para substitution. For both o, p-directing and m-directing substituents, the preceding arguments suggest that the o/p ratio for monosubstituted benzenes should be 2. (The corresponding ratio of partial rate factors, o_f/p_f, is unity.) In fact, however, the ratio can be much smaller or larger than this, as the data in Table 7.IV show [3,12]. Several factors can operate to control the ortho-para ratio.

Table 7.IV Isomer Distribution in the Chlorination and Nitration of Monosubstituted Benzenes [3,12]

Substituent	Chlorination				Nitration			
	o	m	p	o_f/p_f	o	m	p	o_f/p_f
-NO$_2$	17.6	80.9	1.5	5.9	6.4	93.2	0.3	11.0
-CN	23.2	73.9	2.9	4.0	17.1	80.7	2.0	4.3
-Br	39.7	3.4	56.9	0.35	36.5	1.2	62.4	0.3
-Cl	36.4	1.3	62.3	0.29	29.6	0.9	69.5	0.21
-CH$_3$	74.7	2.2	23.1	1.6	58.4	4.4	37.2	0.78
-OCH$_3$	34.9		65.1	0.27	44.0		55.0	0.41

A steric repression of the o/p ratio can be identified in the data of Table 7.II on the nitration of monoalkyl benzenes. The partial rate factors m_f and p_f show that the groups -CH$_3$, -C$_2$H$_5$, -CH(CH$_3$)$_2$, -C(CH$_3$)$_3$ are increasingly activating in the order written. The ortho rate factors, however, decrease in the same order, the ratios o_f/p_f being Me, 0.96; Et, 0.45, iPr, 0.21; t-Bu, 0.07 [13].

Unusually high o/p ratios cannot be accounted for by steric hindrance to attack by the electrophile. Specific intramolecular assistance has been postulated for some of these reactions. For example, methyl benzoate, which is expected to be meta directing, yields almost exclusively the ortho-substituted product upon reaction with thallium (III) trifluoroacetate. This unusual result is rationalized by complex formation between the substituent and the electrophile, followed by its attack on the ortho position [14].

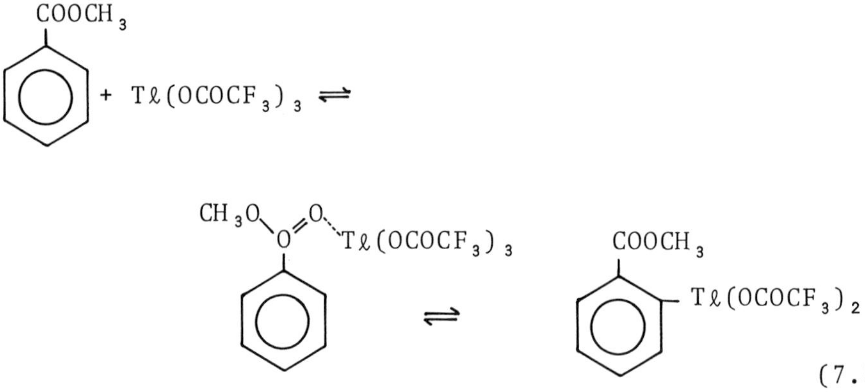

$$(7.5)$$

It has been observed that o/p and m/p ratios may follow similar trends. Since the m/p ratio is electronically controlled, presumably the o/p ratio is also determined electronically in these reactions [15]. The inductive effect is known to fall off rapidly with distance, so the ortho position should be more sensitive to inductive electronic effects than is the para position. This behavior should lead to low o/p ratios for electron-withdrawing substituents and high ratios for electron-releasing groups. Although it is difficult to isolate the inductive effect from other factors, it appears that the expected behavior is observed. This is an initial state argument, and it may be inadequate. The ortho/para orientation has been accounted for in terms of the stability of the transition state, for reactions in which the benzenonium ion intermediate can be taken as a model of the transition state [12]. The argument, which is a rationalization of a quantum mechanical result, avers that 16a makes a greater contribution to the resonance hybrid than does 16b or 16c.

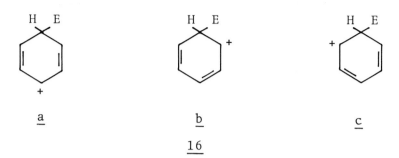

a   b   c

16

The nucleus of <u>16</u> possesses four π electrons, which dis-
tribute themselves in pairs, with the two pairs tending as
far apart as possible, that is, on opposite sides of the
nucleus. Thus <u>16a</u> is the most stable of the three valence-
bond structures, and the charge delocalization results in
greater positive charge density para to the position occupied by E than in the ortho positions.

Such a transition state will be stabilized by an
electron-releasing group para to E, with the result that
the o/p ratio of such a substrate is decreased. An electron-withdrawing group, on the other hand, will provide
greater stabilization from the ortho rather than the para
position, leading (even though it deactivates relative to
benzene) to a high o/p ratio.

A similar argument can be given for those reactions
in which the initial state is a good model for the transition state. A substituent capable of electron withdrawal
by resonance should deactivate the para position more
strongly than the ortho positions, again because structure
17a is more important than 17b.

R⁻   R⁻

a   b

17

It was implied above that the product distribution may depend upon the location of the transition state along the reaction coordinate, that is, by its resemblance to the initial state or intermediate. The nature of the transition state will depend upon the nature of the electrophile as well as the substrate, though we have thus far considered only the substrate. Table 7.V gives rate and product data for substitution reactions of toluene with several electrophiles [3]. Data like these led H. C. Brown to develop the selectivity-reactivity relationship for aromatic substitution that was described in Sec. 5.1. Briefly, this treatment reveals linear relationships between quantitative measures of selectivity and reactivity. The selectivity factor S_f, defined as log (p_f/m_f), is a measure of the intramolecular discrimination exercised by the electrophile. A linear plot of log p_f against S_f (for a single substrate subjected to several electrophilic substitutions) reveals such selectivity-reactivity behavior. By defining a scale of σ^+ substituent constants for reactions in which a positive charge develops at the site of reaction, it was possible to obtain reaction constants, ρ, describing the sensitivity of a reaction to substituent effects (see Sec. 4.2). Table 4.VI gives ρ values for some aromatic substitutions. Plots of log (partial rate factor) against ρ constitute another type of selectivity-reactivity presentation; Stock and Brown show many of these [3].

Table 7.V Relative Rates and Product Distribution for Substitutions on Toluene [3]

Reaction	Relative rate[a]	Product distribution	
		% m	% p
Bromination	605	0.3	66.8
Chlorination	350	0.5	39.7
Benzoylation	110	1.5	89.3
Nitration	23	2.8	33.9
Mercuration	7.9	9.5	69.5
Isopropylation[b]	1.8	25.9	46.2

[a]Relative to benzene.
[b]Friedel-Crafts conditions, with gallium bromide.

In qualitative terms, these demonstrations mean that the more reactive the electrophile, the less discriminating it is. In the limit of an infinitely reactive electro-

phile, presumably the statistical distribution of products
would be observed. The reactivity of the reagents in
Table 7.V increases from top to bottom [16], and the product distribution in the alkylation reaction is clearly
less selective than with the milder halogenations.
     Since aromatic molecules are particularly suited to
theoretical calculations (because many of the approximations, such as neglect of σ electrons, seem to be relatively harmless), aromatic substitutions have been investigated
by these methods. The two approaches described for qualitative interpretation, namely, the initial state method
and the use of the benzenonium ion intermediate as a model
for the transition state, have been adapted. In the initial state (static) method, the π electron density, or a
related quantity called the "free valence," is calculated
at each position [17]. The transition state (dynamic)
approach calculates the energy required to localize a pair
of π electrons on the carbon being attacked by the electrophile. The smaller this "localization energy," the
more reactive the site is predicted to be. These theoretical approaches appear to have been most successful in
accounting for the behavior of polynuclear hydrocarbons.

Mechanisms. Two reasonable mechanisms can be written for
aromatic electrophilic reactions. Equation (7.6) shows a
one-step concerted displacement; in the transition state
18 the C-H bond is partially broken and the C-E bond is
partially formed.

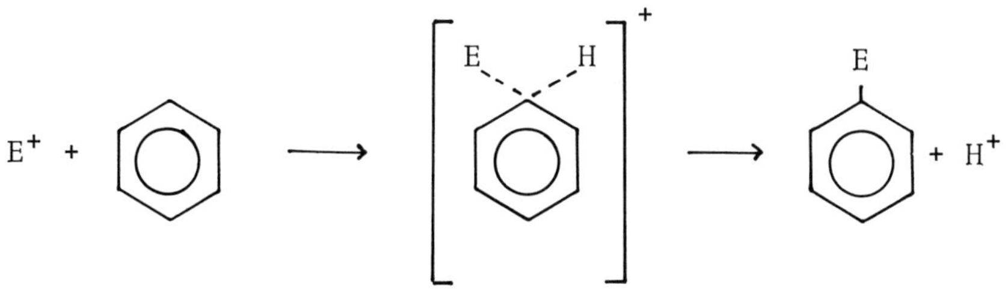

$$18 \hspace{8cm} (7.6)$$

The distinctive feature of this pathway is the retention
of the completely delocalized aromatic system throughout
the reaction.
     The other pathway is a two-step reaction involving
the intermediate 19. 19 is called a benzenium ion, benzenonium ion, Wheland intermediate, or σ-complex (this
last because the electrophile is bonded to the nucleus
with a sigma bond). Pathway (7.7) entails some expenditure

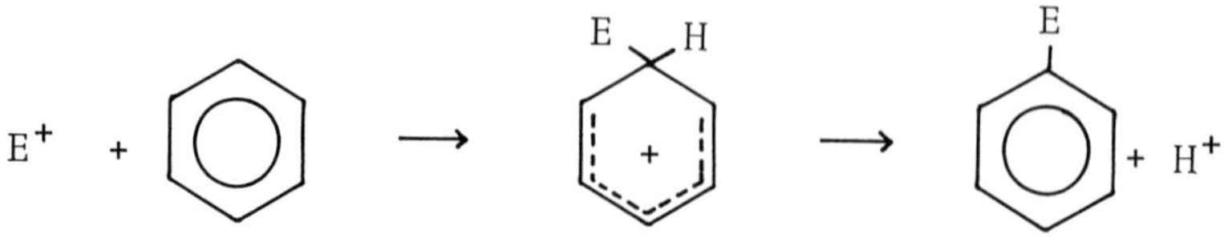

$$E^+ \; + \; \text{(benzene)} \; \longrightarrow \; \underset{\underline{19}}{\text{(intermediate, } E \diagup H)} \; \longrightarrow \; \text{(product, } E) + H^+ \qquad (7.7)$$

of energy to compensate for the disruption of the aromatic system in 19; the loss of aromatic delocalization energy is not complete however, and is partly offset by hyper-conjugation ("no-bond" resonance) of the methenyl hydrogen and the electrophile.

Melander [18] provided the first experimental approach to deciding between these alternatives. Suppose that the hydrogen being displaced were either deuterium or tritium. Then the one-step mechanism would result in a slower reaction rate for the substrate with the heavier isotope, because in the rate-determining step the bond to H, D, or T is partially broken. The two-step mechanism, however, need not lead to a kinetic isotope effect; if formation of the intermediate is rate determining, and if in the transition state for this step the C-H bond is essentially fully maintained, replacement of H by D or T should not alter the rate. If the second step is rate determining, an isotope effect should be observed.

Melander found, for the nitration of benzene and some substituted aromatics that were partially tritiated, that the isotopic rate ratio k_H/k_T was close to unity. This appeared to rule out a one-step mechanism, although the strictest interpretation is simply that no appreciable stretching of the C-H bond occurs in the rate-determining step, a condition that is conceivable, though unlikely, in the one-step mechanism [19]. Certainly the two-step mechanism provides a less strained interpretation for the absence of an isotope effect.

Later work [20] has shown that many, in fact most, substitutions occur with a significant isotope effect. As noted above, the two-step mechanism can accommodate either result. Writing the mechanism schematically (and postulating reversibility in the first step):

$$ArH + E^+ \underset{k_{-1}}{\overset{k_1}{\rightleftharpoons}} ArHE^+ \qquad (7.8)$$

$$ArHE^+ \xrightarrow{k_2} ArE + H^+ \qquad (7.9)$$

Applying the steady-state approximation to the intermediate gives

$$rate = \frac{k_1 k_2 [ArH][E^+]}{k_{-1} + k_2} \qquad (7.10)$$

Thus an overall second-order rate constant k is given by

$$k = \frac{k_1 k_2}{k_{-1} + k_2} = \frac{\alpha k_1}{1 + \alpha} \qquad (7.11)$$

where $\alpha = k_2/k_{-1}$. By the earlier argument, neither k_1 nor k_{-1} should show an appreciable isotope effect, whereas k_2 should be subject to such an effect. The net isotope effect observed for k therefore depends upon the magnitude of α. If $\alpha \gg 1$, then $k \simeq k_1$ and no isotope effect will be observed; this appears to be the case for nitration. If $\alpha \ll 1$, then $k \simeq k_1 k_2/k_{-1}$, and an isotope effect is expected (and is observed for most reactions other than nitration)[20].

It is not surprising that the second step, Eq. (7.9), might be general base catalyzed, and it is easily seen that inclusion of the general base concentration [B] leads to Eq. (7.12) for the apparent second-order rate constant.

$$k = \frac{\alpha [B] k_1}{1 + \alpha [B]} \qquad (7.12)$$

The observation of an isotope effect then depends upon the value of α[B] relative to unity [21]. This quantity is affected by several factors: (1) concentration and efficacy of the catalyst B; (2) effect of the electrophile on the acidity (leaving tendency) of the departing proton; (3) steric crowding in the intermediate.

The present view is that a two-step mechanism (ArS_E2, bimolecular aromatic electrophilic substitution) can account for all kinetic observations. We have gone so far, in Eq. (7.7), as to postulate a structure for the intermediate, but have given little evidence to support this structure. Our next concern is, therefore, with the nature of the intermediate on the substitution reaction path.

The electronic properties of the σ-complex or benzenonium ion type of intermediate 20 were utilized in the preceding section to rationalize, in qualitative terms, the effects of substituents on orientation and reactivity in benzene derivatives. Much evidence has been obtained

20

about the properties of these species, and there is no
doubt of their existence in suitable systems. Olah, who
has been an important contributor in this field, has been
able to prepare isolable compounds whose physical and
chemical properties allow them to be formulated as σ-complexes [22]. These stable σ-complexes are obtained as
salts; Eq. (7.13) is a typical reaction.

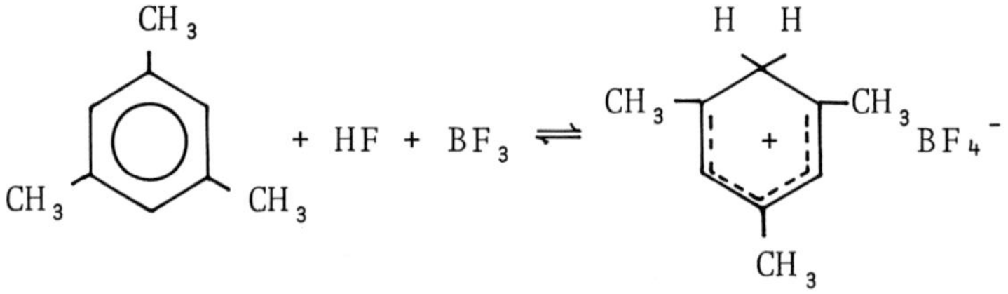

(7.13)

Such salts have definite decomposition temperatures, they
give the correct analyses, they conduct electricity, and
they are colored. Even more interesting, when they are
heated they yield the substitution products that would be
anticipated under similar conditions from appropriate
reactants. Thus alkylation occurs when the isolated salt
21 is heated.

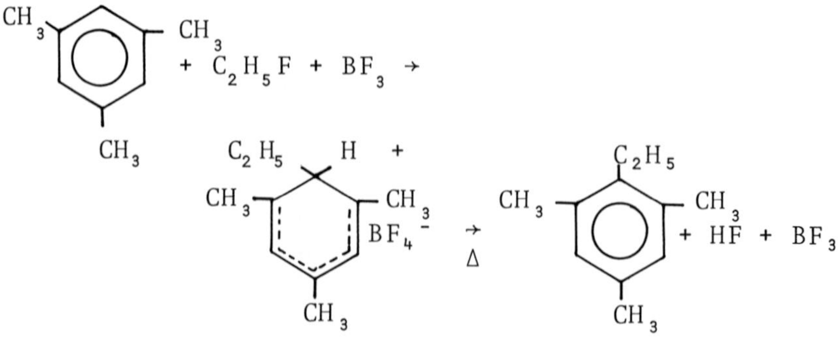

21

Ultraviolet and nuclear magnetic resonance spectral studies
on related compounds provide convincing evidence of the
reality of σ-complexes [23].   None of this is proof, however,   that σ-complexes are the required intermediate in
normal substitution reactions.   It would be most desirable
to detect such an intermediate in an actual substitution
reaction and to correlate its kinetic behavior with that
of the overall reaction.   Since the foregoing kinetic
analyses indicated that the steady-state approximation is
valid for most substitutions, it is likely that the concentration of intermediate at no time reaches high levels;
its detection is therefore expected to be difficult.
     Zollinger and co-workers [24] have, however, described
a reaction in which the steady-state condition does not
apply.   The bromination of G salt, Eq. (7.14), occurs with
buildup of the intermediate, which can be studied kinetically and spectroscopically.

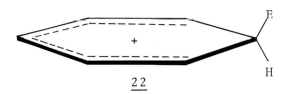

It is now generally accepted that σ-complexes occur
on the reaction path in electrophilic aromatic substitutions.   The intermediate will be like 22, and the reaction

coordinate will have a local minimum corresponding to the
σ-complex, as in Fig. 3.4. The transition state for the
formation of this σ-complex intermediate will have a configuration between that of the fully formed intermediate
22 and the initial state, 23. The Hammond postulate should

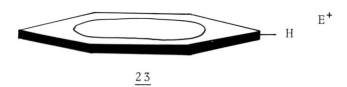

23

apply here. If the electrophile is not highly reactive,
the σ-complex 22 will be very unstable, and the transition
state will resemble closely the σ-complex. If E⁺ is highly reactive, the σ-complex will be more stable, and so the
transition state will occur earlier; in this case the initial state 23 might be a fair model for the transition
state [12,25].
      Another species has been proposed by some authors as
an intermediate in the aromatic substitution process; this
is the π-complex, sometimes represented as 24.

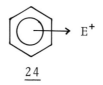

24

This symbolism is meant to represent a dative covalency
from the electron-rich π system of the aromatic nucleus
to the electron-deficient electrophile [26].
      Since the maximum electron density of the π system
is along the periphery, and is unevenly distributed in a
substituted benzene, the dative bond is more localized
than is suggested by 24. This line of thinking leads to
the view that a π-complex may be on the reaction pathway
between the initial state and the σ-complex. Banthorpe
[26] has concluded that the present evidence for π-complexes on the reaction path is very slight, whereas Olah
[27] believes that a π-complex is a reasonable model of a
transition state lying close to the initial state.

Some Electrophiles and Their Reactions. We consider now
a few selected electrophilic substitutions, with particular attention on the nature of the attacking electrophile

(a) <u>Nitration</u> [28]. A large proportion of the mechanistic investigations on electrophilic aromatic substitution has been carried out with the nitration reaction.
In most respects, for example in its response to directive
effects, the nitration reaction is typical of aromatic
substitutions. Its kinetic behavior, as revealed by isotope studies (see the preceding discussion), fits the
general pattern but reveals atypical parameters. Nitration
of aromatic substrates occurs under these conditions:
(1) in nitric acid alone, or with added strong acids such
as sulfuric acid; (2) with nitric acid in some organic
solvents, especially acetic acid and nitromethane; (3)
with acyl nitrates in organic solvents; (4) with nitronium
salts in organic solvents; (5) with dinitrogen pentoxide
in organic solvents; (6) by a process of nitrosation followed by oxidation. In the first five of these, the
attacking electrophile is believed to be the nitronium
ion, NO_2^+. This species has been shown to exist in nitrating media. Thus in sulfuric acid solutions of nitric
acid, cryoscopic measurements give an i-value of about 4.
This is consistent with these equilibria:

$$H_2SO_4 + HNO_3 \rightleftharpoons HSO_4^- + NO_2-OH_2^+$$

$$NO_2-OH_2^+ \rightleftharpoons NO_2^+ + H_2O$$

$$H_2SO_4 + H_2O \rightleftharpoons HSO_4^- + H_3O^+$$

The overall reaction is

$$2H_2SO_4 + HNO_2 \rightleftharpoons NO_2^+ + H_3O^+ + 2HSO_4^-$$

That NO_2^+ is the active electrophile is supported by
kinetic evidence. The rate of nitration in sulfuric acid-
nitric acid mixtures is first order in the aromatic substrate and in nitric acid. This merely says that nitric
acid or some species derived from it in a fast pre-equilibrium is the active reagent. In organic solvents, however,
with reactive substrates (even including benzene), the
rate equation is zero order, being independent of the substrate (with the nitric acid concentration in excess).
Less reactive substrates in the same system give rate
equations first order in substrate. The zero-order rate
process means that the substrate is not involved in the
rate-determining step, which must therefore be a slow
process undergone by nitric acid. This cannot be a proton
transfer, which would be a fast reaction; and it must result in an electrophilic species. The conclusion is that
this rate-determining step is

$$NO_2-OH \rightleftharpoons NO_2^+ + OH^-$$

Since NO_2^+ is known to be present in other nitrating species, it is inferred that this species is the active nitrating agent in these as well as in the systems showing the zero-order kinetics.

The nitronium ion, being a strong electrophile, should coordinate readily with bases; the stronger the base the less powerful the resulting "carrier" of the nitronium ion should be in the substitution reaction. Therefore this series should be progressively less effective as nitrating agents:

NO_2^+, nitronium ion

$NO_2-OH_2^+$, nitric acidium ion

NO_2-ONO_2, dinitrogen pentoxide

NO_2-OCOR, acyl nitrate

NO_2-OH, nitric acid

Nitronium salts, such as $NO_2^+BF_4^-$ and $NO_2^+ClO_4^-$, have been developed into effective nitrating agents for synthetic use [29]. Synthetic procedures for many aromatic nitro preparations have been collected [30].

(b) Diazo Coupling [31]. The general diazo coupling reaction is shown in Eq. (7.15) [32]. We therefore first

$$Ar-N_2^+ + C_6H_5-R \rightarrow Ar-N=N-C_6H_4-R + H^+ \qquad (7.15)$$

consider the diazotization process in which the aromatic diazonium ion $Ar-N_2^+$ is produced from an aromatic amine. Equation (7.16) is the overall reaction for the diazotization of a primary aromatic amine, which is usually carried out in dilute mineral acid solution.

$$Ar-NH_2 + HNO_2 + HCl \rightarrow Ar-N_2^+Cl^- + 2H_2O \qquad (7.16)$$

The reaction sequence [33] involves N-nitrosation [Eq. (7.17)], then a tautomeric shift to give the diazohydroxide, (7.18), and finally acid-assisted loss of hydroxide, Eq. (7.19).

$$Ar-NH_2 + HNO_2 \rightarrow Ar-NH-NO + H_2O \qquad (7.17)$$

$$Ar-NH-NO \rightleftharpoons Ar-N=N-OH \qquad (7.18)$$

$$Ar-N=N-OH + HC\ell \rightarrow Ar-N_2^+C\ell^- + H_2O \qquad (7.19)$$

The sum of these reactions gives Eq. (7.16).

The actual nitrosating agent is not nitrous acid, but a more reactive species derived from it. The most powerful nitrosating agent is NO^+, the nitrosonium ion. This electrophile can combine with bases (analogous to the behavior of the nitronium ion), giving these species (in hydrochloric acid medium).

NO^+, nitrosonium ion

$NO-OH_2^+$, nitrous acidium ion

$NO-C\ell$, nitrosyl chloride

$NO-NO_2$, dinitrogen trioxide (nitrous anhydride)

$NO-OH$, nitrous acid

The nitrosating power decreases in this order, since the NO^+ ion is associated with bases of increasing strength. In any solution the effective reagent will depend upon the relative concentrations and reactivities of several species. Under normal conditions it appears, on the basis of kinetic studies, that the effective species are the nitrous acidium ion, the nitrosyl salt, and dinitrogen trioxide. These are related as in (7.20) [34].

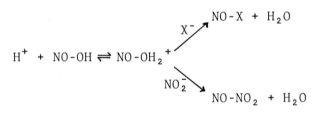

$$(7.20)$$

In aqueous solution the formation of NOX is favored over that of N_2O_3. The bromide, NOBr, is a more effective nitrosating agent than is $NOC\ell$, which is more effective than $NOHSO_4$.

The amine reactant in the diazotization reacts in the unprotonated form. The rate of diazotization is therefore pH dependent; more acidic solutions decrease the rate by placing the substrate in the unreactive protonated form. Opposed to this effect is the equilibrium (7.20), in which formation of the nitrosating agents is favored by high

acidity. The consequence of these opposing effects is
that the rate is maximal at some intermediate pH depending
in part upon the pK_a of the amine.

These principles can be applied for the determination
of primary aromatic amines by titration with standard sodium nitrite into a mineral acid solution of the amine
[35]. Hydrochloric acid gives faster rates than sulfuric
acid, and an alkali bromide can be added to increase the
rate, for the reason given above. Little attention is
usually given to optimization of pH, because the reaction
rate is fast enough for most amines without this additional
effect; control of this variable might by considered when
a further rate increase is desirable [36].

Finally we turn to the electrophilic substitution, the
diazo coupling reaction. The kinetics of this reaction
were studied by Conant and Peterson [37], using phenolic
compounds as the aromatic substrates. They wrote the rate
equation as $v = k'[ArN_2OH][Ar'-OH]$; since this is an electrophilic substitution, however, it seems more reasonable
to express this in the kinetically equivalent form $v =
k[ArN_2^+][Ar'-O^-]$. This is now the accepted formulation
[38]. For amine substrates the corresponding rate equation is $v = k[ArN_2^+][Ar'-NR_2]$. This rate behavior means
that a maximum exists in the pH-rate profile, the location
of the maximum depending upon the acid dissociation constants for the diazonium ion and the aromatic substrate
(usually a phenol or an amine). The rate increases with
pH in the acid region, the rate maximum for most systems
probably lying on the alkaline side of neutrality [39].

Aromatic diazonium ions are relatively unreactive
electrophiles, so coupling takes place only with activated
aromatic compounds, primarily phenols and amines, which
react in their conjugate base forms as noted above. The
reactivity of the diazonium ion can be altered by ring
substitution, with the expected changes occurring as shown
by the data in Table 7.VI, which are approximate relative
rates for the coupling of para-substituted benzenediazonium ions with several phenols [40]. Electron withdrawal,
as by the p-nitro group, intensifies the electrophilic
character of the diazonium group, whereas electron release
decreases electrophilic reactivity by increasing electron
density at the electrophilic site.

Because of the low reactivity of aromatic diazonium
ions, they are quite selective. Manipulation of reactivity, as in the preceding paragraph, will also change
selectivity, both intramolecular and intermolecular. The
ortho/para ratio is sensitive to several factors besides
the reactivity of the electrophile; these include the nature of the reaction medium, temperature, catalysts, and
location of substituents on the substrate. Monosubstituted

substrates appear to undergo nearly exclusive para coupling. 1-Naphthol and 1-naphthylamine couple predominantly
in the 4-position [41].

Table 7.VI   Relative Rates of Coupling by
para-Substituted Benzenediazonium Ions [40]

R in $R-C_6H_4-N_2^+$	Relative rate
OCH_3	0.1
CH_3	0.4
H	1
Br	13
SO_3^-	13
NO_2	1300

     A reaction closely related to diazo coupling can be
included in this section. This process, called oxidative
coupling, was first developed with p-phenylenediamines,
which upon oxidation in the presence of phenols or anilines couple to give indaniline or indamine dyes, respectively. The reaction with 1-naphthol is shown in Eq.
(7.21). This process is called the Nietzki-Fischer reaction.

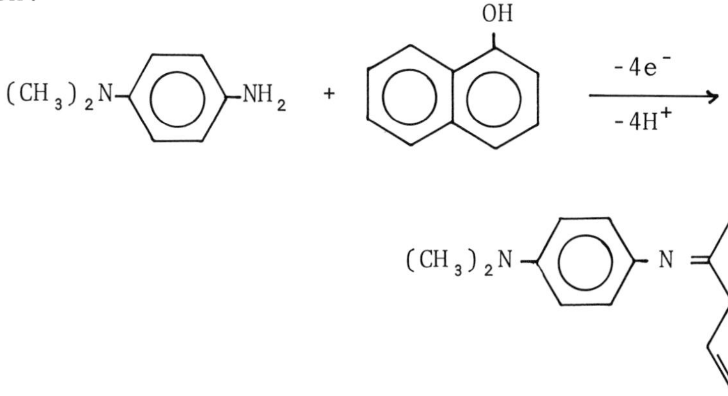

$$(7.21)$$

     The reaction has been generalized, and a series of
hydrazones has been found to be valuable reactants [42].
Equation (7.22) is an example of oxidative coupling of
one of these hydrazones with phenol.

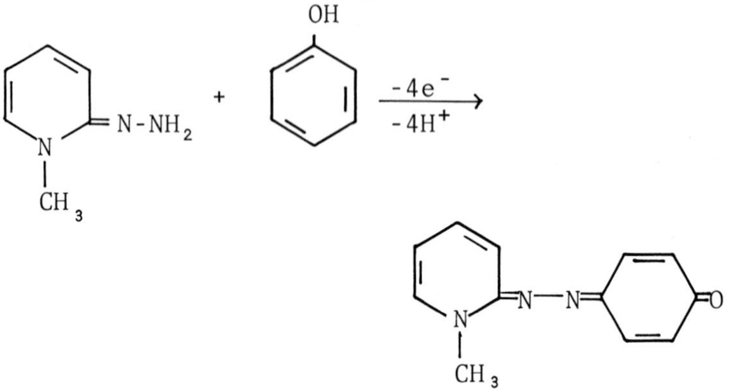

$$(7.22)$$

The actual electrophile in the substitution reaction is believed to be 25.

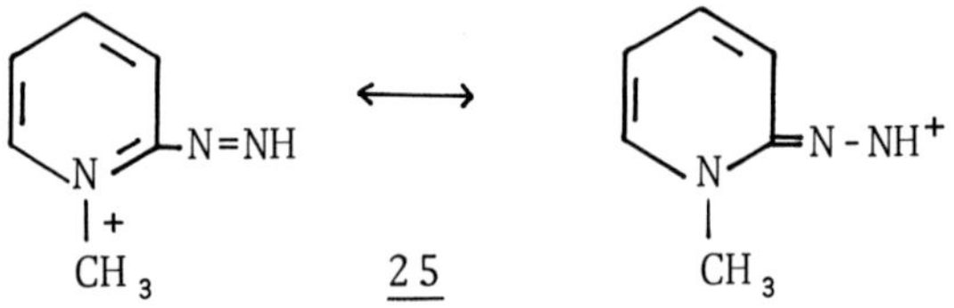

(c) Halogenation. The net result of an aromatic halogenation is the introduction of X^+, but the actual attacking species is not necessarily this one. In aqueous solution this halogenonium ion may be combined with various bases or "carriers." The stronger such a base is, the less potent the resulting halogenating agent. Thus brominating power decreases in this series of species: Br^+; $Br\text{-}OH_2^+$; $Br\text{-}Br$; $Br\text{-}OAc$; $Br\text{-}OH$. Kinetic studies of the halogenation (bromination, chlorination, iodination) of phenols and aromatic amines yield rate equations with terms of the form $k[IOH][PhOH]$, for example. This, of course, is kinetically equivalent to $k'[IOH_2^+][PhO^-]$, which seems mechanistically more reasonable. Moreover, in aqueous solution this term is also indistinguishable from $k''[I^+][PhO^-]$. Similar results are found with aniline as the substrate, giving the corresponding terms $k[IOH][PhNH_3^+]$, $k'[IOH_2^+][PhNH_2]$, and $k''[I^+][PhNH_2]$. The substitution is ortho-para directed. Since the anilinium ion is expected to be meta-directing, the first of these equivalent terms appears to be eliminated [43].

The halogen carriers that are effective electrophiles vary with the particular halogen. Kinetic evidence indi-

cates that these electrophilic species are effective in
the substitutions named, in aqueous media:

Chlorination : Cl^+; $Cl-OH_2^+$; $Cl-Cl$; $Cl-OH$

Bromination : Br^+ or $Br-OH_2^+$; $Br-Br$

Iodination : I^+ or $I-OH_2^+$

In nonaqueous polar solvents, like acetic acid and nitromethane, molecular halogens readily halogenate polymethylbenzenes [44]. The reaction is catalyzed by electrophiles,
which may act by polarizing the molecular halogen and
assisting the transformation of X-X to X^+-X^-. A second
molecule of halogen may function as this electrophile,
giving a rate term second order in halogen. Lewis acids,
such as zinc chloride, are very effective catalysts.
    Orientation and reactivity in halogenations depend
upon the identity of the electrophilic species, and therefore upon the reaction conditions. Norman and Taylor [45]
have reviewed this subject. An interesting substrate for
substitution is the metal chelate of acetylacetonate anion,
26, where M is Cr(III), Co(III), or Rh(III).

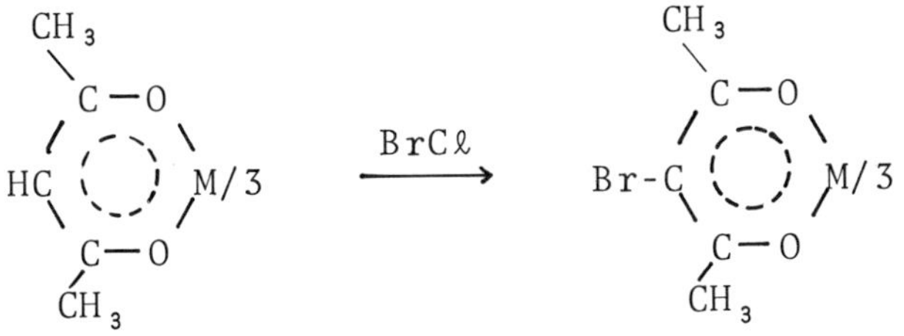

$$(7.23)$$

26

The chelate ring has aromatic character. Numerous substitutions other than halogenation can be carried out [46].

    (d) Alkylation and Acylation. Weakly electrophilic
carbon atoms exist in molecules containing electron-withdrawing groups bonded to carbon. Thus the carbon atom in
methyl chloride is electrophilic. Such molecules are not
strongly electrophilic, and they will not commonly react
with an aromatic substrate. When, however, this electronic polarization is augmented by coordination of the electron-withdrawing atom with a strong acid (Lewis acid or
protonic acid), the electrophilicity may be sufficiently
increased to allow aromatic substitution to occur. Substitution of a carbon fragment to give alkylated or acylated aromatics by this general procedure is called the
Friedel-Crafts reaction. Some similar processes can be

included, as shown in Table 7.VII.

Table 7.VII   Electrophilic Alkylation and
Acylation Reactions

Reaction	Reagent	Acid	Product
Friedel-Crafts alkylation	alkyl halides, R-X	$AlCl_3$, $SnCl_4$, $FeCl_3$, etc.	Ar-R
Friedel-Crafts alkylation	alcohols, R-OH	H_2SO_4, HF, BF_3-H_2O	Ar-R
Friedel-Crafts acylation	acid halides, anhydrides, RCOX	$AlCl_3$, etc.	Ar-COR
Blanc chloro-methylation	CH_2O + HCl	$ZnCl_2$, H_2SO_4	Ar-CH_2Cl
Gatterman-Koch reaction	CO + HCl	$AlCl_3$, Cu_2Cl_2	Ar-CHO
Hoesch reaction	RCN + HCl	$AlCl_3$, $ZnCl_3$	Ar-COR

A general formulation of the Friedel-Crafts reaction includes complex formation between the reagent and the acid, followed by attack of electrophilic carbon on the aromatic substrate to give the usual σ-complex, which loses a proton to yield the product. This scheme is shown in Eqs. (7.24)-(7.26) [47].

$$CH_3Cl + AlCl_3 \rightleftharpoons CH_3Cl \cdot AlCl_3 \qquad (7.24)$$

$$Ar-H + CH_3Cl \cdot AlCl_3 \rightarrow Ar \underset{CH_3}{\overset{H}{<}} + \quad + AlCl_4^- \qquad (7.25)$$

$$Ar \underset{CH_3}{\overset{H}{<}} + \quad + AlCl_4^- \rightleftharpoons Ar-CH_3 + HCl + AlCl_3 \qquad (7.26)$$

If a stable carbonium ion can be generated from the reagent, then this electrophilic species may be produced in a pre-equilibrium (7.27), with the rate-determining step being the reaction of the carbonium ion with the aromatic substrate.

$$(CH_3)_3CCl + AlCl_3 \rightleftharpoons (CH_3)_3C^+ + AlCl_4^- \qquad (7.27)$$

Alkylations often occur with concomitant rearrangement

of the alkyl group or intramolecular migration of the alkyl group to another position on the ring. Alkylation is a relatively unselective process, as indicated by the rate and distribution data for isopropylation in Table 7.V [3]. Strongly deactivated substrates, such as nitrobenzene, do not undergo alkylation. Interestingly, the strongly acti-vating hydroxy, alkoxy, and amino groups do not activate substrates markedly toward alkylation, presumably because coordination between the Lewis acid and the substituent heteroatom reduces electronic delocalization.

Friedel-Crafts acylations are usually carried out with acyl halides or anhydrides, though esters, amides, and the free acids have been used. The electrophilic species may be a polarized 1:1 complex of the acyl reagent and the Lewis acid (27), or the acylium ion 28. There is evidence

$$O \cdots AlCl_3$$
$$\| $$
$$C$$
$$/ \ $$
$$R \quad X$$

$$R - \overset{+}{C} = O$$

27   28

for the existence of both species in solution, though the predominating species depends upon the particular system [48,49].

Friedel-Crafts acylations display several marked con-trasts with alkylations. The acylation reactions are much more selective than are alkylations. Moreover, acylations are cleaner reactions, at least partly because introduction of an acyl group into the substrate deactivates it, so monosubstitution occurs; alkylation, on the other hand, activates the substrate, and further alkylation often takes place. Stoichiometric quantities of the Lewis acid are required in acylation, possibly because the ketone product can complex with the acid; in alkylation only catalytic amounts of the Lewis acid are needed.

7.2 Analytical Reactions

Diazo and Oxidative Coupling. The diazo coupling reaction is the most valuable of the electrophilic aromatic substi-tutions in present analytical practice. The process is summarized in Eqs. (7.28) (diazotization) and (7.29) (coupling).

$$Ar-NH_2 + HNO_2 + HCl \rightarrow Ar-N_2^+Cl^- + 2H_2O \qquad (7.28)$$

$$Ar-N_2^+Cl^- + H-Ar'-R \rightarrow Ar-N=N-Ar'-R + HCl \qquad (7.29)$$

We will consider the kinds of compounds that can be determined with these reactions, some of the specific reagents that have been used, and the significance of the reaction chemistry in its analytical applications. The closely related oxidative coupling reactions are included.

Reaction (7.28) forms the basis of a titrimetric determination of primary aromatic amines as mentioned earlier in this chapter. Our present concern is with the coupling reaction, and this route provides a powerful analytical method for primary aromatic amines, $Ar-NH_2$. Such an amine is diazotized and then coupled to an aromatic substrate (the coupling reagent) chosen for its suitability to the particular analysis. The final measurement is nearly always spectrophotometric.

The second major analytical use of this process is for the determination of the coupling component, $H-Ar'-R$ in Eq. (7.29). In this case the diazonium reagent $Ar-N_2^+$ is selected to suit the sample. Since diazonium ions are not highly reactive electrophiles, the diazo coupling method is applicable primarily to activated aromatic compounds, such as phenols, aromatic amines, and aryl ethers. Table 7.VIII gives a few of the many analytical applications of the diazo coupling reaction for the detection or determination of either a diazotizable amine or an aromatic substrate susceptible to coupling.

The most widely used coupling component for the determination of diazotized primary aromatic amines is N-(1-naphthyl)ethylenediamine, 29, the Bratton-Marshall reagent [50]. Coupling probably occurs predominantly in

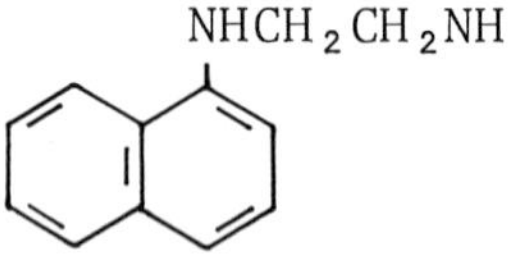

NHCH$_2$CH$_2$NH

29

the 4-position, to give the azo compound 30. Most of the coupling products with the Bratton-Marshall reagent exhibit an absorption maximum near 545 nm. In selecting

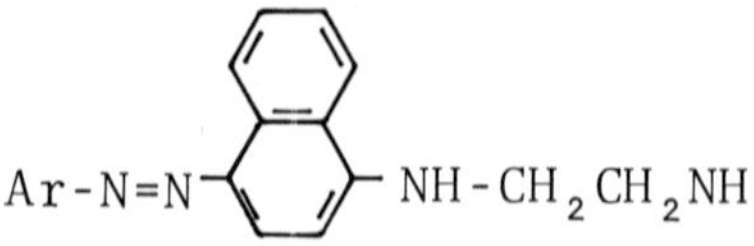

Ar-N=N⟨ ⟩NH-CH$_2$CH$_2$NH

30

Table 7.VIII   Examples of Diazo Coupling Analyses[a]

The sample as diazonium ion, $Ar-N_2^+$

Sample	Coupler	Ref.
Sulfonamides, $H_2N-C_6H_4-SO_2NHR$	N-(1-naphthyl)-ethylenediamine	b
Procaine, $H_2N-C_6H_4-CO_2CH_2CH_2NEt_2$	N-(1-naphthyl)-ethylenediamine	c
p-Aminobenzoic acid	resorcinol	d
Reduced folic acid	N-(1-naphthyl)-ethylenediamine	e
Aniline	2,6-Diaminopyridine	f

The sample as coupling component, $H-Ar'-R$

Sample	Diazonium ion	Ref.
Aniline	azobenzene-4-diazonium fluoborate	g
Tyrosine	diazotized arsanilic acid	h
Thiamine	diazotized p-aminoacetophenone	i
Aminotriazole	diazotized H-acid	j
8-Hydroxyquinoline	p-nitrobenzenediazonium chloride	k

[a]See Eqs. (7.28) and (7.29).
[b]J. T. Woods and G. H. Schneller, Pharmaceutical Analysis (T. Higuchi and E. Brochmann-Hanssen, eds.), Wiley (Interscience), New York, 1961, Chap. 5.
[c]F. J. Bandelin and C. R. Kemp, Ind. Eng. Chem., Anal. Ed., 18, 470 (1946).
[d]W. L. Koltun, J. Am. Chem. Soc., 79, 5681 (1957).
[e]R. A. Kaselis, W. Leibmann, W. Seaman, J. P. Sickels, E. I. Stearns, and J. T. Woods, Anal. Chem., 23, 746 (1951).
[f]L. J. Dombrowski and E. L. Pratt, ibid., 43, 1042 (1971) (fluorometric finish).
[g]E. Sawicki, T. W. Stanley, and T. R. Hauser, Chemist-Analyst, 48, 30 (1959).
[h]M. Tabachnick and H. Sobotka, J. Biol. Chem., 234, 1726 (1959).
[i]O. Mickelsen and R. S. Yamamoto, Methods Biochem. Anal., 6, 221 (1958).

[j]B. B. L. Agrawal and E. Margoliash, Anal. Biochem., 34, 505 (1970).
[k]M. A. Leonard and M. P. Murray, Anal. Letters, 3, 67 (1970) (photometric titration).

N-(1-naphthyl)ethylenediamine as the preferred coupling agent for sulfanilamide determination, Bratton and Marshall compared 16 potential reagents for their speed of coupling, stability of color, and solubility of the coupling product. The effects of reaction conditions on analytical results with the Bratton-Marshall coupler have been studied by Dux and Rosenblum [51] and Kaselis et al [52]. Optimum color intensity is achieved in the pH range 1.0-1.6. At the absorption maximum of 545 nm, the apparent molar absorptivity of the coupling products of sulfanilamide and sulfaquinoxaline with 52 are 5.15×10^4 and 5.25×10^4, respectively [51]. Beer's law is not obeyed over wide ranges of concentration. If the final colored solution should be too concentrated for accurate spectrophotometric measurement, it is not possible simply to dilute the solution [52]. High salt concentrations decrease the color intensity.

Sulfonamides of the general structure $H_2N-C_6H_4-SO_2NHR$ are determined according to the route described by Eqs. (7.28) and (7.29). Some of the medicinal sulfonamides are acylated aromatic amines, $R'CONH-C_6H_4-SO_2NHR$; after preliminary hydrolysis of the amide function the primary aromatic amine can be diazotized and coupled in the usual way. Aromatic nitro compounds that can be reduced to the amine can similarly be determined; thus nitroimidazoles can be reduced, diazotized, and coupled. Folic acid, 31, is reduced by zinc amalgam to p-aminobenzoylglutamic acid, 32, which is diazotized and coupled with the Bratton-Marshall reagent [52,53].

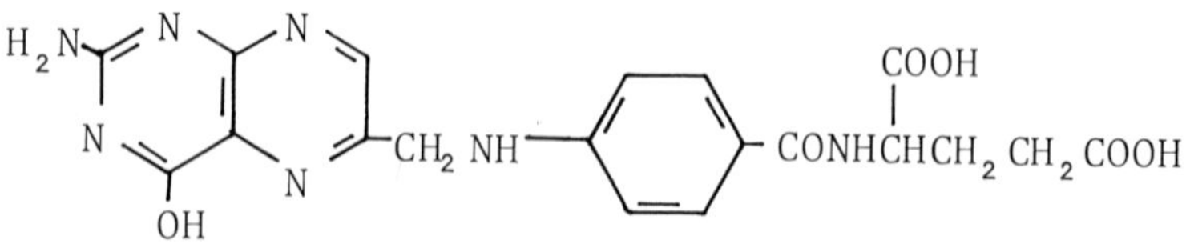

31

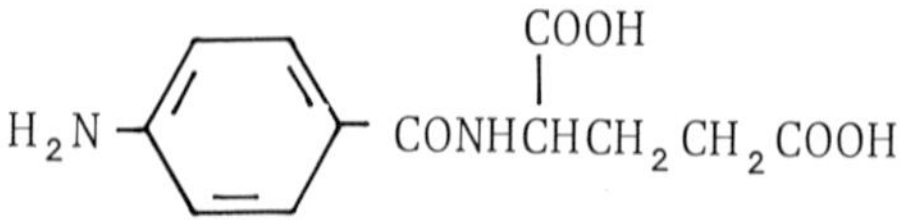

32

Aryl diazonium salts can themselves be determined by coupling. Koltun [54] measured the extent of coupling between diazotized p-aminobenzoic acid and insulin by adding excess resorcinol (the coupling component) and measuring the resulting p-carboxyphenylazoresorcinol spectrophotometrically, using the predetermined molar absorptivity $\epsilon_{385} = 2.15 \times 10^4$. A titrimetric method for diazonium salts is based on the coupling reaction with 1,3-diaminotoluene.

$$\text{Ar-N}_2^+ \; + \; \underset{\underset{\text{NH}_2}{}}{\overset{\overset{\text{CH}_3}{}}{\bigcirc}}\text{-NH}_2 \; \rightarrow \; \text{Ar-N=N-}\underset{\underset{\text{NH}_2}{}}{\overset{\overset{\text{CH}_3}{}}{\bigcirc}}\text{-NH}_2 \; + \; \text{H}^+$$

The excess coupler is titrated with a standard solution of p-toluenediazonium chloride [55].

Some rather indirect analyses make use of the coupling reaction for the final measurement. Thus toluene 2,4-diisocyanate is hydrolyzed to the diamine, which is diazotized and coupled with N-(1-naphthyl)ethylenediamine [56]. Acid anhydrides react with aromatic amines and nitrite to give a diazonium ion that can be coupled for colorimetric analysis [57], possibly as shown below (the species RCO_2NO is an acyl nitrite):

$$(RCO)_2O + NO_2^- \rightarrow RCO_2NO + RCOO^-$$

$$RCO_2NO + Ar\text{-}NH_2 \rightarrow Ar\text{-}N_2^+ + RCOO^- + H_2O$$

$$Ar\text{-}N_2^+ + Ar'\text{-}H \rightarrow Ar\text{-}N=N\text{-}Ar' + H^+$$

Even nitrous acid has been determined by means of the diazo coupling reaction. Urea can be quantitatively decomposed with an excess of nitrous acid.

$$H_2NCONH_2 + 2HONO \rightarrow 2N_2 + CO_2 + 3H_2O$$

The unreacted nitrous acid is measured by adding sulfanilic acid, $H_2N\text{-}C_6H_4\text{-}SO_3H$, which is diazotized by the nitrous acid. α-Naphthylamine is added as a coupler to develop a color [58]. Some N-1-substituted nitroimidazoles have been determined by alkaline hydrolysis to give nitrite ion, acidification, reaction with sulfanilamide, and coupling of the diazotized sulfanilamide with the Bratton-Marshall reagent [59]. A similar method for pentaerythritol tetranitrate uses p-nitroaniline and azulene as the reactants [60].

We turn now to analyses in which the sample takes the part of the coupling component (aromatic substrate) and the aryl diazonium ion is the analytical reagent. Most of the diazonium ions thus used are substituted with electron-withdrawing groups, which tend to increase the electrophilicity of the diazonium function. Among the reagents that have been used are the diazonium ions derived from p-aminobenzoic acid [54], p-nitroaniline, p-arsanilic acid [61], sulfanilic acid, sulfanilamide, etc. Urbanyi and Mollica [62] preferred diazotized 4-amino-6-chloro-m-benzenedisulfonamide as the diazo reagent, among 27 aromatic amines studied, for the determination of estradiol. Usually the diazonium ion is prepared in situ. It is possible, however, to prepare some solid isolable diazonium salts, the aryldiazonium fluoborates [63], and these have been used analytically [64].

An alkaline solution in dimethylformamide or dimethylsulfoxide of the azo dye formed from a primary aromatic amine and azobenzene-4-diazonium fluoborate develops a color intensity considerably greater than in neutral solution [64a]. The coupling product of azulene with this diazonium reagent, Eq. (7.30), exhibits a striking change in its visible absorption spectrum upon acidification, which Sawicki et al. [64b] attribute to delocalization resulting from monoprotonation and then diprotonation on nitrogen.

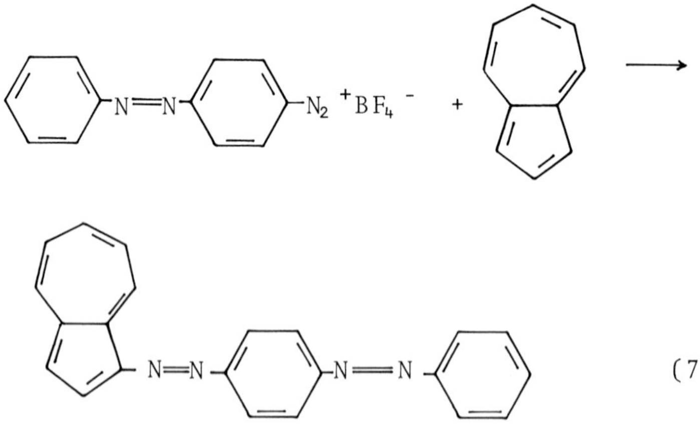

$$(7.30)$$

33

The orange-red neutral compound 33 has λ_{max} 480 nm, ε_{max} 3.5 × 10^4. In moderate concentrations of hydrochloric acid a red-violet color is produced, with λ_{max} 557 nm, ε_{max} 4.8 × 10^4, and in concentrated HCl the color is blue,

λ_{max} 620 nm, $\varepsilon_{max} \simeq 1.1 \times 10^5$. It would be of interest to know the positions of protonation; conceivably C-protonation occurs with enhancement of quinoidal delocalization in the protonated forms.

The environment of tyrosyl and histidyl residues in proteins, particularly those residues implicated in the active sites of enzymes, has been investigated by measuring the reactivity of these residues toward diazonium reagents [61,65]. p-(Arsonic acid)-benzenediazonium fluoborate and diazonium 1H-tetrazole (DHT, 34) have been used [66].

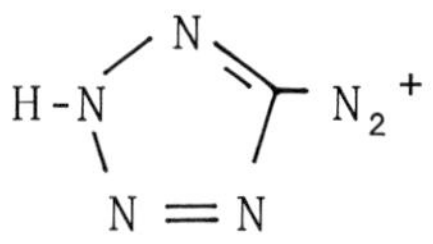

$$H-N \begin{array}{c} N \\ \diagdown \\ \diagup \\ N = N \end{array} N_2^+$$

<u>34</u>

The oxidative coupling of phenols provides several analytical methods, as shown in Eq. (7.31) for the reagent 3-methyl-2-benzothiazolinone hydrazone (MBTH) [67]. The Gibbs test for phenols unsubstituted in the para position,

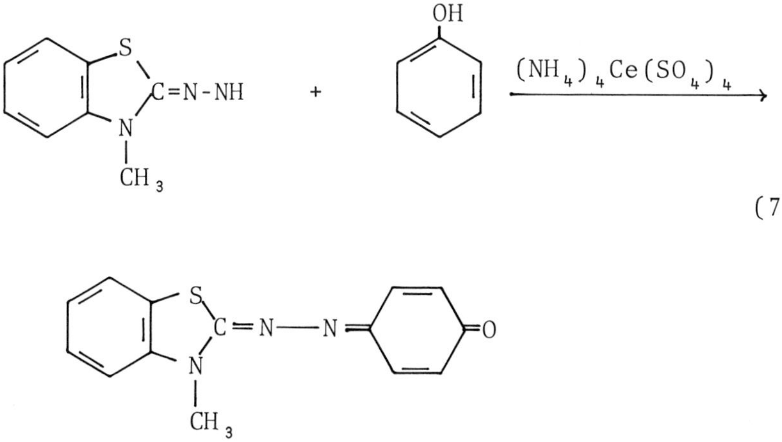

$$\underset{\underset{CH_3}{|}}{\overset{S}{\underset{N}{\bigcirc}}} C = N-NH \quad + \quad \overset{OH}{\bigcirc} \quad \xrightarrow{(NH_4)_4 Ce(SO_4)_4}$$

(7.31)

$$\underset{\underset{CH_3}{|}}{\overset{S}{\underset{N}{\bigcirc}}} C = N - N = \bigcirc = O$$

based on their alkaline coupling with 2,6-dibromoquinone chloroimide [68], may be considered an oxidative coupling reaction. The oxidative coupling of phenols with 4-amino-antipyrine, Eq. (7.33), forms the basis of a sensitive analytical method [68c,d,69]. Phenols and arylamines have been oxidatively coupled with N,N-dimethyl-p-phenylene-diamine [70].

Thiamine, <u>35</u>, which is nonfluorescent, is converted

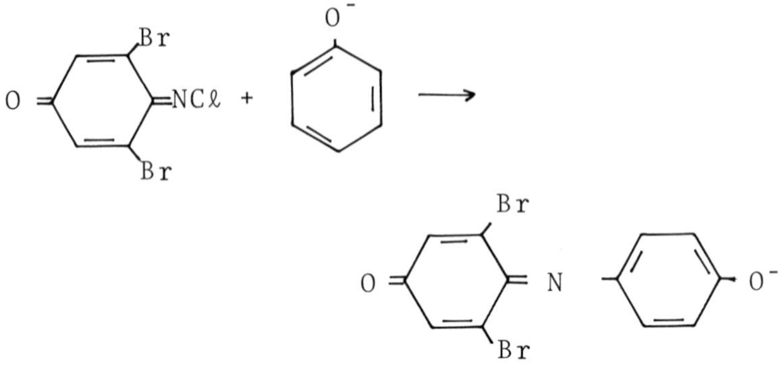

$$(7.32)$$

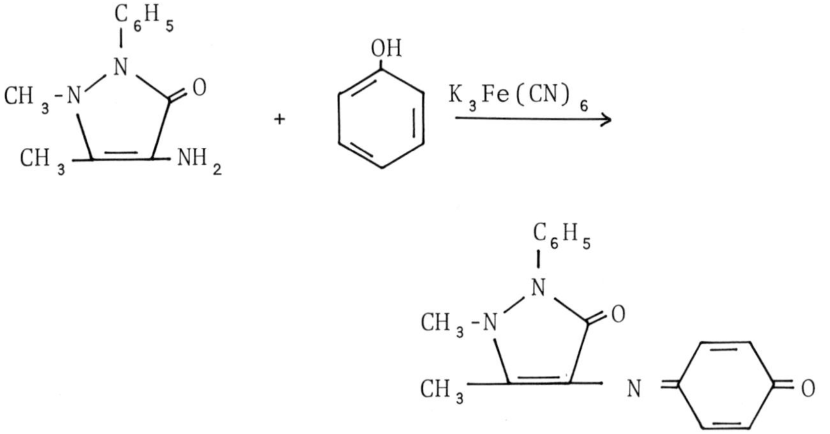

$$(7.33)$$

upon oxidation with ferricyanide or other oxidizing agents [71] to thiochrome, 36, which fluoresces strongly.

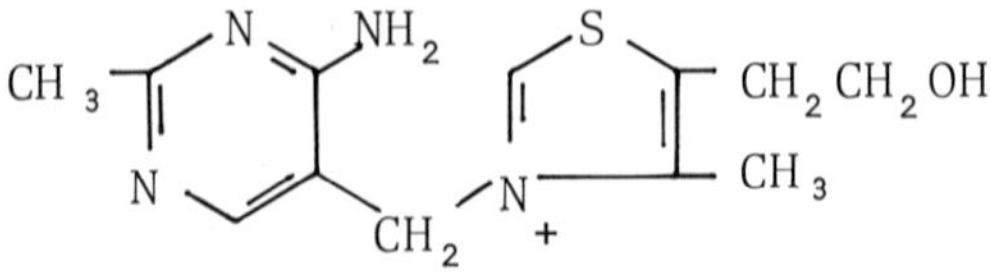

35

$$\underline{36}$$

This reaction appears to be an intramolecular oxidative coupling.

The primary reason that diazo coupling reactions have found wide analytical use is because of the desirable spectral properties of the azo products. Absorption bands appear in the visible region, and intensities are high; molar absorptivities greater than 1×10^4 are the rule. It is fortuitous that diazonium ions are relatively mildly electrophilic in aromatic substitution reactions, with the consequence (the selectivity-reactivity relationship) that they are unusually selective reagents for this type of reaction. This selectivity results in good reproducibility in analytical systems, for this is a reflection of coupling product distribution. Moderate variations in reaction conditions do not appreciably alter the pattern of electrophilic substitution in these reactions, and mono-substitution seems to predominate. Because of the low reactivity of aryl diazonium electrophiles, the analytical applications have emphasized activated aromatic substrates. In few instances, however, is it probable that optimal coupling reaction conditions are employed, and the scope of these reactions might be extended with more quantitative data on the coupling kinetics in analytical reactions.

Halogenation. Bromination and iodination have been applied to the qualitative and quantitative analysis of activated aromatic compounds. Substitution by bromine is considered first.

Phenols and aryl ethers are brominated with bromine in polar solvents to give isolable crystalline derivatives suitable for characterization [72]. Mono-, di-, tri-, and even tetra-bromination may occur. Phenol yields 2,4,6-tribromophenol.

Quantitative determinations based on bromination require close control of reaction conditions to achieve reproducible bromine consumption. Most procedures are titrimetric [73]. A typical procedure for phenol determination uses a standard solution of potassium bromate in excess potassium bromide, which upon acidification generates bromine according to the stoichiometry shown by

$$BrO_3^- + 5Br^- + 6H^+ \rightarrow 3Br_2 + 3H_2O \qquad (7.34)$$

The electrophilic substitution occurs in aqueous solution; the reaction for phenol is shown by

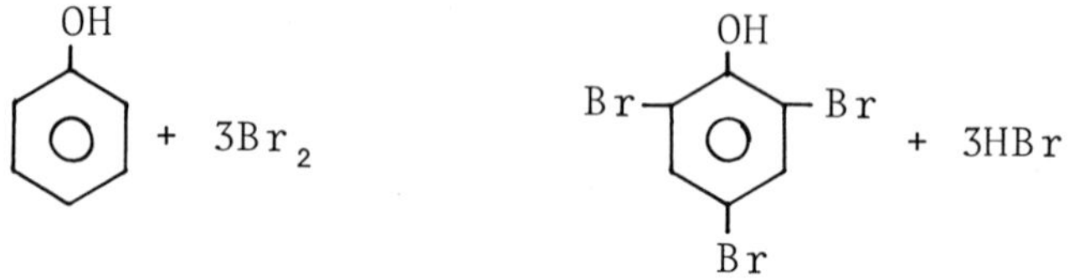

$$+ 3HBr \qquad (7.35)$$

The unreacted bromine is determined by adding excess potassium iodide and titrating the iodine released with standard thiosulfate.

Cheronis and Ma [73] summarize reaction conditions for the quantitative determination of phenols. Bromine monochloride also has been used to brominate phenols and aromatic amines. An indirect use of the bromination procedure allows the determination of anhydrides [74]. The anhydride is reacted with an excess of 2,4-dichloroaniline. After acylation, Eq. (7.36), the unreacted 2,4-dichloroaniline is determined by bromination.

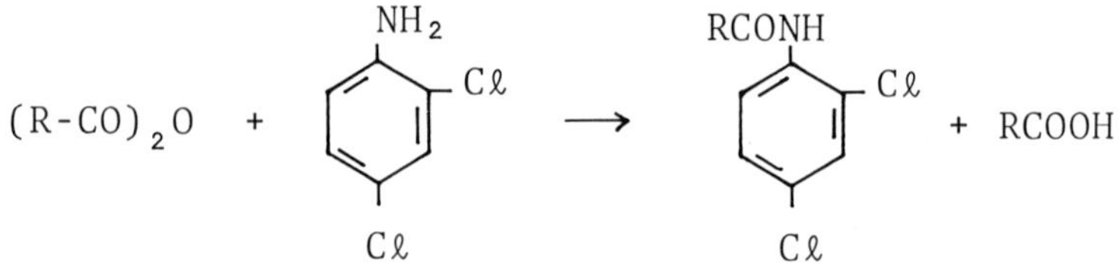

$$(R-CO)_2O \; + \qquad\qquad \longrightarrow \qquad\qquad + \; RCOOH \qquad (7.36)$$

Phenols have been determined by ultraviolet spectrophotometry following their bromination [75].

Iodination of phenols has formed the basis of titrimetric, gravimetric [76], and spectrophotometric analyses. Resorcinol, 37, is quantitatively iodinated in less than one minute at pH 5.0 [77]. After iodination the excess iodine is titrated with standard thiosulfate. Tri-iodina-

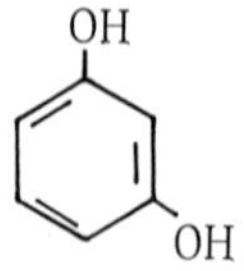

37

tion occurs, and no interference is observed from alkyl and aryl phenols. This selectivity is kinetically controlled, and probably has its origin in the circumstance that three of the four open positions on the resorcinol nucleus are ortho or para to two hydroxy groups.

Willard and Wooten [78] have described colorimetric methods for certain substituted phenols based on reaction with iodine. An ortho-alkyl or aryl phenol reacts with iodine in alkaline medium to form a colored polymeric substance, presumably a para-linked quinone. Another color reaction is selective for o-dihydroxy and m-dihydroxy benzenes; resorcinol and catechol together form, in the presence of iodine, a colored product that dissolves in acetone with an absorption maximum at 725 nm. This reaction can be used to determine whichever reactant is present in smaller amount. The nature of the reaction is not known; possibly both substitution and oxidation by the iodine are required.

Iodination of amino acid side chains in proteins reveals marked differences in reactivity among several like residues in the same protein. For example, of the four tyrosine residues in α-chymotrypsin, the carboxy-terminal tyrosine of the B chain is most reactive toward iodine, being iodinated about three times as rapidly as the other tyrosines. Loss of enzymatic activity is associated with tyrosine iodination [79]. Similar phenomena are seen with ribonuclease A, whose iodination yields three iodinated tyrosine residues and one iodinated histidine; the initial loss of enzymatic activity appears to be due to tyrosyl iodination [80]. Lysozyme undergoes iodination in water giving two iodinated tyrosyl residues, whereas in 8 $\underline{M}$ urea all three of the tyrosyl residues are iodinated (diiodo-tyrosine is the major product). It is inferred that in the native protein one of the tyrosyl residues is "buried," and therefore inaccessible to the reagent [81].

As an analytical reagent for substitution reactions, iodine has the disadvantage of being also an oxidant, which may lead to unwanted side reactions. These have been observed with peptide and protein substrates [82].

Nitration; Nitrosation; Friedel-Crafts Reactions. Aromatic hydrocarbons, aromatic halides, and aryl ethers yield crystalline nitro substitution products upon treatment with a nitric acid-sulfuric acid nitrating mixture [83]. The nitration product depends upon the reaction conditions, and di-nitro or tri-nitro derivatives are common.

Some quantitative use has been made of the nitration reaction. Nitration does not proceed with a quantitative yield, so an empirical working curve must be employed. An indirect application of nitration is used to analyze biphenyl, which is nitrated, the nitro group is reduced, and the resulting aromatic amine is diazotized and coupled to the Bratton-Marshall reagent [84]. Pyrene and related compounds containing active methylene can be nitrated, the nitrated product being dissolved in an alkaline medium

to give a color suitable for quantitative measurement [85].
Inorganic nitrate and nitric acid esters were determined
by hydrolysis to nitric acid, which nitrates 3,4-xylenol
to give, after separation and treatment with alkali, a
colored product [86].

Tetranitromethane, $C(NO_2)_4$, is a nitrating agent that
is fairly selective for tyrosyl residues in proteins, o-
nitrotyrosine being produced [87]. The reaction with phe-
nols has been studied kinetically; the rate law is $v = k[ArO^-][C(NO_2)_4]$. Bruice et al. [88] propose a free-
radical mechanism for the substitution, with the rate-
determining step being an electron-transfer mediated by a
charge-transfer complex.

Phenols can be determined by nitrosation (substitution of NO). β-Naphthol is titrated in acid solution with
standard sodium nitrite [89]. A colorimetric method is
based on formation of a nitrosophenol in acid solution;
upon treatment with alcoholic ammonia a colored compound,
λ_{max} 420 nm, is formed. This may have a quinone-type
structure [68c].

A qualitative color test for aromatic compounds utilizes a Friedel-Crafts reaction with chloroform and aluminum chloride [90]. The main product of the reaction with
chloroform is triphenylmethane, which arises from multiple
substitutions.

Friedel-Crafts acylations yield derivatives of aromatic hydrocarbons and halogenated aromatics [91]. Acetyl
chloride and aluminum chloride yield aryl methyl ketones,
which can be characterized via the usual carbonyl derivatives. Reaction with phthalic anhydride, Eq. (7.37), gives
crystalline o-aroylbenzoic acids.

$$\text{ArH} + \text{phthalic anhydride} \xrightarrow{Al\,Cl_3} \text{o-aroylbenzoic acid} \qquad (7.37)$$

Comment. Diazo coupling provides an excellent example of
an electrophilic aromatic substitution exploited for analysis. The azo product has a high molar absorptivity,
giving good sensitivity. Selectivity is achieved in two
ways: (1) the azo absorption maximum is near 550 nm, far
removed from that of many possible interferences; (2)
kinetic selectivity results from the mildly electrophilic
character of the diazonium ion. The reaction rate is dependent upon pH, control of which provides a further means
to selective analysis. Diazo coupling occurs in aqueous

solution, which is often an advantage.

The analytical possibilities of this reaction probably could be extended, by the use of reactive diazonium ions and optimal reaction conditions, to less activated substrates than the phenols and amines that are the usual samples (though decomposition of the diazonium ion may limit such uses). The simplicity of diazo coupling suggests its wider use as the finish step in analyses requiring preliminary reactions.

Friedel-Crafts acylation seems to be fairly selective; thus ρ = -9.1 for acetylation with acetyl chloride and aluminum chloride. This approach may deserve analytical attention, particularly since the product is a ketone, which is readily susceptible to further reactions.

The synthetic literature, especially recent results, should stimulate new analytical methods. If a synthetic process provides a yield of over 50%, it is likely that, with optimization of conditions for analytical use, a quantitative yield can be forced (unless the reaction is severely equilibrium limited). A potentially useful example, in the present context, is provided by thallation of aromatic compounds. The reaction of benzyl alcohol is shown as an example:

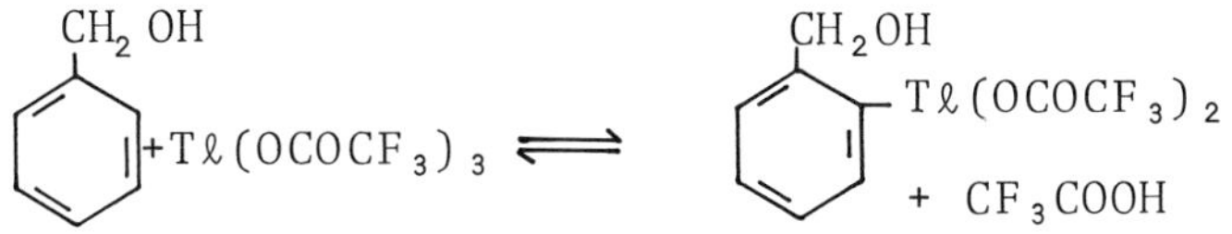

Treatment of the arylthallium ditrifluoroacetate with aqueous potassium iodide yielded over 99% of the o-iodobenzyl alcohol. The isomer distribution in this thallation route to aromatic iodides can be influenced by the reaction conditions. When the reaction is under thermodynamic control (reflux conditions), meta substitution predominates. With kinetic control, ortho substitution is observed when the substituent can interact with the electrophile directly, delivering it to the adjacent position as shown in Eq. (7.5); otherwise para substitution occurs [14]. Typical yields of the aromatic iodides were over 80%. Substituted phenols and aromatic nitriles have been prepared in good yields by suitable treatment of the arylthallium ditrifluoroacetate [92].

Aromatic substitution with the proton as a leaving group is, in effect, a functional group reaction of the phenyl ring, and it offers the possibility of a general method for the determination of aromatic compounds.

Several aromatic substances have been converted to o-nitro-
sophenols by means of a cupric ion-hydrogen peroxide-
hydroxylamine reagent [93]. Simple hydroxylation of aromatic compounds to give phenols is an effective analytical
approach. The reagent is the Hamilton hydroxylating system, consisting of hydrogen peroxide, catechol, and ferric
ion [94]. Phenols thus produced by hydroxylation of aromatic compounds are detected by oxidative coupling with
4-aminoantipyrine [95].

Problems

1. The rate of chlorination of fluorobenzene relative to
    the rate for benzene is 0.74; 11% of o-chlorofluoro-
    benzene and 89% of p-chloro-fluorobenzene are formed.
    Calculate the partial rate factors.

    Answer: o_f = 0.24, p_f = 3.95.

2. The rate of nitration of biphenyl relative to benzene
    is 35, and the isomer distribution is 69% ortho, 31%
    para. What are the partial rate factors?

    Answer: o_f = 36, p_f = 33.

3. It has been suggested that nitration by nitronium salts
    $NO_2^+X^-$ in organic solvents occurs via a π-complex,
    giving a σ-complex, which leads to a second π-complex
    from which the proton is eliminated, finally giving
    the product. Sketch the presumed reaction coordinate
    diagram.

    Answer: See R. O. C. Norman and R. Taylor, Ref. [5],
    p. 67.

4. In the diazo coupling analysis of a primary aromatic
    amine, the excess nitrous acid is destroyed by reac-
    tion with sulfamic acid before addition of the coupling
    agent.

    $$HNO_2 + H_2NSO_3H \rightarrow H_2SO_4 + N_2 + H_2O$$

    Suggest a reason.

5. Devise an analytical approach to the quantitative de-
    termination of both solutes in an aqueous solution of
    aniline and phenol.

6. Suggest a method for revealing the aromatic compounds after a thin-layer chromatographic separation of the naturally occurring amino acids.

References

1. C. K. Ingold, Structure and Mechanism in Organic Chemistry, Cornell Univ. Press, Ithaca, N. Y., 1953, Chap. VI; E. Berliner, Progr. Phys. Org. Chem., 2, 253 (1964); R. O. C. Norman and R. Taylor, Electrophilic Substitution in Benzenoid Compounds, Monograph 3 in Reaction Mechanisms in Organic Chemistry, Elsevier, Amsterdam, 1965; L. M. Stock, Aromatic Substitution Reactions, Prentice-Hall, Englewood Cliffs, N. J., 1968.
2. C. K. Ingold, Ref. [1], pp. 246-247.
3. L. M. Stock and H. C. Brown, Advan. Phys. Org. Chem., 1, 35 (1963).
4. Usually the ortho directive effect is quantitatively different from the para effect, and sometimes the ortho partial rate factor is comparable with or smaller than the meta factor, as for the acetylation of toluene shown in Fig. 7.1. It will be appreciated that anomalies in ortho position behavior are not surprising. This question is considered subsequently.
5. G. W. Wheland, J. Am. Chem. Soc., 64, 900 (1942); R. O. C. Norman and R. Taylor, Electrophilic Substitution in Benzenoid Compounds, Elsevier, New York, 1965.
6. G. S. Hammond, J. Am. Chem. Soc., 77, 334 (1955); see also Sec. 3.3.
7. L. M. Stock, Aromatic Substitution Reactions, Prentice-Hall, Englewood Cliffs, N. J., 1968, p. 48.
8. J. H. Ridd, Aromaticity, Chemical Society Special Publication No. 21, London, 1967, pp. 149-162.
9. R. O. C. Norman and R. Taylor, Ref. [5], p. 53.
10. M. J. S. Dewar, T. Mole, and E. W. T. Warford, J. Chem. Soc., 1956, 3581.
11. A. Albert, Heterocyclic Chemistry, Essential Books, Fair Lawn, N. J., 1959, pp. 62-69.
12. R. O. C. Norman and G. K. Radda, J. Chem. Soc., 1961, p. 3610.
13. Further examples of steric effects are given by G. S. Hammond and M. F. Hawthorne, Steric Effects in Organic Chemistry (M. S. Newman, ed.), Wiley, New York, 1956. See also, for steric effects in polymethylbenzenes, E. Baciocchi and G. Illuminati, Progr. Phys. Org. Chem., 5, 1 (1967).

14. E. C. Taylor, F. Kienzle, R. L. Robey, and A. Mc-Killop, J. Am. Chem. Soc., $\underline{92}$, 2175 (1970). Thallation is a reversible reaction.

15. P. B. D. de la Mare and J. H. Ridd, Aromatic Substitution, Butterworths, London, 1959, p. 82; M. Charton, J. Org. Chem., $\underline{34}$, 278 (1969).

16. The rates relative to benzene decrease, meaning that the more reactive reagent also exhibits less intermolecular discrimination. The limiting relative rate would be 0.833 (5/6) for a monosubstituted benzene.

17. B. Pullman and A. Pullman, Progr. Org. Chem., $\underline{4}$, 31 (1958).

18. L. Melander, Arkiv Kemi, $\underline{2}$, 211 (1950).

19. G. S. Hammond, J. Am. Chem. Soc., $\underline{77}$, 334 (1955).

20. E. Berliner, Progr. Phys. Org. Chem., $\underline{2}$, 253 (1964); H. Zollinger, Advan. Phys. Org. Chem., $\underline{2}$, 163 (1964).

21. Inclusion of general base catalysis in the one-step mechanism leads to the rate equation $v = k[ArH][E^+][B]$, giving a linear dependence on $[B]$, unlike the prediction by Eq. (7.12).

22. G. A. Olah and S. J. Kuhn, J. Am. Chem. Soc., $\underline{80}$, 6535, 6541 (1958); G. A. Olah, A. E. Pavlath, and J. A. Olah, ibid., $\underline{80}$, 6540 (1958).

23. E. Berliner, Ref. [20], pp. 262-270. The parent benzenonium ion, $C_6H_7^+$, has been demonstrated by NMR measurements on a solution of benzene in $SO_2C\ell F$-SO_2F_2; G. A. Olah, R. H. Schlosberg, D. P. Kelly, and Gh. D. Mateescu, J. Am. Chem. Soc., $\underline{92}$, 2546 (1970).

24. M. Christen and H. Zollinger, Helv. Chim. Acta, $\underline{45}$, 2057, 2066 (1962); M. Christen, W. Koch, W. Simon, and H. Zollinger, ibid., $\underline{45}$, 2077 (1962).

25. E. Berliner, Ref. [20], p. 303.

26. The role of π complexes in organic reactions has been critically surveyed by D. V. Banthorpe, Chem. Revs., $\underline{70}$, 295 (1970).

27. G. A. Olah, Accounts Chem. Res., $\underline{4}$, 240 (1971).

28. C. K. Ingold, Ref. [1], pp. 269-288; J. H. Ridd, Accounts Chem. Res., $\underline{4}$, 248 (1971).

29. G. A. Olah and S. J. Kuhn, J. Am. Chem. Soc., 83, 4564 (1961); $\underline{84}$, 3684 (1962); G. A. Olah, S. J. Kuhn, and S. H. Flood, ibid., $\underline{83}$, 4571, 4581 (1961).

30. R. B. Wagner and H. D. Zook, Synthetic Organic Chemistry, Wiley, N. Y., 1953, Chap. 28; S. R. Sandler and W. Karo, Organic Functional Group Preparations, Academic, New York, 1968, pp. 437-444; L. F. Fieser and M. Fieser, Reagents for Organic Synthesis, Wiley, New York, 1967.

31. H. Zollinger, Azo and Diazo Chemistry, Wiley (Interscience), New York, 1961; P. A. S. Smith, The Chemistry of Open-Chain Organic Nitrogen Compounds, Vol

II, Benjamin, New York, 1966, Chap. 11.

32. The product R-N=N-R´ is an azo compound. When one
     of the bonds from nitrogen is to a nonorganic moiety,
     the compound is a diazo compound.

33. J. H. Ridd, J. Soc. Dyers Colourists, 75, 285 (1959).

34. H. Zollinger, Ref. [31], pp. 27-36.

35. S. Siggia, Quantitative Organic Analysis via Function-
     al Groups, 3rd ed., Wiley, New York, 1963, pp. 446-
     449; N. D. Cheronis and T. S. Ma, Organic Functional
     Group Analysis by Micro and Semimicro Methods, Wiley
     (Interscience), New York, 1964, p. 236.

36. Secondary aromatic amines, and aliphatic primary and
     secondary amines, react with nitrous acid to give N-
     nitroso products, and, for some primary aliphatic
     amines, alcohols.

37. J. B. Conant and W. D. Peterson, J. Am. Chem. Soc.,
     52, 1220 (1930).

38. R. Wistar and P. D. Bartlett, J. Am. Chem. Soc., 63,
     418 (1941); C. R. Hauser and D. S. Breslow, ibid.,
     63, 418 (1941).

39. H. Zollinger, Ref. [31], pp. 227-228.

40. L. P. Hammett, Physical Organic Chemistry, McGraw-
     Hill, New York, 1940, p. 314.

41. H. Zollinger, Ref. [31], pp. 253-255; H. S. Turner,
     J. Chem. Soc., 1949, p. 2282.

42. S. Hünig, J. Chem. Educ., 46, 734 (1969).

43. E. Berliner, J. Am. Chem. Soc., 72, 4003 (1950); C.
     K. Ingold, Ref. [1], p. 290. This argument is
     slightly weakened by the recent data of Ridd on
     directing effects in anilinium ions [8]; see Table
     7.III.

44. E. Baciocchi and G. Illuminati, Progr. Phys. Org.
     Chem., 5, 1 (1967).

45. R. O. C. Norman and R. Taylor, Ref. [5], pp. 134-152.

46. J. P. Collman, Reactions of Coordinated Ligands and
     Hamogeneous Catalysis (Adv. Chem. Ser., 37), American
     Chemical Society, Washington, D. C., 1963, p. 78.

47. Reaction (7.25) could be viewed as an aliphatic
     nucleophilic substitution in which $AlCl_4^-$ is displaced
     by the nucleophile Ar-H.

48. G. A. Olah, Science, 168, 1298 (1970).

49. The acetylium ion, CH_3CO^+, may be the effective aci-
     dic species in acidimetric titrations with perchloric
     acid in acetic anhydride medium.

50. A. C. Bratton and E. I. Marshall, Jr., J. Biol. Chem.,
     128, 537 (1939). This reagent is available as the
     dihydrochloride salt, which crystallizes from 6 N
     HCl as colorless prisms, m.p. 188-190°.

51. J. P. Dux and C. Rosenblum, Anal. Chem., 21, 1524
     (1949).

52. R. A. Kaselis, W. Leibmann, W. Seaman, J. P. Sickels,
     E. I. Stearns, and J. T. Woods, Anal. Chem., 23, 746
     (1941).
53. The spectrophotometric calibration curve is prepared
     with p-aminobenzoylglutamic acid rather than with
     folic acid. Very few spectrophotometric methods have
     been described in which the standard substance is
     different from the sample compound.
54. W. L. Koltun, J. Am. Chem. Soc., 79, 5681 (1957).
55. S. Siggia, Quantitative Organic Analysis via Func-
     tional Groups, 3rd ed., Wiley, New York, pp. 55-59,
     548.
56. K. Marcali, Anal. Chem., 29, 552 (1957).
57. H. F. Liddell and B. Saville, Chem. Ind., 16, 493
     (1957).
58. W. Brandt, Mikrochemie, 22, 181 (1937).
59. E. P. K. Lau, C. Yao, M. Lewis, and B. Z. Senkowski,
     J. Pharm. Sci., 58, 55 (1969); J. A. F. De Silva,
     N. Munno, and N. Strojny, ibid., 59, 201 (1970).
     The sensitivity of this approach has been increased
     by incorporating a double solvent extraction; G. R.
     Macchi and B. S. Cescon, Anal. Chem., 42, 1809 (1970).
60. V. Hankonyi and V. Karas-Gasparec, Anal. Chem., 41,
     1849 (1969).
61. L. Wofsy, H. Metzger, and S. J. Singer, Biochemistry,
     1, 1031 (1962).
62. T. Urbanyi and J. A. Mollica, Jr., J. Pharm. Sci.,
     57, 1257 (1968).
63. A. Roe, Organic Reactions (R. Adams, ed.), Vol. V,
     Wiley, New York, 1949, pp. 193-228.
64. (a) E. Sawicki, T. W. Stanley, and T. R. Hauser,
     Chemist-Analyst, 48, 30 (1959); (b) E. Sawicki, T.
     W. Stanley, and W. Elbert, Proceedings, 1961 Intern.
     Symp. Microchem. Tech. (N. D. Cheronis, ed.), Wiley
     (Interscience), New York, 1962, p. 633.
65. M. Tabachnick and H. Sobotka, J. Biol. Chem., 235,
     1051 (1960).
66. T. F. Spande, B. Witkop, Y. Degani, and A. Patchornik,
     Advan. Protein Chem., 24, 183 (1970).
67. H. O. Friestad, D. E. Ott, and F. A. Gunther, Anal.
     Chem., 41, 1750 (1969).
68. (a) H. D. Gibbs, J. Biol. Chem., 72, 649 (1927); (b)
     J. C. Dacre, Anal. Chem., 43, 589 (1971); (c) E. F.
     Mohler, Jr., and L. N. Jacob, ibid., 29, 1369 (1957);
     (d) D. Svobodova, J. Gasparic, and L. Novakova, Coll.
     Czech. Chem. Commun., 35, 31 (1970); D. Svobodova and
     J. Gasparic, ibid., 35, 1567 (1970).
69. D. Svobodova and J. Gasparic, Mikrochim. Acta, 1971,
     p. 384.

70.   D. N. Kramer and L. U. Tolentino, Anal. Chem., 43,
      834 (1971).
71.   O. Mickelsen and R. S. Yamamoto, Methods Biochem.
      Anal., VI, 223-228 (1958).
72.   R. L. Shriner, R. C. Fuson, and D. Y. Curtin, The
      Systematic Identification of Organic Compounds, 5th
      ed., Wiley, New York, 1964, pp. 275, 297, 348-9,
      374-6.
73.   N. D. Cheronis and T. S. Ma, Organic Functional Group
      Analysis by Micro and Semimicro Methods, Wiley (Inter-
      science), New York, pp. 340, 447-451.
74.   H. Roth, Microchim. Acta, 1958, 767.
75.   F. Pellerin and R. Chasset, Ann. Pharm. Fr., 27, 571
      (1969).
76.   M. Francois and L. Seguin, Bull. Soc., Chim. Fr.,
      53, 711 (1933).
77.   H. H. Willard and A. L. Wooten, Anal. Chem., 22, 585
      (1950).
78.   H. H. Willard and A. L. Wooten, Anal. Chem., 22, 423,
      670 (1950).
79.   S. K. Dube, O. A. Roholt, and D. Pressman, J. Biol.
      Chem., 239, 1809, 3347 (1964).
80.   I. Covelli and J. Wolff, J. Biol. Chem., 241, 4444
      (1966).
81.   I. Covelli and J. Wolff, Biochemistry, 5, 860 (1966);
      J. Wolff and I. Covelli, ibid., 5, 867 (1966).
82.   L. A. Cohen, Ann. Rev. Biochem., 37, 683 (1968).
83.   R. L. Shriner, R. C. Fuson, and D. Y. Curtin, Ref.
      [72], pp. 284, 348-349, 352-353, 357-358.
84.   R. B. Bruce and J. W. Howard, Anal. Chem., 28, 1973
      (1956).
85.   E. Sawicki and R. R. Miller, Anal. Chem., 30, 109
      (1958); E. Sawicki and T. W. Stanley, Chemist-Analyst,
      49, 77 (1960).
86.   A. C. Holler and R. V. Huch, Anal. Chem., 21, 1385
      (1949).
87.   G. R. Stark, Advan. Protein Chem., 24, 289 (1970).
88.   T. C. Bruice, M. J. Gregory, and S. L. Walters, J.
      Am. Chem. Soc., 90, 1612 (1968).
89.   M. Matrka, Coll. Czech. Chem. Commun., 25, 964 (1960).
90.   R. L. Shriner, R. C. Fuson, and D. Y. Curtin, Ref.
      [72], p. 116. For an unusual color test for benzene
      based on the formation of p-phenylazobenzene by elec-
      trophilic substitution see this reference, p. 115.
91.   R. L. Shriner, R. C. Fuson, and D. Y. Curtin, Ref.
      [72], p. 283.
92.   E. C. Taylor, H. W. Altland, R. H. Danforth, G.
      McGillivray, and A. McKillop, J. Am. Chem. Soc., 92,
      3520 (1970).

93. J. Bartos, Ann. Pharm. Fr., <u>27</u>, 759 (1969).
94. G. A. Hamilton, J. P. Friedman, and P. M. Campbell,
     J. Am. Chem. Soc., <u>88</u>, 5266 (1966); G. A. Hamilton,
     J. W. Hanifin, and J. P. Friedman, ibid., <u>88</u>, 5269
     (1966).
95. K. A. Connors and K. S. Albert, Anal. Chem., <u>44</u>, 879
     (1972).

Chapter 8.   NUCLEOPHILIC AROMATIC SUBSTITUTION

For certain aromatic substrates the nucleophilic displacement reaction (8.1) can occur.

$$N^- + \underset{X}{\text{(ring)}} \longrightarrow \underset{N}{\text{(ring)}} + X^- \qquad (8.1)$$

N^- is a nucleophile, which may be neutral or negatively
charged.   This substitution of one nucleophile by another
is not so characteristic of aromatic substrates as is the
electrophilic substitution process treated in Chap. 7.
One of the important reasons for this difference in behavior is that the departing nucleophile, X in Eq. (8.1),
must take with it an electron pair formerly involved in
bonding with the aromatic nucleus.   Hydrogen is therefore
not a good leaving group in nucleophilic aromatic substitution.   Several mechanisms have been identified in these
reactions, and these are briefly discussed in Sec. 8.1.
More extensive reviews are available [1].

8.1   Mechanisms of Nucleophilic Substitution

The Bimolecular Mechanism.   The majority of aromatic nucleophilic substitutions appear to proceed via an addition-
elimination pathway closely analogous to that described
for electrophilic substitution.   An intermediate 1 is
formed.

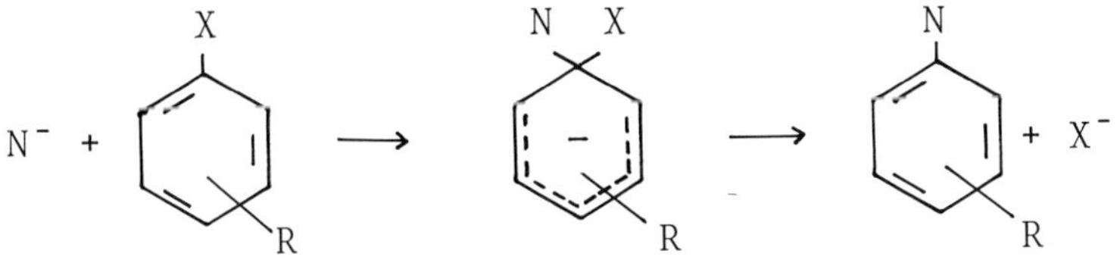

$$N^- + \underset{R}{\overset{X}{\text{(ring)}}} \longrightarrow \underset{R}{\overset{N\ X}{\text{(ring)}}} \longrightarrow \underset{R}{\overset{N}{\text{(ring)}}} + X^- \qquad (8.2)$$

1

This route overcomes in part the disadvantage noted above in the loss of X together with an electron pair; this heterolysis is made easier by the prior bonding of N⁻ with its electron pair. The reaction is facilitated by "good" (i.e., weakly nucleophilic) leaving groups, by substituents on the aromatic substrate that stabilize the intermediate, and by strongly nucleophilic reagents. A typical substitution is illustrated by the formation of 2,4-dinitroanisole from 2,4-dinitrochlorobenzene,

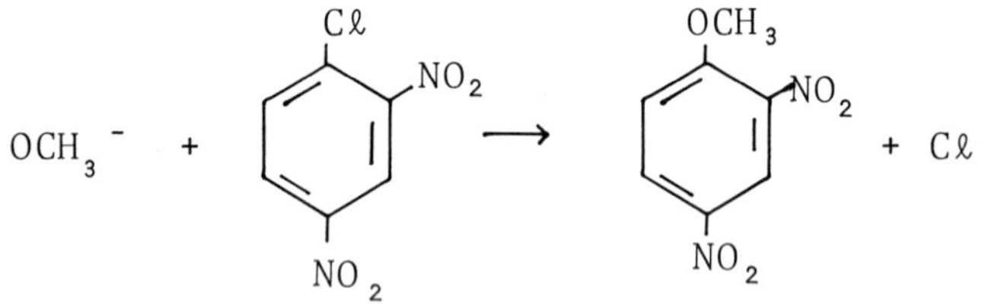

$$OCH_3^- \quad + \quad \text{[structure]} \quad \longrightarrow \quad \text{[structure]} \quad + \quad C\ell \qquad (8.3)$$

The benzenanion intermediate for this reaction is stabilized by conjugative interaction with the electron-withdrawing o,p nitro groups, which delocalize the charge as illustrated in 2.

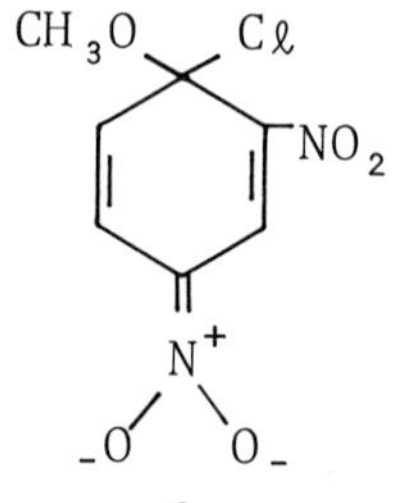

2

The existence of an addition intermediate on the reaction pathway is supported by several lines of evidence [2]. If the aromatic substrate is activated by electron-withdrawing groups and if the leaving group is strongly nucleophilic, evidently the intermediate will be relatively stable. Meisenheimer [3] found that treatment of 2,4,6-trinitroanisole, 3, with potassium ethoxide appeared to give the same salt as did 2,4,6-trinitrophenetole, 5, upon treatment with potassium methoxide. This salt can be formulated as 4, potassium 1-methoxy-1-ethoxy-2,4,6-trinitrocyclohexadienylide. Addition compounds of this type, often called Meisenheimer complexes, have since been well studied [4]. Their formation is an equilibrium process for which rate and equilibrium constants can be determined.

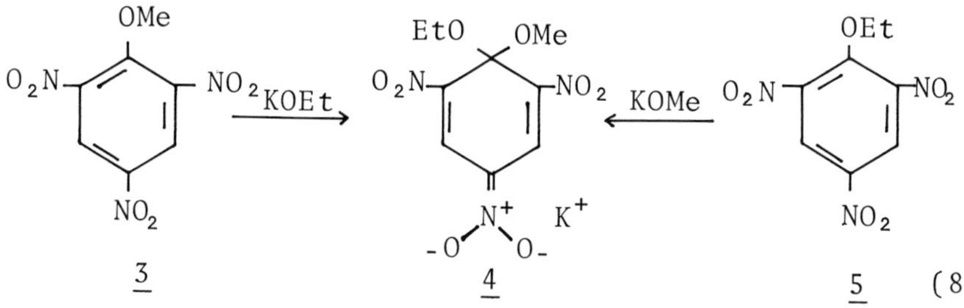

$$(8.4)$$

Table 8.I   Rate and Equilibrium Constants for
Meisenheimer Complexes of 2,4,6-Trisubstituted
Anisoles [5,6][a],[b]

R_1	R_2	R_3	$k_1/M^{-1}\text{-sec}^{-1}$	$10^3 k_{-1}/\text{sec}^{-1}$	K/M^{-1}
NO_2	NO_2	NO_2	17.3	1.04	17,000
CN	NO_2	NO_2	18.8	7.20	2,600
NO_2	NO_2	CN	6.1	22.0	280
CN	CN	NO_2	∿12	373	34
CN	NO_2	CN	2.0	198	10

[a]Cf. Eq. (8.5).
[b]At 25.0° in methanol.

Thus the stabilizing influences of substituent groups are
revealed by the data in Table 8.I on the formation and
decomposition of Meisenheimer complexes of isomeric 2,4,6-
cyanodinitroanisoles [5] and 2,4,6-dicyanonitroanisoles
[6].   The reaction is formulated as Eq. (8.5), the equili-

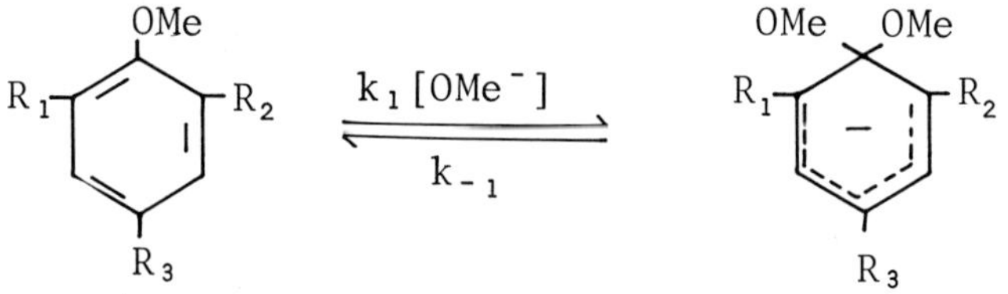

$$(8.5)$$

brium constant being $K = k_1/k_{-1}$.   Two immediate inferences from the data in Table 8.I are that the nitro group has a greater stabilizing effect on the complex than does the cyano group, and that the stabilizing power of a group is greater in the para position that in the ortho position. This latter conclusion follows from the observation that replacement of p-NO_2 by p-CN decreases K by a greater factor than does replacement of o-NO_2 by o-CN. It may be concluded that a para electron-withdrawing substituent is more activating in the aromatic nucleophilic substitution reaction than is the same substituent in an ortho position [7]. Note that in these reactions the relative stabilities are controlled primarily by k_{-1} (decomposition) rather than k_1 (formation).

Bernasconi [8] studied the Meisenheimer complexes formed from several amine nucleophiles with the substrate 1,3,5-trinitrobenzene.

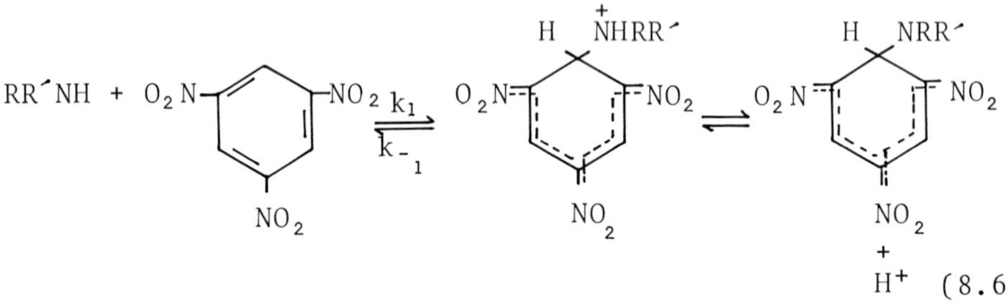

$$ \text{(8.6)} $$

In 10% dioxane-90% water, at 25°, k_1 increases with increasing basicity of the amine, as expected; thus for n-butylamine, pK_a = 10.68 and k_1 = 123 M^{-1}-sec^{-1}; piperidine, pK_a = 11.12, k_1 = 3 × 10^4 M^{-1}-sec^{-1}; pyrrolidine, pK_a = 11.30, k_1 = 8.1 × 10^4 M^{-1}-sec^{-1}. It is therefore surprising that k_{-1} is roughly the same (in the range 1.5 × 10^4 to 2.5 × 10^4 sec^{-1}) for the three amines; n-butylamine, being the weakest nucleophile, is expected to be the best leaving group [9].

Synthetic, kinetic, and spectroscopic studies on Meisenheimer complexes leave no doubt that addition compounds exist that can be formulated as shown above. These demonstrations strengthen the proposal that such species may be intermediates in nucleophilic substitution, but they do not require such a conclusion. Bunnett et al. [10] used the following argument to establish the involvement of an intermediate. In a substitution on the activated aromatic substrates 6, where the first atom in X is changed from one element to another, the rate should be markedly dependent upon the nature of this element if C-X bond breaking has made appreciable progress in the transi-

tion state of the rate-determining step. This would be
the expected result for a one-step displacement, Eq. (8.7),
analogous to the aliphatic S_N2 mechanism.

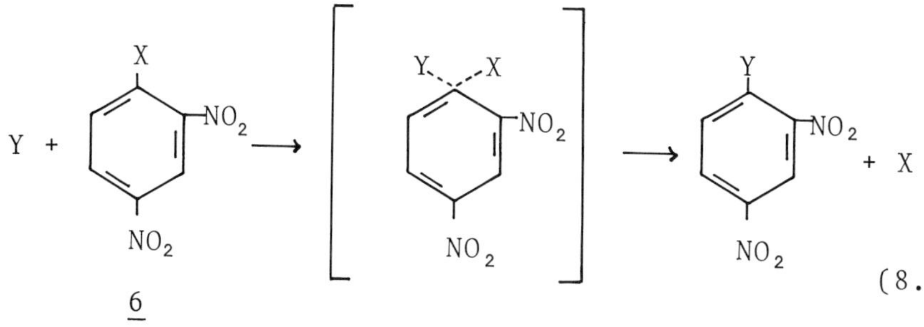

$$(8.7)$$

6

An insensitivity of rate to the element in the C-X bond
would rule out such a transition state. The reaction
studied was the substitution by piperidine of nine 1-sub-
stituted-2,4-dinitrobenzenes. The relative rates are
shown in Table 8.II.

Table 8.II   Relative Rates of Substitution by Piperidine
of 1-X-2,4-Dinitrobenzenes [10][a]

X	Relative rate
F	3300
NO_2	890
$OSO_2C_6H_4CH_3$-p	100
SOC_6H_5	4.7
Br	4.3
Cℓ	4.3
$SO_2C_6H_5$	3.2
$OC_6H_4NO_2$-p	3.0
I	1.0

[a]At 0.0° in methanol.

The results are conclusive. For the last six substituents, representing five elements bonded to carbon, the
maximum rate variation was fivefold. The simplest interpretation is in terms of a two-step mechanism, with the

first step rate determining, and with no significant
breaking of the C-X bond in the transition state for this
step. The minor variations in rate for the last six substituents in Table 8.II are reasonable, for the ease of
formation of a bond to the attacking nucleophile should be
dependent in part upon the nature of X. This effect is
presumably also responsible for the very facile substitution reactions undergone by the first three substrates in
Table 8.II; all of these substituents are very electronegative (activating) and should facilitate attack by the
nucleophile [11].

   Although other kinds of intermediates can be imagined
(charge-transfer complexes for example), the Meisenheimer
complex structure can best account for the available data
on nucleophilic substitutions of activated aromatic substrates. Most kinetic investigations of these substitutions have dealt with systems in which the steady-state
condition applies to the intermediate. Even in this situation several types of kinetic behavior can be distinguished. The two-step mechanism can be schematically
written as

$$Y + Ar-X \underset{k_{-1}}{\overset{k_1}{\rightleftharpoons}} ArYX \xrightarrow{k_2} Ar-Y + X \qquad (8.8)$$

Applying the steady-state approximation to the intermediate gives

$$v = \frac{k_1 k_2 [Ar-X][Y]}{k_{-1} + k_2} \qquad (8.9)$$

showing the bimolecular nature of the kinetics. The experimental second-order rate equation is therefore

$$k = \frac{k_1 k_2}{k_{-1} + k_2} \qquad (8.10)$$

Now, if $k_2 \gg k_{-1}$, Eq. (8.10) becomes simply $k = k_1$; the
formation of the intermediate is rate determining. The
data in Table 8.II for the substitution of several 1-
substituted-2,4-dinitrobenzenes by piperidine exemplify
this kinetic behavior. In these systems the group X
leaves the intermediate with greater facility than does
Y, which is the strong nucleophile piperdine.

   If $k_{-1} \gg k_2$, Eq. (8.10) becomes $k_1 k_2 / k_{-1}$. Displacement of fluoride by iodide (X = F, Y = I) illustrates
this case [12,13]. When k_{-1} and k_2 are comparable, the

full Eq. (8.10) describes the kinetics, and both steps
are partially rate controlling.

Another kind of kinetic distinction can be made between substitution by an anion, as in

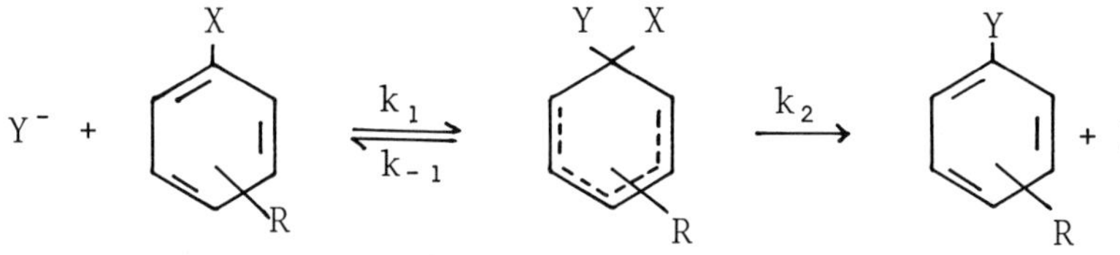

$$(8.11)$$

and by an amine, as in

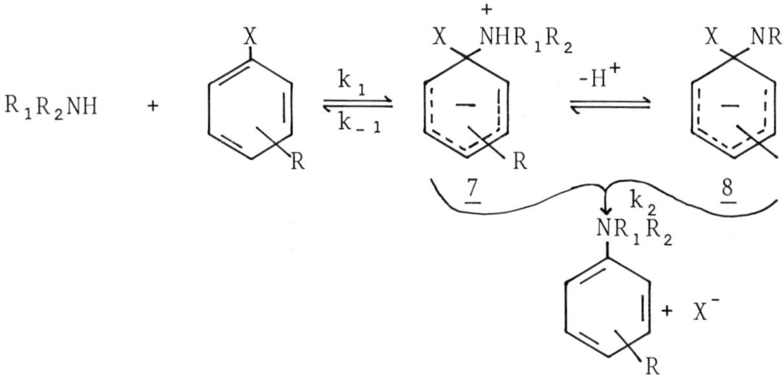

$$(8.12)$$

This latter process involves a proton transfer, which may
occur before, or concurrently with, expulsion of the leaving group. It may therefore be susceptible to acid-base
catalysis. Such catalysis will clearly be limited to the
second step (k_2), so it will be observable only in those
systems in which k_2 appears in the expression for the
second-order rate constant.

Bunnett and Randall [14] observed general base catalysis in the reaction of N-methylaniline with 2,4-dinitro-
fluorobenzene, whereas the reactions of 2,4-dinitrochloro-
benzene and 2,4-dinitrobromobenzene were not base catalyzed. This result is consistent with the above picture,
the ease of bond breaking [11,12] being Br>Cℓ>>F. Thus
for the fluoro compound k_{-1}>>k_2, and for the bromo and
chloro substrates k_2>>k_{-1} [15].

Three mechanisms have been suggested for the general
base catalysis:

a. Rate-determining transfer of a proton from 7 to

the conjugate base B of the catalyst, giving 8, followed
by rapid loss of X^-.
    b. Concerted breaking of the N-H and C-X bonds under
the influence of the base B.
    c. Rapid reversible reaction of 7 with the general
base B to give 8 and the conjugate acid BH^+, followed by
rate-determining loss of X^- from 8 catalyzed by the general acid BH^+. This mechanism is kinetically equivalent to
the general base catalytic mechanisms in (a) and (b).

    The detailed kinetic behavior of an unstable inter-
mediate cannot be evaluated under steady-state conditions,
because the concentration of the intermediate is essentially invariant. Orvik and Bunnett, however, have studied
a non-steady-state system, the reaction of 2,4-dinitro-1-
naphthyl ethyl ether with primary aliphatic amines in dimethyl sulfoxide solution, and have been able to observe
both steps of the substitution reaction [16]. A combination of kinetic and spectral evidence reveals that the
process can be written as follows:

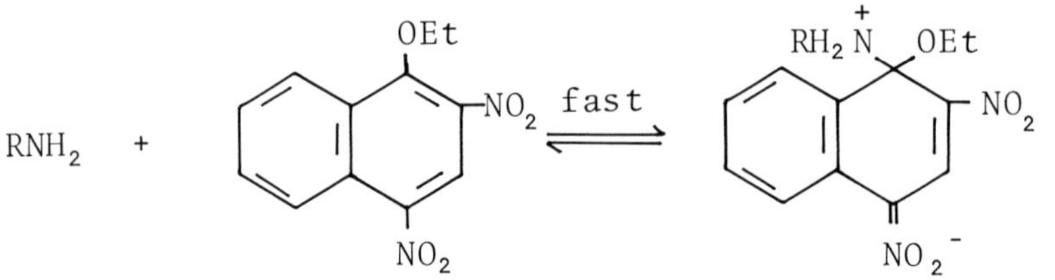

(8.13)

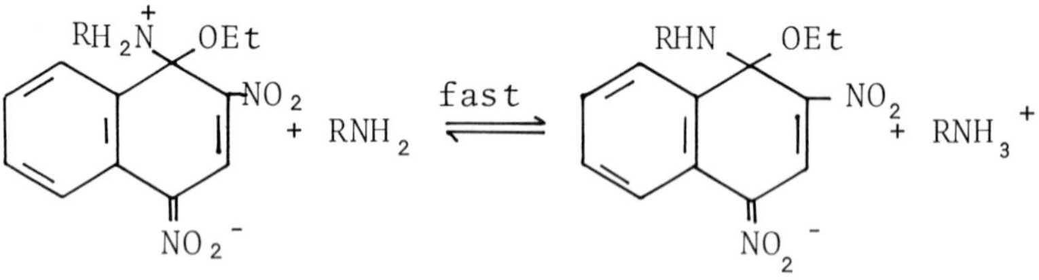

9   10   (8.14)

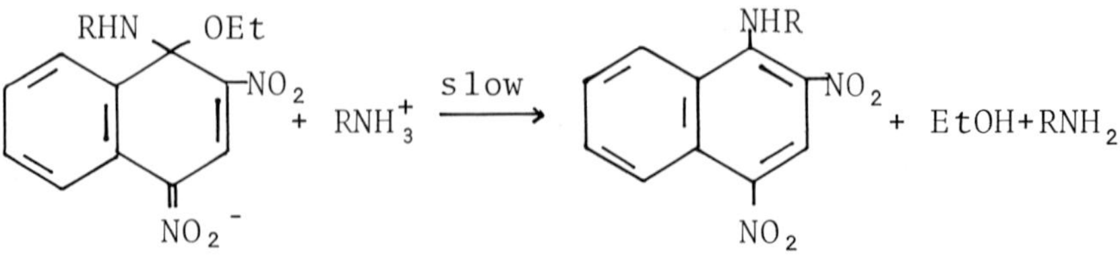

11   (8.15)

Formation of the Meisenheimer complex is very rapid. The
zwitterionic form 9 loses a proton to the general base,
with the equilibrium (8.14) lying far to the right. The
anionic intermediate 10 then is converted to product 11
in a slow step catalyzed by the conjugate acid of the
base. Thus mechanism (c) describes the general base catalysis in this reaction.
      The reaction of n-butylamine with 2,4-dinitrochloro-
benzene is catalyzed by n-butylamine. Since intermediate
formation is probably rate controlling in this reaction,
Ross [17] has suggested that the catalysis occurs in this
step with a proton being removed to give the anionic intermediate directly.

The Unimolecular Mechanism. Aryldiazonium ions react with
nucleophiles in aqueous solution showing first-order kinetics, with rates being approximately independent of the
nature and concentration of the nucleophile. This behavior has been interpreted in terms of a rate-determining
formation of an aryl cation, followed by fast reaction to
give the substituted product.

$$Ar\text{-}N_2^+ \xrightarrow{\text{slow}} Ar^+ + N_2 \qquad (8.16$$

$$Ar^+ + X^- \xrightarrow{\text{fast}} ArX \qquad (8.17)$$

This mechanism, formally equivalent to the aliphatic S_N1
mechanism, is not favored for most aromatic substitutions
because of the instability of aryl carbonium ions [18].
Presumably the formation of such a species could occur
with diazonium ions because of the concurrent production
of the highly stable N_2 molecule. Consistent with this
formulation of the mechanism are kinetic substituent
effects, shown in Table 8.III, for hydrolysis in aqueous
solution of monosubstituted benzenediazonium ions.

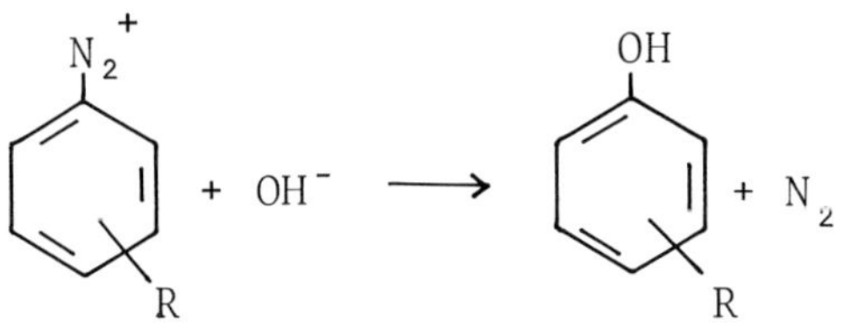

$$(8.18)$$

In general, the reaction rates are increased by electron-
donating substituents (with some remarkable exceptions)

and decreased by electron-withdrawing groups. These effects are consistent with the S_N1 mechanism, for electron-
donation would stabilize the aryl cation. The meta substituent effects are entirely normal in this interpretation.
Ortho and para substituents capable of direct resonance
interaction are deactivating, and this effect has been
attributed to delocalization as in 12, which would
strengthen the C-N bond.

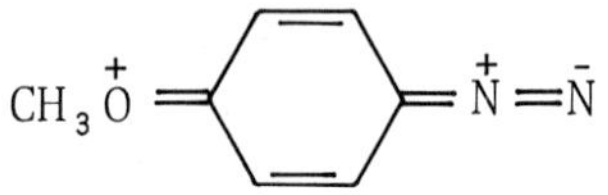

12

Table 8.III   Relative Rates of Decomposition of
Benzenediazonium Ions [18][a,b]

Substituent	ortho	meta	para
OH	0.0092	12	0.0013
OCH$_3$	--	4.6	0.00015
C$_6$H$_5$	1.5	2.3	0.050
CH$_3$	5.0	4.6	0.12
H	(1.0)	(1.0)	(1.0)
COOH	0.19	0.55	0.12
SO$_3^-$	0.12	0.21	0.057
Cℓ	0.00019	0.042	0.0019
NO$_2$	0.00050	0.00093	0.0042

[a]From J. F. Bunnett and R. E. Zahler, Chem. Revs., 49
 273 (1951).
[b]At 28.8° in water.

    Lewis and his co-workers have made the simple S_N1
interpretation doubtful by their recent observations on
the kinetics of reaction with nucleophiles [19], and the
rearrangement of nitrogen atoms accompanying hydrolysis
[20]. This rearrangement

$$Ar - {}^{15}N^+ \equiv {}^{14}N \rightarrow Ar - {}^{14}N^+ \equiv {}^{15}N \tag{8.19}$$

occurs at a rate that is about 0.026 times the rate of
the concurrent hydrolysis for several substituted benzenediazonium ions. The fact of rearrangement requires that

aryl cation formation as in Eq. (8.16) be reversible.
Such reversibility had been postulated to account for a
rate dependence on nucleophile concentration [21]; however the extent of rearrangement does not correlate simply
with the nucleophile kinetic effect. Moreover, in reaction systems containing both anionic nucleophiles X^- and
water, in which a mixture of products ArX and ArOH is
possible, the ratio of yields can be correlated with the
rate ratios [19]. Lewis et al. [19] prefer a rate-determining reaction of the diazonium ion with water in the
hydrolysis reaction. Whatever the mechanism of nucleophilic substitution on aryldiazonium ions may be, the S_N1
route with production of a free aryl cation now seems
unlikely.

Aromatic diazonium ions can be converted to aryl
halides by a displacement reaction, known as the Sandmeyer
reaction, using cuprous chloride (for example) in hydrochloric acid solution. The reaction involves the complex
ion $CuCl_2^-$.

$$Ar-N_2^+ + CuCl_2^- \rightarrow Ar-Cl + CuCl + N_2 \qquad (8.20)$$

The Sandmeyer reaction appears to be a free radical reaction [22]. The kinetics show first-order dependence on
the diazonium ion and on cuprous chloride. Bromides,
cyanides, and other aryl derivatives can be prepared.

The Elimination-Addition (Benzyne) Mechanism. As we have
seen, activated aromatic substrates tend to undergo nucleophilic substitution by a bimolecular mechanism via a
cyclohexadienylide intermediate; the overall substitution
process can be viewed as an addition-elimination route.
Aryldiazonium ions are substituted in a manner formally
analogous to an S_N1 mechanism, though the reaction is not
well understood. A third type of nucleophilic substitution is known, with these features: Nonactivated aryl
halides when reacted with metallic amides (for example,
KNH_2 in liquid ammonia at room temperature) yield arylamines with a mixture of positional isomers. The entering
substituent is within one position of the original substituent. Thus o-chlorotoluene yields o-methylaniline and
m-methylaniline; m-chlorotoluene gives the o-, m-, and p-
isomers; and p-chlorotoluene gives m-methylaniline and p-
methylaniline. The reaction requires an unsubstituted
position adjacent to the halogen. These observations,
and many others like them [23], are best accounted for by
a sequential elimination-addition reaction with an intermediate (a benzyne, 13) carrying a formal triple bond in
the aromatic ring.

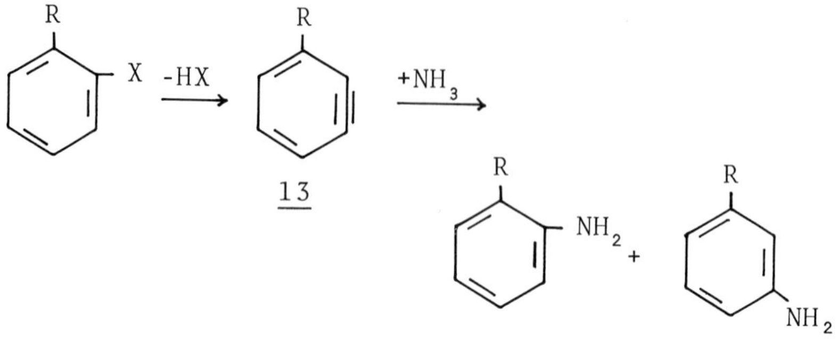

(8.21)

The benzyne mechanism is favored by substrates that
are not activated enough to undergo substitution by the
normal bimolecular mechanism, and that have a hydrogen
ortho to an electronegative substituent. The strongly
basic metal amide functions by removing the proton. Substituted anilines are obtained by reaction with the corresponding amide; for example, lithium diethylamide in ether
leads to the formation of N,N-diethylanilines. Benzyne
intermediates have proved to be valuable synthetic tools.

8.2   Analytical Reactions

Substitutions on Activated Aromatics. Most analytical
examples of nucleophilic aromatic substitutions utilize
an activated aromatic halide as an analytical reagent.
Historically and practically the most important of these
reagents is 1-fluoro-2,4-dinitrobenzene (FDNB), 14, which
Sanger employed in determining the amino acid sequence of
insulin [24].   Upon reaction with a protein in alkaline

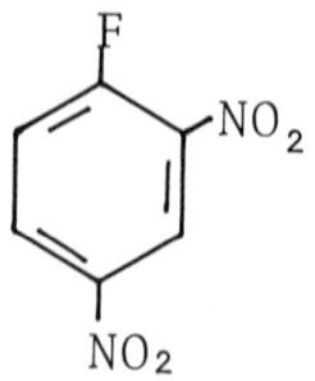

14

solution, FDNB arylates terminal amino groups [Eq. (8.22)], as well as exposed ε-amino groups (lysine), imidazoles (histidine), phenols (tyrosine), and thiols (cysteine).

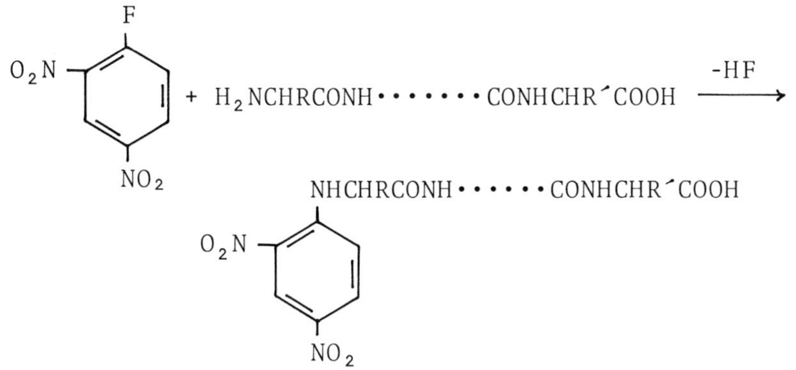

$$O_2N \overset{F}{\underset{NO_2}{\bigcirc}} + H_2NCHRCONH\cdots\cdots CONHCHR'COOH \overset{-HF}{\longrightarrow}$$

$$NHCHRCONH\cdots\cdots CONHCHR'COOH$$

$$O_2N \overset{}{\underset{NO_2}{\bigcirc}}$$

(8.22)

After reaction is complete, the excess reagent (and dinitrophenol produced by concurrent hydrolysis of FDNB) is extracted and the dinitrophenylated protein is hydrolyzed with 6 N HCl into its constituent amino acids. Upon extraction with ether the N-terminal DNP-amino acids (dinitrophenylamino acids) pass into the organic phase; they can be separated, identified, and determined by chromatographic procedures. The aqueous phase retains any ε-DNP-lysine, O-DNP-tyrosine, and Im-DNP-histidine, and the amino acids that did not react with FDNB. In this way the N-terminal amino acids of a protein can be determined. Such "end group" analysis of insulin, for example, reveals one phenylalanine and one glycine free α-amino group, showing that this protein has two peptide chains [25].

Besides its use in end group analysis, FDNB has been employed as a reagent for active site labeling of enzymes and for studying protein tertiary structure [26].

Several reagents similar to FDNB have been introduced for end group analysis and as enzyme probes. 2,4,6-Trinitrobenzenesulfonic acid (TNBS, 15) does not react with proline, histidine, or phenols, but it reacts with lysine more rapidly than does FDNB. Two reagents that give water-soluble derivatives are 4-fluoro-3-nitrobenzenesulfonic acid, 16, and 2,6-dinitrobenzene-1,4-disulfonic acid, 17. With 17 the 1-sulfonate is displaced; this reagent is used as a blocking group prior to attachment of a peptide to a resin in solid-phase peptide syntheses [27]. 1,5-Difluoro-2,4-dinitrobenzene, 18, yields

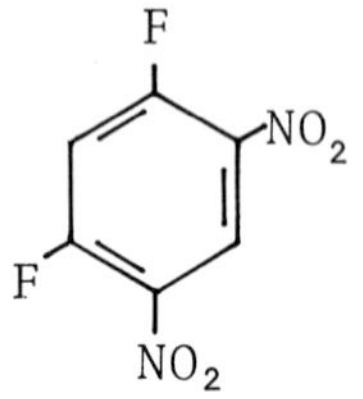

18

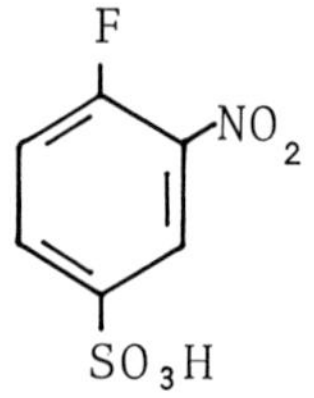

16

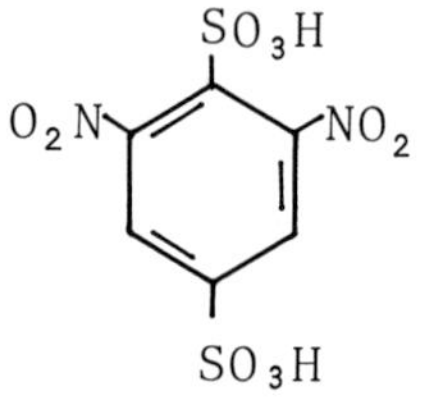

17

cross-linked lysine residues upon reaction with ribonuclease [28].

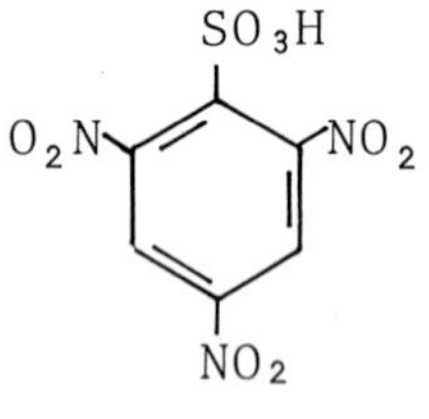

15

McIntire et al. [29] developed a colorimetric method
for amines based on their dinitrophenylation with FDNB.
A bicarbonate solution of FDNB and the amine is heated to
speed the reaction, then excess FDNB is hydrolyzed to
dinitrophenol with strong alkali. The DNP-amines are extracted into cyclohexane for spectrophotometric measurement. It was noted that most primary amine derivatives
have absorption maxima in the range 325-335 nm, whereas

secondary amine derivatives absorb maximally near 350-360
nm. Dubin [30] increased the sensitivity and simplified
the procedure by acidification prior to spectrophotometric
measurement, the extraction step being eliminated. He
found that the absorbance ratio A_{350}/A_{390} of the DNP-amino
acid is a good criterion of the class of amine; for primary amines this ratio is greater than 2, being about 2.2
for many amines, whereas with secondary amines it is less
than 0.8.

DNP-amines can be identified by thin-layer chromatography and by their fragmentation behavior in the mass
spectrometer [31], and by gas chromatography [32].

In 1935 Mannich [33] described a gravimetric method
for morphine, 19, a phenolic alkaloid, which upon reaction

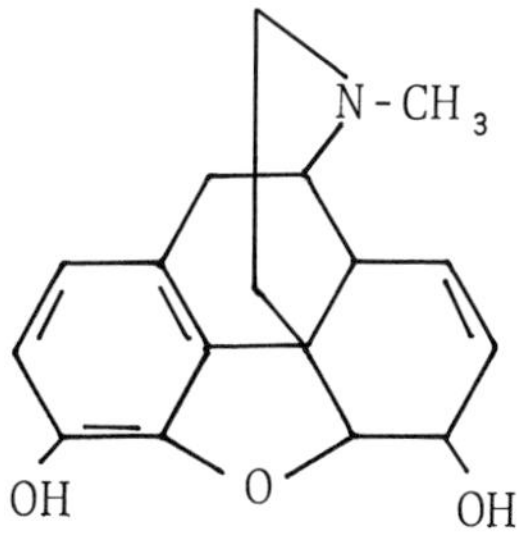

19

with 1-chloro-2,4-dinitro-benzene yields the 2,4-dinitro-
phenyl ether. FDNB is now the preferred reagent for this
analysis [34], and for the gravimetric determination of
other phenols [35]. Phenols can be characterized as their
crystalline 2,4-dinitrophenyl ethers, prepared by reaction
with FDNB [36].

Baernstein developed a method for aromatic compounds
that appears to involve a stable Meisenheimer complex
[37]. When a m-dinitrobenzene having an unsubstituted
position para to one of the nitro groups is treated with
butanone and alkali, blue or red colors develop. This
colorimetric procedure was applied to the analysis of
mixtures of benzene and toluene, an analysis that illustrates some key ideas on both electrophilic and nucleophilic aromatic substitution. The procedure requires
three steps, shown for the two sample compounds in

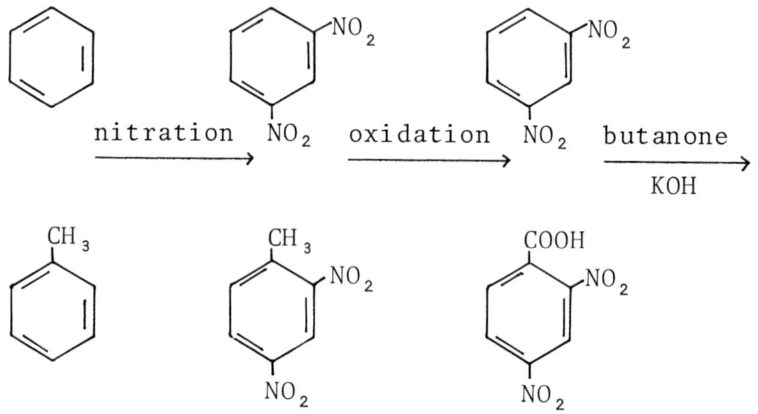

nitration NO_2 oxidation NO_2 butanone   color

                                                    KOH   (8.23)

no color

First the mixture is nitrated. The first nitro group to enter benzene has a deactivating m-directing influence, so m-dinitrobenzene is formed. Toluene, carrying the activating o,p-directing methyl group [38], gives 2,4-dinitrotoluene. Upon oxidation this compound gives 2,4-dinitrobenzoic acid, but the m-dinitrobenzene does not react. When treated with alkali and butanone, the product derived from benzene yields the color, presumably due to the Meisenheimer complex, whereas the toluene derivative does not give the color [39].

This color production probably occurs as in the Janovsky reaction [40], which produces blue-violet colors when m-dinitrobenzenes are treated with alkaline acetone. Lehmann [41] has developed a general colorimetric method for phenols by using two nucleophilic aromatic substitutions. First the phenol is arylated with FDNB,

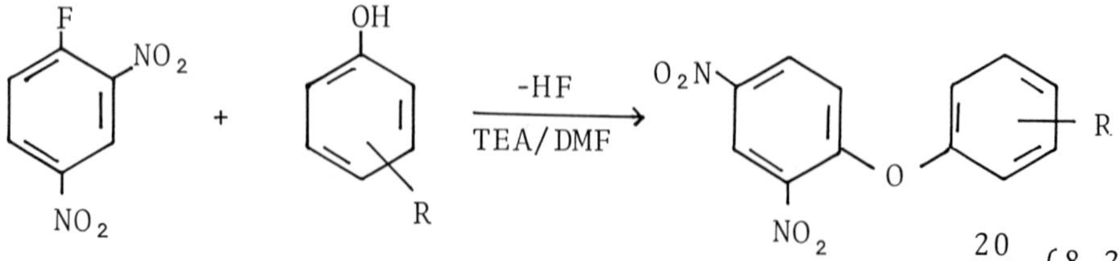

          -HF
         ────────→   R
         TEA/DMF

                                                    20 (8.24)

The 2,4-dinitrophenyl ether, 20, is then treated with acetone and NaOH or KOH; a colored Meisenheimer complex 21 is formed by attack of the acetone carbanion

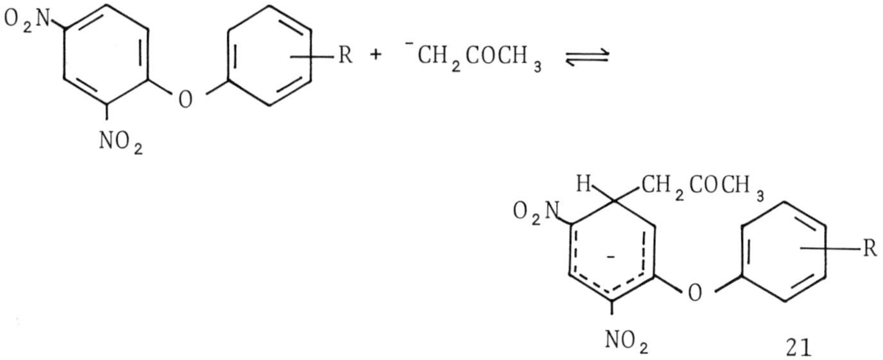

The initial nucleophilic attack appears to be at the 3-position (between the nitro groups), with subsequent transannular migration to the 5-position [42].

Dialkylphosphites produce a red color upon reaction with s-trinitrobenzene in alkaline solution. These compounds undergo a tautomeric shift analogous to enolization:

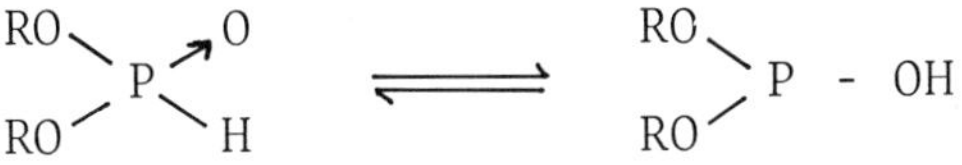

$$\underline{22}$$

Trialkylphosphites, phosphates, and phosphonates do not give the color. It was proposed, by analogy with Baernstein's method, that a Meisenheimer complex is formed by the anion of 22 [43].

Heterocyclic aromatic halides have been used as analytical reagents. A ring nitrogen atom has an activating effect less than that of a nitro group; the relative rates of substitution by ethoxide on 23, 24, 25, and 26 are 414, 40, 82, and 1, respectively [44].

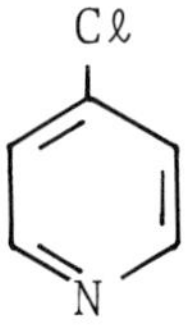

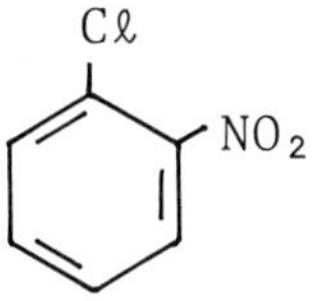

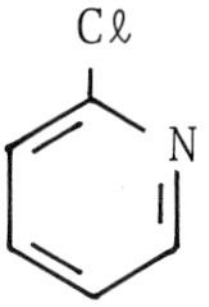

Further activation is provided by the fused ring systems
in 27 and 28, which are 7.5 and 710 times more reactive
than 24, respectively. Compound 28, 9-chloroacridine,

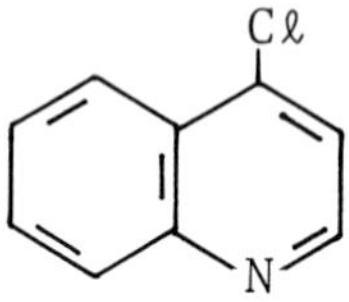

27 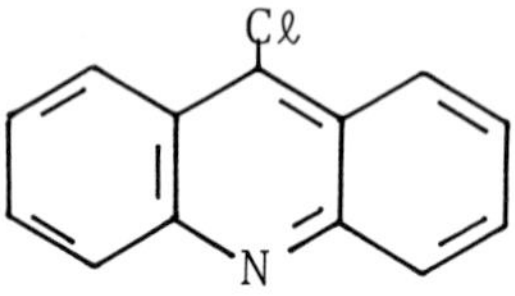   28

is an effective analytical reagent for primary aromatic
amines [45]. The product of the reaction absorbs maximally
at 435 nm, and amines other than primary aromatics do not
interfere. It is possible that this peculiar selectivity
is a consequence of the formation of a new chromophore in
the product, as in 29; this structure can be achieved only
by primary amine nucleophiles, and only aromatics will
markedly alter the chromophoric properties by conjugation
with the acridine moiety.

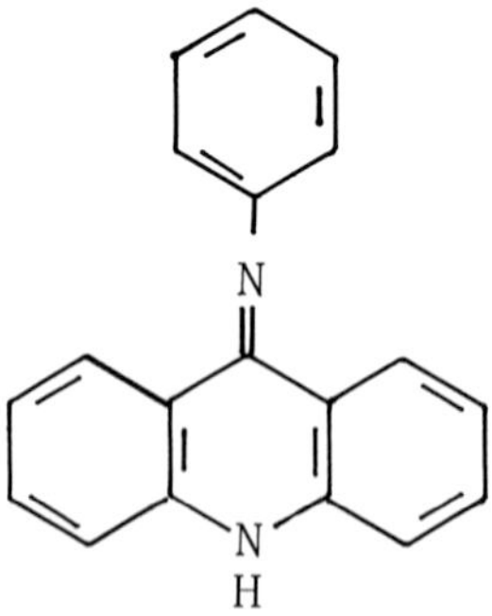

29

     Other heterocyclic halides that have been introduced
for the structural study of proteins are 2-fluoropyridine
N-oxide, 30 [46], and cyanuric fluoride, 31 [47].
     Several analytical methods involving reactions be-
tween quinones and amines may be nucleophilic substitu-

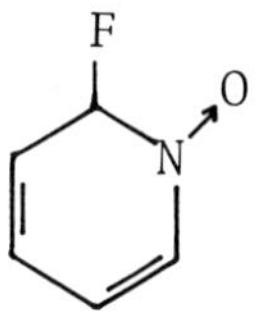

30 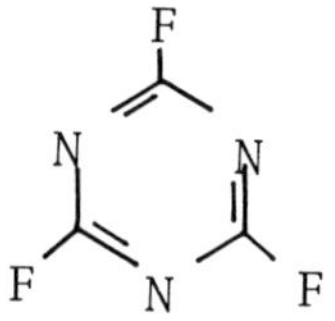   31

tions, though the mechanisms are uncertain. Hydroquinone
can be determined spectrophotometrically by air oxidation
to benzoquinone 32 and reaction with n-butylamine [48].
By analogy with the reaction of methylamine with benzoquinone [49], the product was written as 33 [48]. Because
the aromatic compound does not posses a good leaving group,

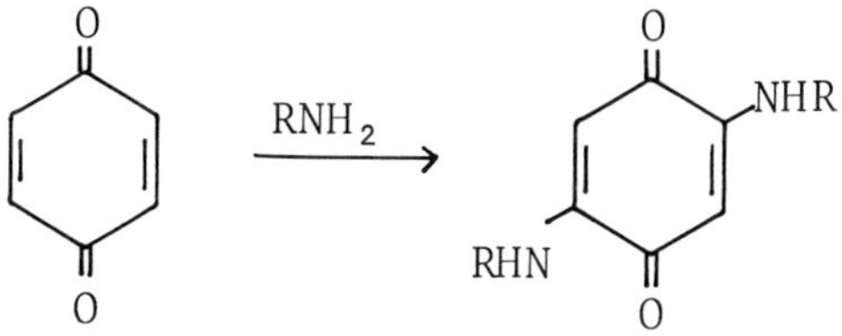

$$(8.27)$$

32   33

this reaction seems more likely to be an addition followed
by dehydrogenation rather than a nucleophilic substitution.
    Other quinones with good leaving groups may react in
a normal substitution process. Primary aromatic amines
(also reactive methylene compounds) give colored solutions
with β-naphthoquinone-4-sulfonic acid; Feigl [50] writes
this reaction as

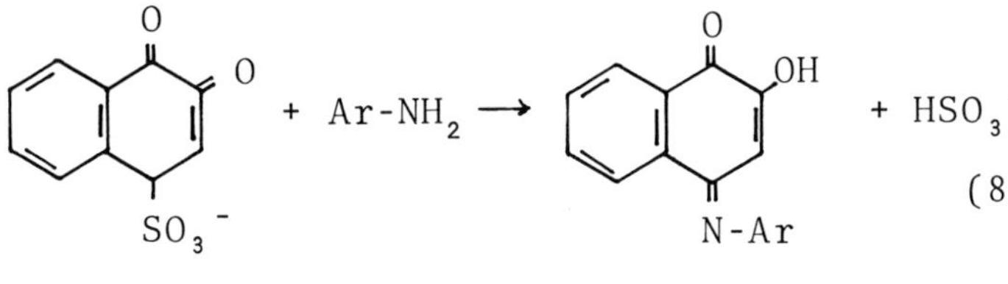

$$(8.28)$$

This reagent has been applied to norepinephrine [51] and
to ethylenimine [52], 34, which may attack via carbon
rather than nitrogen.

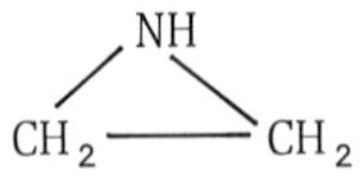

$$\begin{array}{c} \quad\nearrow^{NH}\searrow \\ CH_2 \longrightarrow CH_2 \end{array}$$

34

    Chloranil, 35, reacts with tertiary aliphatic amines
in toluene to give colored products suitable for quantitative measurement. Sass et al. [53] give Eq. (8.29) for
this reaction.

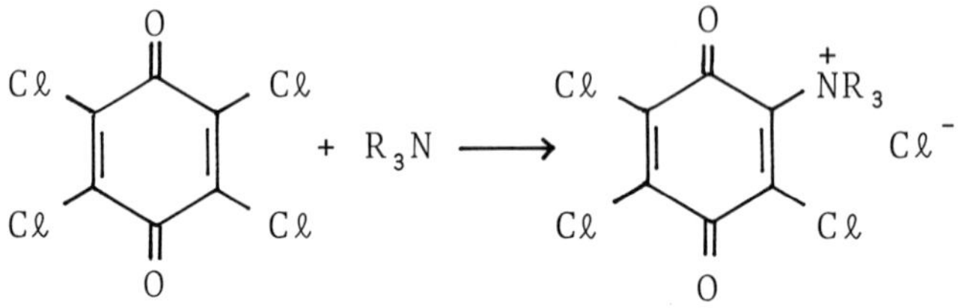

35

The reaction with triethylamine, however, was shown by
Buckley, Dunstan, and Henbest [54] to proceed through
diethylvinyl amine, which is produced by dehydrogenation
of triethylamine, Eq. (8.30).

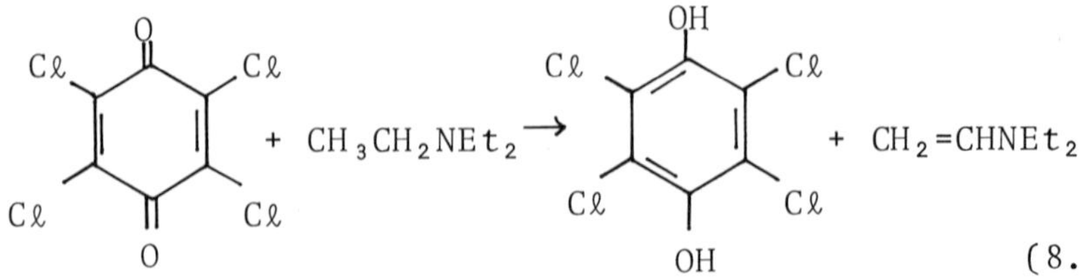

Compound 36 is blue. Reaction (8.31) represents nucleophilic attack by the negatively charged carbon, the polarization being as in 37.

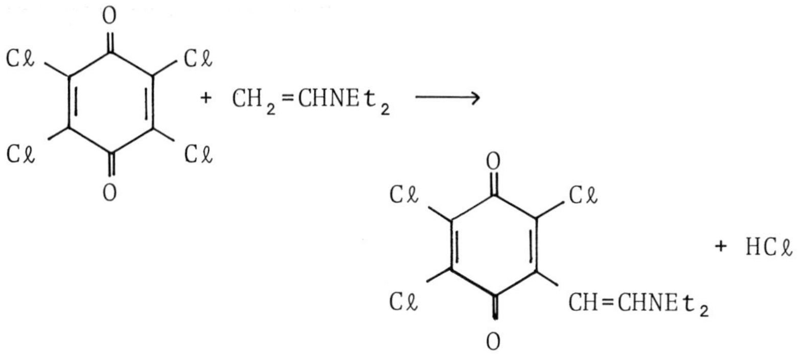

$$+ \text{HCl} \qquad (8.31)$$

$$\underline{36}$$

$$CH_2 = CH - NEt_2 \longleftrightarrow {}^-CH_2 - CH = \overset{+}{N}Et_2$$

$$\underline{37}$$

Aromatic primary amines have been determined by reaction with chloranil; products analogous to $\underline{33}$ are postulated [55].

Substitutions on Nonactivated Aromatics. Diazonium ions can be quantitatively determined by the Sandmeyer reaction,

$$Ar-N_2^+ + CuCl_2^- \rightarrow Ar-Cl + CuCl + N_2 \qquad (8.32)$$

This is a free radical reaction. Siggia describes a procedure in which the volume of nitrogen evolved is measured [56].

Arylsulfonic acids undergo substitution to give phenols when fused with alkali.

$$Ar-SO_3K + KOH \overset{\longrightarrow}{\Delta} Ar-OH + K_2SO_3 \qquad (8.33)$$

Siggia et al. [57] developed a quantitative method based on this nucleophilic substitution. The extent of reaction was established either by gas chromatographic determination of the phenol, or by measuring, titrimetrically, the sulfite produced.

Comment. Since hydrogen is not an effective leaving group in nucleophilic aromatic substitutions, these reactions usually do not yield mixtures of isomers or polysubstituted products, in contrast with electrophilic substitutions.

Reactions of activated aromatics with good leaving groups
characteristically occur quantitatively. Most analytical
applications have been for the determination of amines,
and the extension to other nucleophiles may be worthwhile.
Some modest advantages may be realized in adapting reagents
introduced for protein studies.

Several specific areas for potential analytical
development are:

1. Further application of Meisenheimer complexes.
These species absorb strongly in the visible region. The
principle requirements for the production of a stable
Meisenheimer complex are a powerful nucleophile and a
highly activated aromatic substrate without a good leaving
group.

2. Catalytic effects of micelles upon nucleophilic
aromatic substitution. Bunton and Robinson [58] found
that the reactions of aniline with fluorodinitrobenzene
and with chlorodinitrobenzene are catalyzed by cationic
micelles and, to a smaller extent, by anionic micelles.
Substitutions by the nucleophiles glycinate and glycylgly-
cinate were assisted by cationic and inhibited by anionic
micelles. Such effects possibly could be exploited to
achieve interesting analytical selectivity.

3. Benzyne intermediates. The benzyne mechanism for
nucleophilic substitutions on nonactivated aromatics
appears not to have been used analytically. Though the
reagents are powerful, these reactions are usually conducted at room temperature. Use could possibly be made
of the reaction to determine suitable aromatics by reaction via the benzyne, or to determine nucleophiles by
reaction with a benzyne-generating aromatic compound [59].

Problems

1. Give an expression for k_{obs}, the pseudo-first-order
    rate constant for the Meisenheimer complex equilibrium
    in Eq. (8.5).

    Answer: $k_{obs} = k_1[OMe^-] + k_{-1}$

2. Suggest a transition state for the rate-determining
    step of the amine catalyzed reaction of n-butylamine
    with 2,4-dinitrochlorobenzene [16].

3. Devise analytical approaches to the analysis of these
    mixtures:

    a. Nitrobenzene and chlorobenzene,
    b. Nitrobenzene and benzaldehyde,
    c. n-Butylamine and aniline.

4. Which of the substituent constants, σ, σ^+, or σ^-,
    would be expected to give the best correlation of the
    rates of decomposition of m-substituted benzene-
    diazonium ions? Make these plots.

5. Predict the products obtained by subjecting each of
    the four isomers of monochoro-m-ethyltoluene to
    potassium amide in liquid ammonia.

References

1. J. F. Bunnett and R. E. Zahler, Chem. Revs., 49, 273
    (1951); J. F. Bunnett, Quart. Revs., 12, 1 (1958); S.
    D. Ross, Progr. Phys. Org. Chem., 1, 31 (1963); L. M.
    Stock, Aromatic Substitution Reactions, Prentice-Hall,
    Englewood Cliffs, N. J., 1968, Chap. 5; J. Miller,
    Aromatic Nucleophilic Substitution, Monograph 8 in
    Reaction Mechanisms in Organic Chemistry, Elsevier,
    Amsterdam, 1968.
2. S. D. Ross, Progr. Phys. Org. Chem., 1, 31 (1963).
    This bimolecular mechanism is operative primarily with
    "activated" aromatic substrates, that is, compounds
    substituted with electron-attracting groups capable
    of charge delocalization.
3. J. Meisenheimer, Ann., 323, 205 (1902).
4. J. Miller, Aromatic Nucleophilic Substitution, Else-
    vier, Amsterdam, 1968, pp. 11-14; M. R. Crampton,
    Advan. Phys. Org. Chem., 7, 211 (1969); M. J. Strauss,
    Chem. Revs., 70, 667 (1970).
5. J. H. Fendler, E. J. Fendler, and C. E. Griffin, J.
    Org. Chem., 34, 689 (1969).
6. E. J. Fendler, J. H. Fendler, C. E. Griffin, and J.
    W. Larsen, J. Org. Chem., 35, 287 (1970).
7. J. Miller, Ref. [1], Chap. 4, has reviewed substituent
    effects on the bimolecular mechanism.
8. C. F. Bernasconi, J. Am. Chem. Soc., 92, 129 (1970).
9. In this reaction system two additional intermediates
    are formed. One of these is the Meisenheimer complex
    resulting from attack by hydroxide ion, and the other
    is a species resulting from amine attack on nitrogen
    of a nitro groups [8].
10. J. F. Bunnett, E. W. Garbisch, Jr., and K. M. Pruitt,
    J. Am. Chem. Soc., 79, 385 (1957).
11. In aliphatic substitution reactions (see Chap. 9)

fluorine is less easily displaced than are the other halogens, in agreement with the proposal that in aromatic substitution the C-F bond is not markedly weakened in the transition state.

12. It was earlier [11] pointed out that the C-F bond is not as easily broken as are other carbon-halogen bonds.

13. The stable Meisenheimer complexes described above are extreme examples of this kinetic behavior.

14. J. F. Bunnett and J. J. Randall, J. Am. Chem. Soc., 80, 6020 (1958).

15. In the reaction of piperidine with these same sub-strates, Table 8.II, evidently $k_2 \gg k_{-1}$ for all of the substrates, because piperidine is a much more powerful nucleophile than is N-methylaniline. Base catalysis can be seen in substitutions by piperidine if the leaving group is sufficiently poor, as in 2,4-dinitrophenyl phenyl ether; cf. J. F. Bunnett and R. H. Garst, J. Am. Chem. Soc., 87, 3879 (1965); J. F. Bunnett and C. Bernasconi, ibid., 87, 5209 (1965).

16. J. A. Orvik and J. F. Bunnett, J. Am. Chem. Soc., 92, 2417 (1970).

17. S. D. Ross, Tetrahedron, 25, 4427 (1969).

18. J. Miller, Ref. [1], Chap. 2.

19. E. S. Lewis, L. D. Hartung, and B. M. McKay, J. Am. Chem. Soc., 91, 419 (1969).

20. E. S. Lewis and J. M. Insole, J. Am. Chem. Soc., 86, 32 (1964); E. S. Lewis and R. E. Holliday, ibid., 91, 426 (1969); E. S. Lewis and P. G. Kotcher, Tetrahe-dron, 25, 4873 (1969).

21. E. S. Lewis and J. E. Cooper, J. Am. Chem. Soc., 84, 3847 (1962).

22. E. S. Gould, Mechanism and Structure in Organic Chemistry, Holt, Rinehart and Winston, New York, 1959, p. 729.

23. Some evidence has, however, been presented implicat-ing a free radical mechanism in substitutions on aryl iodides; cf. J. K. Kim and J. F. Bunnett, J. Am. Chem. Soc., 92, 7463 (1970).

24. F. Sanger, Biochem. J., 39, 507 (1945).

25. The FDNB method has been reviewed by H. Fraenkel-Conrat, J. I. Harris, and A. L. Levy, Methods Biochem. Anal., 2, 359 (1955).

26. S. J. Singer, Advan. Protein Chem., 22, 1 (1967); L. A. Cohen, Ann. Rev. Biochem., 37, 683 (1968).

27. G. R. Stark, Advan. Protein Chem., 24, 261 (1970). See also T. F. Spande, B. Witkop, Y. Degani, and A. Patchornik, Advan. Protein Chem., 24, 97 (1970).

28. P. S. Marfey, M. Uziel, and J. Little, J. Biol. Chem., 240, 3270 (1965).

29. F. C. McIntire, L. M. Clements, and M. Sproull, Anal.
     Chem., 25, 1757 (1953).
30. D. T. Dubin, J. Biol. Chem., 235, 783 (1960).
31. A. Zeman and I. P. G. Wirotama, Z. Anal. Chem., 247,
     155 (1969).
32. I. C. Cohen and B. B. Wheals, J. Chromatog., 43, 233
     (1969).
33. C. Mannich, Arch. Pharm., 273, 97 (1935).
34. D. C. Garratt, C. A. Johnson, and C. J. Lloyd, J.
     Pharm. Pharmacol., 9, 914 (1957). This paper also
     describes the purification and properties of FDNB.
35. H. Zahn and A. Würz, Z. Anal. Chem., 134, 183 (1951);
     N. D. Cheronis and T. S. Ma, Organic Functional Group
     Analysis by Micro and Semimicro Methods, Wiley (Inter-
     science), New York, 1964, p. 453. 2,4-Dinitrophenyl
     phenyl ethers have been determined by gas chromato-
     graphy; cf. I. C. Cohen, J. Norcup, J. H. A. Ruzicka,
     and B. B. Wheals, J. Chromatog., 44, 251 (1969).
36. P. A. Lehmann, Anal. Chim. Acta, 54, 321 (1971).
37. H. D. Baernstein, Ind. Eng. Chem., Anal. Ed., 15,
     251 (1943).
38. The terms activating and deactivating are here used
     in the sense of Chap. 7, that is, with respect to
     electrophilic attack.
39. Although 2,4-dinitrobenzoic acid has an open position
     para to a nitro group, in alkaline solution this
     substrate will be deactivated toward nucleophilic
     attack.
40. J. V. Janovsky and L. Erb, Chem. Ber., 19, 2155
     (1886).
41. P. A. Lehmann, Rev. Latinoamer. Quim., 1971, 112.
42. P. A. Lehmann, personal communication.
43. S. Sass and J. Cassidy, Anal. Chem., 28, 1968 (1956).
44. N. B. Chapman and D. Q. Russell-Hill, J. Chem. Soc.,
     1956, 1563.
45. J. T. Stewart, T. D. Shaw, and A. B. Ray, Anal. Chem.,
     41, 360 (1969); J. T. Stewart, A. B. Ray, and W. B.
     Fackler, J. Pharm. Sci., 58, 1261 (1969); J. T.
     Stewart and D. M. Lotti, ibid., 59, 838 (1970).
46. D. Sarantakis, J. K. Sutherland, C. Tortorella, and
     V. Tortorella, J. Chem. Soc., C, 1968, 72.
47. K. Kurihara, H. Horinishi, and K. Shibata, Biochim.
     Biophys. Acta, 74, 678 (1963); M. J. Gorbunoff,
     Biochemistry, 6, 1606 (1967); Arch. Biochem. Biophys.,
     138, 684 (1970). Cyanuric chloride is a reagent for
     N-acylated glycine; presumably the methylene group
     is involved. Cf. S. Suzuki, Y. Hachimori, and U.
     Yaoeda, Anal. Chem., 42, 101 (1970).
48. R. J. Lacoste, J. R. Covington, and G. J. Frisone,
     Anal. Chem., 32, 990 (1960).

49. W. K. Anslow and H. Raustrick, J. Chem. Soc., <u>1939</u>, 1446.
50. F. Feigl, Spot Tests in Organic Analysis, 7th ed., Elsevier, Amsterdam, 1966, p. 153. See also J. Bartos and M. Pesez, Bull. Soc. Chim. Fr., <u>1970</u>, 1627.
51. M. E. Auerbach, Drug Standards, <u>20</u>, 165 (1952).
52. D. H. Rosenblatt, P. Hlinka, and J. Epstein, Anal. Chem., <u>27</u>, 1290 (1955).
53. S. Sass, J. J. Kaufman, A. A. Cardenas, and J. J. Martin, Anal. Chem., <u>30</u>, 529 (1958).
54. D. Buckley, S. Dunstan, and H. B. Henbest, J. Chem. Soc., <u>1957</u>, 4880.
55. Y. Tashima, H. Hasegawa, H. Yuki, and K. Takiura, Bunseki Kagaku (Japan Analyst), <u>19</u>, 43 (1970).
56. S. Siggia, Quantitative Organic Analysis via Functional Groups, 3rd ed., Wiley, New York, 1963, p. 544.
57. S. Siggia, L. R. Whitlock, and J. C. Tao, Anal. Chem., <u>41</u>, 1387 (1969).
58. C. A. Bunton and L. Robinson, J. Am. Chem. Soc., <u>92</u>, 356 (1970).
59. For examples of such reagents see L. F. Fieser and M. Fieser, Reagents for Organic Synthesis, Wiley, New York, 1967.

Chapter 9.   NUCLEOPHILIC ALIPHATIC SUBSTITUTION

9.1   Nature of the Reaction

Survey of the Reaction.   Perhaps the most carefully
studied class of organic reactions is that in which a
nucleophile Y displaces a group X, together with an electron pair, from a saturated carbon atom, as in

$$Y^- + \overset{\diagdown/}{\underset{|}{C}}{-}X \rightarrow \overset{\diagdown/}{\underset{|}{C}}{-}Y + X^- \tag{9.1}$$

These are aliphatic S_N (nucleophilic substitution) reactions.   Equation (9.1) is written for the substitution of
a neutral substrate by an anionic nucleophile, but the
charge types are not central to the classification.   Many
apparently diverse reactions are related by their conform-
ity to this general scheme.   A few examples indicate the
scope of aliphatic S_N reactions.   The hydrolysis of an
alkyl halide, Eq. (9.2), clearly is such a reaction.

$$OH^- + C_2H_5Br \rightarrow C_2H_5OH + Br^- \tag{9.2}$$

Closely related is the Williamson synthesis of ethers:

$$RO^- + R'\text{-}Br \rightarrow R'\text{-}O\text{-}R + Br^- \tag{9.3}$$

Halide exchange (the Finkelstein reaction) provides a
further example of substitution by an anion:

$$I^- + R\text{-}C\ell \rightarrow R\text{-}I + C\ell^- \tag{9.4}$$

The Menschutkin quaternization reaction involves displacement by a neutral nucleophile:

$$(CH_3)_3N + C_2H_5I \rightarrow C_2H_5N(CH_3)_3^+ + I^- \tag{9.5}$$

The displaced (leaving) group may be polyatomic, as in
the hydrolysis of a p-toluenesulfonate ester [1] shown in

$$OH^- + R\text{-}OSO_2C_6H_4CH_3 \rightarrow R\text{-}OH + CH_3C_6H_4SO_3^- \qquad (9.6)$$

When the attacking species is the solvent the reaction is called solvolysis. Usually a hydroxylic solvent is involved, and sometimes the type of solvolysis is specified; for example, hydrolysis is reaction with water (or hydroxide), ethanolysis with ethanol (or ethoxide), and acetolysis with acetic acid (or acetate).

Substitution on a saturated (sp^3-hybridized) carbon atom is the most important aliphatic S_N reaction, though others are known. For example, S_N reactions occur at silicon atoms, as in [2]

$$OH^- + R_3SiC\ell \rightarrow R_3SiOH + C\ell^- \qquad (9.7)$$

The mechanisms of S_N reactions were elucidated in the 1930's, particularly by Hughes, Ingold, and their co-workers. This work is reviewed by Ingold [3]. Many other reviews have appeared [4].

Kinetics and Mechanisms. S_N reactions can proceed by more than one mechanism. Two mechanistic routes have been clearly identified [3]. One of these is schematically shown by

$$R\text{-}X \xrightarrow{\text{slow}} R^+ + X^- \qquad (9.8)$$

$$R^+ + Y^- \xrightarrow{\text{fast}} R\text{-}Y \qquad (9.9)$$

The first step, which is rate determining, is an ionization to a carbonium ion intermediate, which reacts with the nucleophile in the second step. Since the transition state for the rate-determining step includes R-X but not Y^-, the reaction is unimolecular and is labeled S_N1 (substitution nucleophilic unimolecular) [5]. First-order kinetics are observed, with the rate being independent of the nucleophile identity and concentration. The S_N1 mechanism is also called solvolytic displacement, the dissociation mechanism, or the carbonium ion mechanism.

The second S_N mechanism is the one-step direct displacement reaction,

$$Y^- + RX \rightarrow [Y\cdots\cdots R\cdots\cdots X]^- \rightarrow R\text{-}Y + X^- \qquad (9.10)$$

This bimolecular process is called the S_N2 mechanism. It yields overall second-order kinetics (unless the nucleophile is the solvent, in which case apparent first-order kinetics are seen). The species in brackets represents the transition state. Since this is a one-step process, the rate-determining step is also the product-determining step.

The evidence supporting the duality of mechanisms is of several kinds; these in fact are the experimental criteria that are applied in classifying an aliphatic substitution as S_N1 or S_N2. The kinetic behavior is an obvious feature. This is somewhat more complex than is implied by the preceding treatment. A quantitative description of the S_N1 mechanism requires recognition of the reversibility of the ionization step, thus

$$R\text{-}X \underset{k_{-1}}{\overset{k_1}{\rightleftharpoons}} R^+ + X^- \qquad (9.11)$$

$$R^+ + Y^- \xrightarrow{k_2} R\text{-}Y \qquad (9.12)$$

Since the carbonium ion intermediate is unstable, it will be permissible to apply the steady-state approximation, leading to Eq. (9.13) for the reaction rate.

$$v = \frac{k_1 [R\text{-}X]}{(k_{-1}[X^-]/k_2[Y^-]) + 1} \qquad (9.13)$$

If the carbonium ion reacts with nucleophile Y^- (which may be the solvent) to give product much faster than it does with X^- to revert to reactant, then Eq. (9.13) will tend to the simple first-order form, $v = k_1[R\text{-}X]$. In aqueous solvents tert-butyl bromide exhibits this kinetic behavior. If $(k_{-1}[X^-]/k_2[Y^-])$ is not much smaller than unity, then as the substitution reaction proceeds, the increase in $[X^-]$ will increase the denominator of Eq. (9.13), slowing the reaction and causing deviation from simple first-order kinetics. This mass-law or common-ion effect, that is, a rate constant decrease with increase in concentration of the common anion X^- in R-X, is characteristic of an S_N1 process, though, as already noted, it is not a necessary condition. The common-ion effect (also called "external return") occurs only with the common ion, and must be distinguished from a general kinetic salt effect, which will operate with any ion [6]. An example is provided by the hydrolysis of triphenylmethyl chloride (trityl chloride); the addition of 0.01 M sodium chloride decreased the rate by fourfold [7]. The solvolysis rate of diphenylmethyl chloride (benzhydryl chloride) in 80% aqueous acetone was decreased by lithium chloride but increased by lithium bromide [8].

The S_N2 mechanism will also yield pseudo-first-order kinetics in a solvolysis reaction, but it should not be susceptible to a common-ion rate inhibition.

In the presence of an added nucleophile (which could be the lyate ion), a reaction proceeding by the S_N1 mechanism will reveal overall second-order kinetics, first order in each reactant. The reaction of methyl iodide and sodium ethoxide in ethanol shows this behavior [9]. The solvolytic first-order rate constant of an S_N1 reaction will not generally be affected by the presence of an added nucleophile. For example [7], trityl fluoride gives the same rate constant (7.6×10^{-4} sec^{-1} in 50% water-50% acetone at 25°) in the presence or absence of sodium azide, NaN_3. This behavior is consistent with the S_N1 scheme, Eqs. (9.11) and (9.12).

Kinetic results of the above types are the most generally accessible and applicable evidence for the determination of mechanism, but they are not always definitive. Kinetics is normally supplemented with other data, namely, structural and solvent effects, and stereochemical effects; these will be considered subsequently.

The real world of S_N reactions is not quite as simple as the discussion has so far suggested. The preceding treatment in terms of two clearly distinct mechanisms, S_N1 and S_N2, implies that all substitution reactions will follow one or the other of these mechanisms. This is an oversimplification. The strength of the dual mechanism hypothesis and its limitations are revealed by these relative rates of solvolysis of alkyl bromides in 80% ethanol: methyl bromide, 2.51; ethyl bromide, 1.00; isopropyl bromide, 1.70; tert-butyl bromide, 8600. Addition of lyate ions increases the rate for the methyl, ethyl, and isopropyl bromides, whereas the tert-butyl bromide solvolysis rate is unchanged. The reaction with lyate ions is overall second order for methyl and ethyl, first order for tert-butyl, and first or second order for the isopropyl member, depending upon the concentrations. Similar results are found in other solvents [10]. These data show that the methyl and ethyl bromides solvolyze by the S_N2 mechanism, and tert-butyl bromide by the S_N1 mechanism. The isopropyl bromide does not fit so neatly into one of these classes; depending upon the solvent and the concentrations, it can present features of either mechanistic class. Evidently a change in the alkyl substitution on the α-carbon is responsible for a change in mechanism, with primary alkyl substrates tending to react by the S_N2 mechanism and tertiary substrates by S_N1. At some point along this scale of substituent changes presumably the mechanistic switch occurs, and this appears to be, roughly, with the secondary

substrate. This behavior has given rise to the concept
of "borderline" reactions, which do not fit unambiguously
into either the S_N1 or S_N2 classes [11].
     Several proposals have been made to fit the border-
line reactions into a well-defined mechanistic scheme.
Most of these adopt one of two viewpoints: either (1)
borderline substrates undergo concurrent S_N1 and S_N2 processes, with the particular system determining which mechanism, if either, predominates; or (2) all S_N reactions
are related by essentially the same mechanism, which dif-
fers from case to case in the detailed disposition of
electrons in the transition state. In this view pure S_N1
and S_N2 processes are merely the extreme limiting forms
of a single mechanism, and the borderline mechanism is a
merged process having some features of both.
     Concurrent S_N1 and S_N2 processes have been implicated
in several nucleophilic substitutions. The clearest demonstrations have been with added nucleophiles. For example,
p-methoxybenzyl chloride undergoes concurrent S_N1 solvolysis and S_N2 displacement by azide ion [12]. Heat capacities of activation have been interpreted in terms of concurrent S_N1 and S_N2 processes for the solvolysis of p-
methylbenzyl chloride in 50% aqueous acetone [13]. Substitution reactions of 1-hydroxymethyl-1-(3,5-ditrifluoro-
methylphenyl)-cyclopropane mesylate in 70% aqueous dioxane
have been reported to occur by concurrent competitive S_N1
and S_N2 mechanisms [14].
     Interpretations of borderline reactions in terms of
one mechanism rather than two have been more widely accepted. Winstein, Grunwald, and Jones [15] studied several
solvolyses with the aid of the solvent polarity parameter
$\underline{Y}$ [16]. The solvolysis rate should be a function of two
solvent properties: one is its ionizing power, and the
other is its nucleophilicity. An S_N1 process should be
promoted by high ionizing power, and an S_N2 mechanism by
high solvent nucleophilicity. Assuming that $\underline{Y}$ is a good
measure of ionizing power, it is possible to measure the
solvolysis rate in two solvents of equal ionizing power
but greatly different nucleophilicity. For example, ethyl
p-toluenesulfonate solvolyzes in the weakly nucleophilic
solvent acetic acid at about one-hundredth its rate of solvolysis in aqueous ethanol of equal $\underline{Y}$ value. This is taken
as evidence of covalent participation by the solvent in
the rate-determining step. Linear free energy plots of
log k against $\underline{Y}$, according to

$$\log \frac{k}{k_{EtOH}} = m\underline{Y} \qquad (9.14)$$

yield m values, which evidently are measures of substrate
sensitivity to solvent ionizing power.

Winstein et al. [15] proposed a classification of
mechanisms according to the covalent participation by the
solvent in the transition state of the rate-determining
step. If such covalent interaction occurs, the reaction
is assigned to the nucleophilic (N) class; if covalent
interaction is absent, the reaction is in the limiting
(Lim.) class. At their extremes these categories become
equivalent to S_N2 and S_N1, respectively, but the dividing
line between S_N2 and S_N1 does not coincide with that between N and Lim. For example, a mass-law effect, which
is evidence of an intermediate and therefore of the S_N1
mechanism, can be observed for some isopropyl compounds,
but these appear to be in the N class in aqueous media.

The N-Lim. classification does not eliminate the
possibility of borderline cases between these two categories, but it leads to the suggestion that no sharp distinction can be made between the possible intermediates
in these mechanisms, and that perhaps all solvolyses proceed via an intermediate. The mechanistic category of a
particular solvolysis then depends upon the relative
weights of the canonical structures 1, 2, and 3 to the
transition state resonance hybrid [15].

$$Y^-: R-X \longleftrightarrow Y-R\ :X^- \longleftrightarrow Y^-:\ R^+ :X^-$$

$$\underline{1} \qquad\qquad \underline{2} \qquad\qquad \underline{3}$$

The greater the contribution of 2 to the transition state,
the more firmly the system is placed in the N category;
likewise a large contribution from 3 is characteristic of
the Lim. category. A high m value indicates high sensitivity to solvent ionizing power, a large contribution
from 3, and therefore a Lim. reaction; tert-butyl halides
fit this class. Primary halides have low m values; their
solvolysis rates are insensitive to Y but are sensitive
to nucleophilic power of the solvent. These reactions are
in the N class. The smooth increase in m values from (for
example) methyl bromide (m = 0.258), ethyl bromide (0.343),
isopropyl bromide (0.544), to tert-butyl bromide (0.940)
suggests a continuous change of transition state character.
Streitwieser [17] has given a molecular orbital structural
interpretation of this idea.

Recently Sneen and Larsen [18] have proposed a new
mechanism involving an ion-pair intermediate; this mechanism is considered to be general in the sense of accounting for all S_N reactions. The kinetic scheme, for competitive solvolysis and nucleophile attack, is given by

$$R\text{-}X \underset{k_{-1}}{\overset{k_1}{\rightleftharpoons}} R^+X^- \quad \begin{array}{l} \xrightarrow{\quad k_S \quad} R\text{-}S \\ \\ \xrightarrow{\quad k_Y[Y^-] \quad} R\text{-}Y \end{array} \qquad (9.15)$$

By postulating the generality of this mechanism, which includes the intermediate R^+X^- for all reactions, it follows that conventional S_N1 behavior results when the formation of the ion-pair is rate determining, S_N2 behavior is a consequence of a rate-determining decomposition of the ion pair by nucleophilic attack, and borderline cases are the result of comparable rates of formation and destruction of the ion pair. At the S_N1 (or Lim.) end of the mechanistic spectrum, this proposal is not greatly different from earlier views. In fact, two ion-pair intermediates have been postulated to account for kinetic salt effects in the acetolysis of sulfonate esters [19]. These intermediates are related as in

$$R\text{-}X \rightleftharpoons R^+X^- \rightleftharpoons R^+(s)X^-(s) \rightleftharpoons R^+ + X^- \qquad (9.16)$$

$$\underset{4}{\phantom{R\text{-}X}} \qquad\qquad \underset{5}{}$$

Species 4 is called an internal or intimate ion pair, and 5 is an external or solvent-separated ion pair, sometimes symbolized $R^+||X^-$.

The ion-pair mechanism has been applied to some S_N1 and borderline reactions, for which it appears to provide a consistent picture. Its extension to S_N2 reactions introduces a distinct break with the classical view of the S_N2 transition state. In the ion-pair mechanism, nucleophilic attack by solvent or added nucleophile is postulated to occur on the ion pair, never directly on tetracovalent carbon. This generalization to S_N2 reactions is not supported by Kurz and Harris [20], who compare observed and calculated $pK_a^{\pm}$ values for the transition states of substitutions on bromoacetic acid; the observed values are consistent with those expected for covalent transition states, but not for ionic transition states. More direct tests of the generality of the ion pair mechanism can be expected.

Additional kinds of kinetic evidence, some of it supporting a continuity of mechanisms, have been used to characterize transition states in S_N reactions. Streitwieser [21] plots log k for solvolyses of arylmethyl substrates, $ArCH_2X$, against log k for a standard reaction, the solvolysis of $ArCH_2Cl$ in 80% aqueous ethanol; linear

free energy relationships are observed. The slopes of
these lines depend upon the reaction medium, with high
ionizing power tending to give high slope values. These
plots present essentially the same information as the
Winstein-Grunwald m$\underline{Y}$ correlations. As with m, the greater
the slope of the Streitwieser plot, the greater the carbonium ion character of the transition state [22].

Kohnstam [23] and Robertson [24] have made careful
kinetic measurements of solvolyses to extract $\Delta H^{\ddagger}$, $\Delta S^{\ddagger}$,
and $\Delta C_{p}^{\ddagger}$ values. These quantities, though difficult to
interpret, provide some insight into the role of the sol-
vent in the transition state; in fact, Robertson defines
the transition state in terms of relative extents of
solvent-solvent and reactant-solvent interaction.

Stereochemistry. When the α-carbon atom of an aliphatic
substrate is asymmetric, nucleophilic substitution has
stereochemical consequences. Ingold [25] has reviewed
the development of the modern ideas. It is important to
recall that the configuration about an asymmetric carbon
atom bears no necessary relationship to the sign of optical rotation. Nevertheless, many correspondences of configuration and sign of rotation have been established,
and these are invaluable in determining the stereochemical
course of a substitution reaction. Three extreme stereochemical outcomes are possible: inversion of configuration, retention of configuration, or racemization. Which
of these occurs, alone or in combination, is determined
by the mechanism of the substitution.

The situation is (with the advantage of the present
state of knowledge) simplest for the pure S_N2 mechanism.
For this bimolecular mechanism it can be stated that
direct displacement is always accompanied by inversion of
configuration. This change in configuration is called
the Walden inversion. Two kinds of evidence support the
assertion that the S_N2 reaction always occurs with inversion [25,26]. Equation (9.17) shows a typical stereochemical cycle. In this three-step process the dextro-
rotatory alcohol $\underline{6}$ can be converted to its enantiomer $\underline{7}$.

$$
\begin{array}{ccc}
\overset{\displaystyle CH_3}{\underset{\displaystyle |}{PhCH_2-CH-OH}} \xrightarrow{TsC\ell} & \overset{\displaystyle CH_3}{\underset{\displaystyle |}{PhCH_2-CH-OTs}} \xrightarrow{OAc^-} & \overset{\displaystyle CH_3}{\underset{\displaystyle |}{PhCH_2-CH-OAc}} \\
\underline{6}\ (+) & & \Big\downarrow OH^- \qquad (9.17) \\
& & \overset{\displaystyle CH_3}{\underset{\displaystyle |}{PhCH_2-CH-OH}} \\
& & \underline{7}\ (\ )
\end{array}
$$

This result is consistent with an inversion of configuration at one or three of the steps. The first reaction--
an esterification--occurs on oxygen, not on the asymmetric
carbon, and the same is true for the third reaction, as is
shown in Chap. 13; therefore no configuration change
occurred in these reactions. It follows that inversion
occurred in the displacement of tosylate by acetate, an
S_N2 reaction at the asymmetric carbon atom.

   That every act of substitution is accompanied by inversion is shown by an isotopic exchange experiment [25].
The kinetics of isotopic exchange and of racemization in
the reaction of (+)-2-iodooctane and radioactive iodide,

$$I*^- + C_6H_{13}-\overset{\overset{I}{|}}{C}H-CH_3 \rightarrow C_6H_{13}-\overset{\overset{I*}{|}}{C}H-CH_3 + I^+ \qquad (9.18)$$

showed that the S_N2 mechanism was operative (in acetone),
and that the rate of racemization is twice the rate of
iodide exchange. That is, when half the molecules have
been substituted with radioactive iodine, so, too, have
half of them suffered inversion of configuration, resulting in complete loss of observable optical activity.
Therefore each bimolecular substitution occurs with inversion.

   The mechanistic interpretation of this result is very
simple. "Back-side" attack of the nucleophile Y^-, with
respect to the leaving group, will result in the energetically favorable transition state 8, which re-hybridizes
to the inverted product.

$$Y^- + \quad \overset{R_2}{\underset{R_3}{\overset{R_1}{\diagdown}}}C-X \rightarrow Y----\overset{R_2 \, R_1}{\underset{R_3}{|}}C----X \rightarrow Y-\overset{R_1}{\underset{R_3}{C}}R_2 \qquad (9.19)$$

8

The groups R_1, R_2, R_3 and the C atom lie approximately in
a plane in 8, and the Y-C-X bonds are perpendicular to
this plane.

   The stereochemical result of an S_N1 reaction is not
so well defined as for bimolecular substitutions. A
limiting case is that in which the carbonium ion intermediate is sufficiently long-lived (as when it is stabilized by resonance) so that the departing group diffuses

far enough away that the free carbonium ion assumes a
planar form, presumably stabilized by solvation [27], as
in 9.

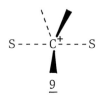

9

Attack by a nucleophile is then equally probable from
either side of the planar ion, so racemization should be
the stereochemical result. The unimolecular substitution
of 1-phenylethyl chloride occurs with nearly complete
racemization, in accordance with this description of the
mechanism.
     If nucleophilic attack occurs on the incipient car-
bonium ion before the departing group has been replaced
by a solvating species, then attack should be favored on
the back side, relative to the departing group. This
process, which does not sound significantly different from
an S_N2 displacement, should lead to inversion of configuration. The distinction between an S_N2 reaction and an S_N1
reaction with inversion can be made on the basis of kinetics and the absence or presence of concurrent racemization. The conceptual difference is one of detailed electronic distribution in the transition state. An ion pair
will fit the description given for the S_N1 intermediate
leading to inversion.
     The third type of stereochemical phenomenon, reten-
tion of configuration, requires a "neighboring group" that
blocks inversion by restricting attack from the back side.
The classic example is provided by the alkaline hydrolysis
of α-bromopropionate [28]. When conditions are chosen [29]
so that the unimolecular route predominates, then retention
of configuration is observed. This is rationalized by an
intramolecular interaction between the carboxylate ion and
the asymmetric carbon; this may be either electrostatic
[28] or covalent (the latter giving the α-lactone 10).
One view is that the asymmetry of the carbon atom is preserved in the carbonium ion intermediate by a zwitterionic
intramolecular participation of the carboxylate; attack by
hydroxide then yields the substituted product with the
original configuration. Another interpretation is that

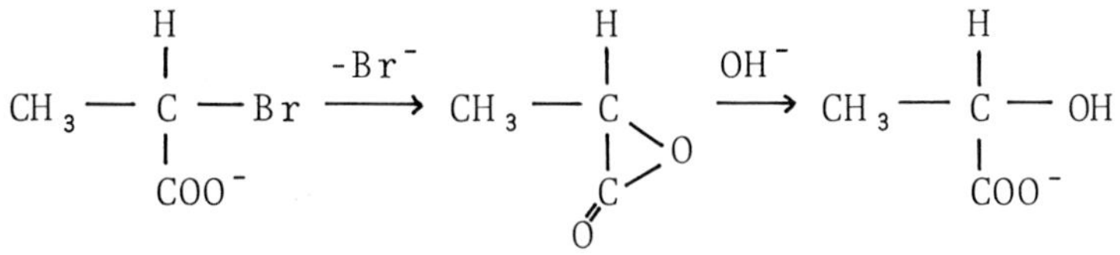

$$
\underset{\underset{COO^-}{\overset{H}{|}}}{CH_3 - C - Br} \quad \xrightarrow{-Br^-} \quad CH_3 - \underset{\underset{C}{\overset{H}{|}}}{C} \overset{}{\underset{O}{\diagdown}} O \quad \xrightarrow{OH^-} \quad \underset{\underset{COO^-}{\overset{H}{|}}}{CH_3 - C - OH} \qquad (9.20)
$$

<u>10</u>

back-side intramolecular attack by carboxylate gives the α-lactone intermediate <u>10</u> with inversion, and subsequent back-side attack by hydroxide gives a second inversion; the observed retention is therefore the consequence of two inversions [30].

Several other neighboring groups, such as acetoxy and phenyl, can influence the stereochemical course of S_N1 reactions [31].

A rather special type of substitution reaction has been identified in which the leaving group can act intra-molecularly to control the stereochemistry of the substitution. Both S_N1 [32] and S_N2 processes may be involved. The clearest example is given by the reaction of alcohols with thionyl chloride to give alkyl chlorides, a reaction that occurs via an alkyl chlorosulfite, <u>11</u>.

$$
R\text{-}OH + SOCl_2 \rightarrow R\text{-}O\text{-}\overset{\overset{\textstyle O}{\parallel}}{S}\text{-}Cl \rightarrow R\text{-}Cl + SO_2 \qquad (9.21)
$$

<u>11</u>

In dioxane solution this reaction occurs with retention of configuration at an asymmetric carbon atom; in toluene the same reaction takes place with inversion [33]. It is believed to occur with the formation of an ion pair, which may be represented by <u>12</u> [34].

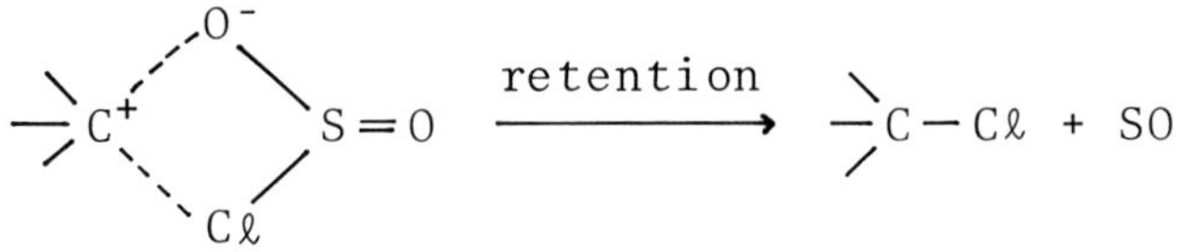

$$
\underset{}{\diagup}C^+ \overset{O^-}{\underset{Cl}{\diagdown \diagup}} S=O \quad \xrightarrow{\text{retention}} \quad \underset{}{\diagup}C\text{-}Cl + SO \qquad (9.22)
$$

<u>12</u>

The chloride is delivered intramolecularly by the ion pair, with retention of configuration. This "internal return" is called the S_Ni mechanism. Presumably it is aided by a good solvating species (like dioxane) for the carbonium

ion. If this is absent, an ordinary S_N2 displacement by
external halide leads to inversion of configuration.

Nucleophiles and Leaving Groups. The discussion of kinetic behavior in S_N reactions showed that the rates of S_N1
reactions are independent of the substituting nucleophile,
whereas S_N2 reactions show a first-order dependence on the
concentration of the nucleophile. The efficacy of the
substituting reagent will be reflected in the bimolecular
rate constant for an S_N2 reaction. This kinetic pattern
can be used as a criterion of mechanism [35]. Moreover,
it is possible to observe a change in mechanism for a substrate subjected to the action of a series of reagents in
turn, the reagents being selected for their varied nucleophilic power. The most powerful nucleophiles will carry
out substitution by the S_N2 route, and the rate will decrease with decreased nucleophilicity of the reagent [36].
At some point (or region) along the scale of decreasing
nucleophilicity, the nucleophilic assistance will be inadequate to produce a clear bimolecular reaction, and subsequent decreases in reagent nucleophilicity will reveal
independence of rate on nucleophilicity [35].
        It is implied that a series of nucleophiles can be
ordered according to their nucleophilicity, and that this
order has meaning outside the defining reaction conditions.
Within limits, some of which were described in Sec. 4.3,
this can be done. Since the nucleophile functions by
coordinating, with an unshared electron pair, at the carbon atom carrying the leaving group, it is reasonable to
suggest that nucleophilicity in an S_N2 reaction may be
related to basicity, which is a measure of equilibrium
affinity by coordination with the solvated proton. In a
series of nucleophiles with the same "attacking atom,"
the orders of nucleophilicity and basicity are often the
same. Thus these oxygen nucleophiles are listed in order
of decreasing basicity and decreasing nucleophilicity
[37]:

$$OH^- > OPh^- > CO_3^{2-} > OAc^- > p\text{-}Me\text{-}C_6H_4\text{-}SO_3^- > p\text{-}Br\text{-}C_6H_4SO_3^- > H_2O > C\ell O_4^-$$

Amine nucleophilicity also is correlatable with basicity,
though the correspondence may be interrupted by steric
effects or even by comparisons of different classes of
amines. This is, at least in part, because the steric
requirements for protonation and for substitution at
tetrahedral carbon are so different that basicity fails
as a universal model property for the S_N2 reaction.
        An important series of nucleophiles in which basicity
is not an effective measure of nucleophilicity is the

halide series.   The "normal" S_N2 nucleophilicity sequence is

$$I^- > Br^- > C\ell^- >> F^-$$

Three factors besides basicity are commonly invoked in treating such sequences. One of these is the polarizability, which is a measure of the ease of distortion of the valence electron cloud. Larger atoms generally are more polarizable than small ones. Polarizability changes in the order given above for the halides. Another factor is ion solvation. The solvation sphere of a nucleophile must be disrupted when it engages in a substitution on carbon, so the more tightly solvated anion should be less nucleophilic. The possible formation of ion pairs may also affect reactivity by decreasing the concentration of the dissociated anion [38].

Because of the several effects that can operate to control reactivity, the net nucleophilicity can depend strongly upon reaction conditions. This is clearly shown in Table 9.I, where nucleophilic sequences are given for halide ions in several solvents [39]. Solvation by protic solvents appears to produce the normal reactivity sequence, whereas polar aprotic solvents reverse the order. The order in acetone, for example, follows the basicity sequence.

Table 9.I   Relative Nucleophilicities of Halide Ions in S_N2 Reactions [39][a]

Solvent	Nucleophilicity sequence
H_2O, MeOH	$I^- > Br^- > C\ell^- >> F^-$
Acetone	$C\ell^- > Br^- > I^-$
DMF[b]	$C\ell^- > Br^- > I^-$
DMF + H_2O	$I^- > Br^- > C\ell^-$
DMSO[c]	$C\ell^- > Br^- > I^-$
DMSO + H_2O	$I^- > Br^- > C\ell^-$
Liquid SO_2	$I^- > Br^- > C\ell^-$

[a]Corrected for ion pairing.
[b]N,N-dimethylformamide.
[c]Dimethylsulfoxide.

A quantitative measure of nucleophilicity in S_N2

reactions is provided by the Swain-Scott approach [40], which was considered in Sec. 4.3. This is a conventional linear free energy relationship

$$\log \frac{k}{k^0} = sn \qquad (9.23)$$

where n is a nucleophilic constant defined by Eq. (4.64). The substrate constant s, which is defined to be unity for the standard substrate, methyl bromide, expresses the sensitivity of a substrate to a change in reagent nucleophilicity in an S_N2 reaction. The nucleophilic constants were evaluated in aqueous media. Since n is defined on a logarithmic scale, the n values of 5.04 for iodide and 3.04 for chloride mean that, in an S_N2 reaction, iodide is one hundred times more nucleophilic than is chloride. With the s and n values tabulated, rate constants for many S_N2 reactions can be generated with fair accuracy.

Since the bond from carbon to the leaving group has undergone partial breaking in the transition state for both the S_N1 and the S_N2 mechanisms, it is not surprising that reaction rates depend upon the leaving group in both mechanisms. Table 9.II gives approximate relative rates of S_N2 displacement reactions for several leaving groups [41]. The halide order (which is also observed in S_N1 solvolyses) is $I^- > Br^- > C\ell^- >> F^-$.

Table 9.II   Relative Rates of Displacement for Several Leaving Groups [41]

Leaving group	Typical reaction	Average relative rate
I^-	$EtI + EtO^-$ in $EtOH$	3
Br^-		(1.00)
$C\ell^-$	$MeC\ell + S_2O_3^{2-}$ in H_2O	0.02
F^-	$PhCH_2F + EtO^-$ in $EtOH$	10^{-4}
$PhSO_3^-$	$MeOSO_2Ph + OH^-$ in H_2O	6
NO_3^-	$EtONO_2 + OH^-$ in $EtOH-H_2O$	10^{-2}

In a first approximation it may be expected that the less nucleophilic a group is, the better it should be as a leaving group. For many groups this is a good rule. For example, hydroxide is not a good leaving group--in fact, it is not known to be displaced at all--and it is a

good attacking nucleophile. When a hydroxy group is protonated, changing it from -OH to $-OH_2^+$, the leaving group
now is H_2O, which is but weakly nucleophilic and is therefore a good leaving group. The halide order, which is the
same as the order of nucleophilicity, is therefore surprising. The explanation probably involves several factors.
One of these is the strength of the carbon-halogen bond,
which increases in the order $I < Br < C\ell < F$ [42]. Another
is the point that the "normal" halide nucleophilicity order may not be the proper one for such a comparison, for
Table 9.I shows that it can be reversed by a solvent
change.

In the S_N1 mechanism the leaving group will influence
the rate, but, in the presence of competing nucleophiles,
should have no influence on the product distribution.
This is a test for the unimolecular mechanism. A complication in many S_N1 solvolyses is a competing elimination
of a β-proton to give an olefin:

$$\text{>CH}\!-\!^+\text{C}\!< \xrightarrow{\ -H^+\ } \text{>C}=\text{C}\!< \qquad (9.24)$$

A further test for the S_N1 mechanism is that the ratio of
elimination to substitution be independent of the leaving
group [43]. Examples are known, however, of solvolyses
apparently proceeding through a carbonium ion intermediate
in which the percent of elimination product is strongly
dependent upon the leaving group. This is rationalized
by the intervention of a product-influencing ion pair [44].

$$RX \rightleftharpoons R^+X^- \begin{array}{l} \nearrow \text{ substitution} \\ \searrow \text{ elimination} \end{array} \qquad (9.25)$$

Effects of Substrate Structure. So far in this chapter
the mechanisms available to a substitution process have
been described, but little has been said to indicate why
an aliphatic substrate will react by one mechanism rather
than another. Two guidelines can be briefly given: (1)
The more powerful the attacking nucleophile, the greater
the tendency for the substitution to occur via the nucleophilic assisted route, that is, S_N2 or N; (2) the more
stable the carbonium ion that can be produced by ioniza
tion of the substrate, the greater the tendency to react
by the unimolecular (S_N1 or Lim.) route.

We first consider reactivity in S_N2 reactions. The
role of the nucleophile in promoting the bimolecular mechanism was treated earlier in discussions of the Winstein-
Grunwald mY and Swain-Scott sn correlations. Of a series

of substrates in a fixed solvent, Dewar and Sampson [45]
conclude that the less reactive compounds should be more
susceptible to nucleophilic assistance than the more reactive ones. This will not be evident unless the series is
defined to exclude complicating variable factors like
steric effects.

Streitwieser [46] summarized the reactivities of many
alkyl halides in direct displacements, finding a fairly
constant order of reactivities for different substrates;
these are given in Table 9.III. The rate decreases observed with increasing α-methyl substitution (Me > Et >
iPr) could be the result of inductive electron release at
the reaction site, which would hinder the approach of the
incoming nucleophile; or they could be examples of steric
hindrance. The sharp rate decreases seen with increasing
β-methyl substitution (Et > Pr > iBu > neoPe) are attributed to severe steric effects. The transition state for
reaction of a neopentyl substrate, 13, indicates how the
β-methyl groups can hinder attack by the nucleophile Y
(this will increase $\Delta H^{\ddagger}$); they also, by interaction with
Y and X, limit the free rotation about the C_{α} - C_{β} bond,
thus decreasing $\Delta S^{\ddagger}$ [47].

Table 9.III   Average Relative Rates of R-X
in S_N2 Reactions [46]

R	Relative rate
Methyl	30
Ethyl	1.0
Propyl	0.4
Butyl	0.4
Isopropyl	0.025
Isobutyl	0.03
Allyl	40
Benzyl	120
Neopentyl	10^{-5}

In the allyl and benzyl systems the substrate unsaturation is conjugated with the reaction site, permitting transition state stabilization as shown in 14 and 15
[48]. m- and p-substituted benzyl substrates have reactivities similar to that of the unsubstituted compound;
small rate increases are usually observed [48].

A profound effect of structure on reactivity is seen

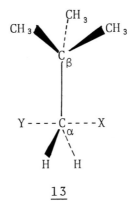

13

14   15

with cyclic substrates carrying the potential leaving
group on a "bridgehead" carbon atom. 1-Chloroapocamphane,
16, yielded none of the parent alcohol after refluxing
for 21 hr with 30% potassium hydroxide in 80% ethanol [49].

16   17

The noncyclic analog, α,α-diethyl-neopentyl chloride 17,
lost 98% of the theoretical halide in 8 hr at 25°. The
interpretation of this reactivity comparison is that, under these S_N2 conditions, backside attack by the nucleophile is prohibited by the apocamphyl cage structure,
whereas 17 is exposed (though hindered). Moreover, even
if 16 could be attacked by the nucleophile, it cannot
possibly undergo a Walden inversion. It is therefore
unreactive to direct displacement.
     Turning to reactivity in S_N1 reactions, it will be
recalled that carbonium ion stability is optimized by a
co-planar orientation of the three bonds to the positive
carbon atom. The inertness of 16 to S_N1 reaction conditions (no chloride produced in 48 hr of reflux with silver
nitrate) [49] is understandable in these terms, because
the desired planarity cannot be achieved in this structure. A more moderate example of the steric influence is
seen in solvolysis rates of a series of 2-aryl-2-chloro-
propanes [50]. These S_N1 reactions proceed at "normal"
rates for aryl substituents, like phenyl and 2-naphthyl
(18), in which little steric opposition is offered to the
attainment of a planar carbonium ion.

          18   19

In "1-naphthyl" types of substrates, however (as in 19),
the peri interaction with the hydrogen of the adjacent
ring restricts the attainment of planarity, and these compounds react relatively slowly. Similar reactivity behavior is observed for solvolyses of arylphenylmethyl
chlorides, $Ar(C_6H_5)CHC\ell$ [51].
     If a planar carbonium ion can be generated by ioniza-
tion, its stability, and therefore the S_N1 reactivity,
will be largely determined by electronic effects at the
α-carbon atom. Alkyl groups promote carbonium ion stability by their electron-releasing inductive effects, with
the consequence that S_N1 solvolysis rates fall in the
order tertiary >> secondary > primary. A phenyl group is
about as effective as two methyl groups in stabilizing
carbonium ions [52]. Benzyl halides and isopropyl halides

are therefore borderline substrates. Benzyl substrates
with electron-releasing substituents, especially with good
leaving groups and in ionizing solvents, will undergo S_N1
solvolysis. Replacement of an α-hydrogen atom by a phenyl
group [53] (as in the series benzyl, benzhydryl, trityl)
increases the unimolecular reactivity by about 10^6.
  Allylic systems provide the possibility of resonance
stabilization of the carbonium ion and of more than one
substitution product. Their reactivity is enhanced by α-
and γ-alkyl substituents. The resonance is indicated by
Eq. (9.26), which shows also how α- or γ-substitution may
occur.

$$R_2C=CH-CHR-X \xrightarrow{-X^-} R_2C=CH-\overset{+}{C}HR \longleftrightarrow R_2\overset{+}{C}-CH=CHR$$

$$\downarrow Y^-$$

$$R_2C=CH-CHY + R_2CY-CH=CHR$$

$$(9.26)$$

  The product mixture of (9.26) suggests a molecular
rearrangement, which is interpreted as shown by α- and
γ-localization of the charge deficiency. Other kinds of
rearrangements are also observed in substitution reactions.
If the carbonium ion initially generated in an S_N1 reaction
can be transformed to a more stable ion by a shift of an
alkyl group (the Wagner-Meerwein rearrangement), the substitution product will appear to be rearranged, as in

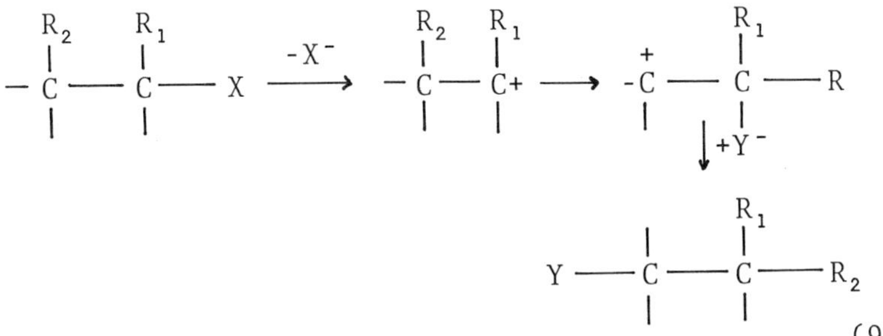

$$(9.27)$$

These rearrangements occur so as to produce tertiary carbonium ions from secondary ions, and secondary from primary [54]. If the rearrangement follows carbonium ion
formation, as in (9.27), the product distribution but not
the reactivity will be affected by the alkyl group migration.
  Many reactions are known, however, in which reaction

rates are accelerated by groups on carbon adjacent to the
leaving group. This behavior is called neighboring group
participation or anchimeric assistance [55]. Neighboring
group participation can affect the product distribution
and stereochemistry as well as the rate; the classic example, the solvolysis of α-bromopropionate, is shown in
Eq. (9.20). As in this case, retention of configuration
is evidence for neighboring group participation. An ele-
gant demonstration of the stereochemical course of an
anchimerically assisted solvolysis was given by Cram in
his study of the acetolysis of the tosylates of threo-
and erythro-3-phenyl-2-butanol [56]. The essential result
is that either of the optically active threo-tosylates
yields racemic acetates, whereas the erythro-tosylates
yield optically active products. This is rationalized by
postulating an intermediate, which may be written as a
cyclic, or bridged, phenonium ion. Equation (9.28) shows
the postulated mechanism for one of the threo enantiomers
[57].

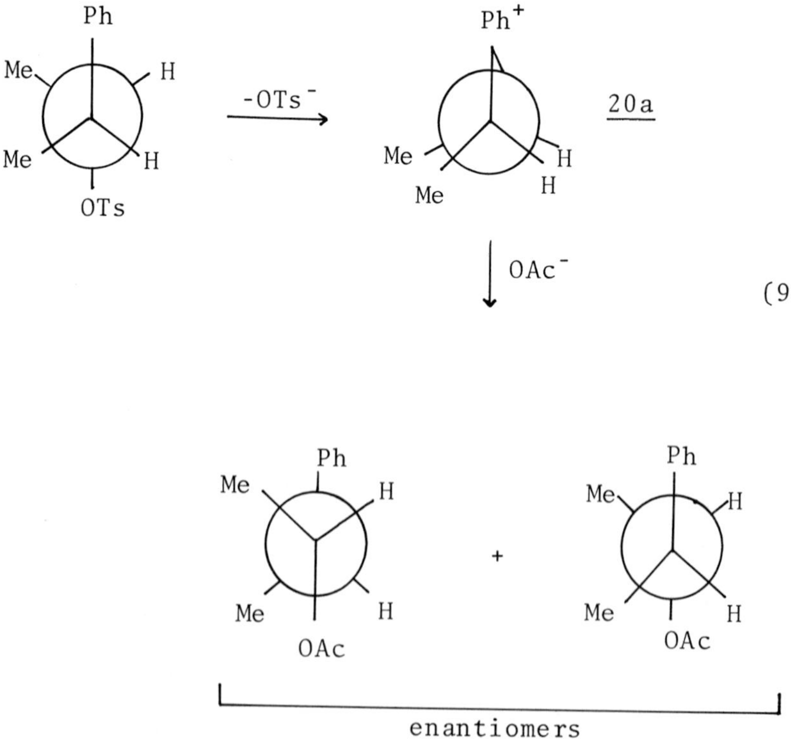

(9.28)

enantiomers

Structure 20a is the bridged ion, produced, with inversion
of configuration, by back-side attack of the neighboring
phenyl group. This ion has a plane of symmetry, and
attack by acetate therefore yields the racemic mixture.
This mechanism also accounts for the stereochemical result
of the solvolysis of an erythro tosylate, shown in Eq.
(9.29). The intermediate 20b does not possess a plane of
symmetry. Attack by acetate at either carbon atom produces
the same optically active acetate.

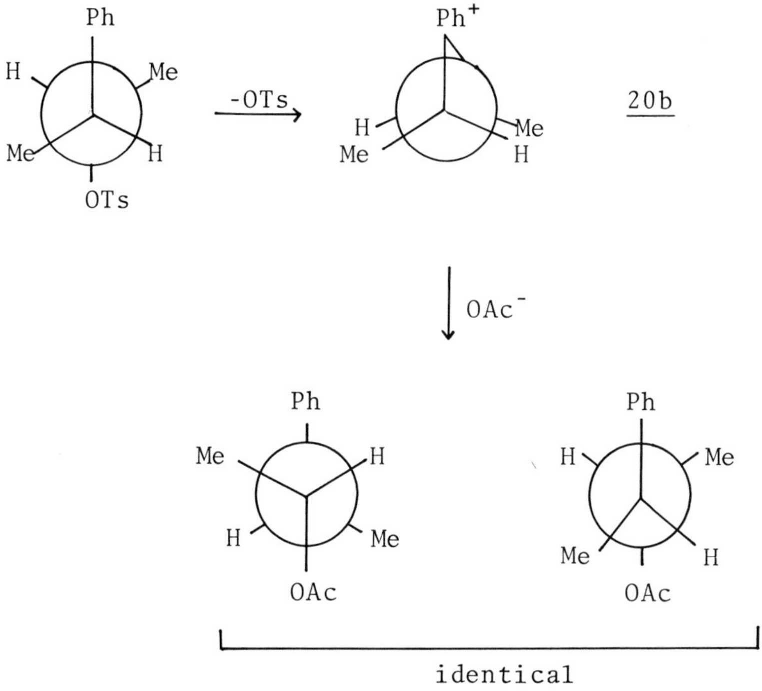

$$(9.29)$$

identical

The nature of the intermediate in solvolyses occurring with such β-aryl participation has been carefully
studied. Three structural proposals have been made: (1)
the "intermediate" is really an equilibrium mixture of
the rapidly equilibrating classical ions 21a, b; (2) it
is the delocalized phenonium ion 22; (3) it is a π-complex
23.
        Attempts to distinguish among these alternatives re-
quire, first of all, a demonstration that β-aryl participation does in fact occur, and this in turn necessitates

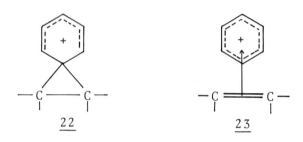

$$21$$

$$22 \qquad\qquad 23$$

an estimate of the solvolytic reactivity of the substrate
supposing the absence of a neighboring group effect. The
problem is usually discussed in terms of two discrete
competing processes. One of these is normal anchimerically unassisted ionization (with rate constant k_S); the
other is anchimerically assisted ionization (k_Δ) [58].
von R. Schleyer has analyzed the separation of these processes, with the conclusions that the k_S process, though
anchimerically unassisted, is strongly solvent assisted,
and that detection of any rate enhancement by the neighboring group, no matter how small, is evidence for strong
β-aryl participation [59]. The success of the kinetic
separation into distinct and competitive routes, with no
crossover possible between the respective intermediates
[59], appears to rule out rapidly equilibrating open cations, for these should be easily converted to the bridged
species. The bridged phenonium ion intermediate is also
supported by stereochemical and isotopic rearrangement
data [60]. Bentley and Dewar [61], however, prefer a π-
complex formulation of the intermediate, based on the observations that the overall kinetic effect of β-aryl participation is very large, but that differences between
different aryl groups are small. The phenonium ion 22
(which, it will be noted, is closely analogous to the σ-
complexes widely accepted as intermediates in electrophilic aromatic substitution) is disfavored by these authors
on the grounds that the rate data for anchimerically
assisted solvolyses of 2-arylethyl p-toluenesulfonates

are not well correlated with localization energies calculated for the formation of σ-complexes [62].
    This discussion of β-aryl participation gives some
idea of the research activity in a small area of study.
Even more effort has been made to account for a related
type of anchimeric assistance in which the intermediate
may be a nonclassical carbonium ion. A nonclassical ion
is one whose ground state has delocalized bonding σ electrons [63]. The possible intervention of such species
has been most carefully studied in bicyclic systems,
especially in norbornane derivatives. Winstein's group
has made many of the major contributions. The nature of
the chemistry is illustrated by the solvolysis of norbor-
nyl arylsulfonates. Exo-norbornyl brosylate [64], 24,
solvolyzes 350 times faster than endo-norbornyl brosylate,
25.

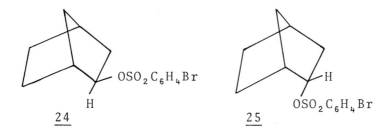

Both of these substrates yield only the exo substitution
product, and substitution is accompanied by complete racemization. These observations were accounted for [65] by
the proposal that the exo isomer is subject to anchimeric
assistance by carbon 6 (see 26) to give the σ-delocalized
nonclassical ion 27. This intermediate has a plane of
symmetry through C_4-C_5-C_6, and atoms 1 and 2 are equivalent. Attack at C_2 gives 28, with the original configuration; attack at C_1 gives its enantiomer 29.
    An alternative view, held especially by Brown [66],
is that the nonclassical ion hypothesis is unnecessary,
the data being equally well accounted for by the postulate
of a pair of rapidly equilibrating classical ions, such
as 30 and 31, which are interconvertible by a 1,2 migration of C_6. It has recently been shown, primarily by nuclear magnetic resonance measurements, that the stable
2-norbornyl cation must be a σ-delocalized nonclassical
ion [67]. This demonstration was achieved in strongly

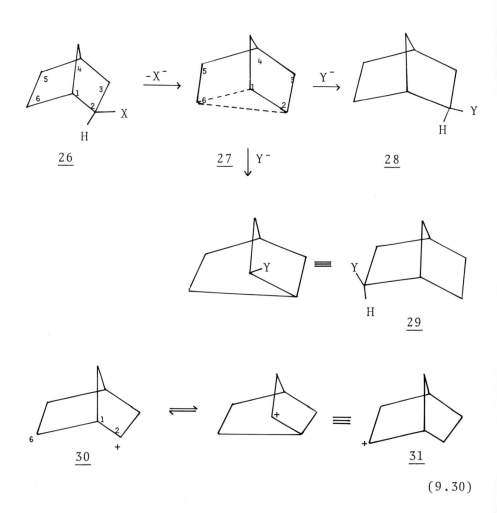

(9.30)

acid solutions, however, and its relevance to the milder
conditions of solvolysis reactions is certain to be
questioned.
      Double bonds are very effective neighboring groups
in solvolyses. Nonclassical ions have been postulated as
intermediates in the solvolysis of the systems 7-norbornyl
(32), syn-7-norbornenyl (33), anti-7-norbornenyl (34),
and 7-norbornadienyl (35) [68]. The remarkable enhancement of solvolytic reactivity is revealed by the relative
rates given below the structures.

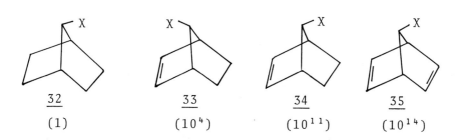

32	33	34	35
(1)	(10^4)	(10^{11})	(10^{14})

Solvent Effects. The role of the solvent in controlling the mechanism of a substitution reaction has been treated in the preceding discussions. Thus we find that S_N1 reactions are promoted by solvents of high ionizing power. The Winstein-Grunwald solvent polarity parameter $\underline{Y}$ is a useful empirical measure of ionizing power. Kosower's $\underline{Z}$ value (see Sec. 4.4) can also correlate nucleophilic displacement reactivities [69]. High nucleophilicity promotes S_N2 displacements. For example, the reaction conditions can be altered for substitutions on aryl methyl chlorides so that the reaction proceeds by the limiting S_N1 mechanism in moist formic acid (a poorly nucleophilic, highly ionizing solvent), or by a transitional mechanism in 80% ethanol-20% water, or by a fully assisted S_N2 displacement by iodide ions in anhydrous acetone [45].

The anticipated effect of a change in solvent polarity on the rate of a substitution reaction can be qualitatively predicted with the aid of the transition state argument of Sec. 3.4: namely, if the transition state is more polar than the initial state, a change to a more polar solvent will increase the reaction rate, whereas a less polar solvent will give a decreased reaction rate. When this approach is applied to substitution reactions several outcomes are predicted, depending upon the mechanism and the charge types of the reactants. The results [70] are shown in Table 9.IV. This treatment is very general, and ignores specific effects such as concomitant alterations in solvent nucleophilicity, which we have seen to be important. Within its limitations, however, it is of some utility [70].

Much of the early kinetic work on aliphatic substitutions was carried out on aqueous alcohol solutions. These introduce some of the complications referred to above. In water-alcohol mixtures equilibrium (9.31) is

$$OH^- + ROH \rightleftharpoons RO^- + HOH \qquad (9.31)$$

Table 9.IV Predicted Solvent Effects on Rates of S_N Reactions [70]

Mechanism	Initial state	Transition state	Relative charge in transition state	Rate effect of increase in solvent polarity
S_N2	$Y^- + RX$	$^{-\delta}Y\text{---}R\text{---}X^{-\delta}$	dispersed	small decrease
	$Y + RX$	$^{+\delta}Y\text{---}R\text{---}X^{-\delta}$	increased	large increase
	$Y^- + RX^+$	$^{-\delta}Y\text{---}R\text{---}X^{+\delta}$	reduced	large decrease
	$Y + RX^+$	$^{+\delta}Y\text{---}R\text{---}X^{+\delta}$	dispersed	small decrease
S_N1	RX	$^{+\delta}R\text{----}X^{-\delta}$	increased	large increase
	RX^+	$^{+\delta}R\text{----}X^{+\delta}$	dispersed	small decrease

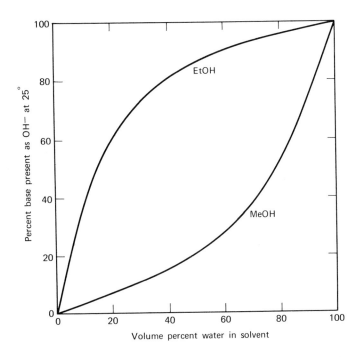

Fig. 9.1. Fraction of total base present as hydroxide as a function of the percent of water in the binary solvent, for ethanol-water and methanol-water mixtures (from Ref. [71]).

established. The position of equilibrium depends upon the identity of the alcohol and the composition of the mixture. Figure 9.1, given by Burns and England [71], shows the profound difference in this equilibrium for ethanol-water and methanol-water mixtures. For example, in a methanol solution containing 20% water, about 6% of total base present is hydroxide, the remainder being methoxide; in the corresponding ethanol-water solution, about 60% of the base is hydroxide. This difference has kinetic consequences because the nucleophilicity of these lyate ions varies in the order $C_2H_5O^- > CH_3O^- > OH^-$ [72,73]. The net effect of solvent composition, in alcohol-water mixtures, upon rates of S_N2 reactions is a consequence of three effects [74]; these are (1) the general transition state effect summarized in Table 9.IV; (2) the change of nucleophile from alkoxide to the less reactive hydroxide

in more highly aqueous solvents; (3) the increased dissociation of ion pairs M^+OR^- to the more reactive dissociated
anion in more highly aqueous solvents. These effects,
which may combine to give several types of observed rate
effects depending upon the system [74], complicate the
mechanistic interpretation of solvent effects in alcohol-
water mixtures. This is especially so in those reactions
(Table 9.IV) for which small changes are predicted.

Much recent work on solvent effects has concerned
S_N2 reactions in polar aprotic solvents; Parker [75] has
been very active in this field. Protic solvents, of which
water, the alcohols, and acetic acid are important examples, are capable of functioning as hydrogen-bond donors.
Polar aprotic solvents are generally taken to be those
with moderate to high dielectric constants; this rules
out dioxane and hydrocarbons. The most extensively studied
polar aprotic solvents are acetone, acetonitrile, N,N-
dimethylformamide (DMF), N,N-dimethylacetamide (DMA), and
dimethylsulfoxide (DMSO).

A few relative rate data are given in Table 9.V [76].
These are typical of the rate effects observed for S_N2
reactions between anions and neutral molecules. The outstanding feature is the rate increase, amounting to several
orders of magnitude, on passing from a protic to a polar
aprotic solvent. This rate effect is more pronounced for
small anions than for large ones, and rates for primary
halides are more sensitive to the solvent change than are
rates for secondary substrates.

Interpretations of the profound rate effect of the
solvent on anion-neutral molecule S_N2 reactions are based
on anion solvation phenomena. It is observed, even with
the few data in Table 9.V, that for a given reaction it
is necessary to establish (to a first approximation) only
two rate classes: slow reactions, which occur in protic
solvents, and fast reactions, which occur in polar aprotic
solvents. One viewpoint is simply that anions are more
strongly solvated (by means of hydrogen bonds) in protic
solvents, and this solvation decreases their effective
nucleophilicity. In aprotic solvents, the solvation is
less extensive, so the inherent nucleophilicity of the
anion can be realized, resulting in a fast reaction. Re-
finement of this idea [77] accounts for the rate variations by postulating that small anions are more strongly
solvated in protic solvents than are large anions, and
that the extent of solvation and the differences both decrease in aprotic solvents. Enthalpies of solvation for
alkali halides in DMSO and in water support this view [77].
The order of solvation in DMSO is $Cl^- > Br^- > I^-$, which
is the same as in water. The reversal in the nucleophilicity sequence (see Table 9.I) is attributed to the smaller

Table 9.V Relative Rates of S_N2 Reactions in Protic and
Polar Aprotic Solvents [76]

	log (relative rate)[a]			
	MeI + Cl⁻	MeI + Br⁻	n-BuBr + N₃⁻	i-PrBr + N₃⁻
Protic solvents				
Water	0.5	-0.3	0.8	--
Methanol	(0.0)	(0.0)	(0.0)	(0.0)
Formamide	1.2	0.7	1.1	--
Polar aprotic solvents				
Acetone	6.2	5.1	3.6	--
Acetonitrile	4.6	--	3.7	--
DMA	6.4	--	3.9	--
DMF	5.9	4.2	3.4	2.7
DMSO	--	--	3.1	--

[a]Logarithm of rate relative to the rate of the same reaction in methanol, at 25°.

differences in solvation in the aprotic solvent.

Undoubtedly solvation phenomena of initial states have much to do with the rate effects being considered. The contribution of transition state solvation to relative rate effects, especially for anions of different size and for substrates of different structure, is less clear. The dependence of the solvent rate effect on substrate structure has been related to the position of the transition state along the spectrum from the pure S_N1 and S_N2 types [78].

9.2   Analytical Reactions

Substitutions on Saturated Carbon. The alkoxyl group is determined by cleavage of the ether linkage with hydriodic acid and measurement of the resulting alkyl iodide [79]; this method, the Zeisel procedure, is based on an S_N2 reaction. It is illustrated in Eq. (9.32) for the methoxyl group.

$$CH_3OR + HI \rightarrow ROH + CH_3I \qquad (9.32)$$

If the ether is unsymmetrical, two different alkyl iodides can be produced. The ratio of these products will depend upon the relative reactivity toward substitution at the two carbons, and upon the leaving group tendencies. When one of the groups is methyl [as in (9.32)] and the other is a much more substituted alkyl group, the principal product is methyl iodide. Even if the initial cleavage gives RI + CH_3OH, the methanol is subsequently converted to methyl iodide by a substitution analogous to (9.32):

$$CH_3OH + HI \rightarrow H_2O + CH_3I \qquad (9.33)$$

The methyl iodide is separated from the reaction mixture by distillation and is determined gravimetrically or titrimetrically [79].

Since alkoxide ions are very powerful nucleophiles, they are poor leaving groups. The substitution is made feasible by the acidic solution, and Eq. (9.32) is better written as Eq. (9.34), showing attack by the nucleophile I^- on an unsubstituted aliphatic carbon carrying the good leaving group ROH.

$$\overset{\text{H}}{\underset{+}{CH_3OR}} + I^- \rightarrow ROH + CH_3I \qquad (9.34)$$

This method is used primarily to determine the

methoxyl group. though higher alkoxyls can also be determined. The N-alkyl group can be determined in a similar
manner, though pyrolysis is required to generate the alkyl
iodide [80].

A simple spectrophotometric determination of tertiary
alcohols, which are difficult to determine by acylation
methods, is based on their nucleophilic substitution by
iodide [81].

$$R_1 - \underset{\underset{R_3}{|}}{\overset{\overset{R_2}{|}}{C}} - OH + HI \longrightarrow R_1 - \underset{\underset{R_3}{|}}{\overset{\overset{R_2}{|}}{C}} - I + H_2O \qquad (9.35)$$

Presumably this is an S_N1 reaction. The tertiary alkyl
iodide is extracted into cyclohexane and the solution absorbance is measured at 268 nm.

Chemical methods for α-epoxides (ethylene oxides,
oxiranes) are based upon ring-opening nucleophilic substitutions, often with halogen acids.

$$-\underset{O}{\overset{|}{C}}-\underset{}{\overset{|}{C}}- + HC\ell \rightarrow -\underset{OH}{\overset{|}{C}}-\underset{}{\overset{\overset{C\ell}{|}}{C}}- \qquad (9.36)$$

The amount of acid consumed is a measure of the epoxide
in the sample [82]. In the presence of water or alcohols,
competing solvolysis reactions can occur. Many variations
of the basic method have been proposed to reduce interferences and to increase the reaction rate [83]. Epoxides
can be titrated directly with HBr in acetic acid solution
[84]. This reagent is unstable, and an interesting modification of the method uses a perchloric acid titrant with
an excess of a quaternary ammonium bromide or iodide in
the sample solution; the halogen acid is generated in situ
[85].

The mechanism [86] of the epoxide hydrolysis reaction
is still not clear. The reaction is acid catalyzed, the
first step being O-protonation. The doubtful point concerns the next step. This could be either a unimolecular
ionization or a bimolecular attack of water on the protonated epoxide. These possibilities are shown in Eqs.
(9.38) and (9.39).

$$-\underset{O}{\overset{|}{C}}-\underset{}{\overset{|}{C}}- + H^+ \rightleftharpoons -\underset{\underset{H^+}{\overset{|}{O}}}{\overset{|}{C}}-\underset{}{\overset{|}{C}}- \qquad (9.37)$$

$$-\underset{\underset{H^+}{\overset{\displaystyle O}{\diagup}}}{C}\underset{}{\overset{\displaystyle |}{-}}C- \xrightarrow{slow} -\underset{\underset{OH}{|}}{\overset{|}{C}}\underset{+}{}-\underset{}{\overset{|}{C}}- \xrightarrow[fast]{H_2O} -\underset{\underset{OH}{|}}{\overset{|}{C}}-\underset{}{\overset{\overset{\displaystyle +OH_2}{|}}{C}}- \qquad (9.38)$$

$$-\underset{\underset{H^+}{\overset{\displaystyle O}{\diagup}}}{C}\overset{\displaystyle |}{\underset{}{}}\overset{\displaystyle \|}{C}- \xrightarrow{slow} -\underset{\underset{OH}{|}}{\overset{|}{C}}-\underset{}{\overset{\overset{\displaystyle +OH_2}{|}}{C}}- \qquad (9.39)$$

Inversion of configuration takes place, which is consistent with the bimolecular mechanism, though it has been pointed out that the asymmetry of the carbon atom might be preserved in a carbonium ion by the adjacent hydroxy group. The rate law for the halogen acid reaction of ethylene oxide includes the halide concentration, which is evidence for the bimolecular mechanism [87]. It seems that either the unimolecular or the bimolecular mechanism may be operative in epoxide reactions, depending upon the structure of the epoxide and the reaction conditions.

A large number of analytical reactions can be classed as alkylations; many of these are aliphatic nucleophilic substitutions. Usually an oxygen or nitrogen atom is alkylated. In the analytical sense, the reagent (the alkylating agent) is the aliphatic substrate, and the analytical sample compound is the attacking nucleophile. A familiar example is provided by the preparation of crystalline esters of acids by reaction of the salt of the acid with a phenacyl halide, as in

$$RCOO^- + C_6H_5-\overset{\overset{\displaystyle O}{\|}}{C}-CH_2Br \rightarrow C_6H_5-\overset{\overset{\displaystyle O}{\|}}{C}-CH_2O_2CR + Br^- \qquad (9.40)$$

The unsubstituted phenacyl bromide $C_6H_5COCH_2Br$ and its p-chloro-, p-bromo-, and p-nitro-derivatives are commonly used for the characterization of acids in this way [88].

Table 9.VI lists many of the alkylating agents that have been introduced as analytical reagents or for allied purposes. Some of these have carbonyl groups alpha to the site of substitution. This substituent retards substitution by an S_N1 mechanism, but it greatly enhances the S_N2 reactivity of a substrate. The rate enhancement is not due solely to inductive electron withdrawal, because other electronegative groups in the alpha position do not have a comparable effect [89]. It appears that some type of direct participation by the carbonyl group with the

attacking nucleophile must be invoked to account for the
rate facilitation.

Koshland's reagent, 2-hydroxy-5-nitrobenzyl bromide
(HNB, 36),

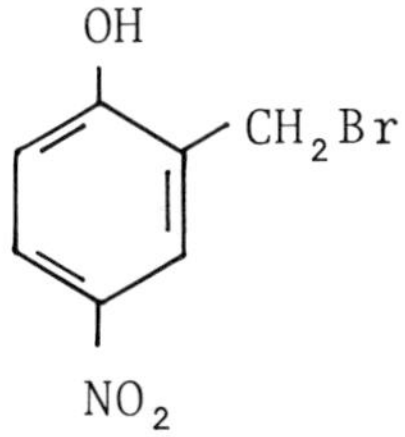

OH

CH$_2$Br

NO$_2$

36

was introduced as a "reporter group," that is, a moiety
that, when covalently bound to an amino acid residue in a
protein molecule, reveals a sensitivity (usually manifested
in its absorption spectrum [90]) to the local conformational environment. HNB displays an unusual reactivity-selec-
tivity pattern. It is highly specific, in neutral or acidic media, for tryptophan and for tryptophan residues in
proteins. On the other hand, it is extremely reactive.
The combination of high reactivity with high selectivity
is abnormal (see Sec. 5.1) and implies the intervention of
a further selectivity-inducing phenomenon; this is thought
to be charge-transfer complexing of the reagent with tryptophan residues, augmented by hydrophobic side-chain interactions [92].

The reactivity of HNB must be a consequence of the
ortho-hydroxy group, because it is much more reactive than
the very similar 2-methoxy-5-nitrobenzyl bromide. Substitutions on HNB probably proceed by an ionization mechanism,
with the carbonium ion being, in a sense, highly stabilized
by loss of a proton to give the o-quinone methide 38, which
may be the active species in the alkylation [93].

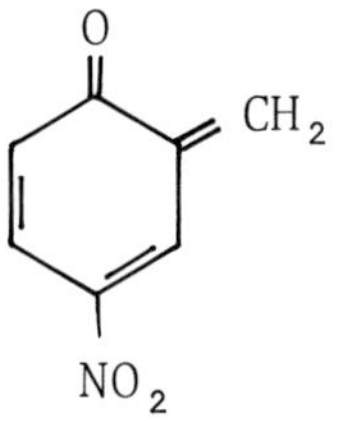

O

CH$_2$

NO$_2$

38

Table 9.VI   Some Alkylating Agents of Analytical Interest

Agent	Sample or reaction

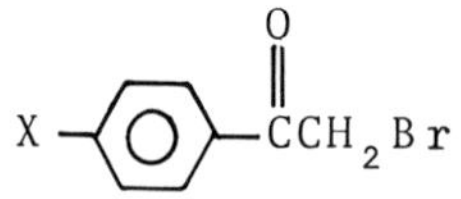

$$X-\underset{}{\bigcirc}-\overset{\overset{O}{\parallel}}{C}CH_2Br$$

Phenacyl halides
Phenacyl bromides

Xanthydrol

Trityl chloride
p-Nitrobenzyl bromide
ICH_2COOH; ICH_2CONH_2

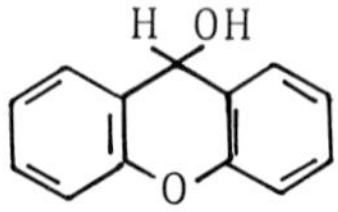

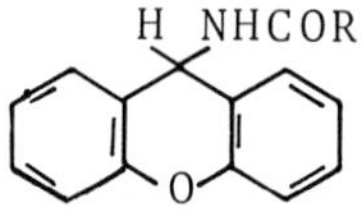

$$R-\overset{\overset{O}{\parallel}}{C}-CH_2C\ell$$

Substituted
   chloromethylketones
$C\ell CH_2COOH$
$CH_2-CH_2-CH_2C\ell$ with $\searrow O \swarrow$

$(CH_3)_2SO_4$
Dimethyl sulfate
CH_3I

CH_3I
Methyl p-nitrobenzene
   sulfonate

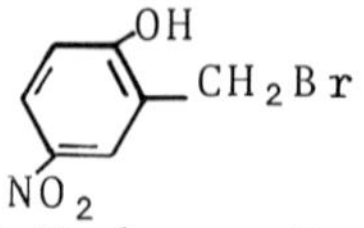

2-Hydroxy-5-nitrobenzyl
   bromide

Sample or reaction:

See Eq. (9.40)

α-chymotrypsin
xanthydrol + $RCONH_2 \rightarrow$

$(C_6H_5)_3CC\ell$ + ROH $\rightarrow$ $(C_6H_5)_3COR$
barbiturates
proteins

enzymes

$ArO^- + C\ell CH_2COOH \rightarrow ArOCH_2COOH$
$ROH + CH_2-CH_2-CH_2C\ell \rightarrow$ with $\searrow O \swarrow$

$ROCH_2CHOHCH_2C\ell$
+
$ROCH(CH_2OH)CH_2C\ell$
theobromine anion + $(CH_3)_2SO_4 \rightarrow$
   caffeine + $CH_3SO_4^-$
$R_3N + CH_3I \rightarrow (R_3NCH_3)^+I^-$

$ArSO_2^- + CH_3I \rightarrow ArSO_2CH_3 + I^-$
α-chymotrypsin

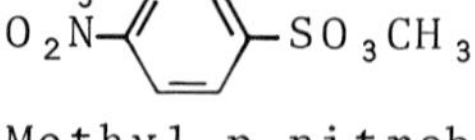

proteins

[a]D. S. Sigman, D. A. Torchia, and E. R. Blout, Biochemistry, 8, 4560 (1969)

Remarks	Ref.
characterization of acids	[88]
S-alkylation of methionine-192	a
characterization of amides	b
characterization of alcohols	c
characterization of barbiturates by N-alkylation	d
N-, S-, and O-alkylation of amino acid residues	e
affinity labeling of active sites	f
characterization of phenols	g
derivatization of primary alcohols prior to gc separation	h
determination of theobromine	i
characterization of tertiary amines (Menschutkin reaction)	j
characterization of arylsulfinates	k
methylation of histidine-57	l
specific for tryptophan residues	m

Table 9.VI (continued)

---

[b] R. F. Phillips and B. M. Pitt, J. Am. Chem. Soc., 65, 1355 (1943).
[c] Ref. [88], p. 243.
[d] K. A. Connors, Pharmaceutical Analysis (T. Higuchi and E. Brochmann-Hanssen, eds.), Wiley (Interscience), New York, 1961, Chap. VI, p. 224.
[e] L. A. Cohen, Ann. Rev. Biochem., 37, 695 (1968).
[f] S. J. Singer, Advan. Protein Chem., 22, 1 (1967).
[g] Ref. [88], p. 297.
[h] J. Novak and J. Reznicek, J. Chromatog., 43, 437 (1969).
[i] P. A. W. Self and W. R. Rankin, Quart. J. Pharm. Pharmacol., 4, 346 (1931).
[j] Ref. [88], p. 261.
[k] Ref. [88], p. 300.
[l] Y. Nakagawa and M. L. Bender, Biochemistry, 9, 259 (1970).
[m] H. R. Horton and D. E. Koshland, J. Am. Chem. Soc., 87, 1126 (1965).

The reporter group has been selectively delivered to tryptophan residues near the active site of α-chymotrypsin (which has eight tryptophans altogether) by allowing the enzyme to hydrolyze the precursor 2-acetoxy-5-nitrobenzyl chloride [94].

Alkylating agents have important roles in biological systems, having been implicated in carcinogenesis and teratogenesis, and they include some agents useful in medicine and in environmental control [95]. The analysis of such agents can be accomplished by reversing the roles of reagent and sample; that is, the alkylating agent now is the analytical sample. Sawicki and Sawicki [96] give many examples of the determination of alkylating agents. Many of these reactions are aliphatic nucleophilic substitutions; the usual strategy is to attach the alkylating agent to a chromophoric reagent, or to develop a new chromophore. For example, small amounts of epoxides are determined by reaction with picric acid to give the strongly absorbing picryl ether [97]:

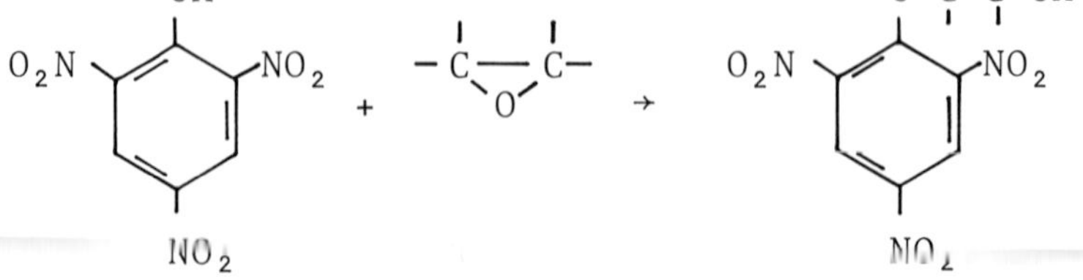

$$(9.41)$$

Alkyl halides and aziridines have been determined colorimetrically by the production of a dye by nucleophilic substitution on 4-p-nitrobenzylpyridine [98]

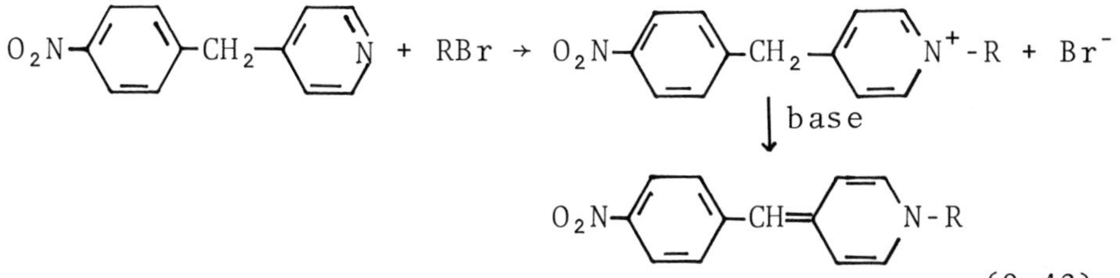

$$O_2N-\text{⟨C₆H₄⟩}-CH_2-\text{⟨pyridine⟩}N + RBr \rightarrow O_2N-\text{⟨C₆H₄⟩}-CH_2-\text{⟨pyridine⟩}N^+-R + Br^-$$

$$\downarrow base$$

$$O_2N-\text{⟨C₆H₄⟩}-CH=\text{⟨pyridine⟩}N-R$$

$$(9.42)$$

Organic halides can be determined by solvolysis reactions. These methods are most successful for the reactive
substrates, such as α-haloacids and tertiary halides. The
analysis is completed by determining the halide ion [99].
Vinyl halides are reacted with methoxide in a similar procedure [99]:

$$CH_2=CHX + OCH_3^- \rightarrow CH_2=CHOCH_3 + X^- \qquad (9.43)$$

Vinson and Fritz [100] took advantage of the rate enhancement of S_N2 reactions in polar aprotic solvents (see the
earlier discussion) to analyze alkyl halides by solvolysis
with potassium hydroxide in DMSO solutions. The proportion
of substitution to elimination is unimportant because the
analysis is completed by measuring the halide produced or
the base consumed. Some selectivity can be achieved
through rate differences; thus the relative rates of solvolysis of n-butyl halides are Cℓ, 1; Br, 38.4; I, 235.

Deamination of aliphatic amines upon treatment with
nitrous acid is a rather special nucleophilic substitution. The overall reaction is

$$R-NH_2 + HNO_2 \rightarrow R-OH + N_2 + H_2O \qquad (9.44)$$

The rate-determining step is the nitrosation of the amine,
which is followed by production of an aliphatic diazonium
ion. This very unstable species rapidly gives the substitution product.

$$RNH_2 \xrightarrow[\text{slow}]{HNO_2} R-NH-NO \xrightarrow{\text{fast}} R-N_2^+ \xrightarrow[\text{fast}]{H_2O} R-OH \qquad (9.45)$$

The last step of Eq. (9.45) is a nucleophilic substitution.

This reaction may proceed through the carbonium ion, though the activation energy for this process is so small that it is unlikely that appreciable solvent stabilization of the ion occurs [101]. Even bridgehead amino groups are readily converted to hydroxy groups by this route. Primary aliphatic amines, especially amino acids, can be determined by nitrous acid deamination, the nitrogen being measured [102]

Substitutions on Atoms Other Than Carbon. One of the most important analytical reactions in modern practice is a nucleophilic substitution on saturated silicon. The substrate (analytical reagent) is nearly always a trimethylsilyl (TMS) derivative, and the nucleophilic atom is oxygen, sulfur, or nitrogen. The general reaction, which is called silylation, is represented by Eq. (9.46); it is clearly analogous to alkylation.

$$RYH + (CH_3)_3Si\text{-}X \rightarrow RYSi(CH_3)_3 + HX \qquad (9.46)$$

$$(Y = O, S, N)$$

Replacement of hydrogen by the silyl group [103] decreases polarity, eliminates hydrogen-bond donor capabilities, and therefore increases the volatility of a compound. Silylation finds its greatest use as a derivatization procedure prior to gas chromatography, and, to a lesser extent, mass spectrometry. The technique has been reviewed by Pierce [104].
    Some of the common silylating agents ("silyl donors") are shown below.

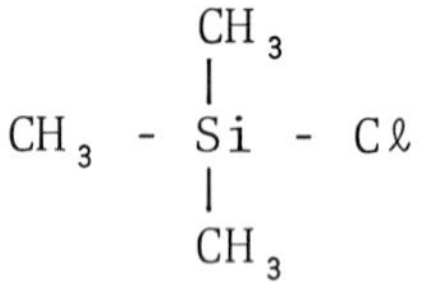

$$\begin{array}{c} CH_3 \\ | \\ CH_3 - Si - C\ell \\ | \\ CH_3 \end{array} \qquad \begin{array}{c} CH_3 \\ | \\ CH_3 - Si - N(C_2H_5)_2 \\ | \\ CH_3 \end{array}$$

Trimethylchlorosilane   N-Trimethylsilyldiethylamine

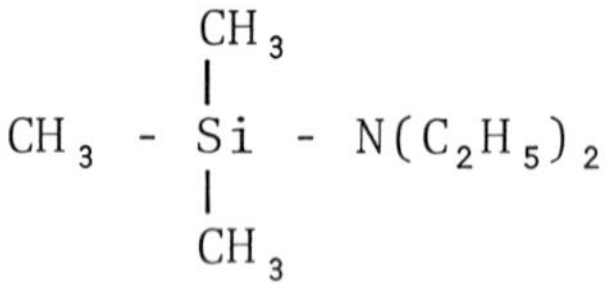

(TMCS)   (TMSDEA)

$$\begin{array}{c} CH_3 \\ | \\ CH_3 - Si - NHCOCH_3 \\ | \\ CH_3 \end{array} \qquad \begin{array}{c} CH_3 \\ | \\ CH_3 - Si - N \\ | \\ CH_3 \end{array}$$

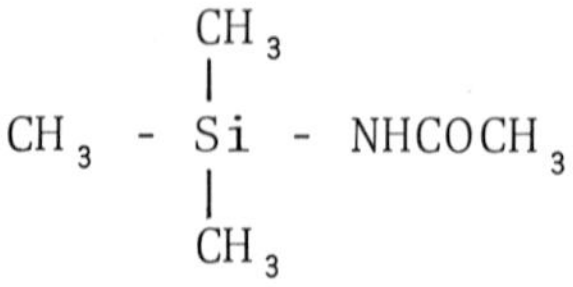

N-Trimethylsilylacetamide 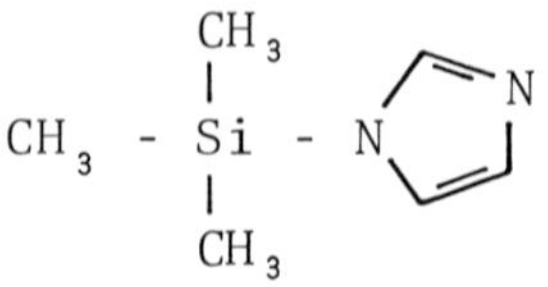   N-Trimethylsilylimidazole

(TSIM)

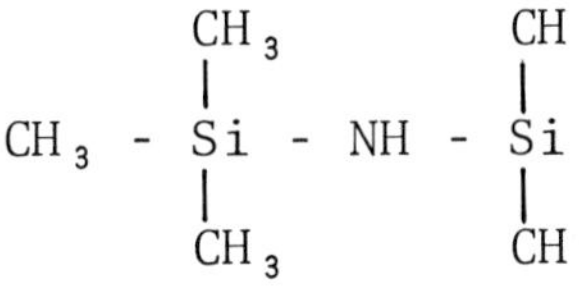

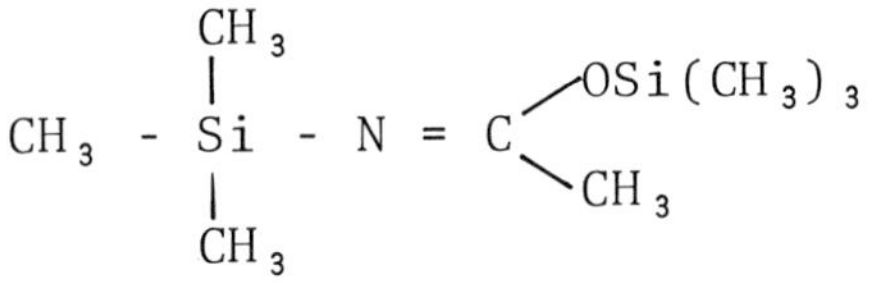

$$CH_3 - \underset{\underset{CH_3}{|}}{\overset{\overset{CH_3}{|}}{Si}} - NH - \underset{\underset{CH_3}{|}}{\overset{\overset{CH_3}{|}}{Si}} - CH_3 \qquad CH_3 - \underset{\underset{CH_3}{|}}{\overset{\overset{CH_3}{|}}{Si}} - N = C \overset{OSi(CH_3)_3}{\underset{CH_3}{}}$$

Hexamethyldisilazane   N,O-Bis(trimethylsilyl)acetamide

     (HMDS)   (BSA)

$$CH_3 - \underset{\underset{CH_3}{|}}{\overset{\overset{CH_3}{|}}{Si}} - N(CH_3) - \overset{\overset{O}{\parallel}}{C} - CF_3$$

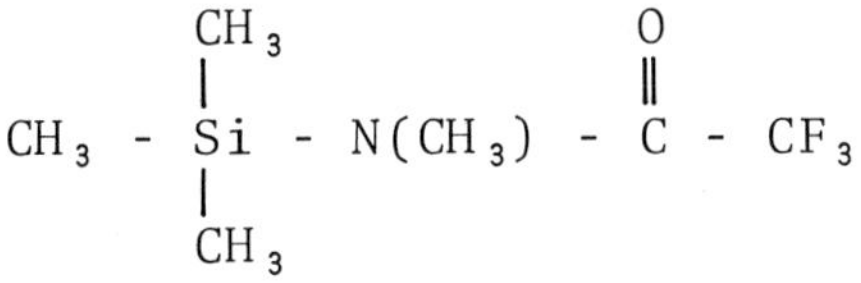

N-Methyl-N-trimethylsilyl trifluoroacetamide

          (MSTFA)

The structure shown for BSA is supported by infrared and nuclear magnetic resonance studies [105]; the alternative is the N,N-substituted structure, $CH_3CON(SiMe_3)_2$. A related agent, bis(trimethylsilyl)trifluoroacetamide (BSTFA), has been formulated as the N,N-bis compound [106], but structural evidence is lacking.

The usual sample compounds ("silyl acceptors") are alcohols, phenols, carboxylic acids, primary and secondary amines, amides, and thiols. The product of the silylation reaction is a silyl ether (from hydroxy groups), silyl ester (from acids), etc. Multifunctional compounds can be fully silylated; thus the usual silylation product of an amino acid is the N,O-di-silyl derivative 39.

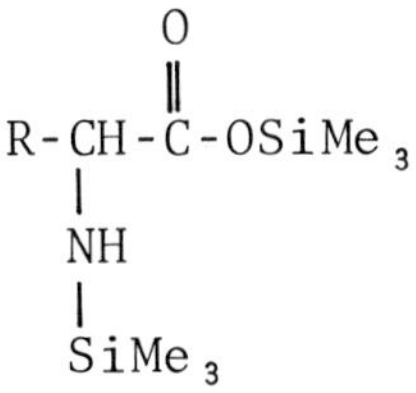

$$R-CH-\overset{\overset{O}{\parallel}}{C}-OSiMe_3$$
$$\underset{\underset{SiMe_3}{|}}{\overset{\overset{|}{NH}}{}}$$

          39

Siliconides of corticosteroids are prepared with the agent dimethyldiacetoxysilane (DMDAS), as in

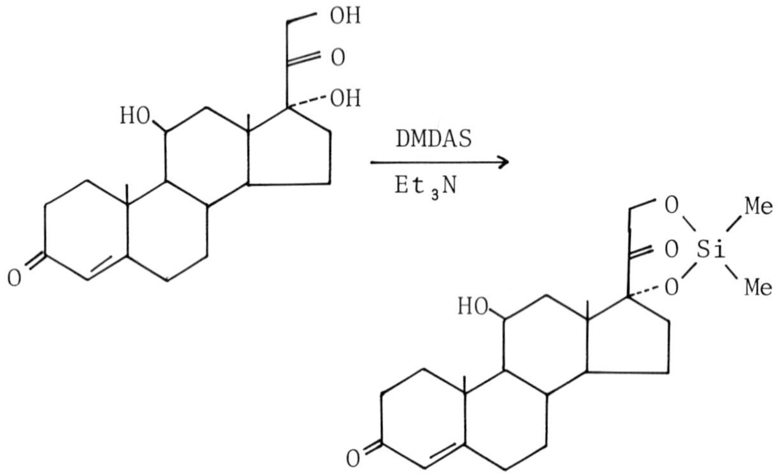

$$(9.47)$$

Silylation reactions must be carried out in aprotic solvents because of the lability of the silylating agents and of the silylated derivatives to solvolysis. Solvolysis methods have, in fact, been employed for the determination of silicon compounds [108].

The chemistry of organosilicon compounds has been studied, though not yet with the thoroughness of carbon compounds. The covalent single-bond radius of silicon is 1.17 Å, considerably larger than the 0.77Å radius of carbon [109]. The trimethylsilyl group is therefore considerably larger than the tert-butyl group. However, less nonbonded repulsion between methyl groups should exist in the TMS group, because the methyls are farther apart than in tert-butyl. Silicon is considerably less electronegative than is carbon [110]. An important difference between silicon and carbon is the availability of 3d orbitals on silicon. This capability provides possible mechanistic routes unavailable to reactions on carbon.

Nucleophilic reactions on silicon, including trimethylsilyl compounds, show second-order kinetics, consistent with a bimolecular mechanism. When the leaving group is good, inversion of configuration (in optically active compounds) is observed; with poor leaving groups retention of configuration occurs [2,111]. Some reactions are accelerated by base, others by acid. The present position is that a range of mechanisms is accessible for nucleophilic substitutions on silicon [112]. For com-

pounds with good leaving groups (these would include most
of the analytical silylating agents), the S_N2 mechanism
is probably operative; this is labeled S_N2-Si by Sommer
to call attention to the probable, or at least possible,
differences from S_N2 reaction at carbon. Silicon compounds with poor leaving groups may react via a quasicy-
clic pyramidal transition state, called the S_Ni-Si mechanism, which leads to retention of configuration [111,112].
An ionization mechanism, S_N1-Si, seems to be less favorable for silicon compounds than for carbon substrates;
this may be partly a consequence of less strain in tetrahedral silicon compounds, as already noted, so there is
less driving force for attainment of a planar ion. The
absence of clearly observable S_N1-Si reactions may also
reflect the availability of more favorable paths.
      Nucleophilic substitutions can occur on divalent
saturated atoms. Edwards [113] has systematized much of
the chemistry of peroxides in terms of a bimolecular nucleophilic substitution process. Edwards points out that
the orbitals containing unshared pairs on oxygen act electronically much as if they were substituents. The presumed transition state for reaction of a nucleophile on a
peroxide is given as 40.

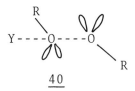

40

Because of the unshared pairs, rates of these reactions
are sensitive to acid, which acts as a catalyst. Second-
order kinetics are widely observed. Relative rates follow
patterns of structural effects in the leaving group and
the nucleophile as expected on the basis of the familiar
substitutions on carbon. The common iodometric method for
the determination of peroxides is an example of a nucleophilic substitution. The overall reaction is

$$ROOR' + 2I^- + 2H_2O \rightarrow I_2 + ROH + R'OH + 2OH^- \qquad (9.48)$$

The first step is the substitution reaction

$$ROOR' + I^- \rightarrow ROI + R'O^- \qquad (9.49)$$

Subsequent reaction with iodide generates the final products.
      A quite different type of substrate exemplifies

another class of nucleophilic substitutions. The substitution of one ligand for another in a metal-ion complex often proceeds by the S_N2 mechanism. Some analytical reactions of mercury (II) derivatives probably occur in this way. Amine halides and quaternary ammonium halides can be titrated as bases in glacial acetic acid if an excess of mercuric acetate is present [114]. The overall reaction is given as

$$2RNH_3Cl + Hg(OAc)_2 \rightarrow HgCl_2 + 2RNH_2 + 2HOAc \quad (9.50)$$

The mercuric acetate and mercuric halide are nonbasic; in effect the amine has been released for titration by a strong acid. The reaction is a nucleophilic displacement by chloride of acetate on mercury.

Mercury-carbon bonds are resistant to nucleophilic displacements, but bonds from mercury to oxygen, sulfur, and nitrogen are more labile. These undergo solvolysis in acetic acid to give mercuric acetate or a substituted mercuric acetate. By adding an excess of methylamine hydrochloride, a stoichiometric amount of methlamine is produced. This method, which is evidently a reversal of the preceding analysis for amine salts, allows mercurials to be determined [115].

A similar substitution occurs in an olefin determination based on the addition of mercuric acetate to the double bond. In one of the modifications of the method [116] the sample is treated with a methanol solution of mercuric acetate. The overall reaction is

$$-CH=CH- + Hg(OAc)_2 + MeOH \rightarrow \underset{\underset{OMe \quad HgOAc}{|}}{-CH-CH-} + HOAc \quad (9.51)$$

The solution is then titrated with standard hydrochloric acid in 1:1 propylene glycol-chloroform. Excess reagent consumes two moles of titrant:

$$Hg(OAc)_2 + 2HCl \rightarrow HgCl_2 + 2HOAc \quad (9.52)$$

whereas the addition product consumes one mole:

$$\underset{\underset{OMe \quad HgOAc}{|}}{-CH-CH-} + HCl \rightarrow \underset{\underset{OR \quad HgCl}{|}}{-CH-CH-} + HOAc \quad (9.53)$$

Both of these reactions are nucleophilic substitutions on mercury (II).

Comment. This survey of aliphatic substitutions in analytical processes is illustrative rather than comprehensive in intent [117]. It may lead one to improvements of
existing methods by combining the mechanistic background
with the analytical requirements, as Vinson and Fritz
[100] did in choosing DMSO as a solvent for the analysis
of halides. Some interesting possibilities are suggested
in the field of alkylations by recent synthetic advances.
Some fluorescent alkylating agents, of which 41 is an example, have been reported; these might be useful analytical tags [118].

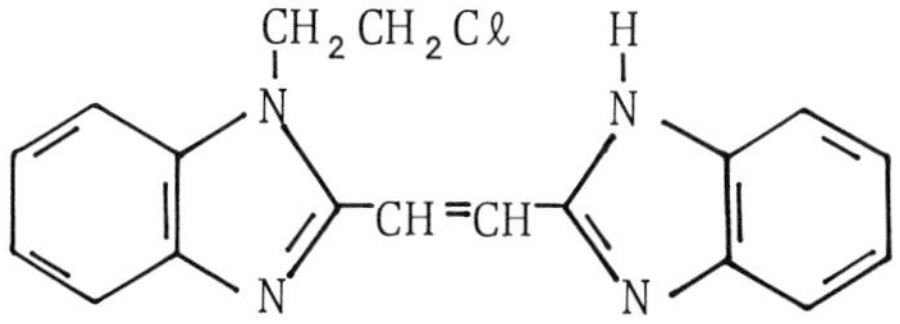

41

Problems

1. Derive Eq. (9.13).

2. Using the data in Table 9.III to construct a model,
   estimate the rate constant for bimolecular displacement by ethoxide ion in ethanol on α-phenylethyl bromide at 25°. The rate constant for benzyl bromide
   [48] under these conditions is 1.84×10^{-3} $\underline{M}^{-1}$-sec^{-1}.

   Answer: The experimental value is 6.2×10^{-5} $\underline{M}^{-1}$-sec^{-1}.

3. In the attempted S_N1 reaction of 1-chloroapocamphane,
   16, is the inertness to be ascribed to stabilization
   of the ground state or destabilization of the transition state?

4. Postulate nonclassical ions derived from the ionization of structures 32, 33, 34, and 35.

   Answer: See Ref. [68].

5. Predict the sign of Bunnett's w value for the hydroly-

sis of ethyl ether in perchloric acid solution (cf.
Table 4.XIV).

Answer: Positive (w = +2.7).

6. Rationalize these comparative rates: Ph₃CX hydrolyzes
    more rapidly than Me₃CX, whereas Me₃SiX reacts faster
    than Ph₃SiX.

7. Design methods for the quantitative analysis of these
    mixtures:
    a.   Chlorobenzene, ethyl chloride, and tert-butyl
    chloride
    b.   n-Butyl chloride and n-butyl iodide.

8. Propose a method for methoxy group determination using
    an S_N reaction with a gas chromatographic finish.

9. a.   Design an analytical method for epoxides based
    upon reaction with an amine.
    b.   Suggest a method for amine determination based
    upon reaction with an epoxide.

References

1. p-Toluenesulfonates are often called tosylates, and
    Eq. (9.6) may be written OH⁻ + R-OTs → R-OH + OTs⁻.
    Notice that substrates for nucleophilic substitutions
    can be viewed as esters, the leaving groups being de-
    rived from acids and the aliphatic portion from alco-
    hols.
2. L. H. Sommer, C. L. Frye, M. C. Musolf, G. A. Parker,
    P. G. Rodewald, K. W. Michael, Y. Okaya, and R.
    Pepinsky, J. Am. Chem. Soc., 83, 2210 (1961).
3. C. K. Ingold, Structure and Mechanism in Organic
    Chemistry, Cornell Univ. Press, Ithaca, N. Y., 1953,
    Chap. VII.
4. See, for example, E. S. Gould, Mechanism and Structure
    in Organic Chemistry, Holt, Rinehart and Winston, New
    York, 1959, Chap. 8; A. Streitwieser, Jr., Solvolytic
    Displacement Reactions, McGraw-Hill, New York, 1962
    [most of which is a reprint of an article published
    in Chem. Revs., 56, 561 (1956)]; C. A. Bunton, Nucleo-
    philic Substitution at a Saturated Carbon Atom, Vol.
    I, Reaction Mechanisms in Organic Chemistry, Elsevier,
    Amsterdam, 1963.
5. The molecularity of a reaction step is defined by
    Ingold (Ref. [3], p. 315) to be the number of molecules

necessarily undergoing covalency change. The participation of solvent molecules as solvating species
(i.e., by noncovalent bonding) is therefore excluded.

6. According to the principle given in Sec. 3.4, we
     would expect a positive salt effect, that is, an in-
     crease in rate constant with ionic strength, for the
     S_N1 mechanism.
7. C. G. Swain, C. B. Scott, and K. H. Lohmann, J. Am.
     Chem. Soc., 75, 136 (1953).
8. F. Spieth, quoted by A. Streitwieser, Jr., Solvolytic
     Displacement Reactions, McGraw-Hill, New York, 1962,
     p. 53.
9. E. A. Moelwyn-Hughes, The Kinetics of Reactions in
     Solution, 2nd ed., Oxford Univ. Press, London, 1947,
     p. 33.
10. A. Streitwieser, Jr., Ref. [8], p. 43.
11. C. K. Ingold, Ref. [3], pp. 320-324.
12. C. A. Bunton, Ref. [4], pp. 21, 164.
13. G. Kohnstam, Advan. Phys. Org. Chem., 5, 121 (1967).
14. K. L. Servis and R. K. Crossland, Abstracts, 159th
     American Chemical Society meeting, Feb. 1970, No.
     0-120. A mesylate is a methylsulfonate.
15. S. Winstein, E. Grunwald, and H. W. Jones, J. Am.
     Chem. Soc., 73, 2700 (1951).
16. S. Winstein and E. Grunwald, J. Am. Chem. Soc., 70,
     846 (1948); this concept is discussed in Sec. 4.4.
17. A. Streitwieser, Jr., Ref. [8], p. 66.
18. R. A. Sneen and J. W. Larsen, J. Am. Chem. Soc., 91,
     362, 6031 (1969).
19. S. Winstein, E. Clippinger, A. H. Fainberg, R. Heck,
     and G. C. Robinson, J. Am. Chem. Soc., 78, 328 (1956).
20. J. L. Kurz and J. C. Harris, J. Am. Chem. Soc., 92,
     4117 (1970); J. L. Kurz, ibid., 85, 987 (1963).
21. A. Streitwieser, Jr., Molecular Orbital Theory for
     Organic Chemists, Wiley, New York, 1961, pp. 367-372.
22. M. D. Bentley and M. J. S. Dewar, J. Am. Chem. Soc.,
     92, 3991 (1970).
23. G. Kohnstam, Advan. Phys. Org. Chem., 5, 121 (1967).
24. R. E. Robertson, Progr. Phys. Org. Chem., 4, 213
     (1967).
25. C. K. Ingold, Ref. [3], pp. 372-386.
26. E. S. Gould, Ref. [4], pp. 263-267; C. A. Bunton,
     Ref. [12], pp. 86-87.
27. A. Streitwieser, Jr., Solvolytic Displacement Reac-
     tions, McGraw-Hill, New York, 1962, p. 68.
28. C. K. Ingold, Ref. [3], pp. 382-386.
29. E. Grunwald and S. Winstein, J. Am. Chem. Soc., 70,
     841 (1948).
30. α-Lactones hydrolyze with alkyl-oxygen cleavage

rather than acyl-oxygen cleavage. Net retention requires either no inversions or two inversions.

31. C. A. Bunton, Ref. [12], pp. 96-100. See also the
later discussion.

32. A. Streitwieser, Jr., Ref. [27], pp. 86-92.

33. C. E. Boozer and E. S. Lewis, J. Am. Chem. Soc., 75,
3182 (1953).

34. E. S. Gould, Ref. [4], pp. 294-295.

35. C. K. Ingold, Ref. [3], pp. 335-338.

36. This is a truism, because reactivity is the measure
of nucleophilicity.

37. E. S. Gould, Ref. [4], p. 259.

38. The anion within the ion pair is expected to be substantially less nucleophilic than is the dissociated
anion. It has been suggested that ion pairs are not
at all nucleophilic; cf. R. A. Sneen and F. R. Rolle,
J. Am. Chem. Soc., 91, 2140 (1969). See also P.
Beronius and L. Pataki, ibid., 92, 4518 (1970).

39. M. S. Puar, J. Chem. Educ., 47, 473 (1970).

40. C. G. Swain and C. B. Scott, J. Am. Chem. Soc., 75,
141 (1953). For some additional n values see A. B.
Ash, P. Blumbergs, C. L. Stevens, H. O. Michel, B. E.
Hackley, and J. Epstein, J. Org. Chem., 34, 4070
(1969).

41. A. Streitwieser, Jr., Ref. [27], p. 30.

42. L. Pauling, The Nature of the Chemical Bond, 3rd ed.,
Cornell Univ. Press, Ithaca, N. Y., 1960, p. 85.

43. A. Streitwieser, Jr., Ref. [27], p. 57.

44. S. G. Smith and D. J. W. Goon, J. Org. Chem., 34,
3127 (1969).

45. M. J. S. Dewar and R. J. Sampson, J. Chem. Soc.,
1957, p. 2946.

46. A. Streitwieser, Ref. [27], pp. 12-15.

47. C. A. Bunton, Ref. [4], pp. 27-29.

48. A. Streitwieser, Jr., Ref. [27], pp. 13-14, 18-20.

49. P. D. Bartlett and L. H. Knox, J. Am. Chem. Soc.,
61, 3184 (1939).

50. M. J. S. Dewar and R. J. Sampson, J. Chem. Soc.,
1957, p. 2952.

51. L. Verbit and E. Berliner, J. Am. Chem. Soc., 86,
3307 (1964). A good linear free energy relationship
is found with the σ^+ substituent constants.

52. A. Streitwieser, Jr., Ref. [27], p. 43.

53. C. A. Bunton, Ref. [4], p. 45.

54. Many examples are given by R. Breslow, Organic Reaction Mechanisms, 2nd ed., Benjamin, New York, 1969,
pp. 93-108.

55. S. Winstein, C. Lindegren, H. Marshall, and L.
Ingraham, J. Am. Chem. Soc., 75, 147 (1953).

56. D. J. Cram, J. Am. Chem. Soc., 74, 2129 (1952).
57. These structures are Newman projections; the circle represents the 2 (or 3) carbon atom, and the direction of view is along the C_2-C_3 bond axis. See M. S. Newman (ed.), Steric Effects in Organic Chemistry, Wiley, New York, 1956, p. 5.
58. A. Diaz, I. Lazdins, and S. Winstein, J. Am. Chem. Soc., 90, 6546 (1968); A. F. Diaz and S. Winstein, ibid., 91, 4300 (1969).
59. P. von R. Schleyer and C. J. Lancelot, J. Am. Chem. Soc., 91, 4297 (1969). This number of the journal includes important communications on this problem from the laboratories of von R. Schleyer, S. Winstein, and H. C. Brown. For a more detailed separation of kinetic behavior into the k_S (solvent assisted), k_Δ (anchimerically assisted), and k_C (unassisted) routes see P. von R. Schleyer, J. L. Fry, L. K. M. Lam, and C. J. Lancelot, ibid., 92, 2542 (1970).
60. R. J. Jablonski and E. I. Snyder, J. Am. Chem. Soc., 91, 4445 (1969).
61. M. D. Bentley and M. J. S. Dewar, J. Am. Chem. Soc., 92, 3996 (1970); M. J. S. Dewar, The Molecular Orbital Theory of Organic Chemistry, McGraw-Hill, New York, 1969, p. 358.
62. At the present time H. C. Brown is the principal proponent of the equilibrating classical ions hypothesis, M. J. S. Dewar supports the π-complex formulation, and most workers follow Cram and Winstein in preferring a bridged phenonium ion intermediate.
63. P. D. Bartlett, Nonclassical Ions: Reprints and Commentary, Benjamin, New York, 1965, page v. Ingold calls nonclassical ions synartetic ions; cf. F. Brown, E. D. Hughes, C. K. Ingold, and J. F. Smith, Nature, 168, 65 (1951).
64. Brosylate ≡ p-bromobenzenesulfonate. Similar results are found with other leaving groups.
65. S. Winstein and D. S. Trifan, J. Am. Chem. Soc., 71, 28 (1949); 74, 1147, 1154 (1952).
66. H. C. Brown, Chem. Soc. Spec. Pub. No. 16, 140 (1962). This is reprinted by P. D. Bartlett, Ref. [63], p. 438.
67. G. A. Olah, A. M. White, J. R. DeMember, A. Commeyras, and C. Y. Lui, J. Am. Chem. Soc., 92, 4627 (1970).
68. S. Winstein and C. Ordronneau, J. Am. Chem. Soc., 82, 2084 (1960); H. Tanida, Accounts Chem. Res., 1, 239 (1968).
69. E. M. Kosower, An Introduction to Physical Organic Chemistry, Wiley, New York, 1968, Part Two.
70. C. K. Ingold, Ref. [3], pp. 345-350.

71. R. G. Burns and B. D. England, Tetrahedron Letters, 1960, No. 24, 1.

72. For example, the relative rates for formation of Meisenheimer complexes with 1,3,5-trinitrobenzene are 1:188:918 for OH^-, CH_3O^-, and $C_2H_5O^-$, respectively; cf. C. F. Bernasconi, J. Am. Chem. Soc., 92, 4682 (1970).

73. A further consequence is that the product distribution is affected. An alkyl halide will yield both an alcohol and an ether upon solvolysis in an alcohol-water mixture, and the product distribution depends upon the lyate ion distribution and the alkoxide nucleophilicity.

74. R. G. Burns and B. D. England, J. Chem. Soc., B 1966, 864.

75. A. J. Parker, Advan. Phys. Org. Chem., 5, 173 (1967).

76. R. Alexander, E. C. F. Ko, A. J. Parker, and T. J. Broxton, J. Am. Chem. Soc., 90, 5049 (1968).

77. R. F. Rodewald, K. Mahendran, J. L. Bear, and R. Fuchs, J. Am. Chem. Soc., 90, 6698 (1968).

78. E. C. F. Ko and A. J. Parker, J. Am. Chem. Soc., 90, 6447 (1968).

79. For reviews and experimental details see G. Ingram, in Comprehensive Analytical Chemistry, C. L. Wilson and D. W. Wilson (eds.), Vol. 1B, Elsevier, Amsterdam, 1960, p. 659; N. D. Cheronis and T. S. Ma, Organic Functional Group Analysis by Micro and Semimicro Methods, Wiley (Interscience), New York, 1964, p. 124; K. G. Stone, Determination of Organic Compounds, McGraw-Hill, New York, 1956, p. 203; A. Elek, Org. Anal., 1, 67 (1953).

80. N. D. Cheronis and T. S. Ma, Ref. [79], p. 223.

81. M. W. Scoggins and J. W. Miller, Anal. Chem., 38, 612 (1966); M. P. T. Bradley and G. E. Penketh, Analyst, 92, 701 (1967).

82. J. L. Jungnickel, E. D. Peters, A. Polgár, and F. T. Weiss, Org. Anal., 1, 127 (1953).

83. G. H. Schenk, Organic Functional Group Analysis, Pergamon, Oxford, 1968, Chap. 5.

84. A. J. Durbetaki, Anal. Chem., 28, 2000 (1956).

85. R. R. Jay, Anal. Chem., 36, 667 (1964); R. Dijkstra and E. A. M. F. Dahmen, Anal. Chim. Acta, 31, 38 (1964). Aziridines (the nitrogen analogs of epoxides) also can be determined by these methods. See also G. Maerker, E. T. Haeberer, L. M. Gregory, and T. A. Foglia, Anal. Chem., 41, 1698 (1969).

86. C. A. Bunton, Ref. [4], p. 157.

87. This rate dependence has been exploited analytically to increase the reaction rate by incorporating excess

halide ions in the reaction mixture [85].
88. R. L. Shriner, R. C. Fuson, and D. Y. Curtin, The
     Systematic Identification of Organic Compounds, 5th
     ed., Wiley, New York, 1964, p. 232.
89. C. A. Bunton, Ref. [4], p. 35.
90. Another example is the spin-label technique, in which
     a small paramagnetic molecule is attached to a macro-
     molecule to study conformational changes by electron
     paramagnetic resonance [91]. The tetramethylpiper-
     idino-1-oxyl compound 37 is a spin label that utilizes
     thy iodoacetamide attaching group and an S_N reaction.

NHCOCH$_2$I

37

91. O. H. Griffith and A. S. Waggonner, Accounts Chem.
     Res., 2, 17 (1969).
92. G. M. Loudon and D. E. Koshland, J. Biol. Chem., 245,
     2247 (1970).
93. H. R. Horton and D. E. Koshland, J. Am. Chem. Soc.,
     87, 1126 (1965).
94. H. R. Horton and G. Young, Biochim. Biophys. Acta,
     194, 272 (1969).
95. Biological Effects of Alkylating Agents, Ann. N. Y.
     Acad. Sci., 163, 589-1029 (1969).
96. E. Sawicki and C. R. Sawicki, Ann. N. Y. Acad. Sci.,
     163, 895 (1969).
97. J. A. Fioriti, A. P. Bentz, and R. J. Sims, J. Am.
     Oil Chem. Soc., 43, 37 (1966).
98. J. Epstein, R. W. Rosenthal, and R. J. Ess, Anal.
     Chem., 27, 1435 (1955).
99. N. D. Cheronis and T. S. Ma, Ref. [79], pp. 460-461.
100. J. A. Vinson and J. S. Fritz, Anal. Chem., 40, 2194
     (1968). This method is also suitable for activated
     aromatic halides.
101. C. A. Bunton, Ref. [4], pp. 103-107.
102. N. D. Cheronis and T. S. Ma, Ref. [79], p. 232.
103. The term "silyl" usually means "trimethylsilyl."
104. A. E. Pierce, Silylation of Organic Compounds, Pierce
     Chemical Co., Rockford, Ill., 1968.
105. A. E. Pierce, Ref. [104], p. 69.
106. D. L. Stalling, C. W. Gehrke, and R. W. Zunwalt,
     Biochem. Biophys. Res. Commun., 31, 616 (1968).
107. R. W. Kelly, J. Chromatog., 43, 229 (1969).

108. N. D. Cheronis and T. S. Ma, Ref. [79], pp. 300, 469.

109. L. Pauling, Ref. [42], p. 224.

110. L. Pauling, Ref. [42], p. 93. The Pauling electronegativities of some atoms are: Si, 1.8; C, 2.5;
N, 3.0; O, 3.5; F, 4.0; Cℓ, 3.0; Br, 2.8.

111. L. H. Sommer, C. L. Frye, and G. A. Parker, J. Am.
Chem. Soc., 86, 3276 (1964); L. H. Sommer, G. A.
Parker, and C. L. Frye, ibid., 86, 3280 (1964).

112. L. H. Sommer, Stereochemistry, Mechanism, and Silicon, McGraw-Hill, New York, 1965.

113. J. O. Edwards in Peroxide Reaction Mechanisms (J. O.
Edwards, ed), Wiley (Interscience), New York, 1962,
p. 67.

114. C. W. Pifer and E. G. Wollish, Anal. Chem., 24, 300
(1952).

115. K. A. Connors and D. R. Swanson, J. Pharm. Sci., 53,
432 (1964).

116. M. N. Das, Anal. Chem., 26, 1086 (1954). This reaction is discussed in Chap. 10.

117. For example, Substitutions on sulfur have not been
covered. The alert reader of chapter titles also
will have noted the absence of a chapter on electrophilic aliphatic substitutions, which proceed through
electron-rich intermediates (carbanions) or transition states; for a review see J. March, Advanced
Organic Chemistry: Reactions, Mechanisms, and Structure, McGraw-Hill, New York, 1968, Chap. 12.

118. K. C. Tsou, D. J. Rabiger, and B. Sobel, J. Med.
Chem., 12, 818 (1969).

Chapter 10.   ADDITION TO CARBON-CARBON MULTIPLE BONDS

10.1   Electrophilic Addition

Classification of reaction type for additions requires
knowledge, or at least an assumption, about the reaction
mechanism [1].   Carbon-carbon double bonds characteristically add reagents that are electrophilic, such as acids
and halogens, suggesting that the olefinic bond participates as a nucleophile.   Thus hydrogen chloride adds to an
olefin to give an alkyl chloride:

$$HC\ell + RCH=CHR \rightarrow RCH_2CHC\ell R \qquad (10.1)$$

Although it is conceivable that this reaction is initiated
by nucleophilic attack by chloride, the weight of chemical
evidence favors the view that the electrophilic proton
attacks the nucleophilic olefin, followed by addition of
chloride ion.   The reaction is therefore an electrophilic
addition (Ad_E).   This mechanism is common to isolated, and
some conjugated, carbon-carbon multiple bonds.   Section
10.1 deals with Ad_E reactions.   An olefinic bond conjugated
with an electron-withdrawing group is capable of acting as
an electrophile, to which nucleophilic reagents may add;
such Ad_N reactions are discussed in Sec. 10.2.

Among the features of Ad_E reactions that must be considered are the number of steps in the reaction (that is,
the possible occurrence of intermediates), the orientation
of addition when both the reagent and substrate are unsymmetrical, the stereochemistry of addition, and possible
rearrangements.

Initiation by Electrophilic Hydrogen.   All of the reactions
to be considered here are additions initiated by a reagent
HX, often a mineral acid.   A reaction sequence invoking an
intermediate formed from the addition of the proton to the
olefin best accounts for several observations.   This intermediate is evidently a carbonium ion, which may add the
nucleophile X^- derived from the reagent, it may add an external nucleophile such as the solvent lyate ion, or it may
rearrange prior to reaction with a nucleophile.   These

possibilities are illustrated by the reaction of hydrogen
chloride with tert-butylethylene in acetic acid solvent
[2]. Three products are formed [3]

$$\text{(10.2)}$$

<u>1</u>   <u>2</u>   <u>3</u>

under conditions of kinetic control. All of these products can be accounted for by the formation of a discrete
carbonium ion intermediate, produced by reaction of tert-
butylethylene with hydrogen chloride [4].

$$Me_3C-CH=CH_2 \xrightarrow{H^+} Me_3C-\overset{+}{C}H-Me \longrightarrow Me_2\overset{+}{C}-CHMe_2$$

$$Cl^- \swarrow \searrow OAc^- \qquad \downarrow Cl^- \qquad \text{(10.3)}$$

<u>1</u>   <u>2</u>   <u>3</u>

Carbonium ion $Me_3C-\overset{+}{C}H-Me$ reacts rapidly with the available
nucleophiles, Cl^- and OAc^-, to give products <u>1</u> and <u>2</u>,
respectively. This ion, which is a secondary ion, can be
transformed into the more stable tertiary carbonium ion
$Me_2\overset{+}{C}HMe_2$ by a methyl group migration (an example of the
Wagner-Meerwein rearrangement); this tertiary ion can capture Cl^- to give product <u>3</u>. The observation of this rearranged product (under conditions of kinetic control, in
which its production in a secondary reaction is not pos-
dible) is the best evidence for an intermediate on the
reaction path.

This reaction of HCl with tert-butylethylene also
serves to illustrate the orientation of addition. This is
described by the Markovnikov rule, which states that when
HX adds to an olefin, group X becomes bound to the more
highly substituted of the two unsaturated carbon atoms.
This normal orientation pattern can be rationalized by
considering the relative stabilities of the two possible
carbonium ions derived from protonation of the olefin.
Since a carbonium ion is stabilized by electron supply to
the electron-deficient carbon, the order of stability is
tertiary > secondary > primary. Secondary and primary
carbonium ions could be generated from tert-butylethylene,
as shown in

$$Me_3C-CH=CH_2 \begin{array}{c} \xrightarrow{\;H^+\;} \\[2ex] \xrightarrow{\;H^+\;} \end{array} \begin{array}{l} Me_3C-\overset{+}{C}H-CH_3 \xrightarrow{\;Cl^-\;} Me_3C-CHCl-CH_3 \\ \text{(more stable)} \qquad \text{(observed)} \\[1.5ex] Me_3C-CH_2-CH_2{}^+ \xrightarrow{\;Cl^-\;} Me_3C-CH_2-CH_2Cl \\ \text{(less stable)} \qquad \text{(not observed)} \end{array} \qquad (10.4)$$

Only the Markovnikov products 1 and 2 are found (aside from rearranged product 3). Note that although orientation patterns are easily accounted for in terms of an intermediate carbonium ion, such a species is not required to explain orientation. An initial state argument leads to the same predictions as does the carbonium ion hypothesis. This argument considers the effects of the olefin structure on the electron densities at the two unsaturated carbons; the electrophilic portion of HX is predicted to add to the more negative carbon. The electron distribution in the double bond of tert-butylethylene, for example, will be very unsymmetrical because of the inductive electron release by the tert-butyl group. The terminal carbon should be more negative (see 4) and will be preferentially protonated.

$$Me_3C \xrightarrow{\quad} CH \overset{\delta^+}{\overset{\frown}{=}} CH_2{}^{\delta^-}$$

4

This normal orientation behavior can be reversed (giving "anti-Markovnikov" orientation) by strongly electron-withdrawing groups. Thus in trifluoromethylethylene the very electronegative $-CF_3$ group polarizes the double bond as in 5, and the anti-Markovnikov product $CF_3CH_2CH_2X$ is formed exclusively [5].

$$F_3C \xleftarrow{\quad} \overset{\delta^-}{CH} \overset{\frown}{=} \overset{\delta^+}{CH_2}$$

5

The stereochemistry of hydrohalogenation is still under active study, and the mechanisms of stereochemical control are not well understood [6]. Both syn (cis)-addition and anti (trans)-addition have been observed [7]. Recent investigations by Pocker et al. [8] and Fahey and co-workers [2,9] have employed product and kinetic studies to elucidate the stereochemistry of some additions and to relate it to mechanism. In nitromethane solvent, DCl adds to 1-methylcyclopentene essentially exclusively in the anti manner [8].

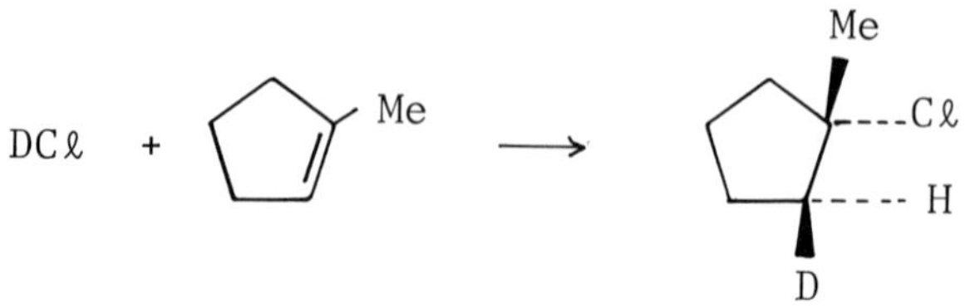

$$\text{(10.5)}$$

The rate equation for addition of HCl (or DCl) in nitromethane is, for several substrates, $v = k[\text{olefin}] [HCl]^2$.
In acetic acid solvent, HCl adds to cyclohexene-1,3,3-d_3
(6), giving three products [9].

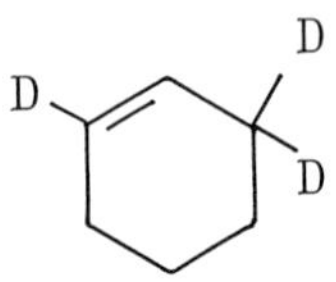

$$\begin{pmatrix} SC, & X = Cl \\ SA, & X = OAc \end{pmatrix} \quad \begin{pmatrix} AC, & X = Cl \\ AA, & X = OAc \end{pmatrix}$$

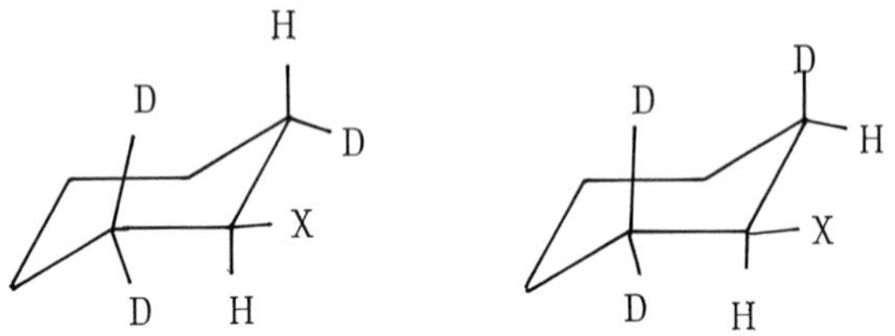

Attack at carbon-1 by a proton could be followed by syn
addition of Cl$^-$ (giving SC) or of OAc$^-$ (giving SA), or by
anti addition of Cl$^-$ (AC) or OAc$^-$ (AA). No SA was observed. The distribution of the other adducts depends
upon the reaction conditions. Acetate formation is an
anti process, whereas chloride addition occurs by syn and
anti routes. In the presence of added chloride ion, the
proportion of anti chloride addition (product AC) is
greatly increased, but the SC production is not increased.
The rate law contains two terms, one a second-order term
in olefin and undissociated acid, the other a third-order
term in olefin, undissociated acid, and dissociated chloride ion. The third-order term leads exclusively to stereospecific anti-HCl addition.
    These results for cyclohexene hydrochlorination are
accounted for by postulating two mechanisms. A termolecular mechanism (AdE3) with a transition state like 9 is
suggested for the production of AC.

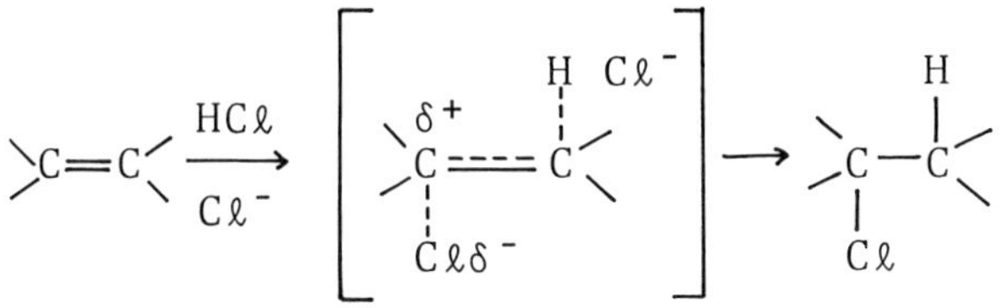

$$
\text{C}=\text{C} \xrightarrow[\text{C}\ell^-]{\text{HC}\ell} \left[\begin{array}{c} \text{H} \quad \text{C}\ell^- \\ \overset{\delta+}{\text{C}} = = = = \text{C} \\ \underset{\text{C}\ell\delta^-}{\big|} \end{array} \right] \longrightarrow \begin{array}{c} \text{H} \\ \text{C} - \text{C} \\ \overset{\big|}{\text{C}\ell} \end{array} \qquad (10.6)
$$

$$\underline{9} \qquad\qquad\qquad\qquad \text{AC}$$

Most of the AA may be produced by the same concerted termolecular mechanism in which the chloride ion is replaced
by acetic acid; this would give rise to a second-order
rate term because acetic acid is the solvent. The syn-HCℓ
adduct SC (and perhaps some AA) could be formed in an Ad$_E$2
reaction via a carbonium ion-chloride ion pair intermediate
10, which would collapse to SC and AA [10].

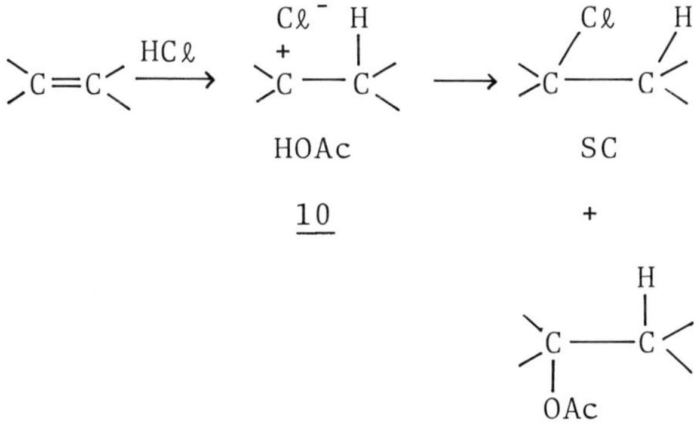

$$
\text{C}=\text{C} \xrightarrow[\text{HOAc}]{\text{HC}\ell} \begin{array}{c} \text{C}\ell^- \;\; \text{H} \\ \overset{+}{\text{C}} - \text{C} \end{array} \longrightarrow \begin{array}{c} \text{C}\ell \quad \text{H} \\ \text{C} \!-\! \text{C} \end{array} \qquad (10.7)
$$

$$\text{HOAc} \qquad\qquad\qquad \text{SC}$$

$$\underline{10} \qquad\qquad\qquad + $$

$$
\begin{array}{c}
\text{H} \\
\text{C} - \text{C} \\
\overset{\big|}{\text{OAc}}
\end{array}
$$

$$\text{AA}$$

Other interpretations of anti addition have been proposed [6]. One of these favors a π-complex intermediate
11 or the similar bridged ion 12, either of which would
allow access by the nucleophile only from the backside,
with resultant anti addition.

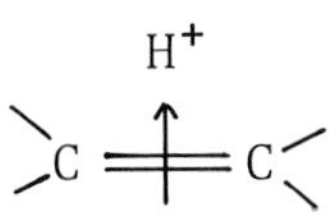

$$
\begin{array}{cc}
\text{H}^+ & \overset{+}{\text{H}} \\
\text{C} = = \text{C} & \text{C} - \text{C}
\end{array}
$$

$$\underline{11} \qquad\qquad\qquad \underline{12}$$

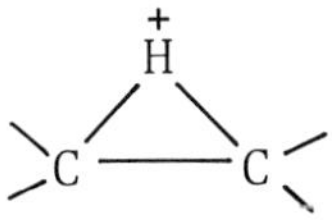

The evidence for these species is slight [11].
   A concerted termolecular mechanism similar to that of
Fahey et al. [9] in which the proton and the nucleophile
are synchronously delivered to the alkene in an anti mode,
had been proposed by Hammond and Collins [12] to account

for the anti addition of HCℓ to 1,2-dimethylcyclopentene.
The recent work [9] supports the concerted mechanism with
consistent kinetic evidence.
    The addition of water to olefins (hydration) is an
acid-catalyzed reversible process. Markovnikov orientation is observed.

$$(CH_3)_2C=CH_2 + H_2O \underset{H^+}{\rightleftharpoons} (CH_3)_2\overset{\overset{\displaystyle OH}{|}}{C}-CH_3 \qquad (10.8)$$

The intermediacy of a free carbonium ion in rapid equilibrium with the olefin [Eq. (10.9)] accounts for these observations [13]. However, it has been observed [14] that

$$\begin{array}{c} \diagup \\ C=C \\ \diagdown \end{array} \diagdown \underset{-H^+}{\overset{+H^+}{\rightleftharpoons}} \overset{+}{C}-\overset{\overset{\displaystyle H}{|}}{C} \underset{-H_2O}{\overset{+H_2O}{\rightleftharpoons}} \underset{+OH_2}{-\overset{\overset{\displaystyle H}{|}}{\underset{|}{C}}-\overset{\overset{\displaystyle H}{|}}{\underset{}{C}}} \underset{+H^+}{\overset{-H^+}{\rightleftharpoons}} \underset{OH}{-\overset{\overset{\displaystyle H}{|}}{\underset{|}{C}}-\overset{\overset{\displaystyle H}{|}}{\underset{|}{C}}-} \qquad (10.9)$$

2-methylbutene-1 and 2-methylbutene-2, both of which are
hydrated to give only tert-amyl alcohol, are not detectably
interconverted under hydration reaction conditions. Addition of a proton to either olefin should lead to the same
carbonium ion intermediate $\underline{13}$.

$$\begin{array}{l} Me \diagdown \\ C=CH-CH_3 \\ Me \diagup \end{array} \searrow^{H^+}$$

$$\begin{array}{l} Me \diagdown \\ \overset{+}{C}-CH_2-CH_3 \\ Me \diagup \end{array} \xrightarrow[-H^+]{H_2O} Me_2\overset{\overset{\displaystyle OH}{|}}{C}-CH_2CH_3 \qquad (10.10)$$

$$\begin{array}{l} Me \diagdown \\ C-CH_2-CH_3 \\ CH_2 \diagup \end{array} \nearrow^{H^+}$$

$$\underline{13}$$

The failure of these substrates to isomerize means either
that the carbonium ion $\underline{13}$ reacts with water much more
rapidly than it loses a proton, or that the added proton
is not structurally equivalent to the protons already present in the olefin.
    Structural effects upon rates of hydration, along with
orientation behavior and rearrangements accompanying hydration, support the view that the transition state must
have carbonium ion character, but the intermediate cannot
be the free carbonium ion in pre-equilibrium with the olefin, as shown above. Various proposals have been made
concerning the nature of the intermediate and the transitio

state for the rate-determining step [15,16]. One of these
invokes a π-complex 14, with its conversion to the carbonium ion being rate determining.

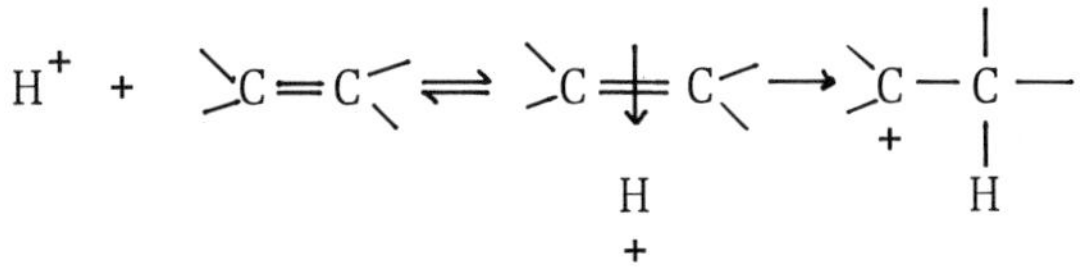

$$H^+ \;+\; \overset{|}{\underset{|}{>}}C{=}C\overset{|}{\underset{|}{<}} \;\rightleftharpoons\; >C{\overset{+}{\pm}}C< \;\rightarrow\; >\overset{|}{\underset{+}{C}}-\overset{|}{\underset{|}{C}}- \qquad (10.11)$$

$$ \underset{+}{H} H$$

$$\underline{14} \underline{15}$$

This mechanism appears to be excluded, at least as a general mechanism, by the finding of general acid catalysis
in the hydration of p-methoxy-α-methylstyrene [17]; if the
rearrangement 14 → 15 is rate determining, the reaction
rate should not be influenced by proton transfer from general acids. Other suggested routes include "encumbered"
π-complexes (like 16) or carbonium ions [16]. It now

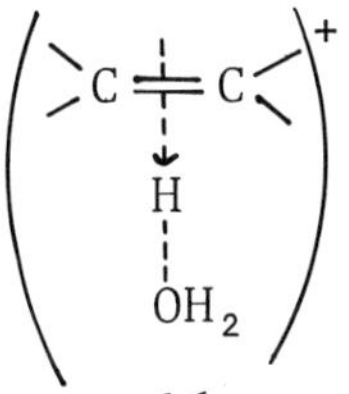

$$\left(>C{\overset{|}{\pm}}C< \atop {\overset{\downarrow}{H} \atop \overset{|}{OH_2}} \right)^{+}$$

$$\underline{16}$$

appears [11,15] that the simplest mechanism consistent with
all observations is a rate-determining proton transfer to
the olefin, as in Eq. (10.12), followed by rapid reaction
of the carbonium ion [18]. Although π-complex formulations

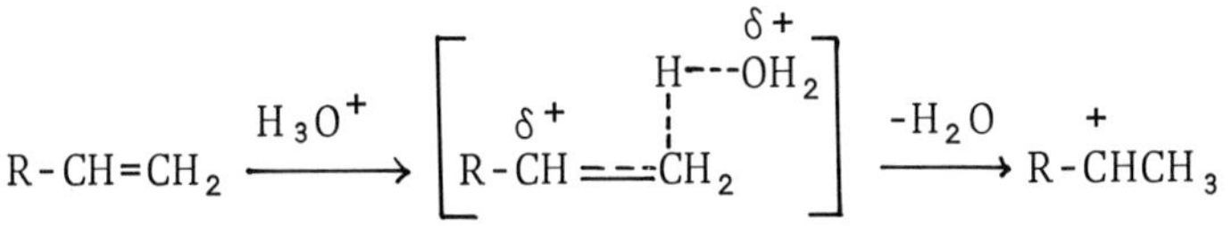

$$R{-}CH{=}CH_2 \xrightarrow{H_3O^+} \left[\overset{\delta+}{R{-}CH} \overset{\overset{\delta+}{H{-}{-}{-}OH_2}}{{=}{=}{=}} CH_2 \right] \xrightarrow{-H_2O} R{-}\overset{+}{CH}CH_3 \qquad (10.12)$$

are consistent, they are not necessary.
     Most of the mechanistic studies on hydrogen-initiated
additions have concerned hydration and hydrohalogenation,
but other related reactions are known. Alcohols, carboxylic acids, and thiols will add to the olefinic bond [19].
Markovnikov orientation is usually observed, and it is believed that the reactions are similar to hydration. Phenols give a product, in the presence of acid catalysts,
that results from attack of the carbonium ion on carbon
rather than on oxygen [19]:

$$R-CH=CH_2 \xrightarrow{H^+} R-CH-CH_3 \xrightarrow{C_6H_5OH} R(CH_3)CH-C_6H_4OH$$

Initiation by Electrophilic Halogen. Halogen molecules add to alkenes to give 1,2-dihalo products. In hydroxylic solvents the solvent lyate species may add in a competing reaction. For example, in aqueous solution the reaction between isobutene and chlorine leads to the dichloro adduct and a chlorohydrin:

$$Me_2C=CH_2 + C\ell_2 \xrightarrow{H_2O} \underset{C\ell}{Me_2C-CH_2C\ell} + \underset{OH}{Me_2C-CH_2C\ell} \quad (10.13)$$

The general mechanistic scheme is similar to that described for hydrohalogenation. The initial, rate-determining step is electrophilic attack by a halogen species on the alkene function, generating an electron-deficient intermediate. This intermediate then captures a nucleophile in a rapid product-determining step. Some of the evidence supporting this scheme is summarized [20]. A large portion of the experimental and speculative effort on halogenation mechanisms has concerned the nature of the electron-deficient intermediate. We first consider kinetic results and structure-reactivity relationships, then observations on products, and finally the intermediate in the reaction. Some oversimplification is introduced by treating all of the halogens together, though differences in behavior will be noted.

The kinetics of halogen addition are relatively uncomplicated in hydroxylic solvents, and most studies have been made in water, alcohols, or acetic acid. In dilute solution the rate law for chlorine and bromine additions is given by

$$- \frac{d[X_2]}{dt} = k[\text{olefin}][X_2] \quad (10.14)$$

Bromine addition in acetic acid, at the concentration region 0.025 M in bromine, can also occur by a third-order process, $v = \bar{k}_3[\text{olefin}][Br_2]^2$. Halide ions can catalyze halogen additions with a rate term of the form $k'[\text{olefin}][X_2][X^-]$. These are the important kinetic forms.

Table 10.I shows some characteristic structure-reactivity effects for bromine addition to aliphatic olefins [21]. These rate constants are the observed second-order rate constants; they certainly include contributions from a term like that in Eq. (10.14) as well as a bromide ion-catalysis term. Nevertheless, the structural effects are so striking that even the observed constants leave no doubt that the reaction is facilitated by substituents that release electrons to the reaction site. In fact, these rate data fit an extended Taft equation (with inclusion of the E_s steric constants), giving $\rho^* = -5.3$. It is apparent that the bromine is functioning as an electrophile.

Table 10.I   Observed Second-Order Rate Constants for Bromine Addition to Olefins [21][a]

Olefin	$10^{-3}k_2/M^{-1}\text{-min}^{-1}$
$CH_2=CH_2$	0.030
$CH_2=CH-Et$	2.90
$CH_2=CH-n-Pr$	2.09
$CH_2=CH-n-Bu$	1.99
$Me-CH=CH-Et$ (cis)	126
$Me-CH=CH-n-Pr$ (cis)	87.5
$Et-CH=CH-Et$ (cis)	195
$Me-CH=C(Me)(Et)$	3,600
$Me_2C=CMe_2$	28,000

[a]In methanol containing 0.2 $\underline{M}$ NaBr; at 25°.

The transition state for the electrophilic attack therefore contains the olefin and the electrophile. It is generally considered to contain the entire halogen molecule, for example, $C\ell-C\ell$ rather than just an electrophilic fragment $C\ell^+$ derived from $C\ell_2$. This conclusion is based on the kinetic effect of added chloride ions, which should markedly reduce the rate by their common ion effect on the equilibrium $C\ell_2 \rightleftharpoons C\ell^+ + C\ell^-$ if $C\ell^+$ were the effective nucleophile. Actually chloride ion has a slight accelerating effect. The evidence presented thus far therefore leads to a mechanism such as that indicated by

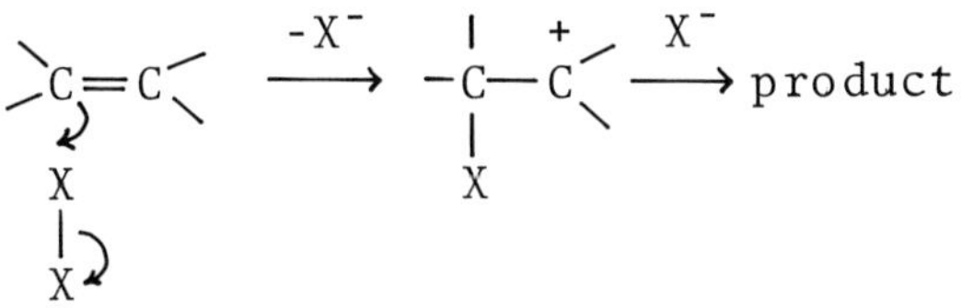

$$\overset{}{\underset{\underset{\displaystyle X}{|}}{\underset{X}{\overset{}{C}}}\!\!=\!\!C \xrightarrow{-X^-} -\overset{}{\underset{\underset{\displaystyle X}{|}}{C}}\!\!-\!\!\overset{+}{C} \xrightarrow{X^-} \text{product} \qquad (10.15)$$

Much attention has been given to the rate effect of bromide ion in brominations. Bromide reacts with bromine to form the tribromide ion

$$Br_2 + Br^- \rightleftharpoons Br_3^- \qquad (10.16)$$

The equilibrium constant, $K = [Br_3^-]/[Br_2][Br^-]$, is 16 M^{-1} in water at 25°. The rate dependence on bromide ion is commonly analyzed in terms of reactions with molecular bromine and with tribromide ion, according to

$$v = k_{Br_2}[olefin][Br_2] + k_{Br_3^-}[olefin][Br_3^-] \qquad (10.17)$$

The experimental rate equation is $v = k_2[olefin][Br_2]_{total}$, where $[Br_2]_{total} = [Br_2] + [Br_3^-]$. Combining these relationships with the expression for K gives

$$k_2(1 + K[Br^-]) = k_{Br_2} + k_{Br_3^-}K[Br^-] \qquad (10.18)$$

The experimental second-order rate constant k_2 is evaluated as a function of bromide ion concentration, and a plot of the left side of Eq. (10.18) against $[Br^-]$ enables k_{Br_2} and $k_{Br_3^-}$ to be estimated.

In this way the rate constants k_{Br_2} and k_{Br_3}- were obtained for the bromination of eight meta- and para-substituted styrenes in acetic acid [22]. Both sets of rate constants correlated well with Brown's electrophilic substituent constants σ^+, giving $\rho_{Br_2} = -4.2$ and $\rho_{Br_3^-} = -2.0$. These values indicate that both processes are electrophilic, but the attack by molecular bromine is much more sensitive to structural effects.

A peculiarity of these results is that $k_{Br_3^-}/k_{Br_2}$ is greater than unity for the less reactive olefins. Variation in this ratio is expected on the basis of the selectivity-reactivity concept, with the ratio tending toward unity for highly reactive substrates, but it is not expected that the anion Br_3^- should ever behave as a more powerful electrophile than Br_2. Table 10.II, selected from the results of Bell and co-workers [23], illustrates the effect [24].

Atkinson and Bell [23a], and later Rolston and Yates [22], discussed the possibility that bromine and tribromide might react with an olefin to produce different intermediates. Since two different rate-determining steps are being compared, it is conceivable that $k_{Br_3^-}$ may be greater than k_{Br_2}. Bell and Pring [23b] suggested that a substrate like diethyl fumarate (see Table 10.II) may react with tribromide by a route other than the electrophilic one characteristic of the more reactive olefins. A further possibility is related to the kinetic analysis. The rate term $k_{Br_3^-}$[olefin][Br_3^-] is kinetically equivalent to k´[olefin][Br_2][Br^-], and the rate constants are related by k´ = $k_{Br_3^-}$K. This interpretation is more accurately described as a bromide ion-catalyzed addition of molecular bromine, so it would be expected that the ratio k'/k_{Br_2} should be greater than unity. Multiplication of the last column in Table 10.II by $K = 16$ gives this ratio, which is greater than unity except for two values of 0.8.

Table 10.II Relative Reactivities of Bromine and Tribromide [23][a]

Olefin	k_{Br_2}[b]	$k_{Br_3^-}$[b]	$k_{Br_3^-}/k_{Br_2}$
Diethyl fumarate	3.40×10^{-5}	3.72×10^{-4}	10.9
Diethyl maleate	3.60×10^{-5}	5.20×10^{-5}	1.45
Ethyl acrylate	1.06×10^{-1}	6.7×10^{-2}	0.64
Ethyl crotonate	2.76	1.04	0.38
cis-3-Chloroallyl alcohol	3.01	0.15	0.05
trans-3-Chloroallyl alcohol	3.08	0.31	0.10
2-Chloroallyl alcohol	65	9.9	0.15
Ethyl cinnamate	220	23	0.10
Allyl cyanide	440	100	0.23
Ethylene	3.9×10^5	2.0×10^4	0.05
Allyl alcohol	6.7×10^5	6.9×10^4	0.10
Propene	4.5×10^6	3.2×10^5	0.07

[a]In water at 25°.
[b]Units are $\underline{M}^{-1}$-sec^{-1}.

Product analyses on brominations in acetic acid yield further information. The two products are the dibromide and the acetoxybromide, and the ratio of these has been interpreted in terms of the dual intermediate hypothesis [22]. The bromide ion-molecular bromine interpretation, which suggests a transition state like 17, has been rejected because it presumably would produce only dibromide [22].

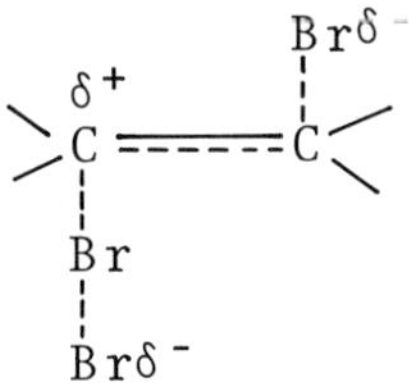

17

The third-order rate term found in more concentrated
solutions of bromine or iodine has the form $k_3[olefin]$
$[X_2]^2$. The role of the second halogen molecule presumably
is to assist the removal of halide ion from the first molecule, thus increasing its electrophilicity. The relative
reactivities of halogens and mixed halogens in acetic acid
by this third-order route are given in Table 10.III [25].

Table 10.III   Relative Reactivities of Halogens
in Acetic Acid [25]

Halogen	Relative Reactivity
I_2	1
IBr	3×10^3
Br_2	10^4
ICl	10^5
$BrCl$	4×10^6

Polarization of mixed halogens is related to the relative
abilities of the atoms to accept electronic deficiency
($I > Br > Cl$) and to their electronegativities ($Cl > Br >$
I). Thus iodine chloride is polarized in the sense
$\delta^+I-Cl\delta^-$, and electrophilic attack should be initiated by
the iodine atom, with chloride ion being the "leaving
group."
    The orientation of halogenation can be studied with
mixed halogens, and the type of orientation seen in hydro-
halogenations is observed. Thus ICl, which we expect to
attack as I^+Cl^-, adds in the Markovnikov manner to propene
[Eq. (10.19)][26].

$$CH_3CH=CH_2 \xrightarrow{ICl} CH_3CHClCH_2I \qquad (10.19)$$

Markovnikov orientation is, however, not so predominant in
halogen-initiated additions as it is in hydrogen-initiated
additions. The difference has been attributed to the
greater capability of the halogen atom to interact with
the adjacent electron-deficient carbon atom, with resulting
decrease in localized carbonium ion character. The extreme
forms of the electron-deficient intermediate are the free
carbonium ion 18 and the symmetrical bridged halogenonium
ion 19.
    The structure 19 was first proposed by Roberts and
Kimball to account for the predominant anti mode of halogen
addition to olefins [7,27]. Formation of this species

18   19

prohibits free rotation about the C-C bond and allows access by the incoming nucleophile only in an anti manner.

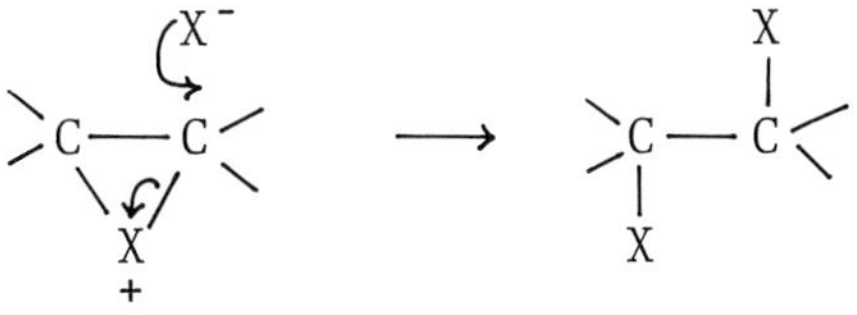

Syn addition has also been observed, and the proportion of syn/anti addition depends upon the identity of the
electrophile and the structure of the alkene. The more
stable the bridged halogenonium ion intermediate, the
greater the fraction of anti addition; conversely, the
more stable the open carbonium ion, the less stereoselective the addition should be, leading in the limit to a
syn/anti ratio of 1:1 [28]. Thus a full range of intermediate structures is considered possible, with 18 and 19
as the limits.
      Until recently the bridged ion has been a convenient
mechanistic idea, without much direct experimental support.
Nuclear magnetic resonance observations on substituted
cumyl cations (21) and the corresponding β-bromo cumyl
cations (22) have now revealed (through large ^{13}C chemical

21   22

shift differences) that 22 is markedly different from 21
(which must be a carbonium ion). 22 is considered to have
some bromonium ion character, either as a single weakly
(unsymmetrical) bridged ion 23, or by equilibration between an open carbonium ion and a strongly bridged bromonium ion [29].
      Cis- and trans-2-butenes give 100% anti addition of
bromine [30] and of BrN$_3$ [31], consistent with strongly

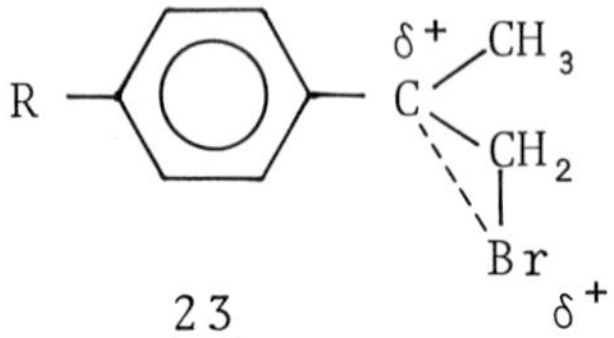

bridged bromonium ion intermediates. With styrene derivatives, however, these reagents give mixed syn- and anti-
addition [30,31]; the implication is that the α-phenylcar-
bonium ion is stabilized relative to the bromonium ion,
perhaps giving a weakly bridged ion [32].

Upon changing the electrophile to iodine (in IN_3), it
is found that addition to a styrene substrate occurs in
the anti manner [31]. Presumably the greater stability of
the iodonium ion compared with the bromonium ion enables
the intermediate to exist in the bridged iodonium form
rather than as the carbonium ion.

π-Complexes have been proposed to account for anti-
addition in halogenations. Although it is known that some
olefins can form charge-transfer complexes with bromine
[33], it appears that it is not necessary to invoke π-
complexes as a general mechanistic device [34]. A special
kind of system may include such species, however, and Dear
[35] has suggested that the anti-addition of BrF to acetylenes may be mediated by π-complexes. The distinction between an olefin-halogen π-complex and a bridged halogen-

$$\text{PhC≡CH} \xrightarrow{\text{BrF}} \text{PhCF=CHBr} \tag{10.20}$$

onium ion may not be significant, and it is possible that
either description is valid.

Hydroboration and Oxymercuration. Electrophilic attack by
hydrogen or halogen accounts for some of the most characteristic reactions of olefins, and has attracted most of
the mechanistic attention. Many other electrophilic addition reactions are known, however [36], and two of these
are now discussed.

The essential features of the hydroboration [37] of
olefins are expressed by

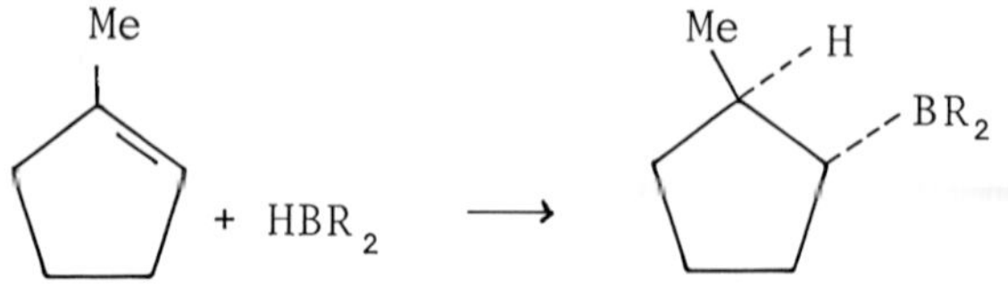

(10.21)

That is, the H-BR$_2$ function adds to the double bond in a syn manner, the orientation indicating that boron is functioning as the electrophile. The initial adding species in most hydroboration procedures appears to be borine, BH$_3$, which is derived, by dissociation, from its dimer diborane, B$_2$H$_6$. Addition to olefins continues until (usually) all of the hydrogens are replaced, giving a trialkylborane, as shown for the overall reaction

$$6RCH=CH_2 + B_2H_6 \rightarrow 2(RCH_2CH_2)_3B \qquad (10.22)$$

In the usual synthetic procedures diborane is prepared in situ by the action of Lewis acids on borohydrides (10.23), or of boron trifluoride with metal hydrides (10.24) [37,38].

$$12RCH=CH_2 + 3NaBH_4 + 4BF_3 \rightarrow 4(RCH_2CH_2)_3B + 3NaBF_4 \qquad (10.23)$$

$$6RCH=CH_2 + 6NaH + 8BF_3 \rightarrow 2(RCH_2CH_2)_3B + 6NaBF_4 \qquad (10.24)$$

Hydroboration is always carried out in ether solvents, usually tetrahydrofuran (THF), diethyleneglycol dimethyl ether (diglyme), or ethyl ether; it is a very fast reaction in these solvents, whereas the reaction is very slow in the absence of ethers.

Hydroboration is very general in scope. Most olefins form trialkylboranes as shown in Eq. (10.22). Highly branched olefins may proceed only to the dialkylborane, as illustrated with bis-3-methyl-2-butylborane (disiamylborane, 24) in Eq. (10.25) [39], or even the monoalkylborane.

$$2Me_2C=CHCH_3 \xrightarrow[0°]{HB} \begin{array}{c} Me_2CHCHCH_3 \\ | \\ BH \\ | \\ Me_2CHCHCH_3 \end{array} \qquad (10.25)$$

$$\underline{24}$$

Steric hindrance by the alkyl substituents is considered to be responsible for limiting these reactions. This steric influence can be utilized to achieve more selective hydroboration by using disiamylborane as the hydroboration reagent rather than the more reactive diborane.

The orientation of hydroboration can be generally accounted for on the basis that the boron-hydrogen bond is polarized in the sense of 25 [40].

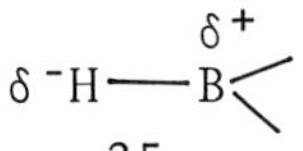

$$\delta^-H\!-\!\overset{\delta^+}{B}\!\!<$$

$$\underline{25}$$

The boron is therefore directed to the olefinic carbon with
the higher electron density; to be consistent with the nomenclature adopted for the addition of HX, it is said that
hydroboration occurs with anti-Markovnikov orientation,
though inspection of 25 shows that the orientation is normal. Table 10.IV gives the isomer distribution for the
hydroboration of some simple olefins. Some of these effects may be ascribed to steric influences, but it is clear
that electronic effects are in primary control of orientation.

Table 10.IV   Orientation in Hydroboration [37]

Olefin	Percent boron added to underlined carbon
Et-CH=CH$_2$	93
n-Bu-CH̄=CH$_2$	94
tert-Bu-C̄H=CH$_2$	94
Et-C(Me)=CH$_2$	99
Et-CH=CH-M̄e (cis)	55
Et-CH=C̄H-Me (trans)	51
Ph-CH=C̄H-Me (trans)	15
Me$_2$C=CH̄(Me)	98
p-MeO-C$_6$H$_4$-CH=CH$_2$	91
C$_6$H$_5$-CH=CH$_2$	80
p-Cℓ-C$_6$H$_4$-CH=CH$_2$	65

Hydroboration appears always to occur with syn stereochemistry, as shown in Eq. (10.21) for 1-methylcyclopentene
[41]. Moreover, hydroboration of norbornene gives 99% of
the exo isomer [Eq. (10.26)], which is the expected course
of the reaction if the reagent approaches from the less
hindered side of the olefin.

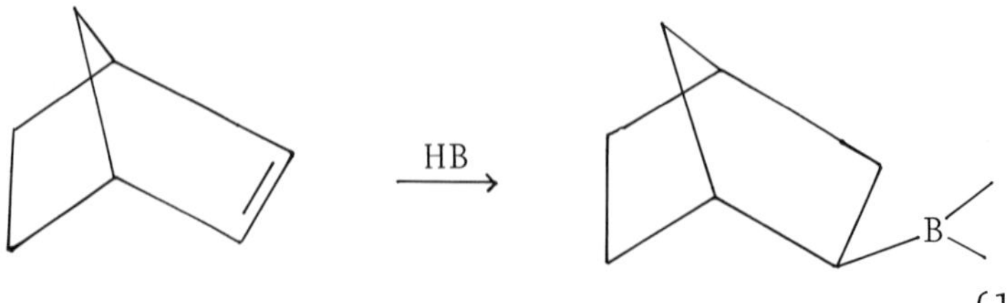

$$\text{(10.26)}$$

It has been implied, to this point, that the study of
hydroboration has involved the structural study of the
product alkylboranes. In fact, however, the results quoted
above are based upon a subsequent reaction in which, upon
treatment of the alkylborane with alkaline hydrogen peroxide, an alcohol is produced.

$$R_3B + 3H_2O_2 + NaOH \rightarrow 3ROH + NaB(OH)_4 \qquad (10.27)$$

This reaction appears to proceed without rearrangement or
stereochemical change. The hydroboration reaction therefore provides a valuable synthetic route for the anti-
Markovnikov hydration of olefins, Eq. (10.28), in contrast
with the acid-catalyzed hydration, which occurs with
Markovnikov orientation.

$$tert\text{-}Bu\text{-}CH=CH_2 \xrightarrow{HB} \xrightarrow[OH^-]{H_2O_2} tert\text{-}Bu\text{-}CH_2CH_2OH \quad (10.28)$$

In addition, hydroboration-oxidation provides a stereospecific syn-hydration route, as in the conversion of
cholesterol to cholestane-3β,6α-diol [42]

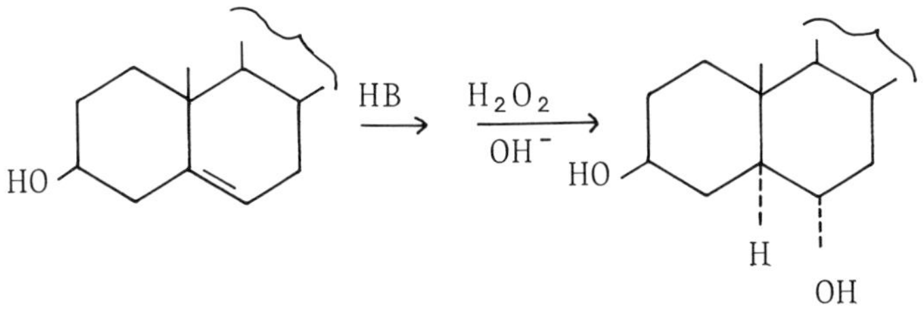

$$(10.29)$$

This syn-hydration occurs on the less hindered side of
the molecule.
    We now consider the mechanism of hydroboration. Reac-
tions of diborane with olefins in ether solvents are too
fast for conventional kinetic measurements, but the selective hydroborating agent disiamylborane 24 reacts at measurable rates [43]. The rate equation for the reaction of
24 with cyclopentene in THF is of the form

$$v = k[olefin][(R_2BH)_2] \qquad (10.30)$$

where $(R_2BH)_2$ is the dimer of 24; disiamylborane appears
to exist mainly as the dimer in THF. By analogy with
other diborane structures, the dimer is proposed to have
the double hydrogen bridge of 26.

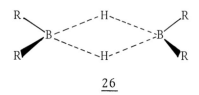

26

Brown and Moerikofer [43] propose a concerted addition by the dimer to account for the kinetics, the transition state being represented by 27.

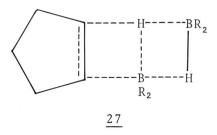

27

Since the stoichiometry is given by Eq. (10.31), it was postulated that the molecule of monomer released upon conversion to the product reacts rapidly with another molecule of olefin.

$$2\ C_5H_8 + (R_2BH)_2 \rightarrow 2\ R_2BC_5H_9 \qquad (10.31)$$

Diborane itself dissociates in THF solution, existing in the form of a 1:1 THF-borine complex [44], $C_4H_8O{:}BH_3$. Brown suggests a four-center concerted addition, for which the transition state is given as 28, and 29 indicates the flow of electrons.

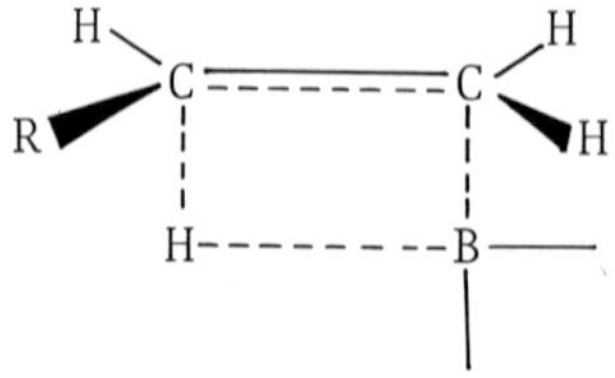

28

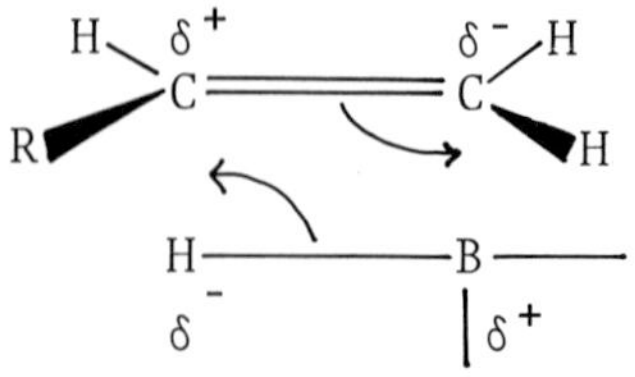

29

This mechanism explains the orientation and the stereo-chemistry of hydroboration. Boron in BH_3 is electron deficient, having a vacant p orbital that the olefinic π electrons can occupy. It is not necessary that the reaction be concerted, and a two-step mechanism [45] (electro-philic boron attack followed by hydride transfer, as in 30) will also account for the experimental results.

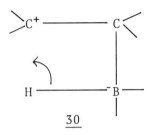

                                    30

     Some aspects of the postulated mechanisms seem doubtful or in need of clarification. For example, the proposal that the monomer of disiamylborane released in the rate-determining step then reacts rapidly with olefin seems in-consistent with the kinetics; if an olefin-monomer reaction were very fast, this route would be expected to predominate over the olefin-dimer pathway, for some of the reagent must always exist in the monomeric form. The four-center con-certed mechanism has not been critically examined, and its possibility is related both to symmetry control [45] (Sec. 10.3) and to the nature of bonding in boron hydrides [46]; a two-step mechanism may be preferable. In this connection some relative rate data show peculiar behavior in hydrobor-ation; the relative rate $k_{cyclopentene}/k_{cyclohexene}$ is 110 for reaction with disiamylborane [43]. Large values for this reactivity ratio have been ascribed to relief of strain on passing to a four-, five-, and six-membered transition state [47], the relief being proportionately greater for the more strained cyclopentene. Table 10.V shows some reactivity data for reactions believed to pass through cyclic transition states. Caution has been urged in using the magnitude of this reactivity ratio as a cri-terion of ring size of cyclic transition states [48]. In any case, the ratio for hydroboration is the largest yet reported, and a satisfactory explanation has not yet been given.

     The remarkable catalytic effect of ethers is also un-explained; coordination of the oxygen unshared pair with the boron vacant p orbital would seem to hinder, rather than facilitate, electrophilic attack by the boron.

     A further unusual observation is that trialkylboranes, though relatively inert toward water, alcohols, alkali,

and mineral acids, are readily attacked by carboxylic
acids [49]. The reaction sequence, shown in Eq. (10.32),
provides a noncatalytic route for the hydrogenation of
alkenes.

Table 10.V Relative Reactivities of Cyclopentene
and Cyclohexene [47,48]

Reaction	Size of cyclic transition state	Relative rate (cyclopentene/cyclohexene)
Diimide reduction	6	15.5
Osmium tetroxide hydroxylation	5	21.9
Ozonolysis	5	3.9
Cr(VI) oxidation	3	1.29
Hydroboration[a]	(4)	110

[a]With disiamylborane [43].

$$3RCH=CH_2 \xrightarrow{HB} 3(RCH_2CH_2)_3B \xrightarrow[\Delta]{R'COOH} 3RCH_2CH_3 \qquad (10.32)$$

Acetylenes are converted in this way to cis-olefins.

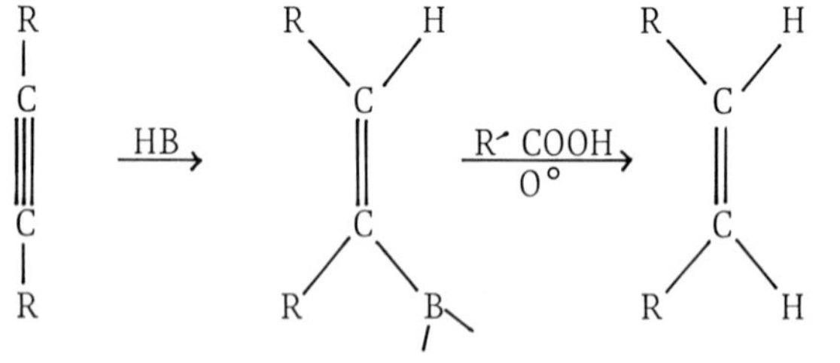

$$(10.33)$$

The effectiveness of carboxylic acids in this "protonoly-
sis" may be due to coordination as in 31, with subsequent
intramolecular proton transfer [49,50].

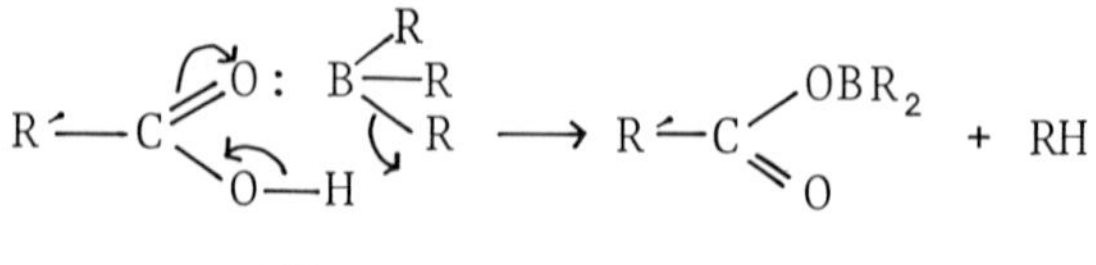

$$(10.34)$$

31

We turn now to the oxymercuration reaction, Eq. (10.35), in which a mercuric salt adds to an olefin in a hydroxylic solvent [51].

$$\begin{array}{c}\diagdown \\ \diagup\end{array}C\!=\!C\begin{array}{c}\diagup \\ \diagdown\end{array} \;+\; HgX_2 \;+\; ROH \;\rightarrow\; \underset{\displaystyle \underset{OR}{|}}{-C}\!-\!\underset{\displaystyle \underset{HgX}{|}}{C}\!- \;+\; HX \qquad (10.35)$$

The rate equation is overall second order [Eq. (10.36)], and the reactions are very fast [52].

$$v = k[\text{olefin}][\text{mercury(II)}] \qquad (10.36)$$

Table 10.VI   Rate Parameters for Hydroxymercuration of Olefins [52]

Olefin	$10^{-3}k/\underline{M}^{-1}sec^{-1}$ [a]	$\Delta H^{\ddagger}/kcal\text{-}mole^{-1}$ [b]	$\Delta S^{\ddagger}/e.u.$ [c]
Ethylene	5.1	9.2	-10
Propylene	about 100		
Isobutene	> 1000		
cis-2-Butene	5.8	9.6	-9
trans-2-Butene	1.7	12.0	-3
Cyclohexene	5		
1-Butene-3-ol	0.26	13.5	-5
1-Butene-4-ol	8.4	8.7	-11
1-Pentene-3-ol	0.14	11.7	-9
1-Pentene-4-ol	6.1	10.7	-5
1-Pentene-5-ol	> 1000		
Allyl alcohol	1.12	10.7	-8
1-Cyclohexene-3-ol	0.43	11.6	-7

[a] At 25°.
[b] ± 0.5 kcal/mole.
[c] Molar concentration scale.

Some rate parameters are given in Table 10.VI for hydroxymercuration [Eq. (10.35) with R = H, X - $C\ell O_4{}^-$]. The structural effects are as expected for an electrophilic addition, with electron-donating groups facilitating the rate; a linear correlation with Taft substituent constants gives $\rho^* = -3.3$.

On the reasonable basis of electrophilic attack by mercury, the orientation of addition is strictly Markovnikovian, as illustrated for propene in [53].

$$CH_3CH{=}CH_2 \;+\; HgCl_2 \;+\; CH_3OH \;\rightarrow\; \underset{\underset{OCH_3}{|}}{CH_3CH}{-}CH_2HgCl \qquad (10.37)$$

The stereochemistry of oxymercuration provides valuable, and perplexing, mechanistic information. Acyclic and un-strained cyclic olefins undergo anti addition, which has been accounted for by the formation of an intermediate mercurinium ion [54], 32, which is analogous to the brom-onium ion discussed earlier.

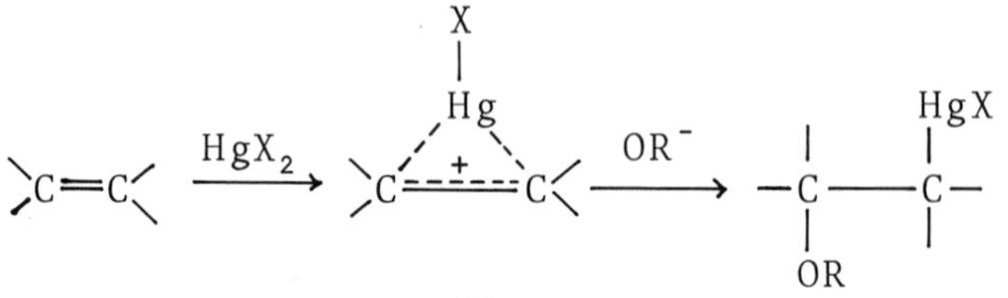

$$\underline{32} \qquad\qquad\qquad\qquad\qquad\qquad\qquad (10.38)$$

Back-side attack by the nucleophile leads to overall anti-addition.

   This simple picture has been complicated by the obser-vation that strained cyclic olefins like norbornene (33) undergo pure syn-addition, giving the exo-cis product with-out any rearrangement [55].

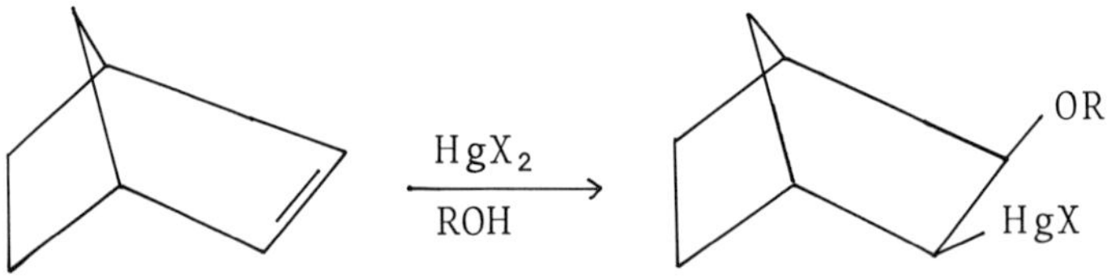

$$\underline{33} \qquad\qquad\qquad\qquad\qquad\qquad\qquad (10.39)$$

Another observation is that other anions, such as acetate, can compete effectively with the solvent in the oxymercur-ation of strained olefins, but not with unstrained sub-strates.

   Traylor and Baker [55] have proposed a single general mechanism for oxymercuration in which a solvated mercur-inium ion 34 is formed in a rapid reversible step. Un-hindered and unstrained substrates can undergo back-side attack by the solvent, giving anti-addition. The transi-tion state for back-side approach on a strained olefin would, however, be energetically unfavorable relative to a front-side approach, which is achieved by collapse of the ion pair to give the acetate [route A in Eq. (10.41)] or the hydroxy compound (route B).

+ Hg(OAc)2
ROH⁻
Cℓ⁻
OR
HgCℓ
+
OAc
HgCℓ

(10.40)

(OH2)n
HgOAc
OAc
A
+ Hg(OAc)2
H2O
OAc
Hg(OH2)n
OAc⁻
34
B
(OH2)n-1
HgOAc
OH

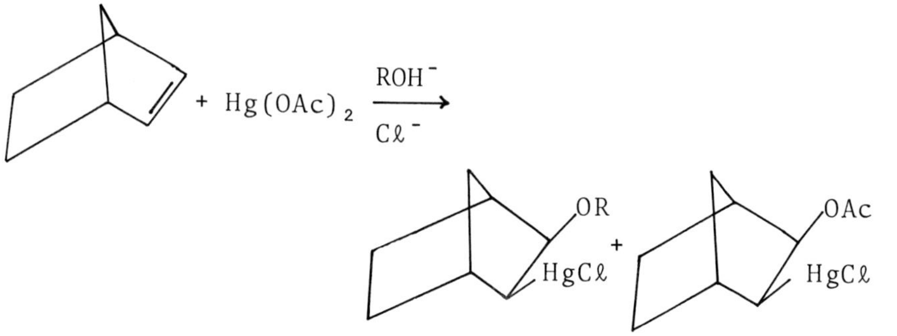

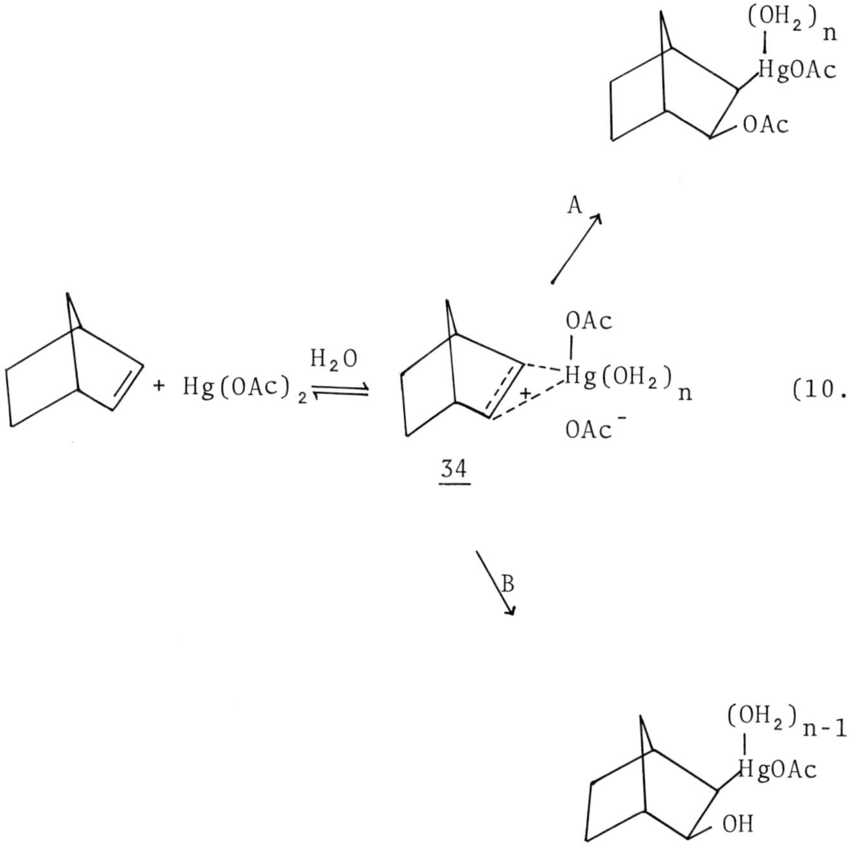

(10.41)

In agreement with this mechanism is the behavior of bicyclo [2.2.2] octene [56], 35, which is less strained than
norbornene, and therefore should be less apt to undergo
syn-addition. The results are shown in [57]

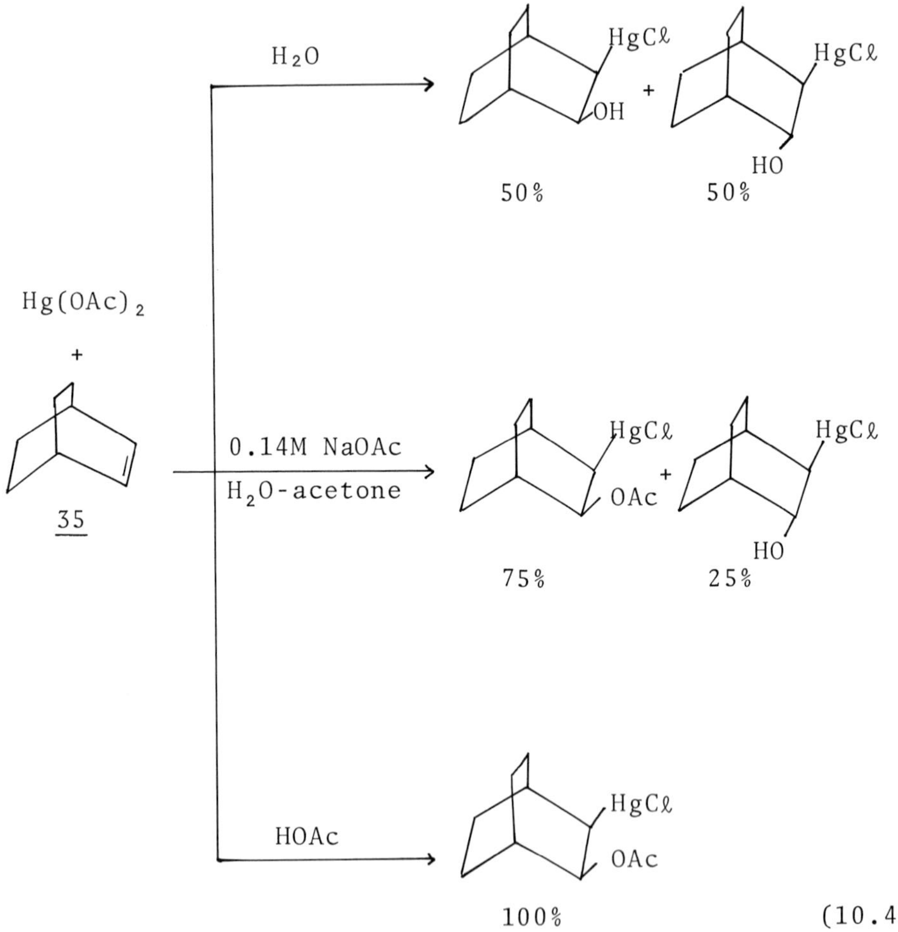

(10.42)

In water, which is a good nucleophile, a 1:1 ratio of
syn-to-anti addition occurred. In aqueous acetone [not
shown in Eq. (10.42)] this changed to 60% syn, 40% anti.
Addition of excess acetate ions to this medium had a small
effect on the production of the trans-hydroxy product, no
cis-hydroxy product was found, but the cis-acetate (and
no trans-acetate) was produced. In the weakly nucleophilic
solvent acetic acid, pure syn-addition occurred. There

results are explicable in terms of the competitive back-
side (anti) intermolecular and front-side (syn) "intramolecular" fates of the mercurinium ion intermediate.
    The extent of free carbonium ion character in the pre-
sumed intermediate is not clear. The absence of rearranged
products in the oxymercuration of norbornene has been taken
as evidence that the intermediate has little carbonium ion
character, and is instead a fairly symmetrical mercurinium
ion [55]. On the other hand, the considerable sensitivity
of the rates to structure is interpreted to mean that the
transition state for the rate-determining step possesses a
strongly localized electron deficiency [52].
    An interesting example of intramolecular participation
is seen with certain hydroxy olefins, in which the internal
hydroxy group participates as the nucleophile [55], giving
a cyclic ether, as illustrated for the reaction of 1-penten-
5-ol [52].

$$HO\text{-}CH_2CH_2CH_2CH\text{=}CH_2 \xrightarrow{HgX_2} \text{(cyclic ether)}\text{-}CH_2HgX \qquad (10.43)$$

    The oxymercuration procedure has been combined with a
borohydride reduction to give a fast, practically quantitative route for the conversion of olefins to Markovnikov-
oriented alcohols [58]. Since the hydroboration procedure
gives anti-Markovnikov hydration products, convenient
routes are available for the synthesis of isomers such as
those shown in

$$Me_3C\text{-}CH\text{=}CH_2 \begin{cases} \xrightarrow{\text{hydroboration}} \xrightarrow[OH^-]{H_2O_2} Me_3C\text{-}CH_2CH_2OH \\ \\ \xrightarrow{\text{oxymercuration}} \xrightarrow{NaBH_4} Me_3C\text{-}CH(OH)CH_3 \end{cases} \qquad (10.44)$$

Addition to Conjugated Systems. Conjugated multiple bonds
introduce further complications in addition reactions.
Although the experimental observations can be easily rationalized, little understanding or predictive capability
are presently at hand. Some of the interesting features
will be pointed out [59].
    The simplest conjugated diene is 1,3-butadiene, whose
reactions illustrate an important experimental problem.
Two products are observed

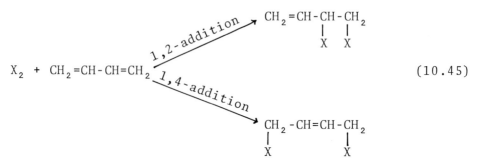

$$X_2 + CH_2=CH-CH=CH_2 \qquad (10.45)$$

If the two products of such a reaction are readily interconvertible under the reaction conditions, the product
distribution will favor the more stable product; the products are said to be <u>thermodynamically</u> <u>controlled</u>. If,
on the other hand, the products are not interconvertible
on the reaction time scale, the product distribution will
favor the more rapidly formed product, and the reaction
is <u>kinetically</u> <u>controlled</u>. Whether a reaction is under
kinetic or thermodynamic control can often be established
by separate experiments in which the interconvertibility
of the products is studied.

The adducts of 1,4-addition of halogens and hydroha-
lides to dienes are generally more stable than are the
1,2-addition products, whereas the 1,2-addition products
are more readily formed [60]. In other words, under kinetic control 1,2-addition predominates, and under thermodynamic control 1,4-addition is favored. This behavior is
not well understood.

A key observation, first made by Mislow and Hellman
[61] for the chlorination of butadiene, is that 1,4-addi-
tion gives only the trans isomer <u>36</u> under conditions of
kinetic control such that cis-trans isomerization could
not have occurred. This result rules out transition

$$
\begin{array}{c}
CH_2C\ell \\
/ \\
CH=CH \\
/ \\
C\ell CH_2
\end{array}
$$

<u>36</u>

states, like <u>37</u> and <u>38</u>, arising from the s-cis conformation of the diene [62]. Reluctance of chlorine to form
chloronium ions presumably is not responsible for this
behavior, because 1,4-addition of bromine also gives
solely the trans adduct [63]. The ratio of 1,2-addition
to 1,4-addition for the bromination of butadiene is

insensitive to the solvent polarity, implying a relatively
nonpolar transition state [63].

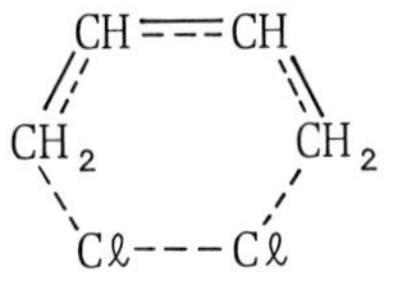

$$\underline{37}$$

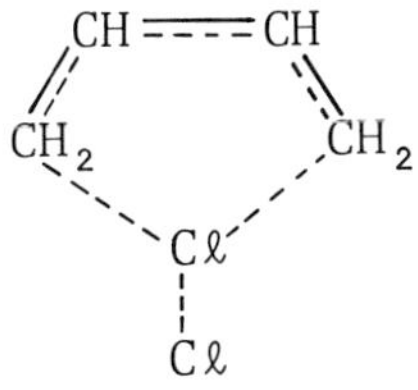

$$\underline{38}$$

The initial electrophilic attack is at a terminal carbon atom, consistent with electron polarizability as shown
in $\underline{41}$, $\underline{42}$, and $\underline{43}$, where R is an electron-releasing group
[64].

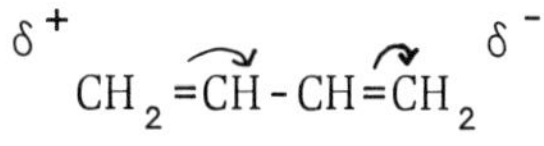

$$\underline{41} \qquad\qquad\qquad \underline{42}$$

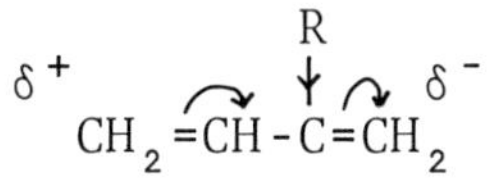

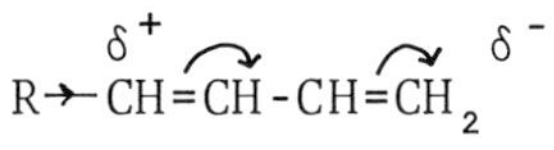

$$\underline{43}$$

Thus of the three possible HCl adducts of butadiene, only
the two that arise from terminal electrophilic attack are
observed [65]

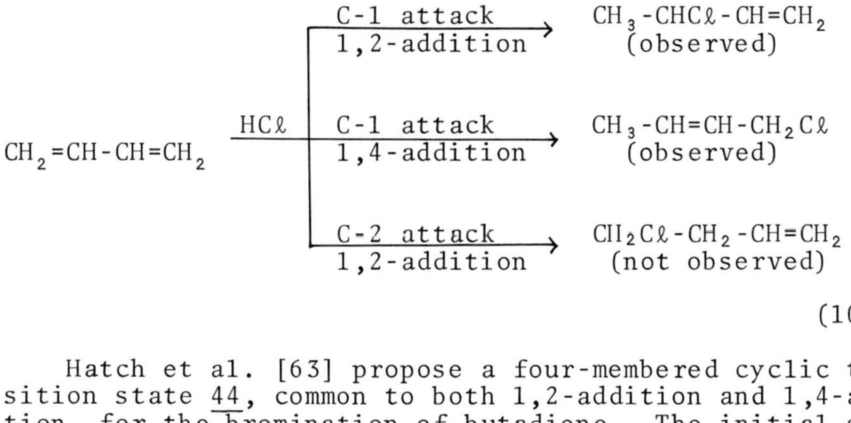

$$(10.46)$$

Hatch et al. [63] propose a four-membered cyclic transition state $\underline{44}$, common to both 1,2-addition and 1,4-addi-
tion, for the bromination of butadiene. The initial step

is presumably similar to that in monoolefins, giving an
intermediate bromonium ion or π-complex.

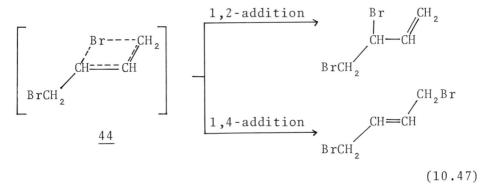

$$(10.47)$$

This mechanism accounts for the formation of trans-1,4-
dibromo-2-butene, and for the independence of rate on solvent polarity.

In contrast to these observations of trans-1,4-addition
products from acyclic dienes are the results of Young et al.
[66] on the bromination of cyclopentadiene and of Hammond
and Warkentin [67] on the addition of deuterium bromide to
cyclohexadiene. Cyclopentadiene gives trans-3,4-dibromo-
cyclopentene (1,2-addition), trans-3,5-dibromocyclopentene
(1,4-addition), and cis-3,5-dibromocyclopentene (1,4-addi-
tion). The 1,2-addition product could easily be the consequence of anti-addition as with a monoolefin. The trans
product of 1,4-addition cannot be accounted for via the
mechanism of Eq. (10.47), since cyclopentadiene is required
to have the s-cis conformation. A transition state like
45 seems plausible, however.

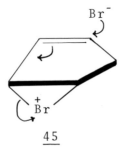

45

This is similar to the transition state 46 proposed to
account for the cis-3,5-dibromo product [66]; this is a
syn-addition by means of lon-pair collapse.

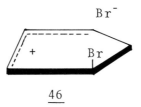

46

The addition of DBr to 1,3-cyclohexadiene gives the
1,2-trans and 1,4-cis products [68]. These additions can
be accounted for as with the preceding reaction.
    A different type of conjugated system of wide occur-
rence is the α,β-unsaturated carbonyl group, -CH=CH-COR.
Because of electron withdrawal from the alkene function,
these are not as susceptible to electrophilic attack as
are simple alkenes, but such attack can be observed. For
example, the chlorination of methyl trans-cinnamate in
acetic acid leads to products expected for electrophilic
attack by chlorine [69], with the initial attachment being
at the carbon alpha to the carboxyl carbon. Four products
are formed: erythro- and threo-dichlorides and acetoxy-
chlorides. These can be accounted for by postulating
the initial formation of an ion pair 47, which is transformed into its conformational isomer 48. The four products can be derived from these intermediates by syn-
addition of the chloride in the ion pair and by anti-
addition of external chloride or acetate ions. Because
of the low dielectric constant of acetic acid, ion pairs
are reasonable species in this solvent.

10.2 Nucleophilic Addition

Survey of the Reaction. Section 10.1 will have made it
clear that a characteristic feature of the olefinic group
is its nucleophilic behavior, in that it is susceptible
to attack by electrophiles. This is a consequence of the
bonding π molecular orbital formed from the two p atomic
orbitals of the olefinic carbons, which are perpendicular
to the plane of the alkene group. This π-electron system
is very polarizable, and if it is acted upon by a powerful
electron-attracting group, sufficient displacement of
charge can occur so that the double bond (or rather one
end of it) may function as an electrophilic site, and be
attacked by nucleophiles. The nature of this polarization
is illustrated by 49 and 50, which show that the β-carbon
is electron deficient relative to the α-carbon.

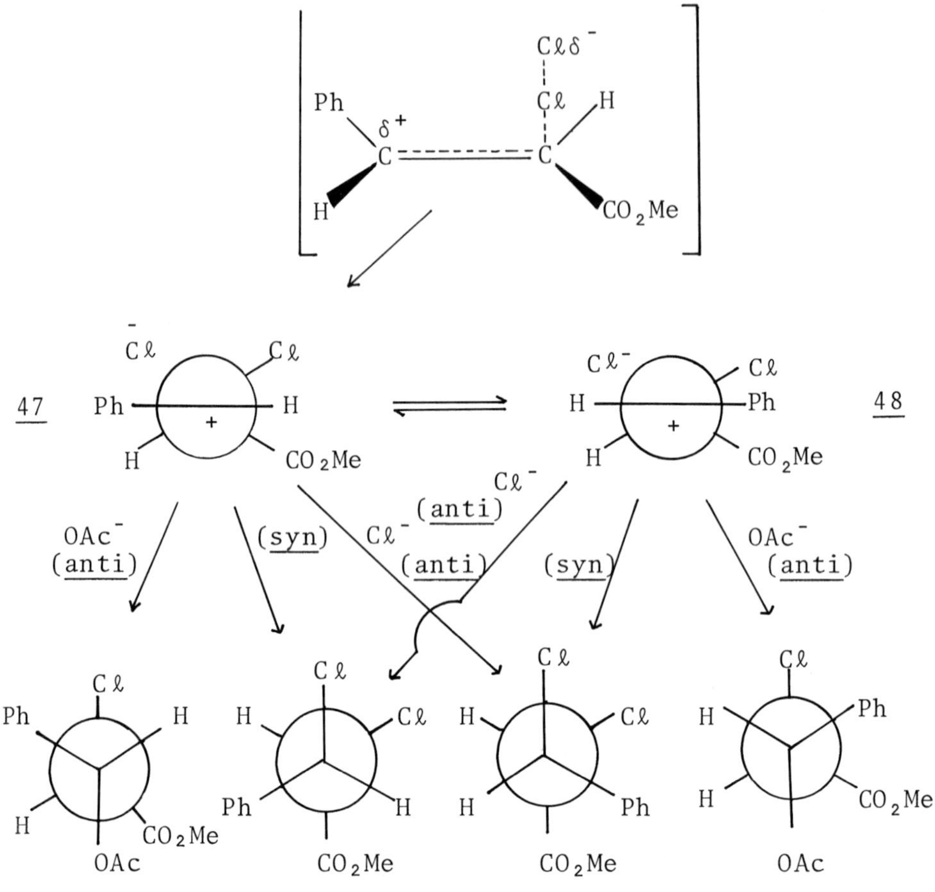

erythro-   threo-   erythro-   threo-
acetoxychloride   dichloride   dichloride   acetoxychloride

(10.48)

$$\overset{\delta^+}{CH_2}=\overset{}{CH}-\overset{\delta^-}{C\equiv N}$$
$$\;\;\beta \;\;\;\;\; \alpha$$

$$\overset{\delta^+}{Ph-CH}=\overset{}{CH}-\overset{\delta^-}{CH}=\overset{}{O}$$
$$\;\;\;\;\;\;\;\;\;\beta \;\;\;\; \alpha$$

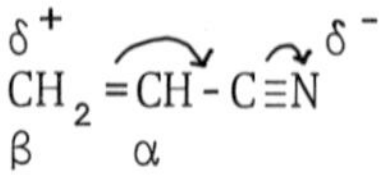

49 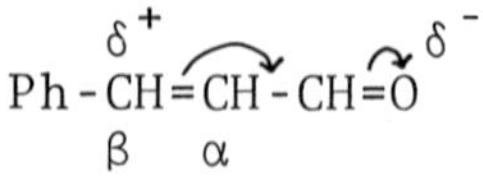   50

    Attack by a nucleophile is therefore expected at the
olefinic carbon β to the electron-withdrawing group. Reaction is completed by subsequent attachment of the electrophilic fragment of the reagent to the α-carbon. This sequence is the reverse of that observed for electrophilic
additions. Letting N⁻ be a generalized anionic nucleophile,

the Ad_N reaction is written as Eq. (10.49), where X is an electron-withdrawing group.

$$\begin{array}{c} R_1 \\ R_2 \end{array}\!\!C\!=\!C\!\!\begin{array}{c} X \\ H \end{array} + \; N^- \; \rightleftharpoons \; \begin{array}{c} R_1 \\ R_2 \end{array}\!\!C\!-\!\bar{C}\!\!\begin{array}{c} X \\ \backslash \\ H \end{array} \;\; \xrightarrow{H^+} \;\; \begin{array}{cc} R & X \\ |^{\,1} & | \\ R_2\!-\!C\!-\!C\!-\!H \\ | & | \\ N & H \end{array} \qquad (10.49)$$

Products other than simple 1,2-adducts are possible, depending on the natures of the nucleophile and the alkene. Among these alternative processes are cyclization [Eq. (10.50)], substitution [Eq. (10.51); this is an overall vinylic substitution], and cis-trans isomerization [Eq. (10.52)].

$$\begin{array}{c} R_1 \\ R_2 \end{array}\!\!C\!=\!C\!\!\begin{array}{c} X \\ H \end{array} + \; NZ^- \; \longrightarrow \; \begin{array}{c} R_1 \\ R_2 \end{array}\!\!C\!-\!\bar{C}\!\!\begin{array}{c} X \\ \\ H \end{array} \;\; \xrightarrow{-Z^-} \;\; \begin{array}{c} R_1 \\ R_2 \end{array}\!\!C\!-\!C\!\!\begin{array}{c} X \\ \\ H \end{array} \qquad (10.50)$$

$$\begin{array}{c} R_1 \\ R_2 \end{array}\!\!C\!=\!C\!\!\begin{array}{c} X \\ H \end{array} + \; N^- \; \longrightarrow \; \begin{array}{c} R_1 \\ R_2 \end{array}\!\!C\!-\!\bar{C}\!\!\begin{array}{c} X \\ \\ H \end{array} \;\; \xrightarrow{-R_1^-} \;\; \begin{array}{c} N \\ R_2 \end{array}\!\!C\!=\!C\!\!\begin{array}{c} X \\ H \end{array} \qquad (10.51)$$

$$\begin{array}{c} R_1 \\ R_2 \end{array}\!\!C\!=\!C\!\!\begin{array}{c} X \\ H \end{array} + \; N^- \; \longrightarrow \; \begin{array}{c} R_1 \\ R_2 \end{array}\!\!C\!-\!\bar{C}\!\!\begin{array}{c} X \\ \\ H \end{array} \;\; \xrightarrow{-N^-} \;\; \begin{array}{c} R_1 \\ R_2 \end{array}\!\!C\!=\!C\!\!\begin{array}{c} H \\ X \end{array} \qquad (10.52)$$

This section discusses some of the important and most characteristic Ad_N reactions [70].

Attack by Nucleophilic Carbon. The Michael reaction is the base-catalyzed addition of an active methylene group to the β-carbon atom of a double bond conjugated with a ketone, ester, nitro group, nitrile, etc. Equation (10.53) is a typical example, showing the addition of ethyl aceto-acetate to acrylonitrile [71].

$$CH_3CO\text{-}CH_2\text{-}CO_2Et + CH_2\!=\!CH\text{-}CN \xrightarrow{\text{base}} \begin{array}{c} CH_2\text{-}CH_2\text{-}CN \\ | \\ CH_3CO\text{-}CH\text{-}CO_2Et \end{array} \qquad (10.53)$$

The addition of ethyl malonate to ethyl cinnamate is another example.

$$CH_2(CO_2Et)_2 + PhCH\!=\!CHCO_2Et \xrightarrow{EtO^-} \begin{array}{c} PhCHCH_2CO_2Et \\ | \\ CH(CO_2Et)_2 \end{array} \qquad (10.54)$$

The reaction is generally agreed to follow the pathway of Eq. (10.49), that is, the base catalyst abstracts a proton from the methylene group (which usually must be flanked by electron-withdrawing groups to increase its acidity), giving a carbanion, which attacks the β-carbon of the activated olefin. A proton then adds to the α-carbon. This sequence is shown below, where B represents the base.

$$CH_2(CN)CO_2Et + B \rightleftharpoons BH^+ + \bar{C}H(CN)CO_2Et \qquad (10.55)$$

$$\bar{C}H(CN)CO_2Et + CH_2=CHCHO \rightleftharpoons \underset{\underset{CH(CN)CO_2Et}{|}}{CH_2-\bar{C}HCHO} \qquad (10.56)$$

$$\underset{CH(CN)CO_2Et}{\overset{CH_2-\bar{C}HCHO}{|}} + BH^+ \rightleftharpoons B + \underset{CH(CN)CO_2Et}{\overset{CH_2-CH_2CHO}{|}} \qquad (10.57)$$

An intramolecular Michael addition can occur as shown for methyl o-mercaptocinnamate-S-acetate [72]:

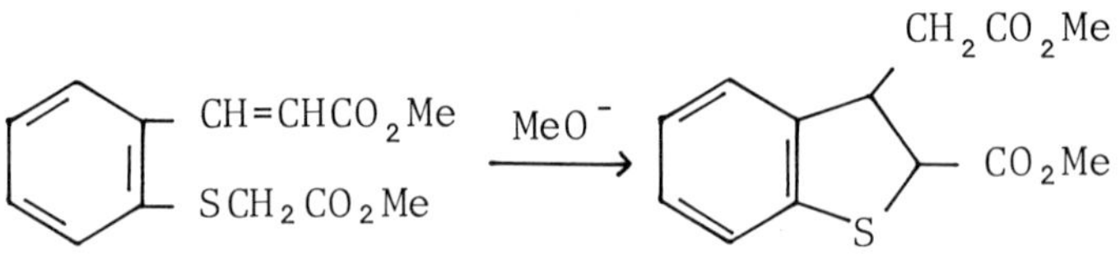

$$(10.58)$$

The Michael reaction is reversible, with the position of equilibrium being dependent upon the temperature (lower temperature favors the addition product), the base concentration (which establishes the amounts of carbanions of the methylene reactant and the product [73]), and the substitution on the olefinic carbons. Such substitution tends to shift the equilibrium in favor of the reactants, at least in part by a steric effect.

Relative reactivities of reaction sites can be studied by positional competition experiments in which two sites are available for reaction in the same molecule. Thus 2-phenylcyclohexanone is cyanoethylated at the 2-position rather than at the 5-position, showing that the 2-hydrogen is more readily dissociated than are those on the 5-carbon [74].

The kinetics of Michael additions are consistent with the formulation of the reaction as Ad_N2. The reaction is first order in the alkene and first order in the nucleophile, the rate equation being $v = k_2[\text{olefin}][\text{carbanion}]$. This behavior was observed in the addition of barbituric acid to β nitrostyrene in aqueous dioxane [75].

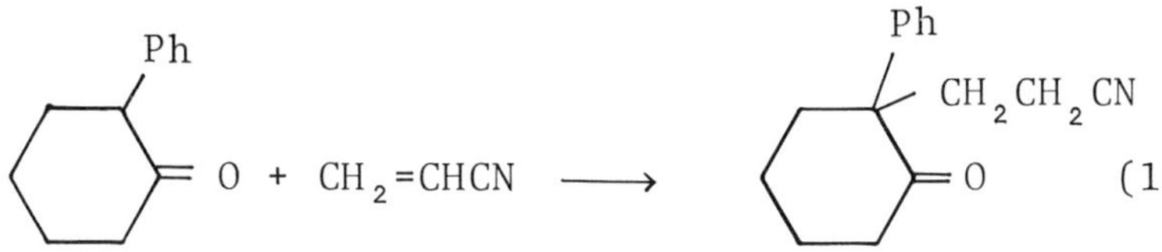

(10.59)

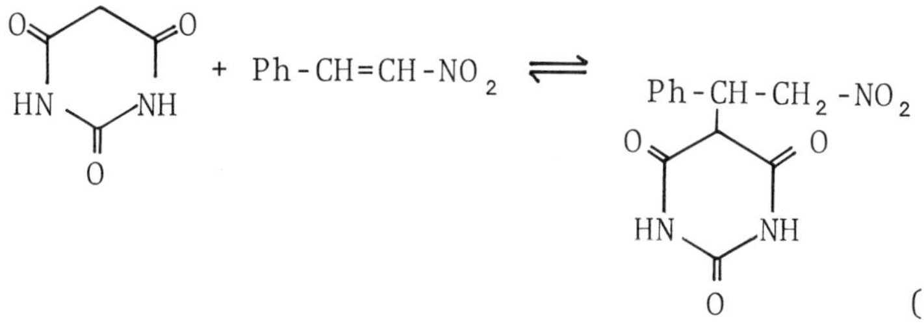

(10.60)

The rate-determining step is probably the attack of the carbanion on the olefin.

The electronic effects of substitution are as expected for the Ad_N2 assignment in the Michael reaction. In the reaction of barbituric acid with a series of substituted β-nitrostyrenes, $R-C_6H_4-CH=CH-NO_2$, the relative reactivities were in the order $p-NO_2 > m-NO_2 > p-Cl > H > p-CH_3O > p-(CH_3)_2N$ [75].

The stereochemistry of the Ad_N2 mechanism can be studied with additions to cyclic alkenes. It is necessary to be alert to the possibility of equilibration between isomers, that is, of thermodynamic control of the product distribution, in interpreting such results. In a reaction like that of Eq. (10.61), syn-addition gives the trans product and anti-addition gives the cis-product. The trans isomer is more stable than the cis, so if the cis is produced by an anti-addition, it is possible for conversion to the thermodynamically preferred trans form to occur. On the other hand, if syn-addition occurs, yielding the trans product, this will not be converted to the less stable cis form. Therefore isolation of a cis product shows that anti-addition must have occurred, whereas isolation of trans product is not conclusive of the mode of addition [76]. The reaction of diethyl malonate with

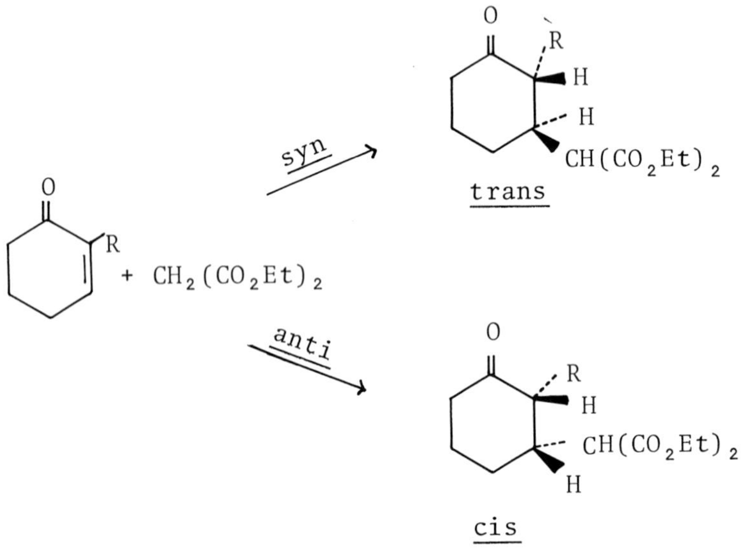

(10.61)

1-cyanocyclohexene gave 72% cis and 28% trans product, showing that the principal route is by anti-addition [77]. Similar results are observed for Ad_N2 reactions other than Michael additions; anti-addition is predominant. However, the steric requirements of the reaction may divert it through a syn-addition route.

The addition of hydrogen cyanide to α,β-unsaturated ketones, esters, and nitriles is closely analogous to the Michael reaction [70,78].

$$Me_2C{=}CH{-}\overset{O}{\overset{\|}{C}}{-}CH_3 \xrightarrow{HCN} Me_2\underset{CN}{\underset{|}{C}}{-}CH_2\overset{O}{\overset{\|}{C}}{-}CH_3 \tag{10.62}$$

The cyanide group always attaches itself to the β-carbon, and the rate is first order in cyanide ion [79]. The reaction sequence is therefore as shown in Eq. (10.49).

Organometallic compounds can add to α,β-unsaturated ketones, esters, etc., in what appears to be a 1,4-addition followed by keto-enol isomerization to give net α,β-addition. Grignard reagents react rapidly [80] as shown in

$$PhCH=CH-\overset{\overset{\textstyle O}{\|}}{C}-NMe_2 \xrightarrow{\quad PhMgBr \quad} Ph_2CHCH_2\overset{\overset{\textstyle O}{\|}}{C}-NMe_2 \qquad (10.63)$$

The carbon fragment is bonded to the β-carbon. Carbonyl
addition can be a competing reaction. The ratio of 1,4-
addition to carbonyl addition is very sensitive to substitution on the carbonyl carbon, with large substituents
favoring formation of the α,β-adduct. Sodium, lithium,
potassium, and calcium organometallics give mainly carbonyl
addition, and mercury, cadmium, beryllium, and manganese
compounds lead preferentially to 1,4-addition. Addition
of a cuprous salt to the Grignard reagent increases the
1,4-addition yield.

A concerted mechanism can account for most of the observations on this reaction. The metal ion, functioning
as a Lewis acid, coordinates with the carbonyl oxygen and
enhances the polarization of the conjugated system, with
the organic moiety of the reagent attacking at the β-carbon,
as in 51.

$$PhCH=CH\overset{\overset{\textstyle O}{\|}}{C}-Et$$
$$+$$
$$CH_3CH_2MgBr$$

$$\longrightarrow$$

$$\left[\begin{array}{c} \overset{\textstyle Br}{\underset{|}{}} \\[-2pt] \cdots Mg \cdots \\ CH_3\overset{..}{C}H_2 \qquad O \\ \qquad \overset{\|}{} \\ Ph-\overset{..}{C}H \qquad \overset{..}{C}-Et \\ \overset{\diagdown}{}CH \end{array}\right]$$

$$51$$

$$\longrightarrow \quad \underset{\quad}{\overset{\textstyle CH_2CH_3 \quad OMgBr}{\underset{|}{} \qquad \underset{|}{}}}$$
$$PhCH-CH\!\!=\!\!C-Et$$

$$\Big\downarrow H_2O$$

$$\overset{\textstyle CH_2CH_3 \quad OH}{\underset{|}{} \qquad \underset{|}{}}$$
$$PhCH-CH\!\!=\!\!C-Et$$

$$\Updownarrow$$

$$\overset{\textstyle CH_2CH_3 \quad O}{\underset{|}{} \qquad \overset{\|}{}}$$
$$PhCH-CH_2-C-Et$$

$$(10.64)$$

Attack by Nucleophilic Nitrogen, Sulfur, and Oxygen. Our
treatment of the Michael reaction, in which a carbon nucleophile attacks the β-carbon of an olefinic group conjugated with an electron-withdrawing group, is generally
applicable to Ad_N reactions with other nucleophiles.
Usually the rate equation is $v = k_2[olefin][conjugate\ base]$,

where the second bracketed term represents the concentration of the conjugate base of the attacking reagent. For amines this is the neutral form, giving a zwitterionic intermediate 52, which then undergoes protonation, perhaps intramolecularly, perhaps from the solvent.

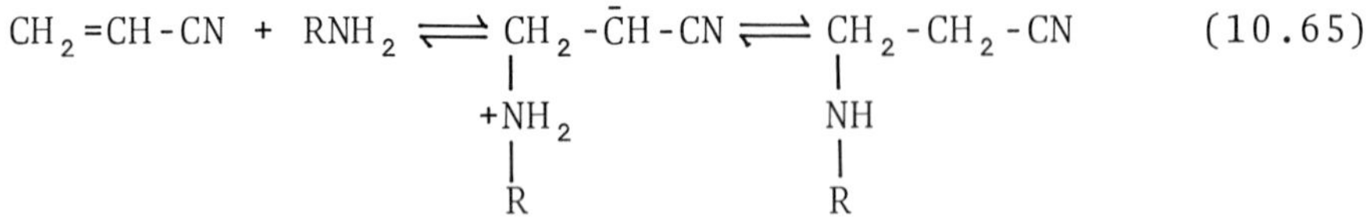

$$CH_2=CH-CN \;+\; RNH_2 \;\rightleftharpoons\; CH_2-\bar{C}H-CN \;\rightleftharpoons\; CH_2-CH_2-CN \qquad (10.65)$$

with the zwitterionic center bearing $+NH_2-R$ and the product bearing $NH-R$.

52

Table 10.VII gives rate data for the addition of morpholine to some activated olefins according to [81]

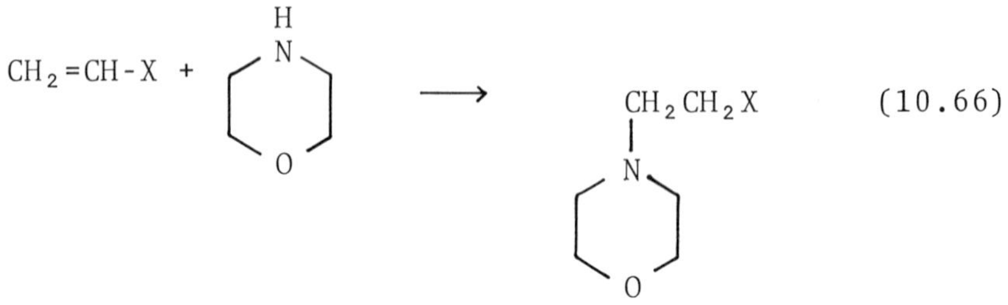

$$(10.66)$$

Table 10.VII   Kinetics of Morpholine Addition to Activated Olefins [81][a]

X in $CH_2=CHX$	Relative rate	$\Delta H^{\ddagger}$/kcal-mole^{-1}	$\Delta S^{\ddagger}$/e.u.[b]
PO(OEt)$_2$	0.014	8.8	-47
CONH$_2$	0.034	8.5	-46
CN	0.536	6.8	-47
CO$_2$Me	(1.00)[c]	5.6	-49
SO$_2$C$_6$H$_4$-Me-p	7.26	6.4	-43
CO$_2$Ph	13.6	5.8	-44
COMe	116	5.9	-40
CHO	123	3.1	-48
SO$_3$Ph	308	5.3	-39
COPh	390	2.4	-47

[a]At 30° in methanol solution.
[b]Molar concentration scale.
[c]$k_2 = 1.04 \times 10^{-2}$ M^{-1}-sec^{-1}.

It is clear that the reactivity differences are determined
almost solely by differences in $\Delta H^{\ddagger}$, the activation entropy
being nearly the same for all olefins. The large negative
value of $\Delta S^{\ddagger}$ may be attributable to the zwitterionic nature
of the transition state, which will immobilize solvent
molecules. The relative reactivities are not successfully
correlated by the usual substituent constants, and it was
suggested that the stabilization of the developing negative
charge at the α-carbon by group X might be the controlling
factor [81]. The acid dissociation of the weak acid CH_3X
was therefore selected as a model process, and reasonable
linear correlations were found for plots of log k_2 for the
addition reaction against pK_a for the model process.

Friedman and Wall [82] have obtained linear free energy
relationships for additions of amino acids to activated
olefins by taking polar and steric effects into account.
It was possible to correlate structural changes in both
reactants.

Thiols react with activated olefins, presumably through
the anionic form. Thiols add to the double bond of maleic
anhydride faster than they attack the anhydride to form the
thiol ester [83].

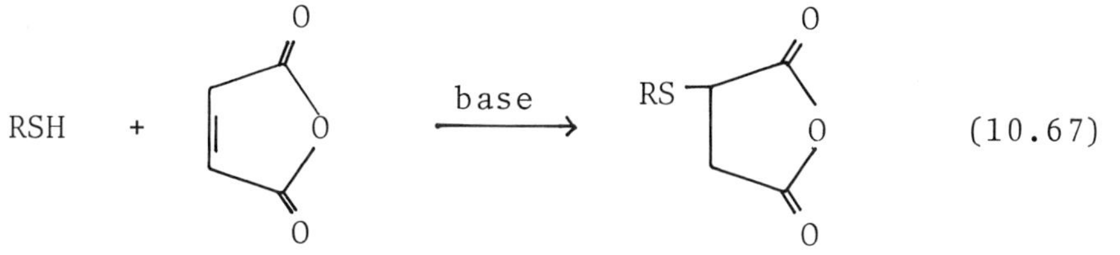

$$(10.67)$$

A similar reaction has been suggested as a basis for the
biological activities of α-methylene γ-lactones [84]. For
example, elephantopin, 53, reacts rapidly with cysteine,
giving the adduct 54.

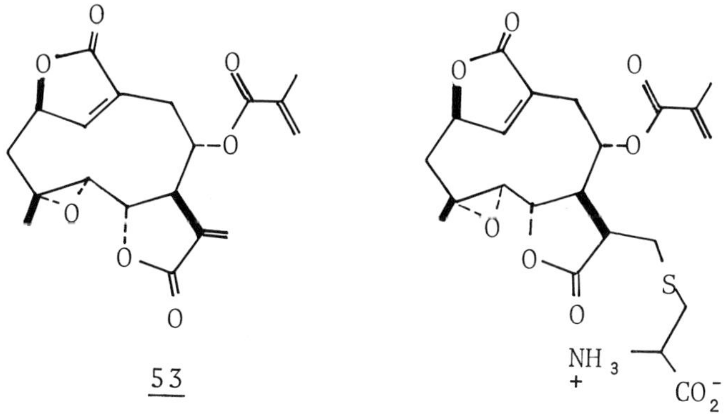

53

54

The addition of thiols to quinones, usually as shown in Eq. (10.68), bears a resemblance to the Ad_N2 reaction, and it could also be viewed as a modified nucleophilic aromatic substitution.

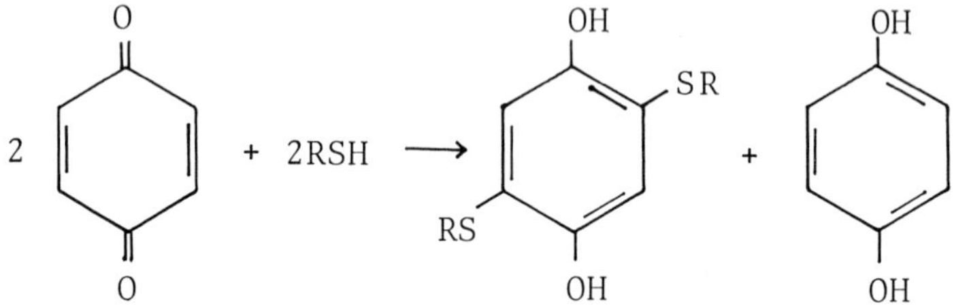

$$ (10.68) $$

Bisulfite adds to activated olefins; Eq. (10.69) shows the reaction with chalcone. With α,β-unsaturated aldehydes, carbonyl addition occurs.

$$PhCH=CHCOPh + NaHSO_3 \rightarrow PhCHCH_2COPh \qquad (10.69)$$
$$\underset{SO_3Na}{|}$$

The stereochemical result of the addition of p-toluene-thiol to 1-p-tolylsulfonylcyclohexene shows that anti-addition occurs, since the product is cis-2-p-tolylmercapto-1-p-tolylsulfonylcyclohexane [85].

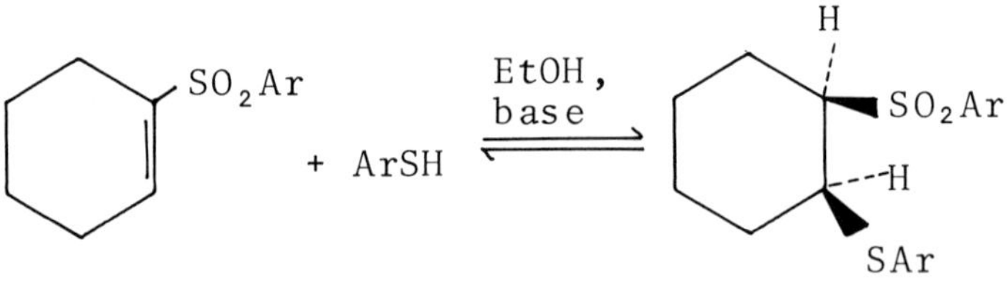

$$ (10.70) $$

No trans adduct was observed. The addition is therefore stereoselective, and it yields the less stable isomer. This was interpreted as follows. The least hindered path is that from the axial direction in the transition state, where tetrahedral geometry is assumed for this state. The arylsulfonyl group will adopt the equatorial position because of its size. Thus the proton is delivered axially by the solvent, giving anti-addition. It was necessary to postulate a concerted process to forestall rearrangement of the cis intermediate anion to the more stable trans anion. This was tested by carrying out the reaction in dioxane with only a limited amount of hydroxylic solvent. Under these conditions largely trans product was isolated, presumably because the limited proton source allowed the usual two-step process to occur, with isomerization of the anion.

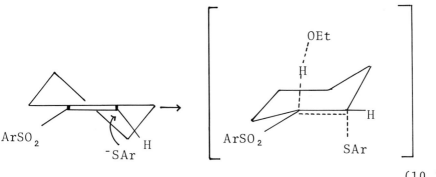

$$(10.71)$$

The epoxidation of an activated olefin, Eq. (10.72), is an Ad_N2 reaction [86].

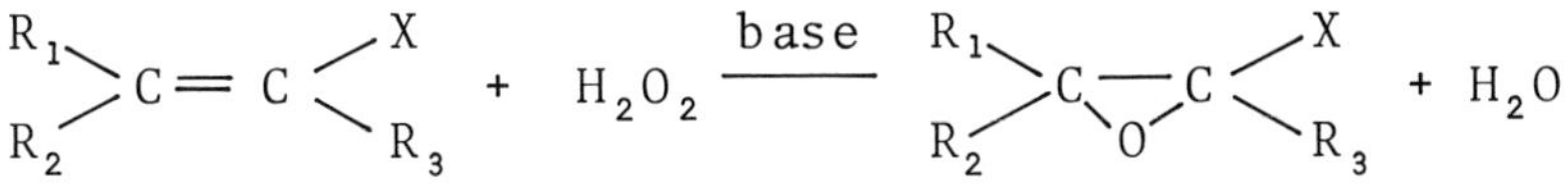

$$\underset{R_2}{\overset{R_1}{>}} C = C \underset{R_3}{\overset{X}{<}} + H_2O_2 \xrightarrow{\text{base}} \underset{R_2}{\overset{R_1}{>}} C \underset{O}{-} C \underset{R_3}{\overset{X}{<}} + H_2O \quad (10.72)$$

The reaction rate law is $v = k_2 [\text{olefin}][HOO^-]$, and the relative rates of epoxidation of mesityl oxide, <u>55</u>, and ethylideneacetone, <u>56</u>, are 1:5.6, respectively.

$$\overset{O}{\underset{\|}{Me_2C=CHC-CH_3}}$$

<u>55</u>

$$\overset{O}{\underset{\|}{MeCH=CHC-CH_3}}$$

<u>56</u>

The rate with <u>55</u> is slower than with <u>56</u> because of the electron-releasing effect of the extra β-methyl group. A reasonable mechanism involves rate-determining attack by peroxide anion followed by ring closure, but a concerted process is not ruled out by the kinetics.

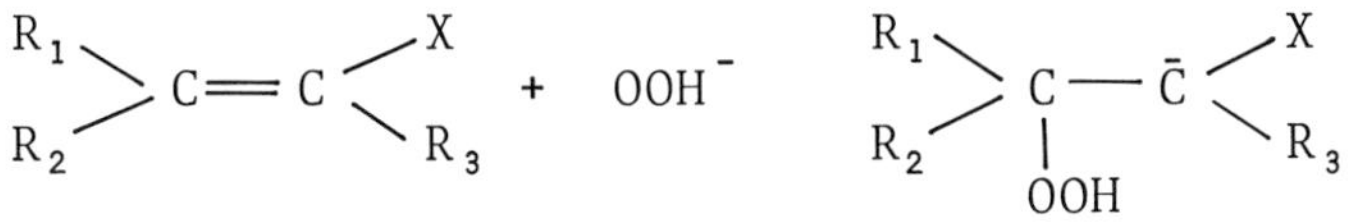

$$\underset{R_2}{\overset{R_1}{>}} C = C \underset{R_3}{\overset{X}{<}} + OOH^- \qquad \underset{R_2}{\overset{R_1}{>}} C - \underset{\underset{OOH}{|}}{\bar{C}} \underset{R_3}{\overset{X}{<}} \qquad (10.73)$$

$$\underset{R_2}{\overset{R_1}{>}} \underset{\underset{O-OH}{|}}{C} - \bar{C} \underset{R_3}{\overset{X}{<}} \qquad \underset{R_2}{\overset{R_1}{>}} C \underset{O}{-} C \underset{R_3}{\overset{X}{<}} + OH^- \qquad (10.74)$$

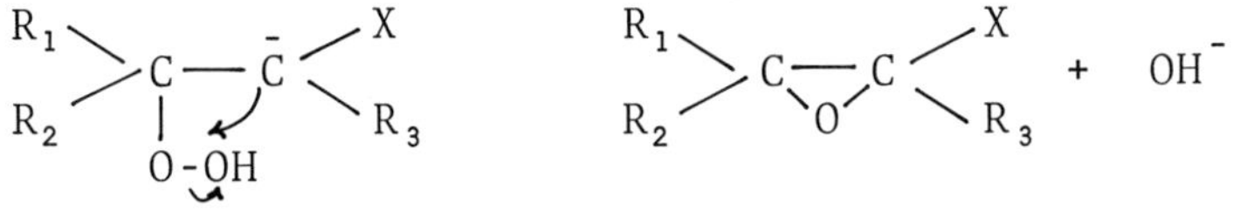

Alcohols add to activated double bonds, giving ethers. These reactions are base catalyzed, as with the other Ad_N2

reactions already considered, the alkoxide ion being the principal nucleophilic species. A kinetic study of the addition of methanol to trans-dibenzoyldethylene has, however, detected an "uncatalyzed" reaction with the un-ionized methanol [87]. The pH-rate profile [88] was fitted over the pH range 4-14 by the equation $k_{obs} = k_1 + k_2[OMe^-]$, with $k_1 = 1.8 \times 10^{-5}$ min^{-1} and $k_2 = 182$ M^{-1}-min^{-1} at 25.0°. It was found that p-substituted trans-dibenzoylethylenes, 57, where X and Y could be the same or different, gave apparent second-order rate constants that could be correlated with the sum of the Hammett substituent constants for the two groups. The direction of the substituent

$$X-\!\!\!\bigcirc\!\!\!-\overset{\overset{\displaystyle O}{\|}}{C}-CH=\!CH-\overset{\overset{\displaystyle O}{\|}}{C}-\!\!\!\bigcirc\!\!\!-Y$$

$$\underline{57}$$

effect is as expected for the Ad$_N$2 reaction. cis-Dibenzoylethylene gave a much slower rate of addition ($k_2 = 4$ M^{-1}-min^{-1}) than its trans isomer; this can be attributed to steric interference in the nonplanar cis system.

The hydration of isolated olefins is an acid-catalyzed reaction, as shown in Sec. 10.1. When the double bond is conjugated with an electron-withdrawing group, it can undergo an Ad$_N$ reaction with hydroxide ion. Moreover, acid catalysis of an Ad$_N$ reaction with water can be observed, in which the function of the acid is to convert the substrate into its conjugate acid, which then undergoes rate-determining attack by water. A possible sequence follows [89].

$$CH_2=CH-CH=O + H^+ \rightleftharpoons CH_2=CH-CH\overset{+}{O}H \longleftrightarrow \overset{+}{C}H_2-CH=CHOH \qquad (10.75)$$

$$\overset{+}{C}H_2-CH=CHOH + H_2O \rightleftharpoons \underset{\underset{\displaystyle +\overset{}{O}H_2}{|}}{CH_2-CH=CHOH} \qquad (10.76)$$

$$\underset{\underset{\displaystyle +\overset{}{O}H_2}{|}}{CH_2-CH=CHOH} \rightleftharpoons \underset{\underset{\displaystyle OH}{|}}{CH_2-CH_2-CHO} + H^+$$

Hydration is a reversible reaction.

The hydration of fumaric acid, giving malic acid, is unusually complex, and is not yet fully understood.

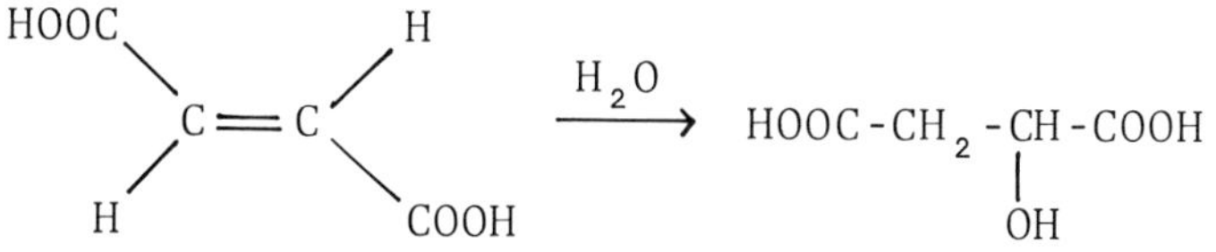

$$
\begin{array}{c}
\text{HOOC} \\
\diagdown \\
\diagup \\
\text{H}
\end{array}
\text{C}=\text{C}
\begin{array}{c}
\text{H} \\
\diagup \\
\diagdown \\
\text{COOH}
\end{array}
\xrightarrow{\text{H}_2\text{O}}
\text{HOOC-CH}_2\text{-}\underset{\underset{\text{OH}}{|}}{\text{CH}}\text{-COOH}
\qquad (10.77)
$$

The reaction is catalyzed by acid and by base [90]. The acid-catalyzed reaction requires a rate term of the form $k_{H^+}^{o}[H_2F][H^+]$, and the base-catalyzed term is $k_{OH^-}^{\prime\prime}[F^{2-}][OH^-]$. Together these do not describe the entire hydration behavior, however, for a bell-shaped pH-rate profile has been observed in the neutral region [91]. This was interpreted in terms of a rate equation including the fumarate monoanion, HF^-:

$$
v = k_{H^+}^{o}[H_2F][H^+] + k'[HF^-] + k_{OH^-}^{\prime\prime}[F^{2-}][OH^-] \qquad (10.78)
$$

The monoanion contribution to the rate was considered to be the consequence of a concerted intramolecular catalysis by a general acid (-COOH) and a general base (-COO⁻) in a one-step process [Eq. (10.79)] or via an intermediate lactone formed by general acid-nucleophilic catalysis [Eq. (10.80)] [91].

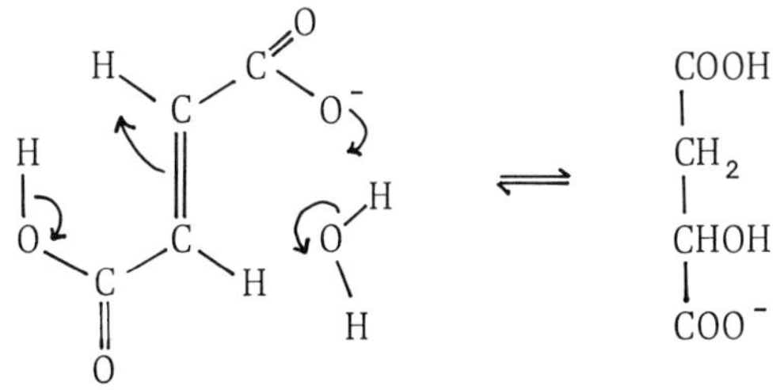

$$
\begin{array}{c}
\text{(structure)}
\end{array}
\rightleftharpoons
\begin{array}{c}
\text{COOH} \\
| \\
\text{CH}_2 \\
| \\
\text{CHOH} \\
| \\
\text{COO}^-
\end{array}
\qquad (10.79)
$$

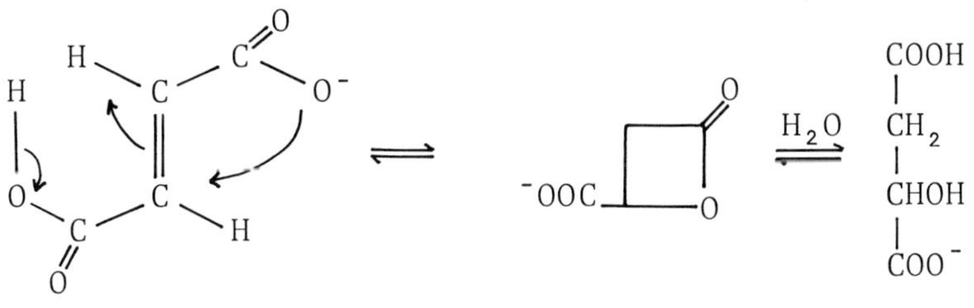

$$
\begin{array}{c}
\text{(structure)}
\end{array}
\rightleftharpoons
\begin{array}{c}
\text{(lactone structure)}
\end{array}
\xrightarrow{\text{H}_2\text{O}}
\begin{array}{c}
\text{COOH} \\
| \\
\text{CH}_2 \\
| \\
\text{CHOH} \\
| \\
\text{COO}^-
\end{array}
\qquad (10.80)
$$

It has recently been suggested [92] that the rate equation should be

$$v = k_H^o [H_2F][H^+] + k_{H_2O}^o [H_2F] + k_{OH}^o - [H_2F][OH^-] +$$
$$k_{OH}^{''} - [F^{2-}][OH^-] \qquad (10.81)$$

This replaces the monoanion term of Eq. (10.78) with an "uncatalyzed" attack by water on un-ionized fumaric acid plus Ad_N2 hydroxide attack on un-ionized fumaric acid. This latter term is kinetically equivalent to the mono-anion formulation, though it is mechanistically very different. Equation (10.81) was preferred to (10.78), partly on the basis that the acid dissociation constants for fumaric acid that are derived from this kinetic treatment [Eq. (10.81)] are in better agreement with pK's extrapo-lated from lower temperatures [93]. Equation (10.81), however, leads to extremely large values for the rate con-stant $k_{OH}^o -$ [94]. One of the reasons for interest in this neutral hydration of fumaric acid is that the enzyme fu-marase catalyzes the hydration reaction under neutral con-ditions, though with much greater efficiency than in the nonenzymatic system. Successful elucidation of the non-enzymatic mechanism might aid in understanding the enzy-matic reaction.

10.3   Cycloadditions

Classification of Cycloadditions. The addition of a mole-cule across a double bond to produce a cyclic structure is a cycloaddition. Several constraints can be added to this description to give a definition that usefully limits the scope of reactions to be considered [95].

    a. Cycloadditions do not involve breaking of σ bonds.
    b. The number of σ bonds is increased in cycloaddi-
       tions.
    c. Cycloadditions are not accompanied by eliminations,
       that is, the product includes all atoms of the
       reactants.
    d. A cycloaddition gives a species corresponding to a
       minimum in the reaction coordinate diagram, that
       is, an intermediate or product, not merely a trans-
       ition state.

A cycloaddition that results in the formulation of one σ bond must be an intramolecular reaction. These have been called electrocyclic reactions, and the stereochemical course of the ring closure (or opening) is now well under-

stood [96]. The reversible inter conversion of cyclobu-
tenes and butadienes is an intramolecular cycloaddition.

$$(10.82)$$

All of the cycloadditions to be considered here involve the formation of two σ bonds. These can be classified in two ways: (1) by the number of atoms contributed
by each reactant to the ring [95]; (2) for addition to a
double bond, by the relative number designation of the
atoms in the addendum that become bonded to the olefinic
carbons [97]. Table 10.VIII illustrates these classification methods with a few examples of cycloadditions to
olefins.

Table 10.VIII   Examples of Olefin Cycloaddition Reactions

Classification		Example
Ring size	Type of addition	
1 + 2 → 3	1,1-addition	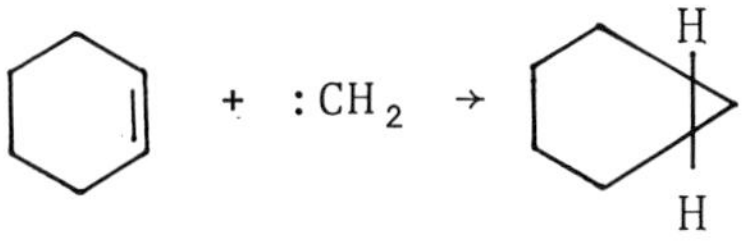
2 + 2 → 4	1,2-addition	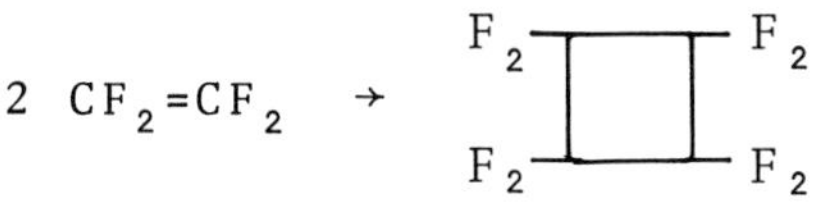 $2\ CF_2=CF_2 \rightarrow$
3 + 2 → 5	1,3-addition	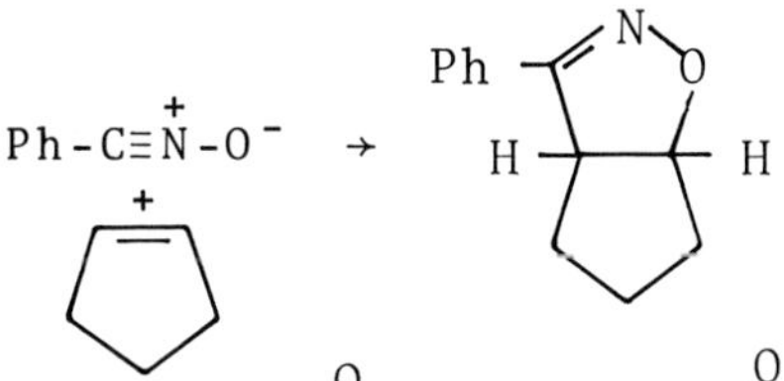 $Ph-C\equiv\overset{+}{N}-O^-$ →
4 + 2 → 6	1,4-addition	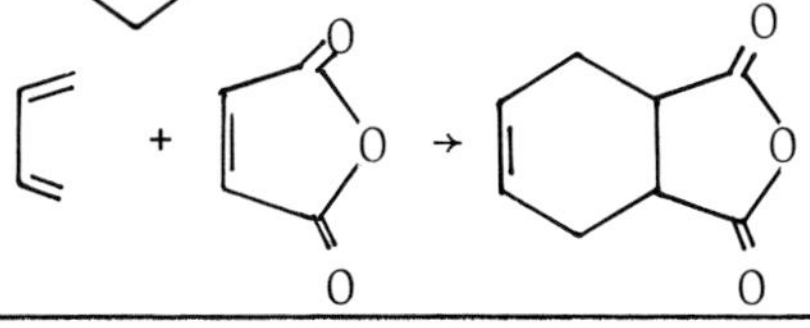

Orbital Symmetry Control in Cycloadditions [98].   Many
potential cycloaddition reactions involve the transformation of two π bonds in the reactants into two σ bonds in
the cycloaddition product.   If one reactant contains m π
electrons and the other contains n π electrons, the reaction is classified as an [m + n] cycloaddition.   (This
classification conforms to that based on ring size, as in
Table 10.VIII, for conjugated reactants.)   Some of these
potential reactions are found to occur readily; others do
not.   The likelihood of occurrence of a concerted (one-
step) cycloaddition can be predicted by means of qualitative arguments based on molecular orbital symmetry [98].
We merely describe the simplest results of these arguments.
     The [2 + 2] cycloaddition of two ethylenes to give cy-
clobutane (and the reverse reaction), Eq. (10.83), is
found to be symmetry-forbidden as a concerted process.

$$CH_2 = CH_2$$
$$+ \qquad \rightleftharpoons \qquad \square \qquad\qquad (10.83)$$
$$CH_2 = CH_2$$

The problem is that in the synchronous transformation of
two π bonds into two σ bonds, electrons in bonding orbitals must pass into antibonding orbitals, a process of low
probability.   It is in fact observed that ethylene does
not yield cyclobutane under thermal conditions; moreover,
even the reverse reaction, which might have been expected
to be facile because of the strain energy of the four-
membered ring, does not proceed readily.   This reluctance
of cyclobutane to isomerize, which had long been a mechanistic puzzle [99], is readily comprehended on the basis
of orbital symmetry control [100].
     On the other hand, the [4 + 2] concerted cycloaddition
of ethylene and butadiene to give cyclohexene is symmetry-
allowed.

$$\| \quad + \quad \rangle \quad \rightleftharpoons \quad \bigcirc \quad\qquad (10.84)$$

The [4 + 2] cycloaddition is one of the best-known reactions in organic chemistry; it is the Diels-Alder reaction, which is discussed later.   The generalized prediction [101] is that a concerted cycloaddition is symmetry-
forbidden if m + n = 4q (q is an integer) and is symmetry-
allowed if m + n = 4q + 2.
     It is observed that tetrahalogenated ethylenes give
cyclobutane derivatives, as in

$$2\ CF_2=CF_2\ \longrightarrow\quad \boxed{\begin{matrix}F_2 & & F_2 \\ & & \\ F_2 & & F_2\end{matrix}} \qquad\qquad (10.85)$$

It follows from the orbital symmetry selection rules that this cycloaddition cannot be concerted. This conclusion had been anticipated on other grounds by Bartlett and his co-workers [102]. For example, the cycloaddition of 1,1-dichloro-2,2-difluoroethylene to butadiene gives only 1,2-addition, with orientation $\underline{58}$ being the only observed product.

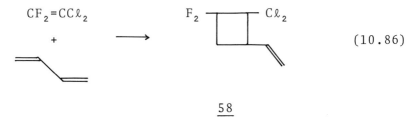

$$\qquad\qquad (10.86)$$

$$\underline{58}$$

Such observations are consistent with a two-step diradical mechanism the intermediate being $\underline{59}$.

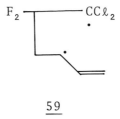

$$\underline{59}$$

It is known that radicals are stabilized more effectively by α-chloro substitution than by fluorine substitution.

For further treatment of molecular orbital symmetry control the cited reviews should be consulted [98]. It is now clear that orbital symmetry conservation imposes constraints that may influence the pathways and even the occurrence of cycloadditions.

Some Cycloaddition Reactions. 1,1-Addition of carbon to olefins gives cyclopropane derivatives. Most of the known reactions seem to involve highly reactive, short-lived divalent carbon species called carbenes or methylenes [103]. The relationship of carbenes to the other carbon intermediates of organic chemistry is shown by the comparisons in Table 10.IX. Carbenes can exist as singlets (the non-bonding electrons having antiparallel spins) or as triplets (parallel spins). Singlet carbenes are electron deficient,

but possess an unshared electron pair; they could therefore function as nucleophiles or as electrophiles.

Table 10.IX   Carbon Intermediates[a]

Intermediate	Number of covalent bonds	Number of valence electrons
Carbanions $\underset{/}{\overset{\textstyle\diagdown}{C}}$:$^-$	3	8
Free radicals $\underset{/}{\overset{\textstyle\diagdown}{C}}$ ·	3	7
Carbonium ions $\underset{/}{\overset{}{C}}$+	3	6
Carbenes $\overset{\textstyle\diagdown}{\underset{/}{C}}$:	2	6

[a]From W. Kirmse, Ref. [103], p. 1. Olah has proposed that
   the trivalent "classical" cations shown above be called
   carbenium ions, and "nonclassical" penta- or tetracoor-
   dinated cations be called carbonium ions, with the generic
   term carbocations applying to both; G. A. Olah, J. Am.
   Chem. Soc., 94, 808 (1972).

The simplest carbene is methylene itself, which can
be produced by the pyrolysis or photolysis of diazomethane,

$$CH_2N_2 \rightarrow N_2 + :CH_2 \qquad\qquad (10.87)$$

or the photolysis of ketene.

$$CH_2=C=O \rightarrow CO + :CH_2 \qquad\qquad (10.88)$$

Methylene adds to the olefinic bond stereospecifically;
that is, a cis-olefin gives a cis-1,2-disubstituted cyclopropane, and a trans-olefin gives a trans-1,2-disub-
stituted cyclopropane. This is consistent with the view
that the addition is concerted (and involves singlet
methylene rather than the diradical triplet state).
   1,1-Cycloaddition to olefins also occurs in the presence of methylene iodide and zinc, but this reaction is
believed to take place via iodomethylzinc iodide or bis-
(iodomethyl)-zinc rather than through methylene.
   Dihalomethylenes have been carefully studied, especially by Hine [103], who showed that they are formed

upon basic hydrolysis of trihalomethanes [104].

$$CHCl_3 + OH^- \rightleftharpoons :CCl_3^- + H_2O \qquad (10.89)$$

$$:CCl_3^- \rightarrow :CCl_2 + Cl^- \qquad (10.90)$$

The addition of dihalomethylenes to olefins is stereo-specific, as illustrated by Eqs. (10.91) and (10.92) [105].

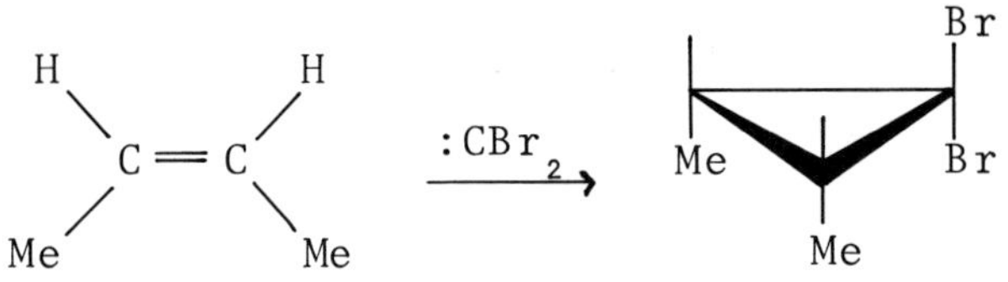

$$(10.91)$$

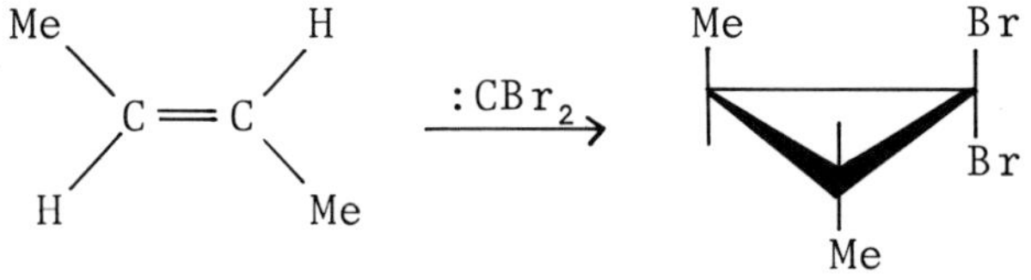

$$(10.92)$$

Relative reactivities indicate that the reagent functions as an electrophile in these cycloadditions, and a concerted three-center mechanism, as in 60, is consistent with the rate and stereochemical results [105,106].

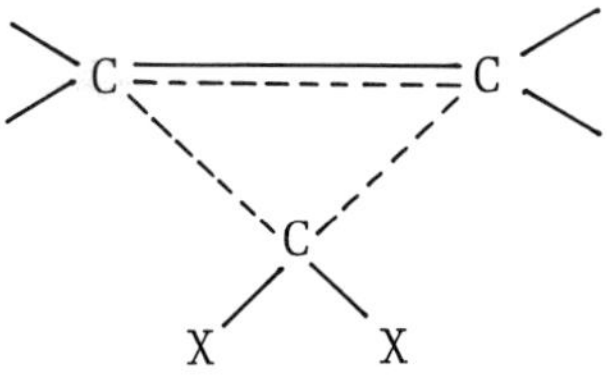

60

1,3-Cycloadditions to olefins include a great range of synthetically useful reactions. cis-Hydroxylations of olefins by osmium tetroxide or by potassium permanganate probably proceed via 1,3-cycloaddition intermediates [107].

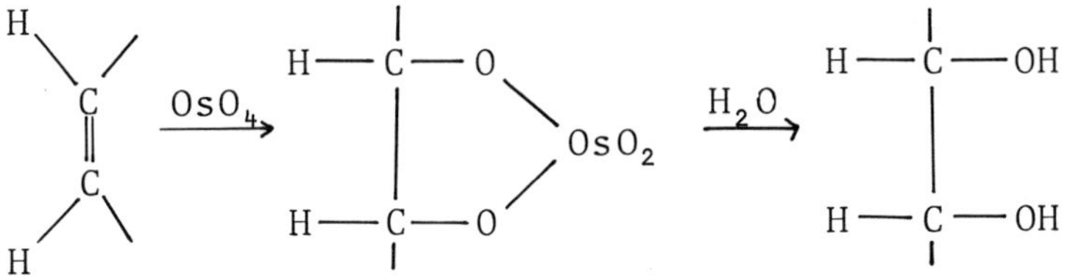

$$(10.93)$$

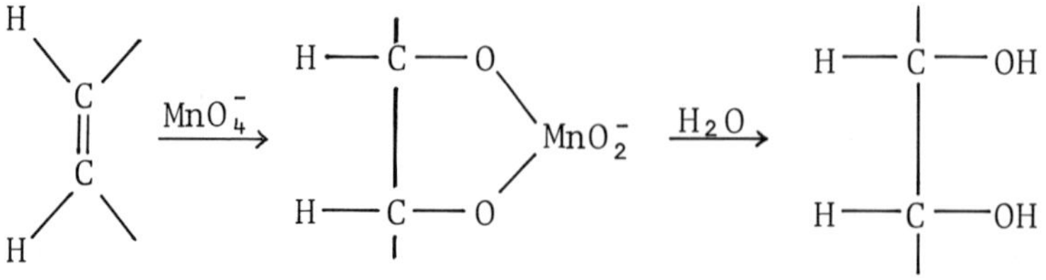

$$(10.94)$$

Huisgen has in recent years exploited, by generalizing it, the 3 + 2 → 5 cycloaddition in which the three-membered reactant carries a formal charge separation. The general reaction is written

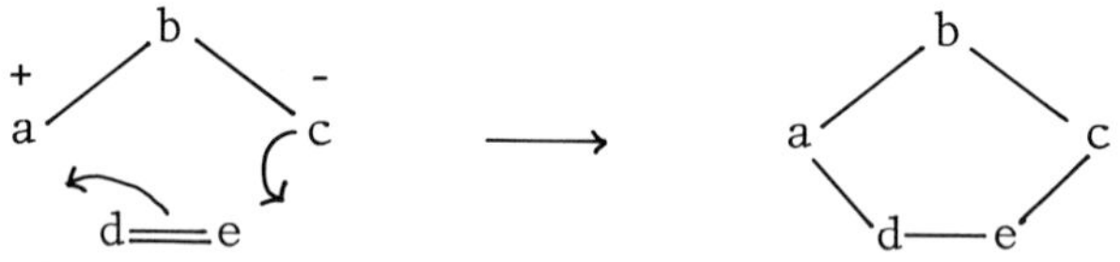

$$(10.95)$$

The a-b-c system is called the 1,3-dipole, the olefinic reactant d=e is the dipolarophile, and the reaction is a 1,3-dipolar cycloaddition [95,108].

In the context of organic synthesis, the important 1,3-dipoles contain carbon, nitrogen, or oxygen as a, b, and c. Atom a is electron deficient in the structure $^+$a-b-c$^-$, but if b carries an unshared pair this deficiency can be relieved as illustrated by the contributing form a=$\overset{+}{b}$-$\bar{c}$. Such species are octet-stabilized 1,3-dipoles. Analogous structures can be written for electron-deficient species containing a double bond:

Without double bond: $\overset{+}{a}$-b-$\bar{c}$ ⟷ a=$\overset{+}{b}$-$\bar{c}$

With double bond: $\overset{+}{a}$=b-$\bar{c}$ ⟷ a≡$\overset{+}{b}$-$\bar{c}$

Some 1,3-diples cannot be stabilized by octet formation; these are highly active species. Table 10.X shows some 1,3-dipoles selected from Huisgen's listings [95].

Additional resonance structures can be written for these 1,3-dipoles, as shown below for a nitrile ylide.

R-$\overset{+}{C}$=N-$\bar{C}$R$_2$ R-C≡$\overset{+}{N}$-$\bar{C}$R$_2$ R-$\bar{C}$=N-$\overset{+}{C}$R$_2$

R-$\bar{C}$=$\overset{+}{N}$=CR$_2$ R-C-N=CR$_2$

Depending upon their relative contributions, of course, these resonance forms suggest that "1,3-dipoles" may not be highly polar because of this balancing of opposing charge distributions.

Table 10.X   1,3-Dipole Structures [95][a]

Octet-stabilized 1,3-dipoles

Left structure		Right structure	Name
$-\overset{+}{C}=N-\overset{-}{C}<$	$\longleftrightarrow$	$-C\equiv\overset{+}{N}-\overset{-}{C}<$	nitrile ylides
$-\overset{+}{C}=N-\overset{-}{\bar{N}}-$	$\longleftrightarrow$	$-C\equiv\overset{+}{N}-\overset{-}{N}-$	nitrilimines
$-\overset{+}{C}=N-\overset{-}{O}$	$\longleftrightarrow$	$-C\equiv\overset{+}{N}-\overset{-}{O}$	nitrile oxides
$\overset{+}{N}=N-\overset{-}{C}<$	$\longleftrightarrow$	$N\equiv\overset{+}{N}-\overset{-}{C}<$	diazoalkanes
$\overset{+}{N}=N-\overset{-}{N}-$	$\longleftrightarrow$	$N\equiv\overset{+}{N}-\overset{-}{N}-$	azides
$\overset{+}{N}=N-\overset{-}{O}$	$\longleftrightarrow$	$N\equiv\overset{+}{N}-\overset{-}{O}$	nitrous oxide
$>\overset{+}{C}-N-\overset{-}{C}<$	$\longleftrightarrow$	$>C=\overset{+}{N}-\overset{-}{C}<$	azomethine ylides
$>\overset{+}{C}-N-\overset{-}{O}$	$\longleftrightarrow$	$>C=\overset{+}{N}-\overset{-}{O}$	nitrones
$\overset{+}{N}-N-\overset{-}{O}$	$\longleftrightarrow$	$-\overset{+}{N}=N-\overset{-}{O}$	azoxy compounds
$\overset{+}{O}-N-\overset{-}{O}$	$\longleftrightarrow$	$O=\overset{+}{N}-\overset{-}{O}$	nitro compounds
$>\overset{+}{C}-\overset{-}{O}-C<$	$\longrightarrow$	$>C=\overset{+}{O}-\overset{-}{C}<$	carbonyl ylides
$>\overset{+}{C}-O-\overset{-}{O}$	$\longleftrightarrow$	$>C=\overset{+}{O}-\overset{-}{O}$	carbonyl oxides
$\overset{+}{O}-O-\overset{-}{O}$	$\longleftrightarrow$	$O=\overset{+}{O}-\overset{-}{O}$	ozone

1,3-Dipoles without octet stabilization

Left structure		Right structure	Name
$-\overset{+}{C}=C-\overset{-}{C}<$	$\longleftrightarrow$	$-C-C=C<$	vinylcarbenes
$-\overset{+}{C}=C-\overset{-}{N}$	$\longleftrightarrow$	$-C-C=N-$	iminocarbenes
$-\overset{+}{C}=C-\overset{-}{O}$	$\longleftrightarrow$	$-C-C=O$	ketocarbenes

---

[a]The name is based on the right-hand electronic structure.

Some of the outstanding features of 1,3-dipolar cycloadditions can be demonstrated by considering a few specific examples. Diphenylnitrilimine, 61, can be generated
in situ from 2,5-diphenyltetrazole. The 1,3-cycloaddition
of diphenylnitrilimine and olefins gives Δ^2-pyrazolines.
Thus ethylene reacts to give 1,3-diphenyl-Δ^2-pyrazoline.

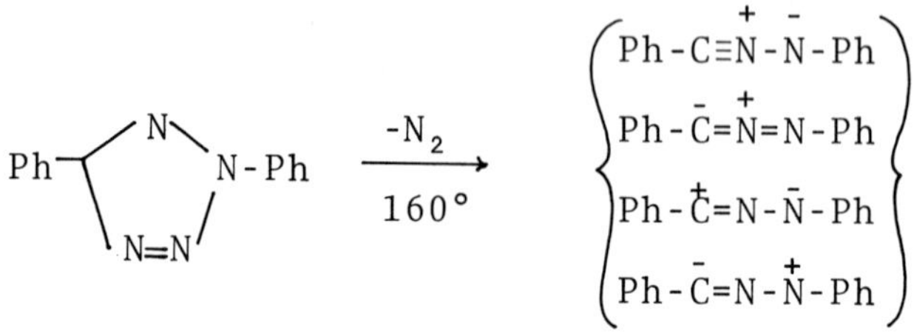

$$\text{Ph} \underset{N=N}{\overset{N}{\diagdown}} \text{N-Ph} \quad \xrightarrow[160°]{-N_2} \quad \left(\begin{array}{c} \text{Ph-C} \overset{+}{\equiv} \text{N-} \overset{-}{\text{N}} \text{-Ph} \\ \text{Ph-} \overset{-}{\text{C}} = \text{N} \overset{+}{=} \text{N-Ph} \\ \text{Ph-} \overset{+}{\text{C}} = \text{N-} \overset{-}{\text{N}} \text{-Ph} \\ \text{Ph-} \overset{-}{\text{C}} = \text{N-} \overset{+}{\text{N}} \text{-Ph} \end{array} \right) \qquad (10.96)$$

$$\underline{61}$$

$$\text{CH}_2 = \text{CH}_2 \ + \ \underline{61} \ \rightarrow \text{Ph} \underset{}{\overset{N}{\diagdown}} \text{N-Ph} \qquad (10.97)$$

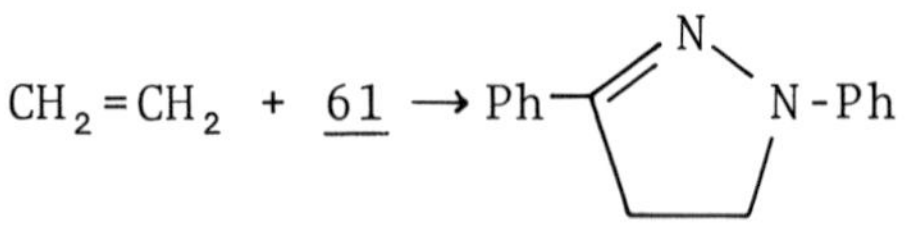

Conjugated olefins react much faster than do unconjugated olefins.

Stereospecific syn-addition is observed with suitably constituted olefins, and this suggests a concerted cyclo-addition. Equations (10.98) and (10.99) show the reaction of diphenylnitrilimine with cis-stilbene and trans-stilbene.

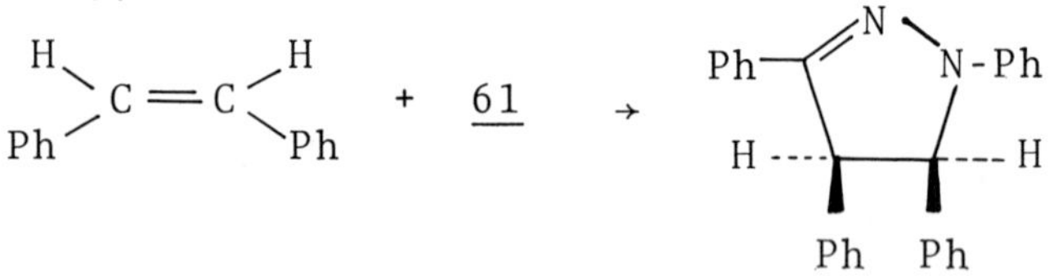

$$\underset{\text{Ph}}{\overset{\text{H}}{\diagdown}} C = C \underset{\text{Ph}}{\overset{\text{H}}{\diagup}} \ + \ \underline{61} \ \rightarrow \quad \text{Ph} \overset{N}{\diagup} \underset{\text{H} \text{---} \underset{\text{Ph}}{\blacksquare} \quad \underset{\text{Ph}}{\blacksquare} \text{---} \text{H}}{} \text{N-Ph} \qquad (10.98)$$

$$\underset{\text{Ph}}{\overset{\text{H}}{\diagdown}} C = C \underset{\text{H}}{\overset{\text{Ph}}{\diagup}} \ + \ \underline{61} \ \rightarrow \quad \text{Ph} \overset{N}{\diagup} \underset{\text{H} \text{---} \underset{\text{Ph}}{\blacksquare} \quad \underset{\text{H}}{\blacksquare} \text{---} \text{Ph}}{} \text{N-Ph} \qquad (10.99)$$

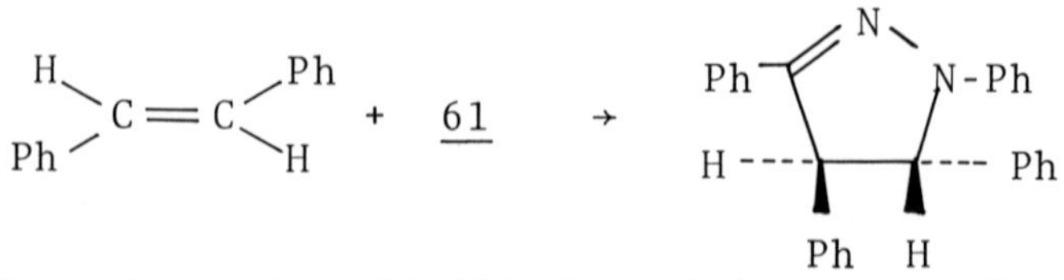

Reaction of a nitrilimine with a monosubstituted ole-fin can, a priori, give two products. In fact, however, the 5-substituted pyrazoline is always produced [95]. Huisgen ascribes this orientation to steric control, sug-gesting that the terminal nitrogen in $\underline{61}$ is more accessible than is the carbon atom. Ulrich, however, suggests that

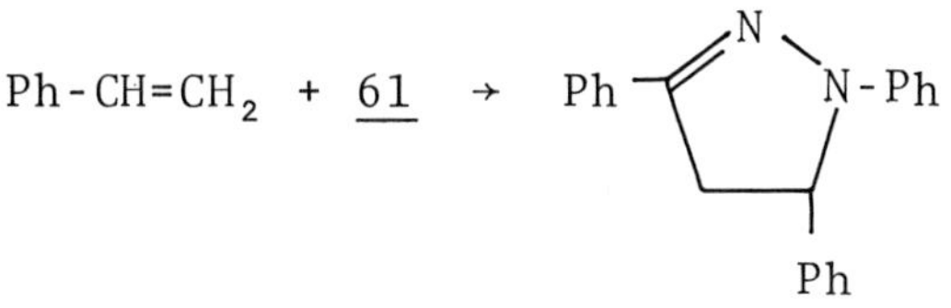

$$Ph\text{-}CH=CH_2 + \underline{61} \rightarrow Ph \overset{N}{\underset{Ph}{\diagdown}} N\text{-}Ph \qquad (10.100)$$

orientation in these reactions may be under orbital symmetry control [109].

Diazoalkanes add to α,β-unsaturated carbonyl compounds and nitroolefins to give pyrazolines. With diazomethane the orientation is as expected for attack of nucleophilic carbon on the electrophilic β-carbon:

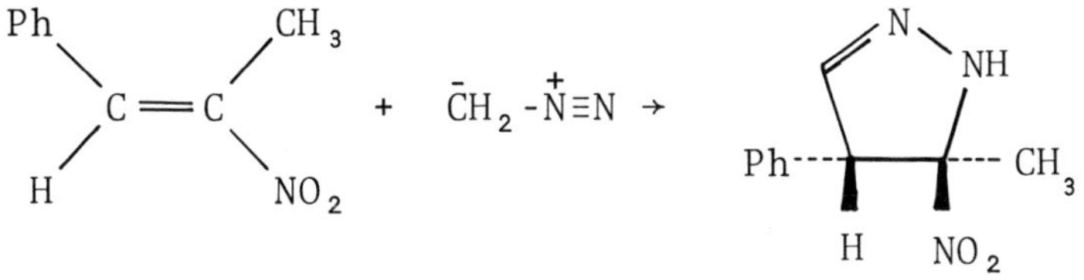

$$\underset{H}{\overset{Ph}{\diagup}}C=C\underset{NO_2}{\overset{CH_3}{\diagdown}} + \bar{C}H_2\text{-}\overset{+}{N}\equiv N \rightarrow \overset{N}{\underset{H \quad NO_2}{\diagdown}}\overset{NH}{\underset{Ph\cdots\quad\cdots CH_3}{}} \qquad (10.101)$$

With diazodiphenylmethane the orientation is reversed. The steric effect may be important here.

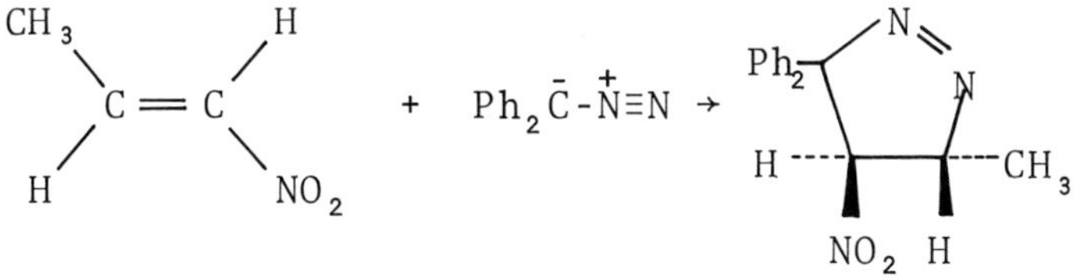

$$\underset{H}{\overset{CH_3}{\diagup}}C=C\underset{NO_2}{\overset{H}{\diagdown}} + Ph_2\bar{C}\text{-}\overset{+}{N}\equiv N \rightarrow \overset{N}{\underset{NO_2 \; H}{\diagdown}}\overset{Ph_2}{\underset{H\cdots\quad\cdots CH_3}{}} \qquad (10.102)$$

The rate data in Table 10.XI show typical relative reactivities of olefins in 1,3-dipolar cycloadditions; these are for reaction with diazodiphenylmethane. Conjugation of the olefinic bond with carbonyl groups is strongly activating. This could be consistent with a primary nucleophilic attack by the 1,3-dipole. Other evidence, however, such as the stereospecificity of the cycloaddition, is better accounted for by a concerted mechanism. The activating influence of conjugated groups has been ascribed to charge dispersal in the cyclic transition state [95]; although the cycloaddition may be concerted, this does not require both of the new σ bonds to form at the same rate, so a net charge separation may exist in the transition state. The greater reactivity of trans versus cis isomers is perhaps due, in part, to their superior capabilities for conjugative interaction.

Table 10.XI   Rate Constants for 1,3-Cycloaddition
of Diazodiphenylmethane to Olefins [95][a]

Olefin	$10^5 k/\underline{M}^{-1}\text{-sec}^{-1}$
Maleic anhydride	5830
Dimethyl fumarate	2450
trans-Dibenzoylethylene	979
Ethyl acrylate	707
Acrylonitrile	434
Dimethyl maleate	68.5
Dicyclopentadiene	3.45
Norbornene	2.86
Ethyl crotonate	2.46
Styrene	1.40
Ethyl cinnamate	1.25
1,1-Diphenylethylene	0.27

[a]At 40° in DMF.

The entropies of activation for 1,3-dipolar cycloadditions are always large negative quantities, of the order -40 e.u. These have been interpreted in favor of a concerted mechanism.

The reaction of ozone with olefins (ozonolysis) may involve two successive 1,3-dipolar cycloadditions [110]. The overall reaction involves the remarkable insertion of ozone between the olefinic carbon atoms to give an ozonide, 62.

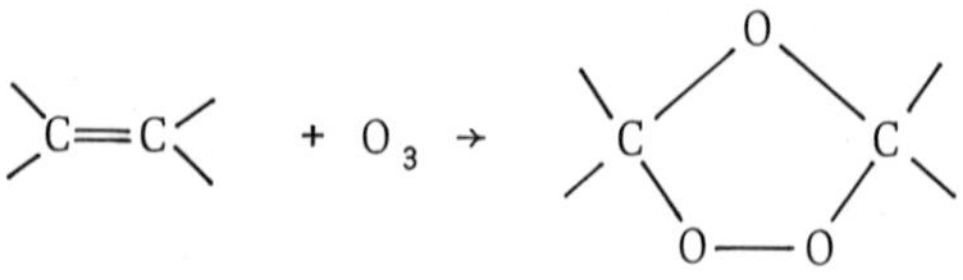

$$\tag{10.103}$$

62

This compound can then be hydrolyzed or reduced to give isolable fragments of the original olefin.

The initial adduct is believed to result from a concerted 1,3-cycloaddition, giving 63, a 1,2,3-trioxolane. Heterolytic cleavage to a carbonyl and a zwitterionic peroxide then can occur:

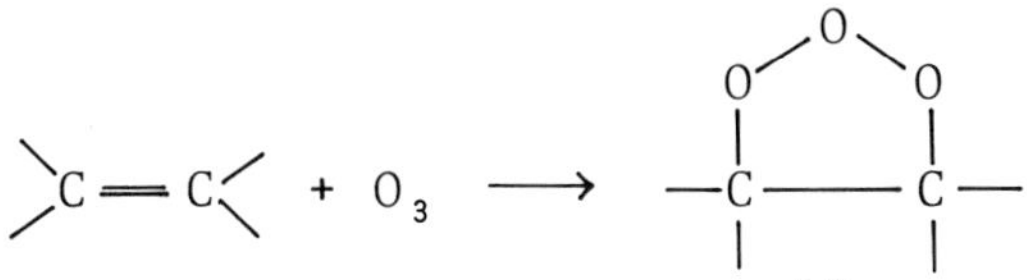

$$\text{>C=C<} + O_3 \longrightarrow \quad \overset{\displaystyle O-O-O}{\underset{\displaystyle -\overset{|}{C}-\overset{|}{C}-}{}} \qquad (10.104)$$

$$\underline{63}$$

$$\overset{O-O}{\underset{-\overset{|}{C}-\overset{|}{C}-}{}} \longrightarrow \overset{O-O^-}{\underset{-\overset{|}{C}-\overset{|}{C}-}{}} \overset{O^+}{} \longrightarrow \overset{O-O^-}{\underset{-\overset{|}{C}^+}{}} + \overset{O}{\underset{C}{\parallel}} \qquad (10.105)$$

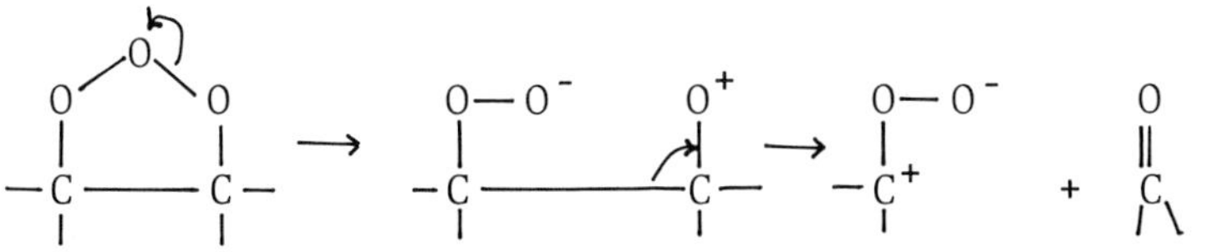

Finally these species can undergo 1,3-dipolar addition to the carbonyl group to give $\underline{62}$.

$$\overset{O}{\underset{-\overset{|}{C}^+}{}} \overset{O=C<}{} \longrightarrow \overset{O}{\underset{\overset{|}{C}}{C}} \qquad (10.106)$$

By postulating the heterolysis (10.105), all of the side products of an ozonolysis reaction can be accounted for.

As the final class of cycloadditions to be considered we turn to 1,4-cycloadditions $(4 + 2 \rightarrow 6)$,

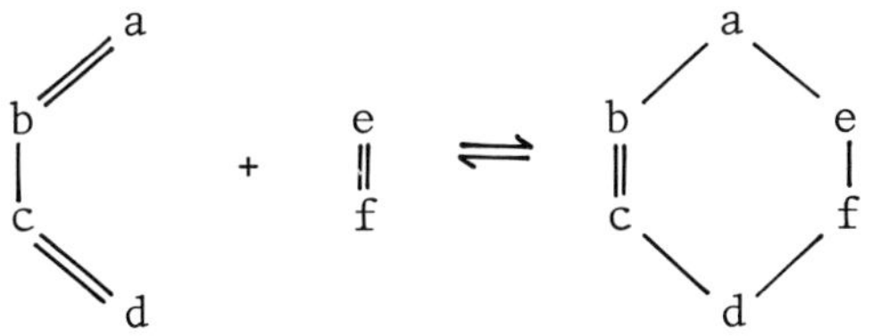

$$\overset{a}{\underset{d}{\overset{b}{\underset{c}{}}}} + \overset{e}{\underset{f}{\parallel}} \rightleftharpoons \overset{a}{\underset{d}{\overset{b}{\underset{c}{}}}} \overset{e}{\underset{f}{}} \qquad (10.107)$$

This is the <u>Diels-Alder</u> reaction [95,111]. In conventional terminology, this is a 1,4-addition of a conjugated <u>diene</u> and a <u>dienophile</u> to form an unsaturated six-membered ring. The Diels-Alder reaction is of great synthetic importance because of its capability for accepting wide variation in the structures of the diene and the dienophile [111,112]. A few examples of the Diels-Alder reaction show its scope. These are written in the general form:

$$\text{Diene + dienophile} \rightleftharpoons \text{adduct}$$

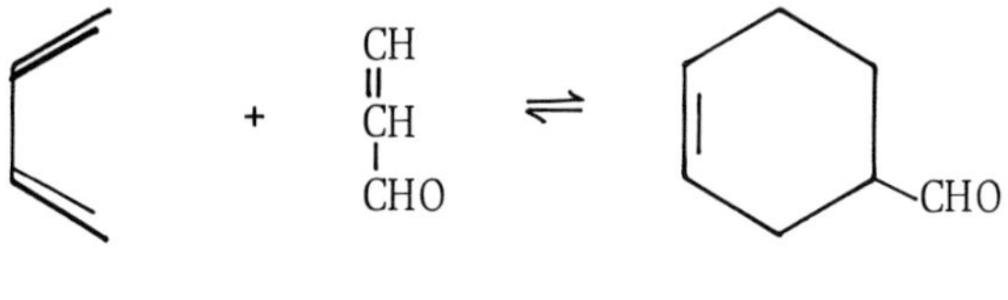

$$\overset{\text{CH}}{\underset{\text{CHO}}{\overset{\text{CH}}{\parallel}}} \rightleftharpoons \overset{\text{CHO}}{} \qquad (10.108)$$

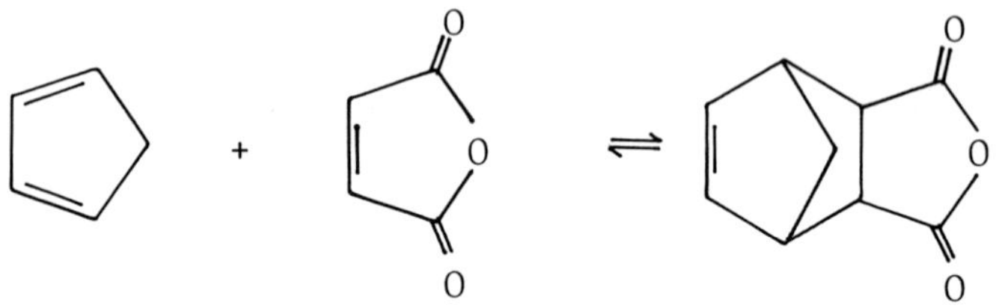

(10.109)

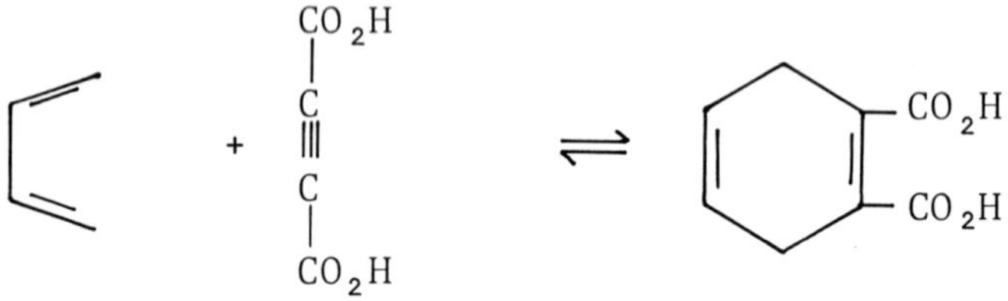

(10.110)

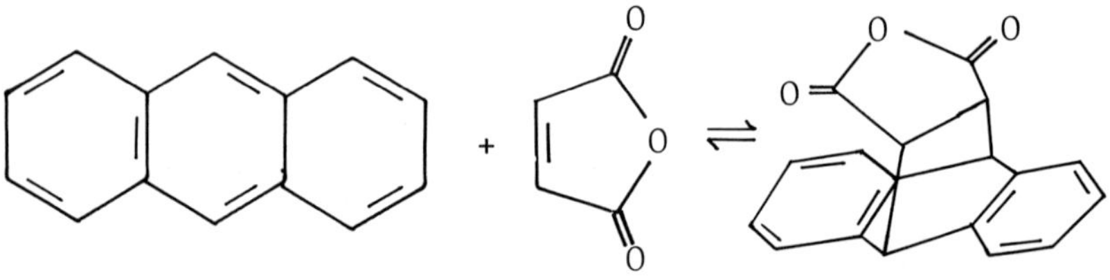

(10.111)

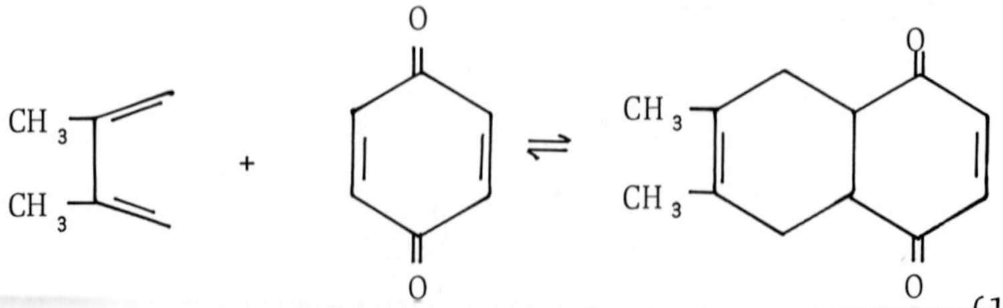

(10.112)

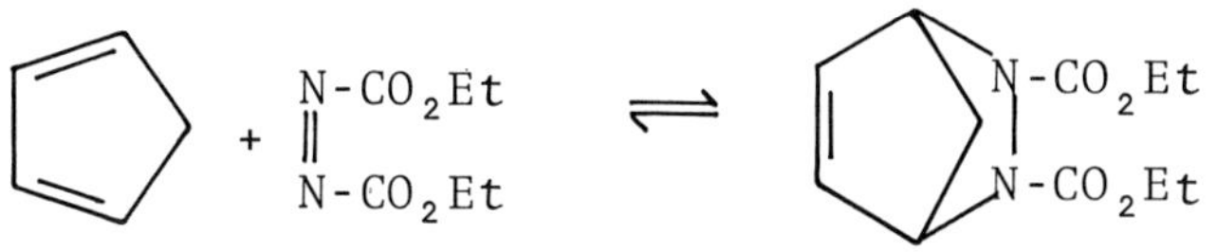

$$(10.113)$$

The Diels-Alder reaction is one of the best-studied organic reactions, but intensive investigation continues and some important questions are still not definitively answered. The weight of evidence favors a concerted mechanism, though the two forming σ bonds may not develop at equal rates. Some of the key observations bearing on the reaction mechanism are reviewed [113].

The reaction is bimolecular, being first order with respect to each reactant. Large negative entropies of activation are observed. The rate is remarkably insensitive to the nature of the solvent, suggesting that the transition state is about as polar as the reactants [114]. The activation volumes of several Diels-Alder reactions of maleic anhydride are more negative than the volume changes of the reactions, indicating that the transition state is smaller than the product [115].

Table 10.XII, listing rate constants for reactions of some dienophiles with cyclopentadiene and of some dienes with maleic anhydride, shows that dienophile activity is enhanced by electron-withdrawing groups, whereas electron-donating groups increase diene reactivity. For these reactions, therefore, the dienophile is functioning as the electrophile and the diene as the nucleophile.

The stereochemical course of the Diels-Alder reaction tells much about the mechanism and, moreover, considerably enhances its synthetic utility [116]. The reaction can only occur with dienes in the s-cis (cisoid) conformation, 64. Any substitutions on an acyclic diene that affect the conformational equilibrium will likewise affect the ease of Diels-Alder cycloaddition. Bulky 2,3-substitution in a 1,3-diene will tend to favor the s-trans conformation 65 and thus to reduce reactivity.

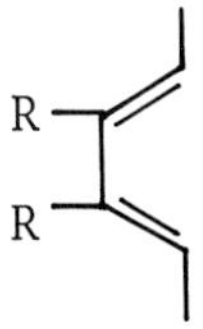

64 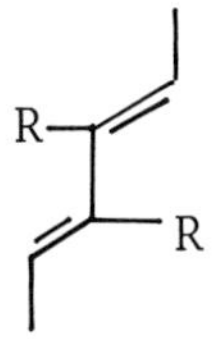   65

Table 10.XII Reactivity in Some Diels-Alder Reactions[a]

Dienophile	$10^5 k^b / M^{-1}\text{-sec}^{-1}$ (reaction with cyclopentadiene)
Tetracyanoethylene	4.3×10^7
1,1-Dicyanoethylene	4.55×10^4
Acrylonitrile	1.04
Dimethyl fumarate	74.2
Dimethyl maleate	0.63

Diene	$10^4 k^c / M^{-1}\text{-sec}^{-1}$ (reaction with maleic anhydride)
Cyclopentadiene	921
2-Phenylbutadiene	6
2-Methylbutadiene	1.54
1-Methoxybutadiene	8.41
Butadiene	0.683

[a] From Ref. [95], pp. 920-921.
[b] At 20° in dioxane.
[c] At 30° in dioxane.

The most outstanding stereochemical feature of the Diels-Alder reaction is that syn-addition always occurs. The evidence for this is that the relative configurations of reactants are retained in the adduct. A typical result is shown by Eqs. (10.114) and (10.115) [117]. The adducts of anthracene with dimethyl maleate and with dimethyl fumarate have, respectively, the cis and trans orientations of the ester groups.

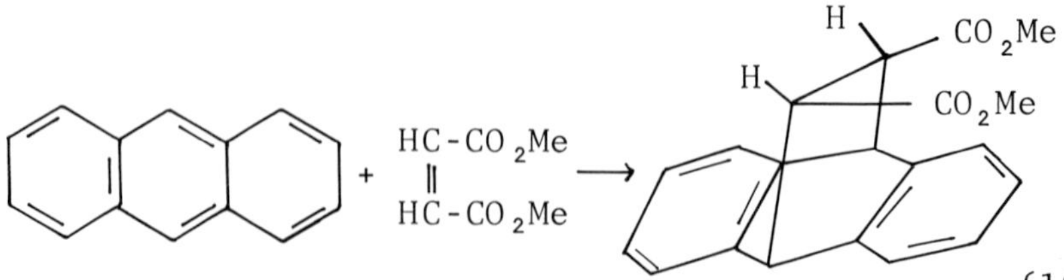

$$ \tag{10.114} $$

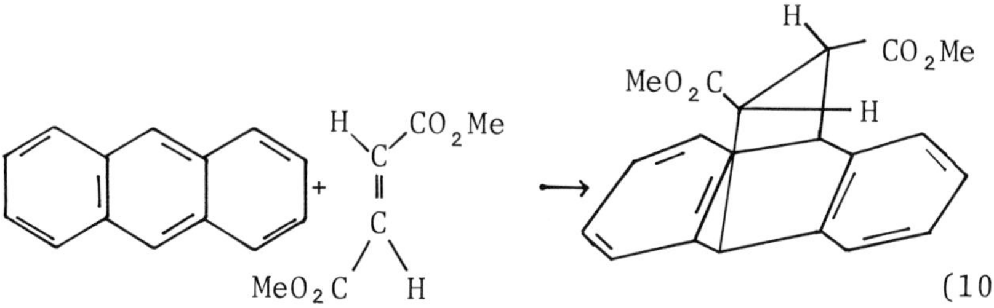

(10.115)

Another type of stereochemical alternative exists with
Diels-Alder reactions between cyclic dienes and dienophiles
lacking a plane of symmetry in the C=C axis. Then the
dienophile substituents can lie either endo or exo in the
bridged bicyclo adduct.

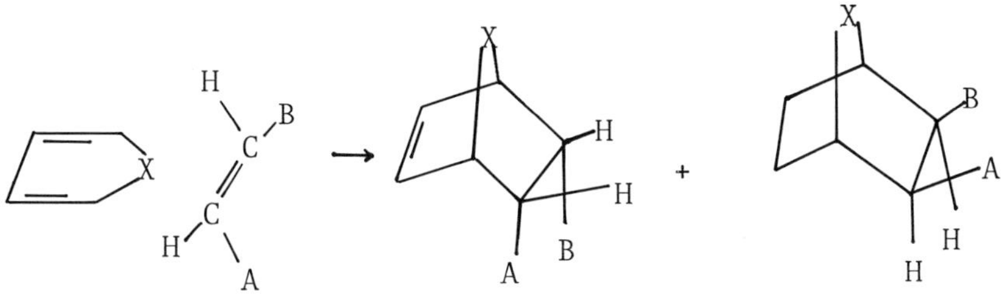

endo   exo   (10.116)

When the product distribution is under kinetic control
(not always an easy condition to assure), the endo adduct
is preferentially, sometimes exclusively, formed. Perhaps
the most thoroughly studied Diels-Alder reaction is that
between cyclopentadiene and maleic anhydride. Only the
endo adduct is formed. Alder proposed that endo addition
was the consequence of a plane-to-plane orientation of
diene and dienophile with "maximum accumulation of double
bonds." Since this same orientation would promote stability in a molecular complex, it has been suggested that
complex formation between the reactants may be responsible
for preferential endo addition [118].
    Woodward and Hoffmann [119], on the other hand, as-
cribe endo addition to interaction of occupied orbitals
with unoccupied orbitals, the endo transition state conformation being favored by orbital symmetry relative to
the exo conformation.
    One further point is of interest. As is implied in

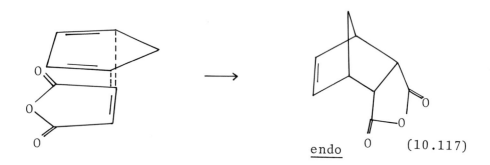

endo   (10.117)

several of the equations given earlier, the Diels-Alder
reaction is reversible, the dissociation of adduct into
the diene and dienophile often being called the retro-
Diels-Alder reaction. (This property has been exploited
synthetically to protect a labile group in a reactant
while a transformation is carried out elsewhere in the
molecule; then the modified diene or dienophile is regenerated in a retro-Diels-Alder reaction.) Since endo
adducts are preferentially formed under kinetic control
in the Diels-Alder reaction, whereas exo adducts are thermodynamically favored, the mechanism of the endo-exo distribution equilibrium has been studied for several adducts.
One possibility is that the interconversion proceeds via
the retro-Diels-Alder dissociation, with recombination in
the alternate conformation. The endo-exo conversion for
the cyclohexadiene-maleic anhydride adduct was shown [120]
to proceed in part by an "internal" mechanism that is different from the dissociation-recombination path. This
internal isomerization might be the consequence of bond-
breaking and -making steps involving other bonds in the
adduct than the σ bonds formed in the Diels-Alder reaction; such endo-exo conversions can be drawn for the cy-
clopentadiene-maleic anhydride adduct [121]. The internal
path for endo-exo isomerization cannot be general for
Diels-Alder adducts, however, because it is absent in the
"isomerization" shown in Eq. (10.118) [121]; in this system, racemization at the asymmetric carbon is equivalent
to isomerization.
      The present view of the Diels-Alder mechanism is that
a concerted four-center transition state best accounts for
the observed features. This is illustrated for the hypothetical reaction, Eq. (10.116), in structure 66.

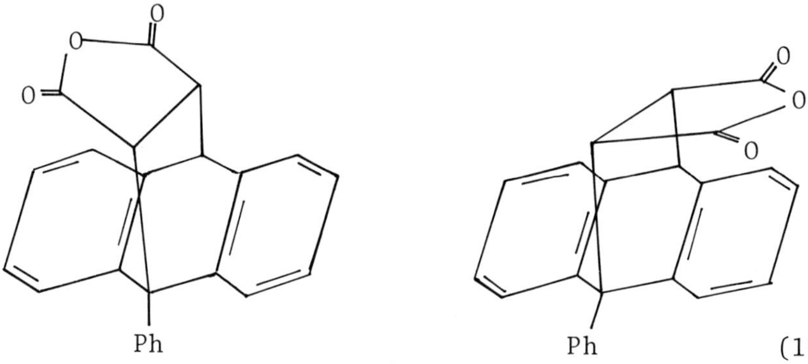

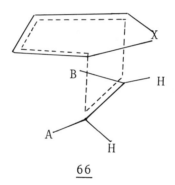

(10.118)

66

This transition state could be symmetrical, in the sense
that the developing σ bonds are about equally formed, or
it could be unsymmetrical, with one bond being formed much
in advance of the other [122]. The structures of the reactants will determine the detailed charge distribution
in the transition state, with strongly dipolar reactants
probably tending to develop unsymmetrical incipient σ
bonds in the cyclic transition state [114].

10.4   Analytical Reactions

Chemical methods for the determination of olefins are
nearly always based on addition reactions; in such analyses
the olefin is of course the sample. A few analyses use
the olefin as the reagent in the determination of substances that add to the double bond. The subject of olefin
determination has been repeatedly reviewed [123-128], and

it is unnecessary to duplicate most of this material.
Enough is described to give a sense of the present analyses based on additions. The organization is by the mechanistic categories used to classify the preceding discussion of the chemistry.

Electrophilic Addition. All analytical halogenations use
bromine or iodine as the electrophilic species that initi-
ates attack on the olefin. Table 10.XIII lists most of
the reagents that have been found useful. A typical procedure calls for the addition of an excess of halogenating
agent to the sample, the elapse of a predetermined reaction time, and the determination of the unreacted halogen
by adding iodide and titrating the resultant iodine with
thiosulfate. Upon reading the literature of the determination of unsaturation by halogenation, one is struck by
the limited applicability of each reagent or procedure.

Table 10.XIII   Analytical Halogenation Reagents
for Olefins

Reagent	Author[a]
ICl in HOAc	Wijs
IBr in HOAc	Hanus
INCS in C_6H_6	Kaufmann and Grosse-Oetringhaus
I_2 in MeOH[b]	Siggia and Edsberg
Br_2 in HOAc	Uhrig and Levin
Br_2 in CCl_4	McIlhiney
Br_3^- in MeOH	Kaufmann
Br_3^- in H_2O	Polgár and Jungnickel
$Br^- - BrO_3^-$ in H_2O[c]	Francis
Br_2 in $CHCl_3$ (aged)	Gal'pern and Vinogradova
$C_5H_5N \cdot H_2SO_4 \cdot Br_2$ in HOAc	Rosenmund and Kuhnhenn
$C_5H_5N \cdot HBr \cdot Br_2$ in HOAc	Rosenmund and Kuhnhenn
$C_5H_5N \cdot CH_3Br \cdot Br_2$ in HOAc	Rosenmund and Grandjean

[a]From A. Polgár and J. L. Jungnickel, Ref. [123].
[b]For vinyl ethers.
[c]Liberates bromine upon acidification.

On the one hand is the possibility of substantially less
than stoichiometric consumption of halogen because of
structural features in the alkene that limit reactivity.
For example, α,β-unsaturated acids are deactivated toward
electrophilic attack, and cannot be analyzed by the usual

halogenation procedures [129]. On the other hand is the
problem of concurrent substitution reactions, in which
hydrogen is displaced by halogen, as in

$$-CH=CH-CH_2- + X_2 \rightarrow -CH=CH-CHX- + HX \qquad (10.119)$$

Substitution is a problem primarily because most halogenation procedures measure the extent of reaction by determining the amount of reagent consumed, rather than the
amount of product formed. Correction procedures based on
an independent estimate of the hydrogen halide produced
have been applied, but these seldom are widely applicable.
Polgár and Jungnickel [123] have described in detail efforts to minimize side-reactions in halogenations [130].
     Despite the inability of any one method to cope with
the full range of alkene samples, limited successes have
been achieved. Moreover, most of the failures, and also
many of the modifications essential for good analytical
results, can be understood in terms of present mechanistic
knowledge. The highly reactive chlorine is not analytically useful because it results in excessive substitution
(and because its solutions are unstable). Iodine by itself is too unreactive for use with unactivated olefins,
but as iodine monochloride and iodine monobromide it is
an effective reagent. Bromine reagents provide the most
widely useful compromise between sufficient reactivity
in addition reactions and minimal concurrent substitution.
     Catalysis of halogenation can be analytically helpful.
An aqueous bromination reagent consisting of bromine and
excess bromide seems to be quite reactive, particularly
for oxygen-containing olefins. This could be an attack
by the tribromide species, or it could be a bromide ion-
catalyzed attack by bromine. The literature contains many
reports of the catalysis of halogenation by mercury (II)
salts. The mechanism of this catalysis apparently has not
been elucidated. Fritz and Hammond [131] suggest facilitation of electrophilic attack by the species $HgBr_2^{2+}$.
Another possibility is an initial electrophilic attack by
mercury (II), as in the oxymercuration reaction. Subsequent events are less obvious. From the analytical point
of view it is not necessary that the overall reaction be
the addition of halogen to the double bond; some process
that results in the equivalent consumption of halogen may
take place. It is known, however, that halogens can catalyze the cleavage of carbon-mercury bonds [132].
     The extent of interference by substitution may be
minimized by direct titration, in which the alkene is not
exposed to a high concentration of halogen for a prolonged
period. Several methods have been developed. Titrations
with coulometrically generated bromine have used ampero-

metric [133] and spectrophotometric [134] end-point detection.   Fritz and Wood [135] carried out direct spectrophotometric titrations of olefins dissolved in 85%
acetic acid-10% water-5% carbon tetrachloride, using as
titrant bromine in glacial acetic acid.   Isolated double
bonds reacted quantitatively, and olefins deactivated
toward electrophilic attack did not react, thus affording
some opportunity for selective analyses.   A few olefins
underwent apparent substitution besides the normal addition.   Conjugated dienes added only a single mole of bromine, but unconjugated dienes added two moles.   Mercury
(II) catalyzes these reactions, but its use tends to increase interference from slower reacting olefins in the
sample.
    An indirect spectrophotometric procedure measures the
decrease in absorbance at 410 nm (this is due to tribromide ion) upon reaction with the olefin sample [135].
About 1 µeq of olefin could be determined.   Greater sensitivity (down to 0.025 µeq) is obtainable by making the
measurement at 270 nm.   A more sensitive method for olefin
impurities in alkanes is based on the addition of [82]Br-
labeled bromine to the alkene; the excess reagent is extracted from the mixture and the organic phase is counted
[136].   A qualitative analysis of the products, after
reaction with a reagent containing higher [82]Br activity
than in the quantitative studies, showed, somewhat dis-
turbingly, that substitution rather than addition had
occurred.   This was attributed to free radical reactions
induced by the radioactive reagent.
    An unusual reaction forms the basis of an analysis for
alkyl vinyl ethers; Siggia and Edsberg [137] treat the
olefin with iodine in aqueous methanol, and write the
reaction as

$$RO\text{-}CH{=}CH_2 + I_2 + CH_3OH \rightarrow RO\text{-}\underset{\underset{OCH_3}{|}}{CH}\text{-}CH_2I + HI \qquad (10.120)$$

The product is an iodoacetal.   Water appears to be necessary for a facile reaction.   It is easy to understand that
Eq. (10.120) describes one of the reactions in this system
but that it is the only reaction is not expected.   Since
iodine addition is reversible [138], the function of the
alcohol (in an analytical sense) may be to effect complete
conversion of the olefin to the stable adduct by diversion
through the onium ion intermediate.   This intermediate may
be stabilized by neighboring group participation as indicated in 67 [139].
    Siggia [140] has analyzed binary mixtures of olefins
by a differential reaction rate method, using bromination

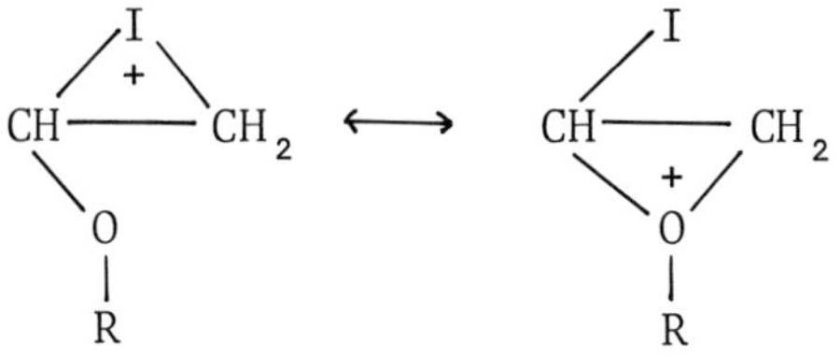

67

in a solvent selected to give a convenient rate; the reaction was followed spectrophotometrically under second-order conditions. It is not surprising that maleic acid and fumaric acid could be analyzed in their mixtures, because it is reasonable that such a major structural difference might have significant kinetic consequences (but see Table 10.II). It is remarkable, however, that mixtures of methyl oleate and ethyl oleate could be analyzed on the basis of their different rates of bromination. The formula of oleic acid is $CH_3(CH_2)_7CH{=}CH(CH_2)_7COOH$. It is not possible for the differential polar effects of the carbomethoxy and carboethoxy groups, transmitted through seven methylene groups, to produce a rate difference of twofold or greater (the difference needed for this determination). Possible reasons for the observed rate difference could be (a) different conformational effects in which the double bond is differently shielded by a carbomethoxy and a carboethoxy group; (b) different extents of addition versus substitution in the two esters; (c) a reaction similar to that between bromine and γ,δ-unsaturated esters, in which a δ-bromo-γ-pentanolactone is formed with the production of an alkyl bromide [141].

Oxymercuration, which was discussed in Sec. 10.1, is a good analytical reaction. Several methods have been developed [142]. Most of these use mercuric acetate as the reagent:

$$\text{-CH=CH-} + Hg(OAc)_2 + MeOH \rightarrow \underset{\displaystyle OMe}{\underset{|}{\text{-CH}}} - \underset{\displaystyle HgOAc}{\underset{|}{\text{CH-}}} + HOAc \qquad (10.121)$$

In one of the procedures the addition reaction is carried out with an excess of mercuric acetate. The unreacted reagent is then converted to mercuric chloride by adding sodium chloride. Finally the acetic acid produced in the addition reaction is titrated with standard alkali. In another modification [143] a known volume of mercuric acetate solution is added to the sample. After the addition is complete, the mixture is diluted with 1:1 propylene glycol-chloroform and is titrated with the standard hydrochloric acid in the same solvent. The excess reagent consumes two moles of titrant acid:

$$Hg(OAc)_2 + 2HC\ell \rightarrow HgC\ell_2 + 2HOAc$$

whereas the adduct consumes one mole:

$$\begin{matrix} -CH\!\!-\!\!CH- \\ |\quad\quad| \\ OMe\ \ HgOAc \end{matrix} + HC\ell \rightarrow \begin{matrix} -CH\!\!-\!\!CH- \\ |\quad\quad| \\ OMe\ \ HgC\ell \end{matrix} + HOAc$$

Marquardt and Luce, who introduced one of the early procedures [144], subsequently modified the method of Martin [145] to produce a very simple method applicable to isolated double bonds and some allylic compounds.

The analytical possibilities of oxymercuration do not seem to have been fully exploited yet. It is usually stated that these methods are applicable (with some exceptions) to isolated double bonds and some cis-alkenes. Slow and incomplete addition reactions limit the technique. Perchloric acid catalyzes the reaction, permitting some extension of the analysis [146]. The data in Table 10.VI show, however, that oxymercuration can be an extremely fast reaction, and if the conditions for fast reaction rate can be combined with those for the final analytical measurement, a very general method for olefins should result [147]. This might be done by spectrophotometric titration, or by a similar spectrophotometric technique, for Halpern and Tinker [52] have demonstrated the feasibility of such titrations.

Table 10.XIV lists some of the other electrophilic addition reactions that have been applied analytically. The hydration reactions give aldehydes or ketones as the ultimate products, and the analysis is completed by measuring this product.

$$RO-CH\!=\!CH_2 + H_2O \rightarrow CH_3CHO + ROH \qquad (10.122)$$

$$\text{(cyclic ether)} + H_2O \rightarrow HO-(CH_2)_3-CHO \qquad (10.123)$$

$$RC\!\equiv\!CR + H_2O \rightarrow R-\overset{\overset{\displaystyle O}{\|}}{C}-CH_2R \qquad (10.124)$$

Conjugated dienes can be determined spectrophotometrically by reaction with aromatic diazonium salts to form colored products [148]. This is not an addition, however, but a substitution, the products having the structure R-CH=CH-CH=CH-N=N-Ar.

Table 10.XIV Some Analytical Electrophilic Additions

Reagent	Sample	Reference
H_2O	vinyl ethers	a
H_2O	α,β-unsaturated epoxides	b
H_2O	alkynes	c
CH_3OH	alkynes	c
HCOOH	dicyclopentadiene	d
HOCℓ	alkenes	e
$(SCN)_2$	isolated double bonds	f
NOCℓ	alkenes	g

[a] Siggia, Ref. [140], p. 401.
[b] Cheronis and Ma, Ref. [142], p. 173.
[c] Siggia, p. 365.
[d] Polgár and Jungnickel, Ref. [123], p. 320.
[e] Polgár and Jungnickel, p. 375.
[f] Polgár and Jungnickel, p. 298.
[g] Polgár and Jungnickel, p. 292.

   An addition reaction of great synthetic and analytical importance is the hydrogenation (reduction) of unsaturation.

$$\diagdown C = C \diagup + H_2 \rightarrow -\underset{H}{\overset{|}{C}}-\underset{H}{\overset{|}{C}}- \qquad (10.125)$$

Catalytic hydrogenation, the usual procedure, utilizes transition metal catalysts in a heterogeneous system. It appears that the olefin is adsorbed to the solid catalyst surface, and the hydrogen is added in a two-step free radical reaction. Syn-addition predominates, presumably because one side of the adsorbed olefin is inaccessible. Catalytic hydrogenation is of great generality, but most procedures are time consuming and require elaborate apparatus and close attention to procedural details. When many analyses are required, catalytic hydrogenation becomes feasible for olefin determination, and is the method of choice. For occasional use, however, probably the most convenient hydrogenation method is that of Brown [149], in which hydrogen is generated from sodium borohydride and the hydrogenation catalyst is prepared in situ.

Nucleophilic Addition.   An unsaturated bond conjugated
with an electron-withdrawing group is susceptible to
attack by a nucleophile, as described in Sec. 10.2.   Addition of a nitrogen, oxygen, or sulfur nucleophile to an
alkene is in effect an alkylation of the heteroatom.   Some
very effective analytical methods are based on the Ad_N
reaction.
     The addition of morpholine is shown in Eq. (10.66).
In the analytical procedure based on the reaction [150],
the olefin is reacted with excess morpholine.   The unreacted morpholine is then acetylated, and the tertiary
amine adduct is titrated with standard acid.   Reaction
rates could be increased by adding acetic acid to the
addition reaction mixture, and it appeared that acetic
acid is more effective than is hydrochloric acid.   Since
the rate of addition will be decreased by protonation of
the morpholine, it is expected that hydrochloric acid
should decrease (not increase) the rate, and that this
effect should be greater than with acetic acid.   The observed rate increase may suggest that a proton-transfer
contributes to the rate-determining step.   Part of the
rate increase could be a medium effect.
     The analytical observation is that if -X in $CH_2=CH-X$
is -CN, -COOH, $-CONH_2$, or -COOR the alkene adds morpholine
readily, whereas if -X is -CHO or -C(O)R the reaction is
too slow for analytical use [150].   These results are not
consistent with the kinetic data of Table 10.VII on morpholine addition to activated olefins [81].   These data
show that the opposite order of addition rates holds in
methanol solution [151].   Possibly the analytical finish
is responsible for this apparent discrepancy, or carbonyl
addition may be occurring.
     Ketenes can be determined by reaction with aniline
[152].

$$>C=C=O + Ph-NH_2 \rightarrow \underset{\underset{\underset{Ph}{|}}{\underset{NH}{|}}}{>CH-C=O} \qquad (10.126)$$

This is formally similar to an Ad_N reaction to the alkene
group, but it is probably more accurate to regard this as
an initial addition to the carbonyl group; and in fact the
orientation shows that the olefinic polarization is opposite to that in a conjugated system [153].
     A colorimetric determination of tertiary amines has
been reported in which upon treatment with cis-aconitic
anhydride, 68, a color develops with maximum absorption
at 500 nm [154].

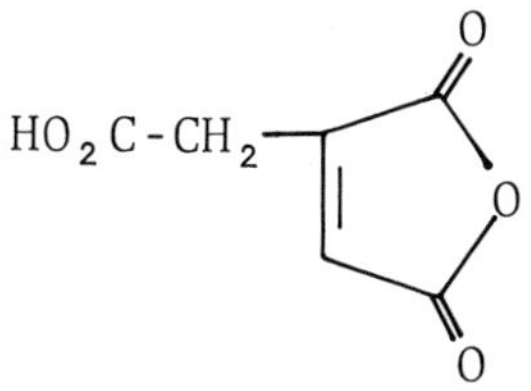

HO$_2$C-CH$_2$-   **68**

Primary and secondary amines do not interfere, perhaps because they undergo Ad$_N$ reaction with the activated double bond. The nature of the reaction with tertiary amines is not known. The reagent is an aged solution of aconitic anhydride in acetic anhydride. This would be an interesting reaction for mechanistic study. A qualitative color test for tertiary amines utilizes a reagent of either citric acid, malonic acid, or aconitic acid heated in acetic anhydride [155]; the chemistry is probably the same as in the aconitic anhydride method.

The Ad$_N$2 reaction of thiols with activated olefins can serve to analyze either the olefin or the thiol. A spectrophotometric method for thiols uses their nucleophilic addition to N-ethylmaleimide (NEM) [156].

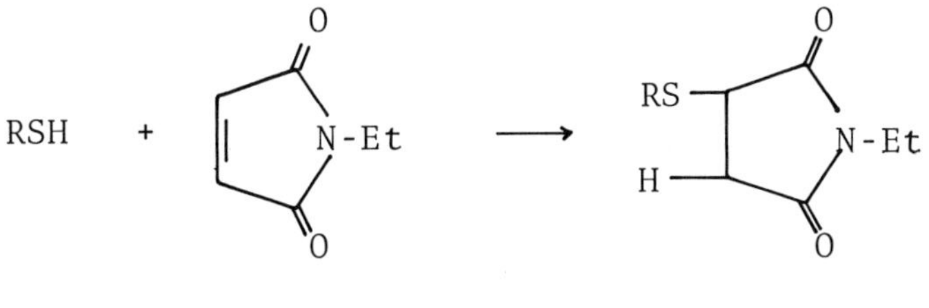

RSH   +   (N-Et) $\longrightarrow$ (RS, H, N-Et)   (10.127)

The absorption of light by NEM is decreased upon the addition of the thiol, which interrupts the system of conjugated double bonds. The NEM absorption (λ_{max} 300 nm, ε_{max} 620) disappears in the product, so the decrease in absorption is a measure of the thiol concentration [157]. An interesting feature of this method is its independence of a sample of the thiol to establish an absorptivity or a standard curve. Only the NEM absorbs at 300 nm, and so NEM serves as the standard substance. A modification of the method increases its sensitivity by carrying out the addition reaction in isopropyl alcohol, which is then made

alkaline. A red color develops whose intensity is proportional to the original thiol concentration [158].
    Acrylonitrile can be determined by reaction with a
thiol [159], and the same reaction can be used to determine thiols [160].

$$RSH + CH_2=CH-CN \rightarrow CH_2CH_2-CN \atop \underset{SR}{|} \qquad (10.128)$$

Sodium bisulfite addition provides a further method for
the determination of activated olefins [161].
    A fluorescence method for determining N-terminal
groups in peptides makes use of an $AdN2$ reaction at one
stage. The primary amino group is converted to the sul-
fonamide 69, which upon treatment with base produces a
strongly fluorescent substance postulated to be 70 [162].

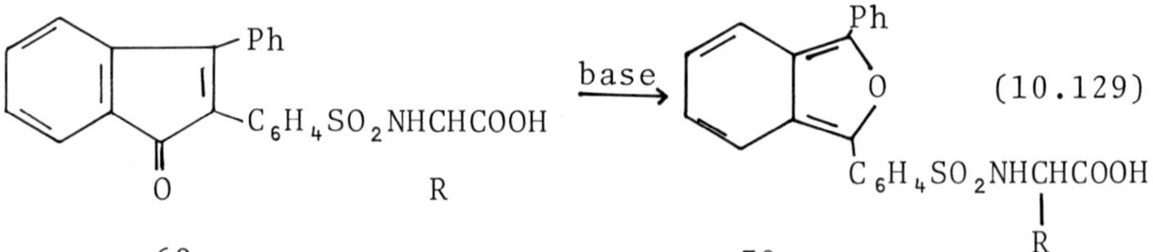

The initial step in this transformation is believed to be
addition of alkoxide to the activated double bond [163].

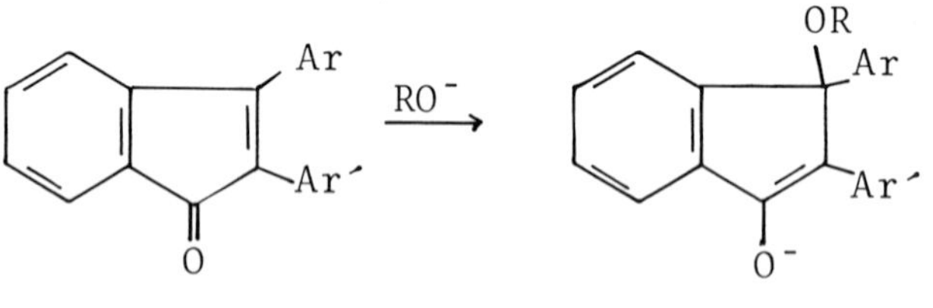

The enolate form of the adduct then undergoes rearrangement to give the fluorophor.
    A final example is the vinyl substitution, Eq. (10.131)
which is used to determine vinyl halides [164].

$$CH_2=CH-X + OCH_3^- \rightarrow CH_2=CH-OCH_3 + X^- \qquad (10.131)$$

The two possible mechanisms for this reaction [165] are an
addition-elimination route,

$$\text{CH}_2\text{=CH-X} \xrightarrow{\text{OMe}^-} \underset{\overset{|}{\text{OMe}}}{\bar{\text{C}}\text{H}_2\text{-CH-X}} \xrightarrow{-\text{X}^-} \text{CH}_2\text{=CH-OMe} \qquad (10.132)$$

or an elimination-addition pathway,

$$\text{CH}_2\text{=CH-X} \xrightarrow{-\text{HX}} \text{HC}\equiv\text{CH} \xrightarrow{\text{MeOH}} \text{CH}_2\text{=CH-OMe} \qquad (10.133)$$

The relative importance of these mechanisms depends upon the nucleophile and the structure of the olefin.

Cycloaddition. Some 1,3- and 1,4-cycloadditions are analytically useful. Ozonolysis is a classical technique for locating carbon-carbon double bonds in an unknown structure [166]. As shown in Eq. (10.104), the initial adduct contains a five-membered ring formed in a 1,3-cycloaddition. This then gives the ozonide 62, which can be decomposed to fragments [aldehydes or ketones, as in Eqs. (10.134) and (10.135)] that represent the two parts of the original alkene molecule.

$$\text{RCH=CHR}' \xrightarrow{\text{ozonolysis}} \text{RCHO} + \text{R}'\text{CHO} \qquad (10.134)$$

$$\text{R}_1\text{R}_2\text{C=CHR}_3 \xrightarrow{\text{ozonolysis}} \text{R}_1\text{R}_2\text{C=O} + \text{R}_3\text{CHO} \qquad (10.135)$$

By combining ozonolysis with gas chromatography it has proved possible to carry out the ozonolysis reaction and the analysis of the carbonyl fragments on microgram quantities of material [167]. Limited quantitative use has been made of the ozonolysis method [168].

Some 1,3-dipolar cycloadditions of the type investigated by Huisgen are used in colorimetric determinations of olefins. Aldrin reacts with phenyl azide to give a phenyldihydrotriazole derivative 71, which then is coupled with an aromatic diazonium salt to give a colored compound [169]. Mattocks [170] described a method for acrylates and acrylamides in which 1,3-dipolar addition takes place with diazomethane, giving a 1-pyrazoline that tautomerizes to the more stable 2-pyrazoline, Eq. (10.137). This secondary amine is then condensed with p-dimethylaminobenzaldehyde (Erlich's reagent) or p-dimethylaminocinnamaldehyde to give a color (possibly due to an imine type of compound).

Hydroxylation of double bonds by permanganate is, in the initial step, a 1,3-cycloaddition [171].

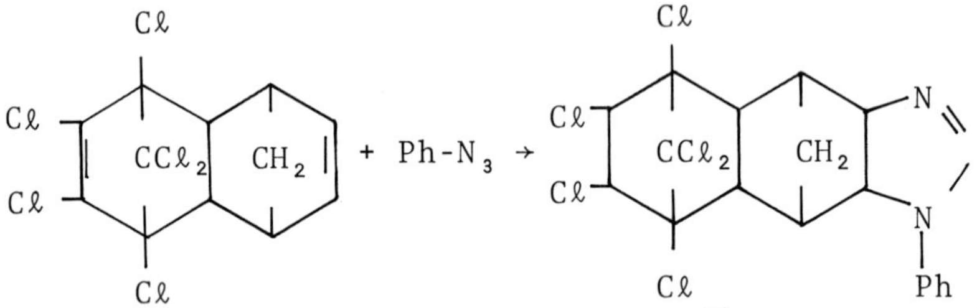

(10.136)

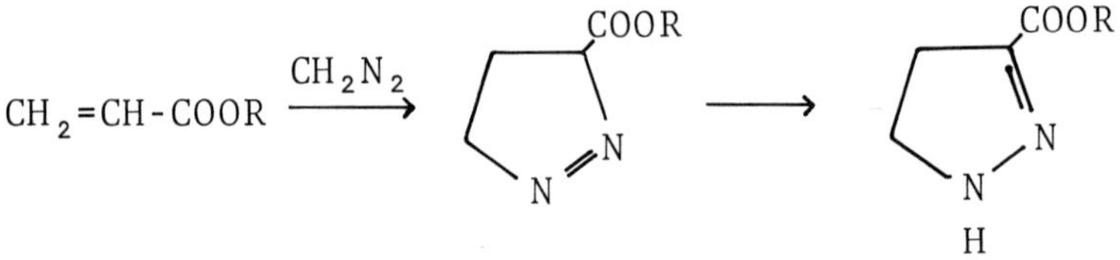

(10.137)

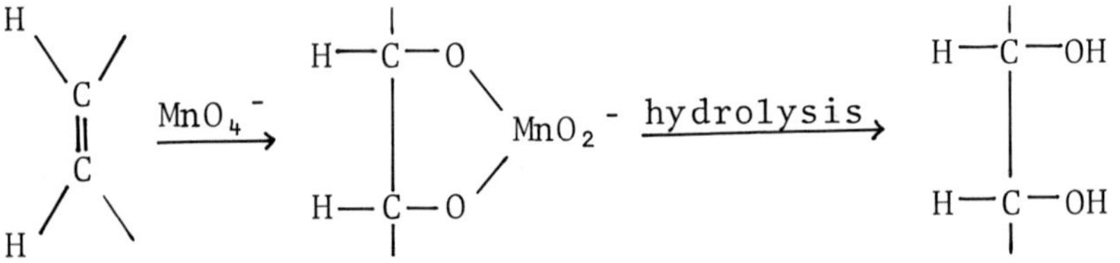

(10.138)

A specific method for terminal double bonds uses $KMnO_4$ hydroxylation, followed by periodic acid oxidation of the glycol. Formaldehyde is one of the products, and this is measured spectrophotometrically by its specific reaction with chromotropic acid [172].

$$\begin{array}{c} CH_2OH \\ | \\ CHOH \\ | \\ R \end{array} + HIO_4 \rightarrow HCHO + RCHO + H_2O + HIO_3 \qquad (10.139)$$

The analytical applications of the Diels-Alder reaction have been minimal relative to its synthetic uses. Schenk has reviewed analyses based on this 1,4-cycloaddition [173]

In all of the published methods the diene takes the role
of the analytical sample; the dienophile is maleic anhydride, chloromaleic anhydride, or tetracyanoethylene
(TCNE). Among the dienes that have been determined are
acyclic 1,3-dienes, anthracene, cyclopentadiene, and dienes
in fats and oils. Most of the methods require separation
of the adduct from the excess reagent; the separation is
often accomplished by solvent extraction. Either the excess dienophile is determined (for example, by titration
of maleic acid with standard base [174]), or the adduct is
determined (by electrophilic halogenation of its isolated
double bond [175]).

A clever modification was introduced by Putnam, Moss,
and Hall [176], who used chloromaleic anhydride to determine 1,3-dienes.

(10.140)

After the addition is complete, the mixture is refluxed
with aqueous silver nitrate. The excess silver is finally
determined by the Volhard titration. The method relies on
the great difference in reactivity of the vinylic chloride
in the dienophile and the tertiary chloride in the adduct
[177].

TCNE was used as the dienophile in a titrimetric method
for dienes [178]. A weighed excess of purified TCNE is
reacted with the sample; after addition is complete the
unreacted TCNE is slowly back-titrated with a solution of
cyclopentadiene. The end point is detected visually or
photometrically with pentamethylbenzene, which forms a
colored molecular complex with TCNE.

Aromatic hydrocarbons have been measured fluorometrically after eliminating interference from anthracene by
forming its Diels-Alder adduct with maleic anhydride [179].
The adduct does not fluoresce under the conditions selected
for the analysis.

Some kinetically controlled Diels-Alder analyses are
of interest. von Mikusch [180] improved a maleic anhydride
method for 1,3-dienes in fats and oils by using iodine as
a catalyst; iodine catalyzes the conversion of the unreactive s-trans isomer to the Diels-Alder-reactive s-cis form.
A separation based on differential rates of Diels-Alder
reaction permitted pure cis-1,3-5-hexatriene to be isolated
from a mixture with its trans isomer [181].

Models showed that the trans isomer could react with a dienophile, whereas the cis isomer could not.

Comment. This survey of addition reactions may help to reveal where significant analytical advances might be achieved, either by clarifying the controlling variables in these reactions or by identifying reaction types that have not yet been exploited analytically. The first point should be considered with some caution in designing sys- tematic research efforts into analytical methods. There is little doubt that a full-scale mechanistic study of most analytical reactions could result in improvments in assay design, such as optimization of conditions for max- imal reactivity, definition of selectivity patterns, etc. However, such improvements may be too modest to justify the effort involved. More interesting is the potential represented by relatively unexplored areas. In the present context these include Michael additions, 1,3-dipolar cyclo- additions, and the Diels-Alder reaction. Photochemical cycloadditions may be interesting; in particular, recall that 1,2-cycloadditions, though symmetry-forbidden in a thermal concerted process, are symmetry-allowed by a photochemical process.

Problems

1. Predict the products in the addition of HOCℓ to
    (a) 2-methylpropene; (b) neurine ($Me_3N^+-CH=CH_2$).

2. For the hydration of styrene,

    $$Ph-CH=CH_2 + H_2O \underset{k_{-1}}{\overset{k_1}{\rightleftarrows}} Ph-CH(OH)CH_2$$

    Schubert et al. [17], report $k_{obs} = 5.66 \times 10^{-5}sec$
    and the equilibrium ratio [styrene]/[alcohol] =
    2.3×10^{-2}. Calculate k_1 and k_{-1}.

3. Rationalize (from Table 10.I) the large increase for bromine addition on going from ethylene to 1-butene, and then the small decrease on going from 1-butene to 1-pentene.

4. In the reaction of methanol with the substituted dibenzoylethylene 57 (X = H, Y = CH_3), two addition products are possible. Which is expected to predominate?

5. (a) Derive an equation relating relative reactivity of two possible pathways (such as syn-addition versus

anti-addition) to the difference in free energies of activation for the two pathways.
(b) In an addition reaction under kinetic control, 99.0% of the cis product and 1.0% of the trans were found. Calculate the quantity $\delta\Delta G^{\ddagger}$ for the two pathways.

Answer: (a) RT ℓn (k_2/k_1) = $\Delta G_1^{\ddagger}$ - $\Delta G_2^{\ddagger}$

6. These are some key observations on the epoxidation of olefins by peracids:

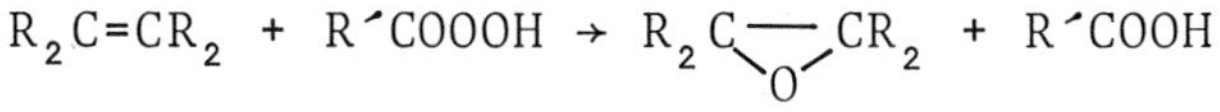

$$R_2C=CR_2 + R'COOOH \rightarrow R_2C\underset{O}{\overset{}{\diagdown\diagup}}CR_2 + R'COOH$$

(a) The rate equation is $v = k_2$ [olefin] [peracid].
(b) These are relative rates of epoxidation by peracetic acid in acetic acid: $CH_2=CH_2$, 1.00; $CH_2=CH-Me$, 22.0; $CH_2=CMe_2$, 48.5; $MeCH=CMe_2$, 6530; $Me_2C=CMe_2$, very rapid.
(c) The order of reactivity of substituted perbenzoic acids, $R-C_6H_4-COOOH$, toward stilbene in benzene solution is $p-NO_2 > p-Cl > H > p-Me > p-OMe$.
(d) The epoxide has the same geometrical structure as the reactant olefin:

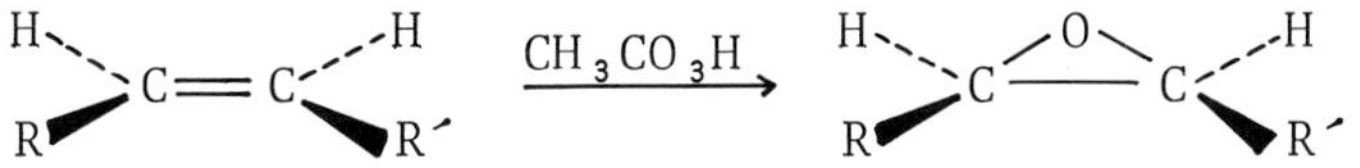

Suggest a mechanism (or mechanisms) consistent with all of these experimental results.

Answer: See P. B. D. de la Mare and R. Bolton, Ref. [130], pp. 154-161.

7. Draw resonance structures to show the ground state polarization depicted in 49.

8. Plan analytical approaches to the quantitative analysis of these mixtures:
(a) 1-Butene and 1,3-butadiene
(b) Methyl crotonate and crotonic acid
(c) Benzene and cyclohexene.

9. Design a quantitative analytical method for active methylene compounds based upon an addition reaction.

References

1.　C. K. Ingold, Structure and Mechanism in Organic Chemistry, Cornell Univ. Press, Ithaca, N. Y., 1953, p. 212.

2.　R. C. Fahey and C. A. McPherson, J. Am. Chem. Soc., 91, 3865 (1969).

3.　2-Acetoxy-2,3-dimethylbutane may also be formed, but under the reaction conditions it is rapidly converted to 3.

4.　In acetic acid, HCl is not extensively dissociated, and molecular HCl (rather than the proton) attacks the olefin to give a carbonium ion-chloride ion pair. For the present purpose we can use the simplified picture of Eq. (10.3). Equation (10.7) gives a more detailed view.

5.　A. L. Henne and S. Kaye, J. Am. Chem. Soc., 72, 3369 (1950). Anti-Markovnikov orientation is also observed in HBr additions when peroxides are not excluded; under such conditions the reaction proceeds by a homolytic (free radical) mechanism.

6.　P. B. D. De La Mare and R. Bolton, Electrophilic Additions to Unsaturated Systems, Monograph 4 in Reaction Mechanisms in Organic Chemistry, Elsevier, Amsterdam, 1966, Chap. 5.

7.　It is preferable to describe the stereochemistry of the process of addition as anti (the two adding groups bonding from opposite sides of the alkene plane) or syn (same side); the terms trans and cis can be reserved to describe the stereochemistry of reactants or products. Most of the literature, however, also uses the cis-trans nomenclature to refer to the stereochemical course of the reaction. The syn/anti terminology is equivalent to the suprafacial/antarafacial nomenclature of R. B. Woodward and R. Hoffmann, J. Am. Chem. Soc., 87, 2511 (1965).

8.　Y. Pocker, K. D. Stevens, and J. J. Campoux, J. Am. Chem. Soc., 91, 4199 (1969); Y. Pocker and K. D. Stevens, ibid., 91, 4205 (1969).

9.　R. C. Fahey, M. W. Monahan, and C. A. McPherson, J. Am. Chem. Soc., 92, 2810 (1970); R. C. Fahey and M. W. Monahan, ibid., 92, 2816 (1970); R. C. Fahey and C. A. McPherson, ibid., 93, 2445 (1971).

10.　Similar mechanisms have been suggested for HCl addition to acetylenes: R. C. Fahey and D.-J. Lee, J. Am. Chem. Soc., 90, 2124 (1968).

11.　D. V. Banthorpe, Chem. Revs., 70, 295 (1970).

12.　G. S. Hammond and C. H. Collins, J. Am. Chem. Soc., 82, 4323 (1960).

13.　L. P. Hammett, Physical Organic Chemistry, McGraw Hill

New York, 1940, pp. 292-293.
14. J. B. Levy, R. W. Taft, Jr., and L. P. Hammett, J.
     Am. Chem. Soc., 75, 1253 (1953).
15. P. B. D. De La Mare and R. Bolton, Ref. [6], pp. 33-
     34.
16. R. H. Boyd, R. W. Taft, Jr., A. P. Wolf, and D. R.
     Christman, J. Am. Chem. Soc., 82, 4729 (1960).
17. W. M. Schubert, B. Lamm, and J. R. Keeffe, J. Am.
     Chem. Soc., 86, 4727 (1964).
18. Some of the evidence contributing to this view con-
     sists of solvent isotope effects and solvent acidity
     dependence; this has been summarized by De La Mare
     and Bolton [15]; see also Ref. [17] and J. L. Jensen,
     Tetrahedron Letters, 1971, p. 7.
19. P. B. D. De La Mare and R. Bolton, Ref. [6], Chap. 4.
20. See also P. B. D. De La Mare and R. Bolton, Ref. [6],
     Chaps. 6,7.
21. J. E. Dubois and G. Mouvier, Tetrahedron Letters,
     1963, p. 1325
22. J. H. Rolston and K. Yates, J. Am. Chem. Soc., 91,
     1483 (1969).
23. (a) J. R. Atkinson and R. P. Bell, J. Chem. Soc.,
     1963, p. 3260; (b) R. P. Bell and M. Pring, ibid.,
     B1966, p. 1119.
24. In a similar study, D. Acharya and M. N. Das, J. Org.
     Chem., 34, 2828 (1969), found $k_{Br_3^-}/k_{Br_2}$ to be less
     than unity for addition to methyl acrylate, methyl
     crotonate, methyl methacrylate, and acrylamide.
25. E. P. White and P. W. Robertson, J. Chem. Soc., 1939,
     p. 1509.
26. C. K. Ingold, Ref. [1], p. 669.
27. I. Roberts and G. E. Kimball, J. Am. Chem. Soc., 59,
     947 (1937).
28. Actually syn/anti ratios in excess of unity are known.
     These probably arise from intimate ion-pair collapse,
     as described for hydrohalogenation reactions. At one
     time it was thought that stereospecific syn addition
     could occur via a concerted four-center cycloaddition,
     as in 20.

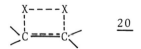

This process is forbidden, however, by orbital sym-
     metry conservation (see Sec. 10.3).
29. G. A. Olah, C. L. Jeuell, and A. M. White, J. Am.
     Chem. Soc., 91, 3961 (1970).
30. J. H. Rolston and K. Yates, J. Am. Chem. Soc., 91,
     1469, 1477 (1969).

31. A. Hassner, F. P. Boerwinkle, and A. B. Levy, J. Am. Chem. Soc., $\underline{92}$, 4879 (1970).

32. J. E. Dubois and A. Schwarcz, Tetrahedron Letters, 1964, p. 2167, find a good correlation with σ^+ (ρ = -4.3) for the bromination of styrenes. They prefer an intermediate of high carbonium ion character. See also the interpretation of M. F. Ruasse and J. E. Dubois, ibid., 1970, p. 1163, of the nonlinear correlation for bromination of stilbenes.

33. R. E. Buckles, R. E. Erickson, J. D. Snyder, and W. B. Person, J. Am. Chem. Soc., $\underline{82}$, 2444 (1960).

34. D. V. Banthorpe, Chem. Revs., $\underline{70}$, 295 (1970).

35. R. E. A. Dear, J. Org. Chem., $\underline{35}$, 1703 (1970).

36. P. B. D. de la Mare and R. Bolton, Ref. [6], Chaps. 8, 9, 10.

37. For reviews see H. C. Brown, Tetrahedron, $\underline{12}$, 117 (1961); Hydroboration, Benjamin, New York, 1962.

38. H. C. Brown, K. J. Murray, L. J. Murray, J. A. Snover, and G. Zweifel, J. Am. Chem. Soc., $\underline{82}$, 4233 (1960).

39. H. C. Brown, who developed the hydroboration procedure, symbolizes the reaction by HB, as in Eq. (10.25) The term "siamyl" is a contraction of s-isoamyl, signifying the group $Me_2CHCH(Me)-$.

40. Hydrogen is more electronegative than boron.

41. H. C. Brown and G. Zweifel, J. Am. Chem. Soc., $\underline{83}$, 2544 (1961).

42. W. J. Wechter, Chem. Ind., 1959, p. 294.

43. H. C. Brown and A. W. Moerikofer, J. Am. Chem. Soc., $\underline{83}$, 3417 (1961).

44. J. R. Elliott, W. L. Roth, G. F. Roedel, and E. M. Boldebuck, J. Am. Chem. Soc., $\underline{74}$, 5211 (1952); B. Rice, J. A. Livasy, and G. W. Schaeffer, ibid., $\underline{77}$, 2750 (1955).

45. R. Breslow, Organic Reaction Mechanisms, 2nd ed., Benjamin, New York, 1969, pp. 125-126. See also K. Fukui, Bull. Chem. Soc. Japan, $\underline{39}$, 498 (1966). A three-center transition state has been suggested: A. Streitwieser, Jr., L. Verbit, and R. Bittman, J. Org. Chem., $\underline{32}$, 1530 (1967). Jones prefers a π complex intermediate, ruling out $\underline{27}$ and $\underline{28}$ on the basis of orbital symmetry; P. R. Jones, J. Org. Chem., $\underline{37}$, 1886 (1972).

46. C. A. Coulson, Valence, 2nd ed., Oxford Univ. Press, London, 1961, pp. 369-382.

47. E. W. Garbisch, Jr., S. M. Schildcrout, D. B. Patterson, and C. M. Sprecher, J. Am. Chem. Soc., $\underline{87}$, 2932 (1965); A. K. Awasthy and J. Rocek, ibid., $\underline{91}$, 991 (1969).

48. R. E. Erickson and R. L. Clark, Tetrahedron Letters, 1969, p. 3997. See also H. C. Brown and K. T. Liu,

J. Am. Chem. Soc., 93, 7335 (1971).

49. H. C. Brown, Ref. [37], pp. 62-66. Diborane is an
     effective reducing agent for many functional groups;
     cf. H. C. Brown, P. Heim, and N. M. Yoon, J. Am.
     Chem. Soc., 92, 1637 (1970).

50. L. H. Toporcer, R. E. Dessy, and S. I. E. Green, J.
     Am. Chem. Soc., 87, 1236 (1965); R. E. Dessy and W.
     Kitching, Advan. Organometallic Chem., 4, 317 (1966).

51. For reviews see J. Chatt, Chem. Revs., 48, 7 (1951);
     R. E. Dessy and W. Kitching, Ref. [50], pp. 298-306.
     The reverse reaction, deoxymercuration, has been ex-
     tensively studied by Kreevoy; see M. M. Kreevoy and
     M. A. Turner, J. Org. Chem., 29, 1939 (1964).

52. J. Halpern and H. B. Tinker, J. Am. Chem. Soc., 89,
     6427 (1967).

53. E. F. Kiefer and W. L. Waters, J. Am. Chem. Soc.,
     87, 4401 (1965).

54. H. J. Lucas, F. R. Hepner, and S. Winstein, J. Am.
     Chem. Soc., 61, 3102 (1939). This intermediate may
     also be formulated as a π-complex. Recent evidence
     for a cyclic mercurinium ion intermediate is given
     by D. J. Pasto and J. A. Gontaz, ibid., 92, 7480
     (1970); G. A. Olah and P. R. Clifford, ibid., 93,
     2320 (1971).

55. T. G. Traylor and A. W. Baker, J. Am. Chem. Soc.,
     85, 2746 (1963). A simple infrared frequency shift
     criterion was developed to facilitate the stereo-
     chemical assignments.

56. T. G. Traylor, J. Am. Chem. Soc., 86, 244 (1964).

57. Chloride is added in order to isolate the products
     as the stable halide salts.

58. H. C. Brown and P. Geoghegan, Jr., J. Am. Chem. Soc.,
     89, 1522 (1967); J. Org. Chem., 35, 1844 (1970).

59. P. B. D. de la Mare and R. Bolton, Ref. [6], Chap.
     12; M. Cais, The Chemistry of Alkenes (S. Patai, ed.),
     Wiley (Interscience), New York, 1964, Chap. 12.

60. E. S. Gould, Mechanism and Structure in Organic Chem-
     istry, Holt, Rinehart, and Winston, New York, 1959,
     p. 531-533.

61. K. Mislow and H. M. Hellman, J. Am. Chem. Soc., 73,
     244 (1951).

62. The conformation of a diene is expressed in terms of
     the orientation of the unsaturated bonds about the
     intervening single bond, so that 39 is the s-cis form
     and 40 is the s-trans form.

39   40

63. L. F. Hatch, P. D. Gardner, and R. E. Gilbert, J. Am.
     Chem. Soc., 81, 5943 (1959).
64. C. K. Ingold, Structure and Mechanism in Organic
     Chemistry, Cornell Univ. Press, Ithaca, N. Y., 1953,
     p. 655.
65. M. S. Kharasch, J. Kritchevsky, and F. R. Mayo, J.
     Org. Chem., 2, 489 (1937).
66. W. G. Young, H. K. Hall, Jr., and S. Winstein, J. Am.
     Chem. Soc., 78, 4338 (1956).
67. G. S. Hammond and J. Warkentin, J. Am. Chem. Soc.,
     2554 (1961).
68. The 1,2-cis product is also found, but this is be-
     lieved to result from a 1,4-cis → 1,2-cis allylic
     rearrangement.
69. M. C. Cabaleiro and M. D. Johnson, J. Chem. Soc.,
     B1967, p. 565. Failure to observe acid catalysis is
     also consistent with electrophilic attack.
70. For reviews see C. K. Ingold, Ref. [64], pp. 690-696;
     S. Patai and Z. Rappoport, The Chemistry of Alkenes
     (S. Patai, ed.), Wiley (Interscience), New York,
     1964, Chap. 8.
71. The Michael addition to acrylonitrile is commonly
     called cyanoethylation.
72. C. F. Koelsch and C. R. Stephens, Jr., J. Am. Chem.
     Soc., 72, 2209 (1950).
73. C. K. Ingold, Ref. [64], p. 692.
74. W. E. Bachmann and L. B. Wick, J. Am. Chem. Soc., 72,
     3388 (1950).
75. M. J. Kamlet and D. J. Glover, J. Am. Chem. Soc., 78,
     4556 (1956).
76. S. Patai and Z. Rappoport, Ref. [70], p. 492.
77. R. A. Abramovitch and J. M. Muchowski, Can. J. Chem.,
     38, 557 (1960).
78. With α,β-unsaturated aldehydes the addition takes
     place across the carbonyl group.
79. W. J. Jones, J. Chem. Soc., 1914, 1547.
80. G. Gilbert and B. F. Aycock, J. Org. Chem., 22, 1013
     (1957); S. Patai and Z. Rappoport, Ref. [70], pp.
     501-506.
81. H. Shenhav, Z. Rappoport, and S. Patai, J. Chem. Soc.,
     B1970, p. 469.
82. M. Friedman and J. S. Wall, J. Am. Chem. Soc., 86,
     3735 (1964); M. Friedman and J. S. Wall, J. Org.
     Chem., 31, 2888 (1966).
83. F. B. Zienty, B. D. Vineyard, and A. A. Schleppnik,
     J. Org. Chem., 27, 3140 (1962).
84. S. M. Kupchan, D. C. Fessler, M. A. Eakin, and T. J.
     Giacobbe, Science, 168, 376 (1970).
85. W. E. Truce and A. J. Levy, J. Am. Chem. Soc., 83,
     4641 (1961).

86. C. A. Bunton and G. J. Minkoff, J. Chem. Soc., 1949, p. 665.
87. T. I. Crowell, G. C. Helsey, R. E. Lutz, and W. L. Scott, J. Am. Chem. Soc., 85, 443 (1963).
88. Equilibrium is reached at 92% reaction, and the kinetics were studied under pseudo-first-order conditions.
89. C. K. Ingold, Ref. [64], p. 696.
90. L. T. Rozelle and R. A. Alberty, J. Phys. Chem., 61, 1637 (1957); L. E. Erickson and R. A. Alberty, ibid., 63, 705 (1959).
91. M. L. Bender and K. A. Connors, J. Am. Chem. Soc., 83, 4099 (1961); 84, 1980 (1962).
92. J. L. Bada and S. L. Miller, J. Am. Chem. Soc., 91, 3948 (1969).
93. Because of the slowness of this reaction, studies were carried out above 100°C in sealed ampuls.
94. The addition of ammonia to fumaric acid has been characterized as Ad_N2, and the similarity of amination and hydration has been invoked to support the Ad_N2 mechanism for the hydration; cf. J. L. Bada and S. L. Miller, J. Am. Chem. Soc., 91, 3946 (1969); 92, 2774 (1970).
95. R. Huisgen, R. Grashey, and J. Sauer, The Chemistry of Alkenes (S. Patai, ed.), Wiley (Interscience), New York, 1964. This chapter is an extensive review of cycloadditions to olefins.
96. R. B. Woodward and R. Hoffmann, J. Am. Chem. Soc., 87, 395 (1965).
97. R. Breslow, Organic Reaction Mechanisms, 2nd ed., Benjamin, New York, 1969, pp. 129-146.
98. R. Hoffmann and R. B. Woodward, J. Am. Chem. Soc., 87, 2046 (1965). For reviews see R. B. Woodward in Aromaticity, Chemical Society Special Publication No. 21, London, 1967, p. 217; R. Hoffman and R. B. Woodward, Science, 167, 825 (1970); J. J. Vollmer and K. L. Servis, J. Chem. Educ., 47, 491 (1970); R. B. Woodward and R. Hoffmann, The Conservation of Orbital Symmetry, Verlag Chemie, Gmbh, Weinheim, 1970.
99. A. Wilson and D. Goldhamer, J. Chem. Educ., 40, 504 (1963).
100. But see M. J. S. Dewar, p. 210 in Aromaticity, Chemical Society Special Publication No. 21, London, 1967, for a typically spirited criticism of the Woodward-Hoffmann approach.
101. This conclusion applies to the thermal (i.e., ground state) reaction. The selection rules are reversed for photochemical (excited state) reactions.
102. P. D. Bartlett, L. K. Montgomery, and B. Seidel, J. Am. Chem. Soc., 86, 616 (1964); L. K. Montgomery,

K. Schueller, and P. D. Bartlett, ibid., 86, 622 (1964); P. D. Bartlett and L. K. Montgomery, ibid., 86, 628 (1964); P. D. Bartlett, Science, 159, 833 (1968).

103. J. Hine, Divalent Carbon, Ronald Press, New York, 1964; W. Kirmse, Carbene Chemistry, Vol. 1 of Organic Chemistry, Academic, New York, 1964.

104. This is called an α-elimination reaction.

105. P. S. Skell and A. Y. Garner, J. Am. Chem. Soc., 78, 3409, 5431 (1956).

106. W. von E. Doering and W. A. Henderson, J. Am. Chem. Soc., 5274 (1958).

107. F. D. Gunstone, Advan. Org. Chem., 1, 103 (1960).

108. R. Huisgen, Proc. Chem. Soc., 1961, 357; R. Huisgen, Angew. Chem., Intern. Ed., 2, 565 (1963).

109. H. Ulrich, Cycloaddition Reactions of Heterocumulenes, Vol. 9 of Organic Chemistry, Academic, New York, 1967, p. 20. Cumulenes are compounds with adjacent double bonds; in heterocumulenes at least one of the atoms is not carbon.

110. R. Criegee, Peroxide Reaction Mechanisms, (J. O. Edwards, ed.), Wiley (Interscience), New York, 1962, p. 29; R. W. Murray, Accounts Chem. Res., 1, 313 (1968). For a recent contrary opinion see P. R. Story, J. A. Alford, W. C. Ray, and J. R. Burgess, J. Am. Chem. Soc., 93, 3044 (1971).

111. M. C. Kloetzel, Org. Reactions, 4, 1 (1948); H. L. Holmes, ibid., 4, 60 (1948); L. W. Butz and A. W. Rytina, ibid., 5, 136 (1949); A. Wasserman, Diels-Alder Reactions, Elsevier, Amsterdam, 1965.

112. J. Hamer (ed.), 1,4-Cycloaddition Reactions. The Diels-Alder Reaction in Heterocyclic Syntheses, Academic, New York, 1967.

113. Recall, from the earlier discussion in this section, that a concerted 1,4-cycloaddition is symmetry-allowed.

114. M. J. S. Dewar and R. S. Pyron, J. Am. Chem. Soc., 92, 3098 (1970).

115. R. A. Grieger and C. A. Eckert, J. Am. Chem. Soc., 92, 2918 (1970).

116. J. G. Martin and R. K. Hill, Chem. Revs., 61, 537 (1961).

117. W. E. Bachmann and L. B. Scott, J. Am. Chem. Soc., 70, 1458 (1948).

118. L. J. Andrews and R. M. Keefer, Molecular Complexes in Organic Chemistry, Holden-Day, San Francisco, 1964, p. 177.

119. R. B. Woodward and R. Hoffmann, The Conservation of Orbital Symmetry, Verlag Chemie, GmbH, Weinheim, 1970, p. 145.

120. J. A. Berson and R. D. Reynolds, J. Am. Chem. Soc.,
      77, 4434 (1955); J. A. Berson, R. D. Reynolds, and
      W. M. Jones, ibid., 78, 6049 (1956).
121. J. A. Berson and W. A. Mueller, J. Am. Chem. Soc.,
      83, 4940 (1961).
122. J. A. Berson and A. Remanick, J. Am. Chem. Soc., 83,
      4947 (1961).
123. A. Polgar and J. L. Jungnickel, Org. Anal., 3, 203
      (1956).
124. K. G. Stone, Determination of Organic Compounds,
      McGraw-Hill, New York, 1956, Chap. 2.
125. S. Siggia, Quantitative Organic Analysis Via Func-
      tional Groups, 3rd ed., Wiley, New York, 1963, Chap.
      7.
126. F. E. Critchfield, Organic Functional Group Analysis,
      Macmillan, New York, 1963, Chap. 6.
127. N. D. Cheronis and T. S. Ma, Organic Functional Group
      Analysis by Micro and Semimicro Methods, Wiley
      (Interscience), New York, 1964, Chap. 10.
128. E. J. Kuchar, The Chemistry of Alkenes (S. Patai,
      ed.), Wiley (Interscience), New York, 1964, Chap. 5.
129. This particular problem can be overcome by converting
      the acid to its anion, which is susceptible to bro-
      mination; F. E. Critchfield and J. B. Johnson, Anal.
      Chem., 31, 1406 (1959).
130. de la Mare and Bolton have concluded that halogena-
      tion methods are so lacking in generality that each
      individual unsaturated sample should be treated as
      a special case for which optimum conditions should
      be established; P. B. D. de la Mare and R. Bolton,
      Electrophilic Additions to Unsaturated Systems,
      Elsevier, Amsterdam, 1966, p. 149.
131. J. S. Fritz and G. S. Hammond, Quantitative Organic
      Analysis, Wiley, New York, 1957, p. 80; these authors
      propose that the net effect of mercuric ion depends
      on this catalytic effect and the opposed inhibiting
      effect of Hg(II)-olefin complexation.
132. R. E. Dessy and W. Kitching, Advan. Organometallic
      Chem., 4, 304 (1966).
133. F. A. Leisey and J. F. Grutsch, Anal. Chem., 28,
      1553 (1956).
134. J. W. Miller and D. D. DeFord, Anal. Chem., 29, 475
      (1957).
135. J. S. Fritz and G. E. Wood, Anal. Chem., 40, 134
      (1968).
136. O. C. Gadeken, R. L. Ayres, and E. P. Rack, Anal.
      Chem., 42, 1105 (1970).
137. S. Siggia and R. L. Edsberg, Anal. Chem., 20, 762
      (1948).
138. P. B. D. de la Mare and R. Bolton, Ref. [130], p. 129.

139. For examples of neighboring group participation by carboxylate, ester, and amide groups see de la Mare and Bolton, Ref. [130], pp. 140-143.

140. S. Siggia, Quantitative Organic Analysis Via Functional Groups, 3rd ed., Wiley, New York, 1963, pp. 670-673.

141. R. T. Arnold, M. de Moura Campos, and K. L. Lindsay, J. Am. Chem. Soc., 75, 1044 (1953).

142. N. D. Cheronis and T. S. Ma, Ref. [127], pp. 374-375. See also Sec. 9.2.

143. M. N. Das, Anal. Chem., 26, 1086 (1954).

144. R. P. Marquardt and E. N. Luce, Anal. Chem., 21, 1194 (1949).

145. R. W. Martin, Anal. Chem., 21, 921 (1949); R. P. Marquardt and E. N. Luce, ibid., 39, 1655 (1967).

146. K. L. Mallik, Anal. Chem., 32, 1369 (1960).

147. One of the advantages of oxymercuration, relative to halogenation, is the low incidence of side reactions.

148. A. P. Altshuller and I. R. Cohen, Anal. Chem., 32, 1843 (1960).

149. C. A. Brown, S. C. Sethi, and H. C. Brown, Anal. Chem., 39, 823 (1967).

150. F. E. Critchfield, G. L. Funk, and G. B. Johnson, Anal. Chem., 28, 76 (1956); F. E. Critchfield, Organic Functional Group Analysis, Macmillan, New York, 1963, p. 116.

151. The rate constants for addition of morpholine to activated olefins are discriminating values for the identification of olefins; for other examples of rate measurements for qualitative analysis see J. R. Robinson and K. A. Connors, J. Chem. Educ., 48, 470 (1971).

152. A. M. Potts, Arch. Biochem., 24, 329 (1949).

153. R. N. Lacey, The Chemistry of Alkenes (S. Patai, ed.), Wiley (Interscience), New York, 1964, Chap. 14.

154. S. Sass, J. J. Kaufman, A. A. Cardenas, and J. J. Martin, Anal. Chem., 30, 529 (1958).

155. F. Feigl, Spot Tests in Organic Analysis, 6th ed., Elsevier, Amsterdam, 1960, p. 281.

156. J. D. Gregory, J. Am. Chem. Soc., 77, 3922 (1955); D. G. Smyth, A. Nagamatsu, and J. S. Fruton, ibid., 82, 4600 (1960).

157. E. Roberts and G. Rouser, Anal. Chem., 30, 1291 (1958); N. M. Alexander, ibid., 30, 1292 (1958).

158. J. Broekhuysen, Anal. Chim. Acta, 19, 542 (1958).

159. A. Polgár and J. L. Jungnickel, Ref. [123], p. 315.

160. N. D. Cheronis and T. S. Ma, Ref. [142], p. 325; L. A. Cohen, Ann. Rev. Biochem., 37, 683 (1968).

161. F. E. Critchfield and J. B. Johnson, Anal. Chem.,

     $\underline{28}$, 73 (1956).
162. C. P. Ivanov and Y. Vladovska-Yukhn, Biochim. Bio-
     phys. Acta, $\underline{194}$, 345 (1969).
163. C. F. H. Allen and J. A. Van Allen, J. Am. Chem.
     Soc., $\underline{70}$, 2069 (1948).
164. N. D. Cheronis and T. S. Ma, Ref. [142], p. 461.
165. S. Patai and Z. Rappoport, The Chemistry of Alkenes
     (S. Patai, ed.), Wiley (Interscience), New York,
     1964, pp. 525-546.
166. P. S. Bailey, Chem. Revs., $\underline{58}$, 925 (1958).
167. V. L. Davison and H. J. Dutton, Anal. Chem., $\underline{38}$,
     1302 (1966); M. Beroza and B. A. Bierl, ibid., $\underline{38}$,
     1976 (1966). M. Beroza, Accounts Chem. Res., $\underline{3}$, 33
     (1970), has reviewed chemical reaction gas chroma-
     tography for structure determination.
168. N. D. Cheronis and T. S. Ma, Ref. [142], p. 389.
169. A. A. Danish and R. E. Lidov, Anal. Chem., $\underline{22}$, 702
     (1950).
170. A. R. Mattocks, Anal. Chem., $\underline{40}$, 1347 (1968).
171. F. D. Gunstone, Advan. Org. Chem., $\underline{1}$, 103 (1960).
172. C. E. Bricker and K. H. Roberts, Anal. Chem., $\underline{21}$,
     1331 (1949).
173. G. H. Schenk, Organic Functional Group Analysis,
     Pergamon, Oxford, 1968, Chap. 6.
174. I. Ubaldini, V. Crespi, and F. Guerrieri, Ann. Chim.
     Applicata, $\underline{39}$, 77 (1949).
175. P. Unger, Analyst, $\underline{80}$, 820 (1955).
176. S. T. Putnam, M. L. Moss, and R. T. Hall, Ind. Eng.
     Chem. Anal. Ed., $\underline{18}$, 628 (1946).
177. Schenk, Ref. [173], p. 76, suggests that the adduct
     undergoes β-elimination, but the reaction conditions
     are just as conducive to S_N1 solvolysis as to E1
     elimination. The adduct is a tertiary chloride (the
     anhydride ring probably opens in the aqueous system),
     and will be susceptible to solvolytic displacement.
178. M. Ozolins and G. H. Schenk, Anal. Chem., $\underline{33}$, 1035
     (1961).
179. G. H. Schenk and D. R. Wirz, Anal. Chem., $\underline{42}$, 1754
     (1970).
180. J. D. von Mikusch, Angew. Chem., $\underline{62}$, 475 (1950).
181. J. C. H. Hwa, P. L. de Benneville, and H. J. Sims,
     J. Am. Chem. Soc., $\underline{82}$, 2537 (1960).

Chapter 11.   β-ELIMINATION

11.1   Nature of the Reaction

Survey of the Reaction. This chapter introduces elimination reactions, in which two atoms or groups are lost,
without replacement, from the substrate molecule. When
both fragments depart from the same atom, the process is
called α-elimination; examples were discussed in Sec.
10.3 in connection with carbene additions to olefins. By
far the most important eliminations are those in which the
two fragments leave adjacent carbon atoms in a heterolytic
process. These β-eliminations are illustrated by

$$R\text{-}CH\text{-}CH\text{-}R' \rightarrow R\text{-}CH{=}CH\text{-}R' + X^+ + Y^- \qquad (11.1)$$
$$\overset{|}{X} \quad \overset{|}{Y}$$

In the most common examples the atom X is a hydrogen atom,
and its departure is aided by a strong base. The group Y,
which leaves with an electron pair, is called the leaving
group. These olefin-forming β-eliminations are exempli-
fied by Eq. (11.2), where Y may be $C\ell$, Br, NR_3^+, SR_2^+, $OCOR$,
etc., and the base may be OH^-, OR^-, SR^-, etc.

$$\overset{\beta\quad\alpha}{R\text{-}CH_2\,CH\text{-}R'} \xrightarrow{\text{base}} R\text{-}CH{=}CH\text{-}R' + HY \qquad (11.2)$$
$$\overset{|}{Y}$$

Notice that olefin-forming eliminations are the formal reverse of olefin addition reactions. Carbonyl-forming eliminations are the reverse of additions to the carbonyl
group, which are described in Chap. 12.
     Examination of Eq. (11.2) from the viewpoint of Chap.
9, which dealt with nucleophilic aliphatic substitution,
leads one to expect that substitution and elimination are
competitive processes, since the base may either abstract
the β proton, giving elimination, or attack the α carbon,
giving substitution. This does in fact occur, and concomitant substitution may be a serious obstacle in syn-

thetic schemes for the formation of olefins by β-elimina-
tion. The similarity of substitution and elimination extends to the mechanistic level, and elimination, like substitution, may occur via unimolecular or bimolecular path-
ways.

Two features of elimination reactions that are of
great synthetic importance and that also yield mechanistic
information are the orientation and stereochemistry of the
reaction. By orientation is meant the site of double bond
formation when a "choice" is available. For example, the
secondary halide 1 gives the mixture of β-elimination products shown in

$$CH_3CH_2CHCH_3 \xrightarrow{\;OH^-\;} CH_3CH=CHCH_3 \; + \; CH_3CH_2CH=CH_2 \qquad (11.3)$$
$$\underset{Br}{|} \qquad\qquad\qquad 81\% \qquad\qquad 19\%$$

1

A theory of β-elimination must account for this phenomenon.
The stereochemistry of elimination is dependent on the
mechanism; in a great many of these reactions anti-elimination occurs, and this observation bears on the structure
of the transition state. These three aspects, mechanism,
orientation, and stereochemistry, are treated in this
section [1].

Among the types of reactions that are not dealt with,
but that may be regarded as eliminations, are 1,1-(α-)
elimination, 1,3-(γ-)elimination, 1,4-(δ-)elimination,
decarboxylation and decarbonylation, and dehydrogenation.

Mechanisms. The general mechanistic features of β-elimination have been recognized and agreed upon for many years
[2], but widespread disagreement currently exists concerning details, such as the timing of bond making and bond
breaking. The dual mechanisms (S_N1 and S_N2) observed in
aliphatic substitutions are seen also in eliminations.
The bimolecular (E2) mechanism has been demonstrated for
many eliminations on the basis of several kinds of evidence, namely: (1) second-order kinetics, the rate being
dependent upon the nature and concentration of the base;
(2) absence of α-hydrogen exchange with the solvent, showing that the reaction occurs in one step; (3) structural
and medium effects on the rate. Elimination from β-phenethyl halides, Eq. (11.4), is a typical E2 reaction.

$$PhCH_2CH_2Br \; + \; EtO^- \; \rightarrow \; PhCH=CH_2 \; + \; EtOH \; + \; Br^- \qquad (11.4)$$

As may be expected, S_N2 substitution often competes with

E2 elimination. The E2 elimination proceeds through a transition state like 2, which is clearly similar to the S_N2 transition state.

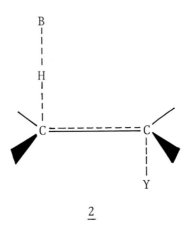

2

The constitutional factors that favor E2 elimination, in particular the quality of the leaving group, will also promote the S_N2 reaction.

A different mechanism, also showing second-order kinetics, can be realized for certain substrates. This is a two-step process involving proton abstraction to give a carbanion intermediate, which then loses the substituent yielding an olefin.

$$R_2CH-CR_2Y + B \rightleftharpoons R_2\bar{C}-CR_2Y + BH^+ \qquad (11.5)$$

$$R_2\bar{C}-CR_2Y \rightarrow R_2C=CR_2 + Y^- \qquad (11.6)$$

This is the E1cb (unimolecular elimination from the conjugate base of the substrate), or carbanion, mechanism. It will be favored by a substrate with an acidic α-hydrogen, a poor leaving group, and a structure that can stabilize the carbanion. 2,2-Dichloro-1,1,1-trifluoroethane undergoes sodium methoxide-catalyzed E1cb dehydrofluorination [3].

$$CHCl_2CF_3 + MeO^- \rightleftharpoons \bar{C}Cl_2CF_3 + MeOH \qquad (11.7)$$

$$\bar{C}Cl_2CF_3 \rightarrow CCl_2=CF_2 + F^- \qquad (11.8)$$

Observation of deuterium exchange with the solvent, at a rate greater than the elimination rate, is consistent with this mechanism [4].

A third elimination mechanism is the two-step E1

process, which shares a carbonium ion intermediate with the competing S_N1 reaction. First-order kinetics are required, with the rate being independent of added base. The E1 mechanism is exemplified by the following equations.

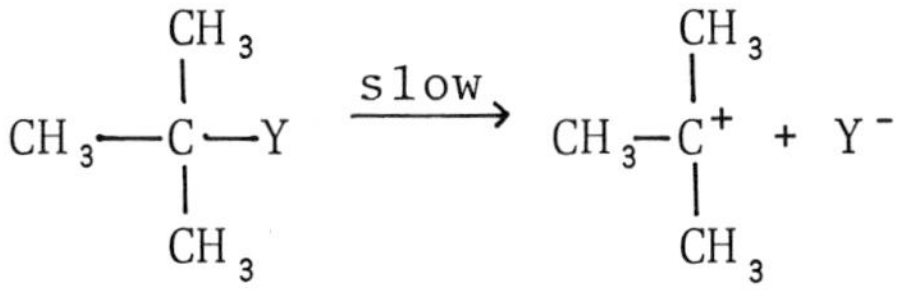

$$CH_3-\underset{\underset{CH_3}{|}}{\overset{\overset{CH_3}{|}}{C}}-Y \quad \xrightarrow{slow} \quad CH_3-\underset{\underset{CH_3}{|}}{\overset{\overset{CH_3}{|}}{C^+}} + Y^- \qquad (11.9)$$

$$CH_3-\underset{\underset{CH_3}{|}}{\overset{\overset{CH_3}{|}}{C^+}} \quad \longrightarrow \quad CH_2=C\overset{\diagup CH_3}{\diagdown CH_3} + H^+ \qquad (11.10)$$

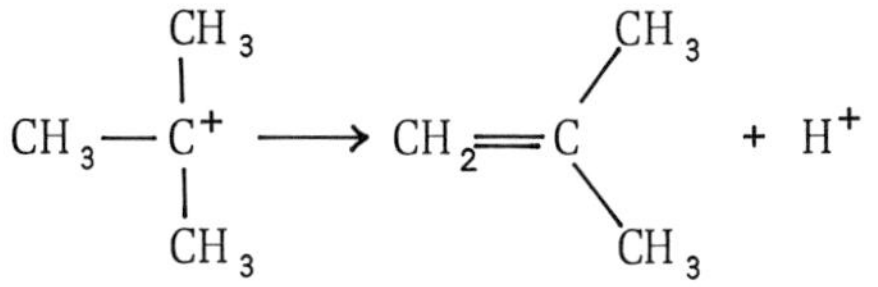

The rate-determining step is the ionization, so different rates are expected for substrates with the same alkyl group but different leaving groups. Since, however, the second step is uninfluenced by the leaving group [5], and moreover the carbonium ion intermediate is common to both the E1 and S_N1 pathways, the ratio of elimination to substitution should be independent of the leaving group. This behavior has been observed; for example, although the overall relative rates of reaction for $Me_3C\text{-}C\ell$ and $Me_3C\text{-}S^+Me_3$ are 7.6:1, the ratio of elimination to substitution (in 80% ethanol) is about 0.36 for both substrates [6,7].

Bunnett [8] has shown how these three mechanisms may be unified in the sense that the E1, E2, E1cb sequence represents breaking of the C-Y bond before (E1), during (E2), or after (E1cb) the breaking of the C-H bond. He has gone further in describing a model of the bimolecular transition state that admits all degrees of relative extents of C-Y and C-H bond breaking. Bunnett's spectrum of mechanisms may be represented as in structures 3 - 7, where dashed lines signify bonds that have been weakened, and dotted lines are bonds whose rupture is significantly advanced [8].

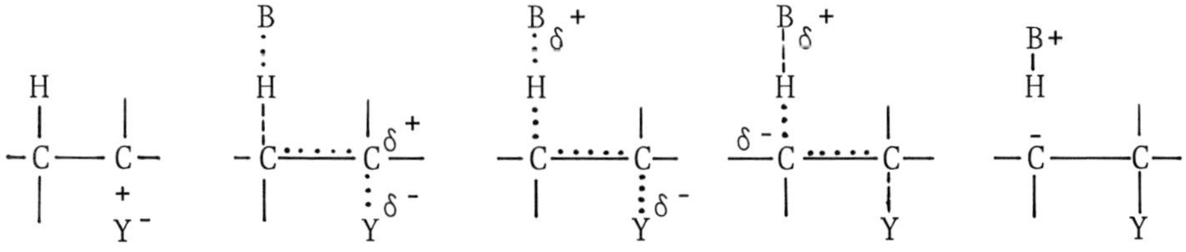

3, E1   4, E1-like   5, "pure" E2 6, E1cb-like   7, E1cb

This continuous range in bond making and bond breaking for E2 reactions, as implied by 4, 5, and 6, is supported by data such as those in Table 11.I [8] for the bimolecular elimination from β-phenethyl compounds.

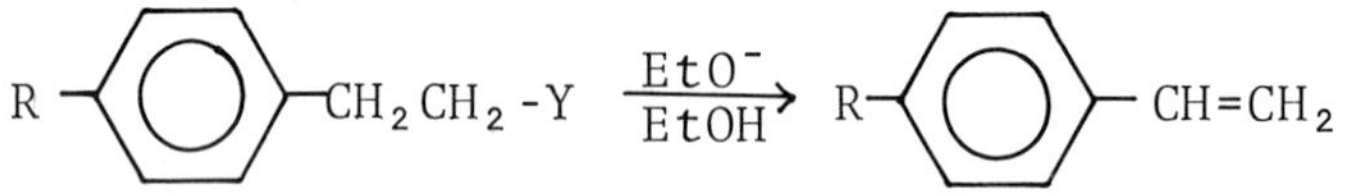

$$R-\langle O \rangle-CH_2CH_2-Y \xrightarrow[EtOH]{EtO^-} R-\langle O \rangle-CH=CH_2 + HY \quad (11.11)$$

Table 11.I   Kinetics of Bimolecular β-Elimination from β-Phenethyl Compounds [8][a]

Leaving group Y	Relative rate (R = H)	ρ	k_H/k_D[c]
I	26,000	+2.07	
Br	4,100	+2.14	7.1
OTs	392	+2.27	5.7
Cℓ	68	+2.61	
SMe$_2^+$	7.7[b]	+2.75	5.1
F	1	+3.12	
NMe$_3^+$		+3.77	3.0

[a]See Eq. (11.11).
[b]A value based on correction for the atypical activation energy.
[c]Ratio (for R = H) of rate for the ordinary compound, Eq. (11.11), to that for the compound with two α-deuteriums.

The positive Hammett rho values show that reaction is favored by para substituents that are electron withdrawing, suggesting that structure 6 is a reasonable picture of the transition state. Moreover, poorer leaving groups are associated with higher ρ values, indicating more E1cb character in their reactions. The variation in k_H/k_D also suggests variation in extent of C-H bond rupture in this series of compounds. The maximum isotope effect should be observed for a reaction in which the C-H bond is half broken at the transition state; the decrease from 7.1 to 3.0 is again consistent with increase in E1cb-like behavior for the lower members of the series in Table 11.I.
      Elimination from benzyldimethylcarbinyl chloride, Ph-CH$_2$-C(Me)$_2$Cℓ, gives an isotope effect of 2.5, indicating that this reaction occurs via a transition state well

removed from the symmetrical one, 5. Since alkyl groups
will tend to stabilize a developing positive charge, an
E1-like transition state, 4, has been suggested for this
reaction [8]. The relative rate of elimination for the
pair benzyldimethylcarbinyl bromide/benzyldimethylcarbinyl
chloride is of the order 10^3, also indicating that C-Y
bond rupture is well advanced in the transition state [9].
   Some alternatives to Bunnett's view of the E2 spectrum
have recently appeared. Parker et al. [10] have suggested
that the base B may attack hydrogen, carbon, or both, giving transition states shown as 8 (E2H, attack on hydrogen),
9 (E2C-like, attack on both hydrogen and carbon), and 10
(E2C, attack on carbon).

8, E2H   9, E2C-like   10, E2C

The following ingenious test was proposed: If the E2H
mechanism is operative, the elimination rate should fit a
Brønsted relationship with the pK_a of the conjugate acid
of B, since B is functioning as a proton-acceptor in this
mechanism. If, on the other hand, the E2C mechanism applies, the elimination rate should be correlated by a
Swain-Scott relationship [11]. For eliminations of cyclo-
hexyl tosylate in acetone a good correlation was found
between log k (E2) and log k (S_N2) for the same bases,
whereas no correlation was observed between log k (E2) and
pK_a of the bases [10]. This was interpreted to mean that
the base attacks carbon in the elimination process. However, the S_N2 and E2C mechanisms do not pass through a
common transition state (the "merged mechanism") as had
earlier been suggested for certain systems [12], as shown
by stereochemical results [13].
   The E2C-like mechanism has been extended to describe
dehalogenations; this so-called E2Haℓ mechanism is shown
as structure 11 [14].

11

The E2C concept is currently a matter of controversy [15].
     The idea that the E1cb and E2 mechanisms are on the
same continuous spectrum has been challenged by O'Ferrall
[16] who claims that E2 and E1cb reactions can occur si-
multaneously, and only when the carbanion is very unstable
is the continuity of transition states a valid picture.
Bordwell et al. [17] point out that elimination via a
carbanion need not generate second-order kinetics, and
that three types of kinetic behavior, dependent upon the
relative rates of carbanion formation, exchange, and elimination, can be demonstrated. These are shown below in
schematic form.

$$\text{B + HCCY} \underset{\text{nil}}{\overset{\text{fast}}{\rightleftharpoons}} \text{BH}^+ + {}^-\text{CCY} \xrightarrow{\text{slow}} \text{C=C} + \text{Y}^- \qquad (11.12)$$

"irreversible" first-order

$$\text{B + HCCY} \underset{\text{fast}}{\overset{\text{fast}}{\rightleftharpoons}} \text{BH}^+ + {}^-\text{CCY} \xrightarrow{\text{slow}} \text{C=C} + \text{Y}^- \qquad (11.13)$$

reversible second-order

$$\text{B + HCCY} \underset{\text{nil}}{\overset{\text{slow}}{\rightleftharpoons}} \text{BH}^+ + {}^-\text{CCY} \xrightarrow{\text{fast}} \text{C=C} + \text{Y}^- \qquad (11.14)$$

"irreversible" second-order

Equation (11.13) illustrates the usual E1cb behavior; this
situation can be detected by solvent deuterium exchange
being greater than the rate of elimination. Equation
(11.12) is a very unusual case; the first-order kinetics
(independence of base concentration) result from the rapid
quantitative production of the carbanion, which then undergoes elimination in the rate-determining first-order step.
An example has been discovered in the sodium methoxide-
catalyzed elimination of 2-phenyl-2-methoxyl-1-nitrocyclo-
pentane [18].

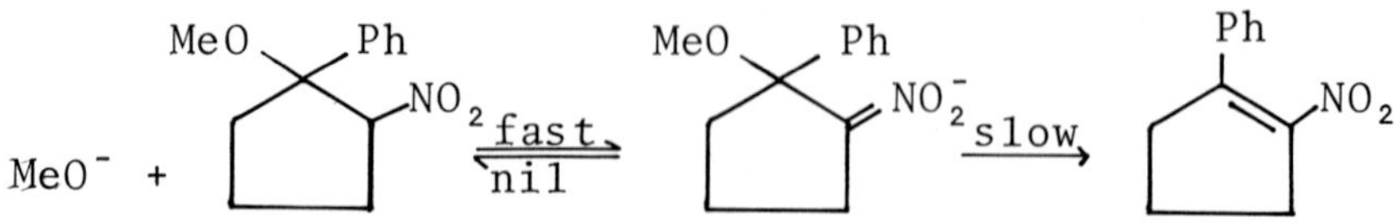

$$(11.15)$$

     Equation (11.14) is an interesting case because here
the carbanion is at steady-state concentration, the

kinetics are second order, and deuterium exchange cannot
be detected. These conditions fit the conventional E2
assignment, and Bordwell et al. [17] have suggested that
many eliminations heretofore assigned to the E2 class may
proceed in two steps via a carbanion, in accordance with
the kinetic scheme of Eq. (11.14).

Stereochemistry. Chlorofumaric acid is dehydrochlorinated
fifty times faster than is chloromaleic acid.

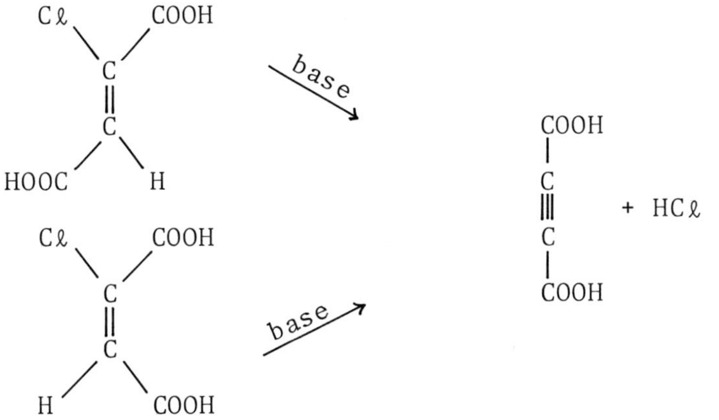

$$+ \ HCl \qquad\qquad (11.16)$$

Meso-1,2-dibromo-1,2-diphenylethane gives cis-bromostil-
bene, Eq. (11.17), whereas the dl stereoisomer gives the
trans olefin.

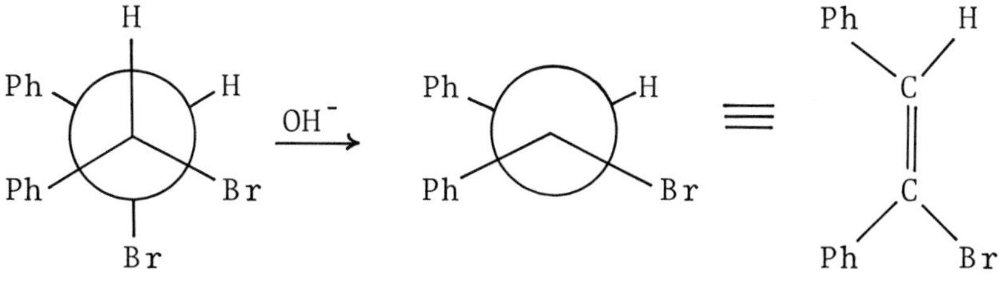

meso   (11.17)

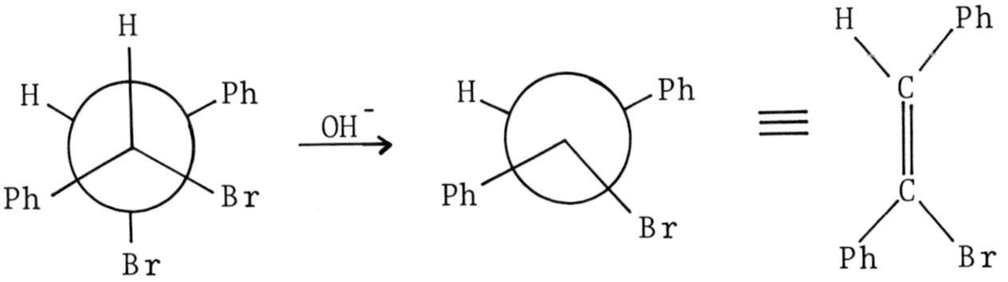

dl (either enantiomer)   (11.18)

Menthyl chloride, 12, yields only one olefin when treated with base, but neomenthyl chloride 13 gives a mixture of two olefins.

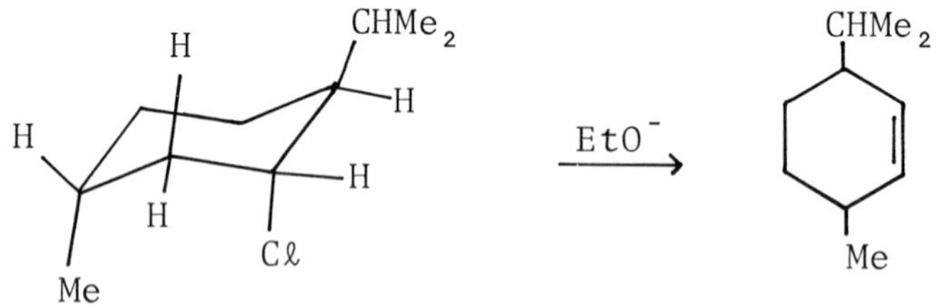

(11.19)

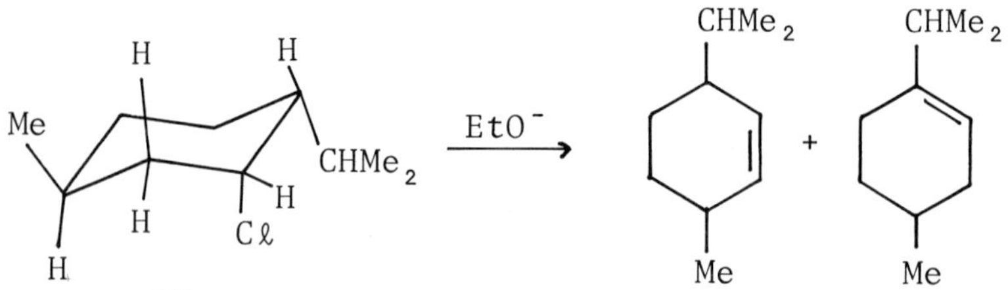

(11.20)

All of these observations suggest that anti-elimination is more facile than is syn-elimination (in E2 reactions). In the large majority of these reactions it appears that elimination is favored by a 180° dihedral angle between the C-H and C-Y bonds, a conformation that is called anti-periplanarity. Including the base, the optimum E2 transition state geometry places five atoms in a plane, as in 14.

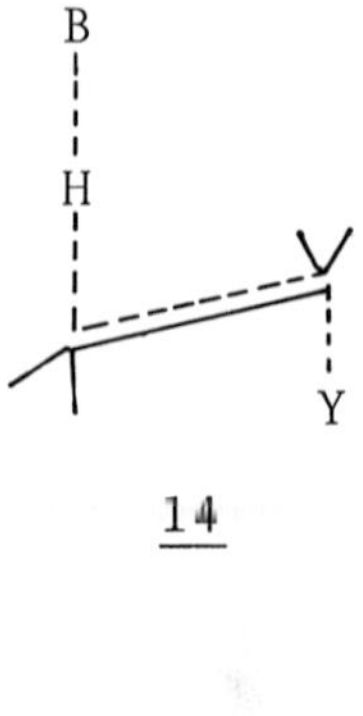

14

The origin of the preference for anti-elimination is not clear. An attractive notion is that the electron pair released as the α-hydrogen is transferred to the base carries out a back-side displacement of Y, analogous to an S_N2 process. However, this electron pair, which eventually contributes to the olefinic double bond, is not available for "attack" in the same electronic or steric sense as is implied by the S_N2 analogy, and this idea has been criticized [19]. In the proposed E2C type of attack [10] stereochemical control is easier to visualize by a direct displacement mechanism.

One factor that probably acts to decrease syn-elimination, and therefore to favor anti-elimination in a relative sense, is the eclipsing effect, which denotes the reduction in reactivity expected from the fully eclipsed syn-periplanar conformation 15 relative to the anti-periplanar form 16.

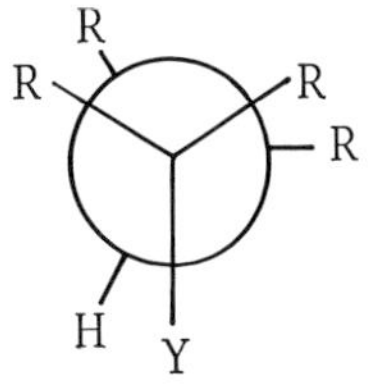

15 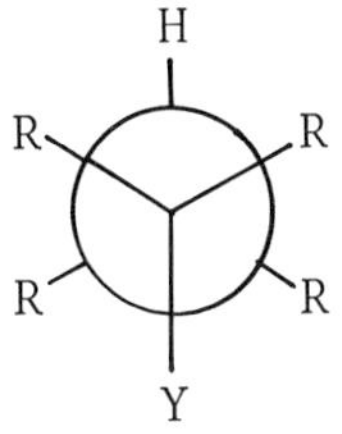   16

Despite these demonstrations and arguments that anti-elimination is favored in E2 reactions, examples are known of syn-eliminations. Some of these occur in systems that cannot, for steric reasons, attain an anti-periplanar conformation. For example, the relative rates of bimolecular elimination (110°, sodium pentoxide in 1-pentanol) for the rigid norbornane derivatives 17 and 18 are 85:1; thus syn-elimination is favored over anti-elimination, presumably because, in 18, an anti-periplanar conformation of C-H and C-Cℓ cannot be achieved, even though these bonds are in a trans relationship [20]. The syn elimination was attributed to a carbanion mechanism [20]. The concerted E2 mechanism is preferred by DePuy et al. [21] for syn-eliminations in which the periplanar conformation is possible. The tosylate 19 readily undergoes the syn-elimination shown.

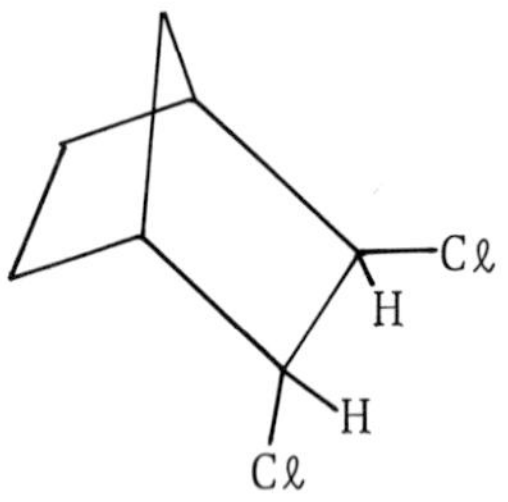

17 (trans) 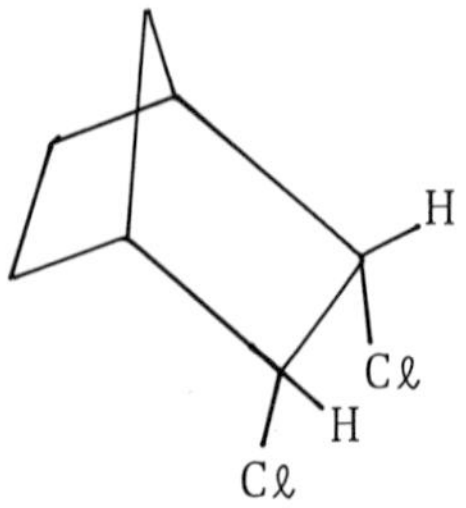   18 (endo-cis)

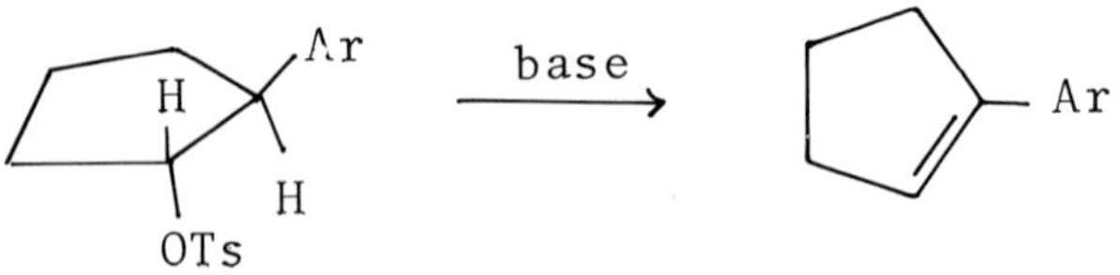

19   (11.21)

The Hammett ρ is +2.34 under conditions for which β-aryl-
ethyl tosylates had ρ = +3.39. Since the acyclic compounds
presumably eliminate in a bimolecular E2 anti fashion, the
cyclic compounds 19, apparently displaying even less
carbanion character, must also react by the E2 mechanism.
It is concluded that ease of elimination is controlled by
the ease with which the periplanar conformation can be
attained, the anti- and syn-periplanar states being favorable to elimination, and deviations from these being increasingly antagonistic as the dihedral angle approaches
90° [22].
        Halide ion promotion of dehalogenations leads primarily
to anti-elimination. This reaction is not well understood,
for the halide ions are surprisingly effective considering
their very low basicity. The E2Haℓ mechanism [14], which
invokes attack on carbon, may help account for this reactivity [23].
        An interesting application of the stereochemical course
of the E2 mechanism is in establishing the stereochemistry
of a substrate. For example, Hammond and Collins [24]
showed that hydrogen chloride adds to 1,2-dimethylcyclo-
pentene predominantly in an anti manner by the following
experiment. The addition product must be either cis-1,2-
dimethylpentyl chloride, 20, or the trans isomer, 21.
Under E2 reaction conditions, so that anti-elimination

occurs, the cis isomer will yield only 2,3-dimethylcyclo-
pentene, whereas the trans isomer can give 1,2-dimethyl-
cyclopentene.

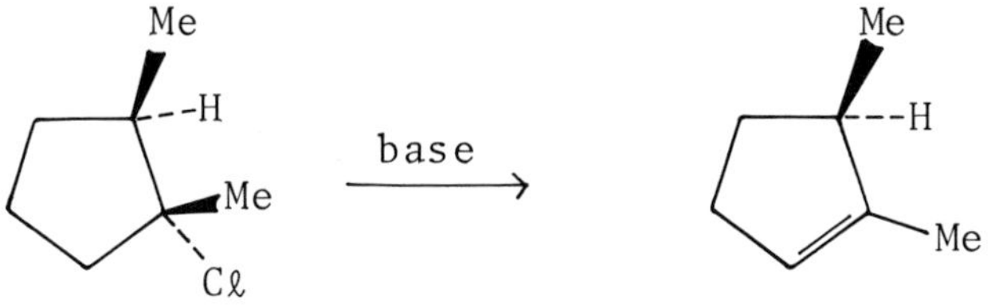

(11.22)

20

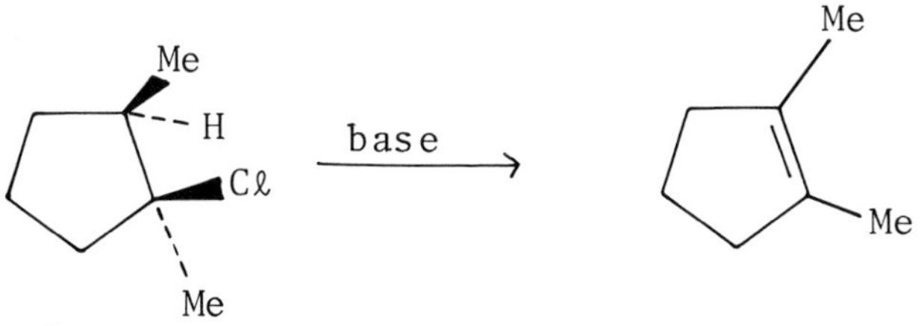

(11.23)

21

(11.23)

The main product of the elimination reaction was the 1,2-
dimethylcyclopentene, showing that 21 was the addition
product and therefore that HCℓ underwent anti-addition to
the olefin.

The stereochemical course of the E1 mechanism is not
so simply established, in part because this is a multistep
pathway for elimination. If the carbonium ion intermediate
is relatively stable and can achieve the optimum planar
(and symmetrical) conformation before elimination occurs,
then anti- and syn-elimination should be equally likely.
If, however, elimination occurs from the various ion-pair
intermediates that have been described for S_N1 reactions,
a definite asymmetry persists at the site of ionization
and anti-elimination may be favored over syn-elimination,
as for E2 reactions. A further complication is provided
by the possibility of neighboring group participation.
Cram [25] attempted to sort out these effects by studying
the E1 reactions of the four stereoisomers of the tosylate
of 3-phenyl-2-butanol [26]. Both a stereospecific anti-
elimination and a nonstereospecific elimination mechanism
were identified; a hydrogen bridge was postulated to account for the stereospecific route.

$$CH_3-\underset{\underset{Ph}{|}}{\overset{\overset{OTs}{|}}{CH}}-CH-CH_3 \xrightarrow{HOAc} \underset{Ph}{\overset{CH_3}{>}}C=C\overset{CH_3}{\underset{H}{<}} + \underset{Ph}{\overset{CH_3}{>}}C=C\overset{H}{\underset{CH_3}{<}} \qquad (11.24)$$

Orientation. When the quaternary ammonium hydroxide 22 undergoes β-elimination, two olefins could be produced, depending upon which alkyl branch contributes the proton.

$$CH_3CH_2-\underset{\underset{Me}{/}\ \underset{Me}{\backslash}}{\overset{+}{N}}-CH_2CH_2CH_3 \xrightarrow{OH^-} CH_2=CH_2 + CH_2=CHCH_3 \qquad (11.25)$$
$$\qquad\qquad\qquad\qquad\qquad\qquad 98\% \qquad\quad 2\%$$

In fact, ethylene is the predominant product. This orientation, of preferential elimination toward the carbon carrying the fewer alkyl substituents (and therefore more hydrogens), is called Hofmann orientation. It is observed for E2 reactions of substrates with ammonium, sulfonium, and sulfone leaving groups.

2-Bromobutane, on the other hand, yields mainly 2-butene rather than 1-butene.

$$CH_3CH_2\underset{\underset{Br}{|}}{CHCH_3} \xrightarrow{OH^-} CH_3CH=CHCH_3 + CH_3CH_2CH=CH_2 \qquad (11.26)$$
$$\qquad\qquad\qquad\qquad\qquad 81\% \qquad\qquad 19\%$$

In this case the hydrogen has been provided by the carbon carrying more alkyl substituents (and therefore fewer hydrogens). This general behavior, called Saytzeff orientation [27], is seen with E2 reactions of halides and esters, and with E1 reactions. Recognition of Hofmann and Saytzeff orientation as general but contrasting modes of elimination behavior was accomplished by Ingold [28], who also provided the first explanations. A key observation is that Hofmann elimination yields the least stable olefin, that is, the one with fewest alkyl substituents attached to the double bond termini; Saytzeff elimination yields the most stable olefin. It is clear that two effects must be operating, with the type of orientation being determined by whichever effect is in control for a particular reaction. Ingold [28] identified these effects as the inductive effect (favoring Hofmann orientation) and the hyperconjugative effect (leading to Saytzeff orientation). The first of these can be illustrated with Eq. (11.25) and structure 23. The argument supposes that the more acidic hydrogen will be preferentially eliminated; of the two possibilities, H_A should be more acidic than H_B, since the latter is adjacent

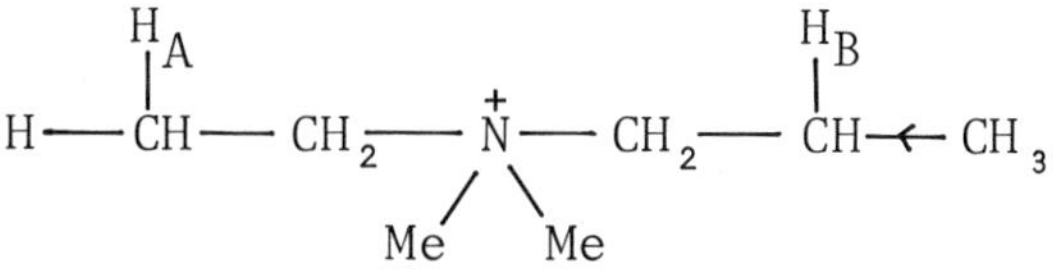

$$H-\overset{\overset{\displaystyle H}{|}{}^A}{\underset{}{CH}}-CH_2-\overset{+}{\underset{\underset{\displaystyle Me\quad Me}{\diagup\quad\diagdown}}{N}}-CH_2-\overset{\overset{\displaystyle H}{|}{}^B}{\underset{}{CH}}\leftarrow CH_3$$

<u>23</u>

to an extra methyl group, which is electron releasing
relative to hydrogen [29].

The argument for Saytzeff elimination assumes that
factors tending to stabilize the final olefin will also
stabilize the developing double bond in the transition
state. Using Eq. (11.26) and structures <u>24</u> and <u>25</u> in
illustration, it is seen that stabilization by hyperconjugation can be more effective in <u>24</u> than in <u>25</u>, because
of the greater numbers of hydrogens beta to the incipient
double bond in <u>24</u>.

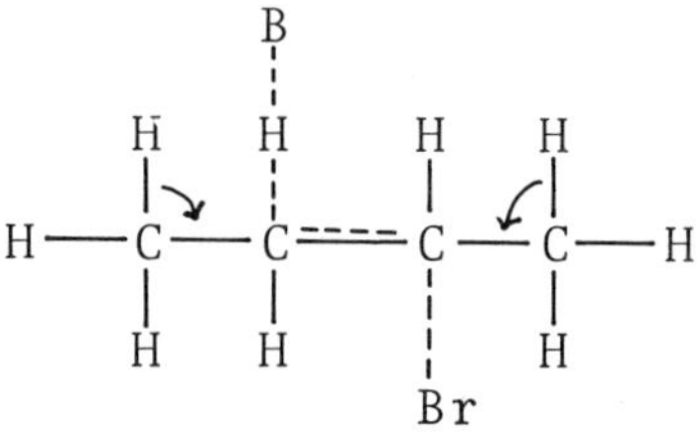

<u>24</u> 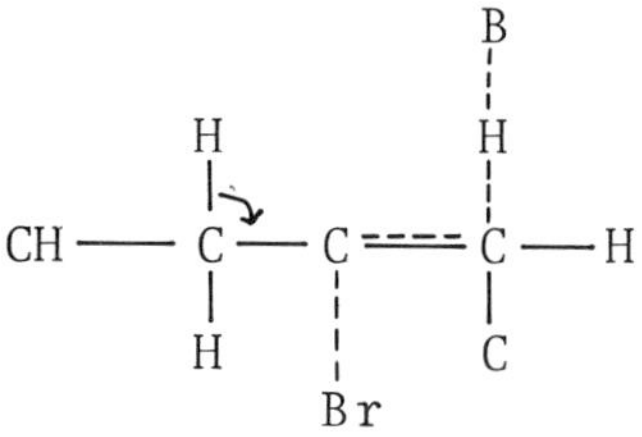   <u>25</u>

This effect is even more pronounced with substituents such
as phenyl groups, which can undergo conjugation rather
than hyperconjugation.

Which of these effects is in control will depend in
part upon structural features. The two effects, inductive
and conjugative, act independently. It is possible to influence (by control of reaction conditions) the elimination mechanism, and for a single substrate to elicit either
Hofmann or Saytzeff orientation. For example, the tert-
amylsulfonium ion <u>26</u> undergoes second-order elimination in
the presence of base, but reacts by first-order kinetics
in the absence of base.

Equation (11.27) and Table 11.II summarize the product
distributions [28].

$$\underset{\underline{26}\quad +SMe_2}{CH_3CH_2-\overset{\overset{\displaystyle CH_3}{|}}{\underset{}{C}}-CH_3}\longrightarrow \underset{(Hofmann)}{CH_3CH_2-\overset{\overset{\displaystyle CH_3}{|}}{\underset{}{C}}=CH_2}\ +\ \underset{(Saytzeff)}{CH_3CH=\overset{\overset{\displaystyle CH_3}{|}}{\underset{}{C}}-CH_3}\qquad (11.27)$$

Table 11.II Dependence of Orientation on Mechanism:
Elimination from tert-Amyldimethylsulfonium Ion [28][a]

Mechanism	Percent olefin	
	Hofmann	Saytzeff
Bimolecular[b]	86	14
Unimolecular[c]	13	87

[a]See Eq. (11.27).
[b]As the ethoxide in 97% ethanol at 24°.
[c]As the iodide in 97% ethanol at 50°.

Clearly the direction of elimination depends upon the
mechanism; in other words, control of orientation depends
upon the nature of the transition state. In terms of the
spectrum of transition states (structures 3 - 7), Hofmann
elimination control is seen with E1cb-like transition
states, and Saytzeff elimination becomes increasingly
manifest as the transition state shifts toward the E1-like
[8]. This recognition gives some predictive ability to
the rationalizations of orientation behavior [30].
    Brown has proposed a theory of orientation behavior
that replaces the inductive effect with steric effects,
the steric natures of the leaving group, the base, and of
substituents in the substrate being taken into account
[31]. There is no doubt that steric factors play a role,
particularly for very large groups, but Brown's postula-
tion of universal control by steric factors, in place of
the inductive effect contribution to Hofmann orientation,
has been severely criticized [8,32].

Some Elimination Reactions. We are now in a position to
consider the effects of reactant structure and reactant
conditions on elimination behavior, and after this discussion some important elimination reactions will be described. It was earlier pointed out that elimination and
nucleophilic substitution are competing reactions, and
the ratio of elimination to substitution products (E/S_N)
is obviously of great concern in any synthetic or analytical scheme based on an elimination. This ratio can be
affected by environmental factors, and some general statements can be made about the directions of these effects.
    Consider, first, a set of circumstances in which only
the unimolecular E1 and S_N1 reactions are important. The
rate-determining step in both of these reactions is the
formation of the carbonium ion intermediate, a step that

is essentially independent of the identity and concentration of base. In the absence of additional nucleophilic
species (other than the solvent), the product ratio $E1/S_N1$
will be independent of the base. If, now, the base concentration is progressively increased so that bimolecular
reactions become important, the rate-determining steps for
elimination and substitution become base dependent. Since,
however, the kinetic dependence on base is identical for
the E2 and S_N2 reactions, the ratio $E2/S_N2$ is independent
of base concentration, just as with the ratio $E1/S_N1$.
However, this does not mean that the product ratios $E1/S_N1$,
$E2/S_N2$ are equal, for a given substrate. Though $E1/S_N1$
is independent of the base identity, $E2/S_N2$ does depend
upon the nature of the base. Consider a base that is a
powerful carbon nucleophile but a weak hydrogen nucleophile (such as thiophenoxide); this nucleophile will favor
substitution over elimination, lowering $E2/S_N2$. A very
strong base (toward hydrogen) will act in the reverse
manner, and so hydroxide and alkoxide ions tend to raise
the $E2/S_N2$ product ratio. Thus a high concentration of a
strong base favors an increased yield of olefin relative
to substitution product.
     The effect of solvent polarity on the elimination/
substitution ratio is usually described in terms of the
Hughes-Ingold theory of transition state charge dispersal
(see Secs. 3.4 and 9.1). Compare the E2 and S_N2 transition states for reaction of hydroxide with an alkyl
halide:

$$HO^{\delta^-} --- H --- \overset{|}{\underset{|}{C}} == \overset{|}{\underset{|}{C}} --- X^{\delta^-} \qquad E2$$

$$HO^{\delta^-} --- \overset{\backslash/}{C} --- X^{\delta^-} \qquad S_N2$$
$$H - \overset{|}{\underset{|}{C}} -$$

Evidently greater charge dispersal occurs in the E2 transition state, so a decrease in solvent polarity, though
promoting both processes, will favor elimination over substitution. The evidence concerning this prediction is
equivocal, as is most evidence on solvent effects, because
it is not possible to vary polarity without concomitantly
altering some other property. Table 11.III gives overall
reactivities (E2 + S_N2) for the bimolecular reactions of
β-phenethyl bromide in alcohol-water mixtures [33].
     According to the above arguments, addition of water to
an alcohol should decrease both the S_N2 and E2 processes,

Table 11.III   Bimolecular Elimination and Substitution of
β-Phenethyl Bromide in Alcohol-Water Mixtures at 25° [33]

Volume percent alcohol	$10^5 k_2/M^{-1}sec^{-1}$ (E2 + S_N2)	Percent olefin
Methanol		
100	3.43	75
90	4.43	77
80	5.58	78
70	6.40	78
60	6.58	79
Ethanol		
100	28.2	100
90	19.2	96
80	15.3	97
70	14.1	92

having the greater effect on elimination. The data in
Table 11.III cannot be accounted for on this basis, and
it was suggested [33], by analogy with earlier results on
substitution reactions as described in Chap. 9, that the
Hughes-Ingold transition state concept is but one of three
important effects, the other two being the alkoxide-hydrox-
ide equilibrium $OH^- + ROH \rightleftharpoons OR^- + HOH$, and the increased
dissociation of ion pairs in more aqueous solvents.   If
the nucleophilicity sequence is $EtO^- > MeO^- > OH^-$, these
results can be accounted for by attributing the rate decrease in ethanol systems to the replacement of ethoxide
by the less nucleophilic hydroxide; the rate increase in
methanol solutions may be a consequence of increased ion-
pair dissociation [33,34].

     A final environmental factor influencing the E/S_N
ratio is the temperature.   In the E2 transition state a
carbon-hydrogen bond is partially ruptured, whereas in
the S_N2 transition state this process is not necessary.
The enthalpy of activation should therefore be greater
for elimination than for substitution (and it usually is,
by 1-2 kcal/mole).   As the temperature is increased, elimination therefore profits more than does substitution.
Putting together these several observations shows why synthetic procedures for olefin production by dehydrohalogenation call for reflux conditions in strong, concentrated
bases.

     Consideration of relative elimination rates for dif-
ferent substrates must take into account the D/C_N competi

tion, for it clearly would be misleading to ascribe to a
substrate a low elimination reactivity solely on the basis
of a low olefin yield. With gas-liquid chromatography,
direct examination of elimination products is greatly
facilitated. The factors available to the experimentalist
for variation (not always independently) are the structure
of the substrate, the base, the solvent, and the temperature; the significant observations are the E/S_N ratio, the
Hofmann-Saytzeff ratio, the cis-trans ratio, and elimination rates. We have already considered some of these. In
extension it can be noted that the effect of a structural
or environmental variation on elimination behavior depends
upon the nature of the transition state. Bunnett [8] has
systematically described these dependencies. An example
is provided by the Hofmann to Saytzeff elimination ratio
from 2-methyl-3-pentyltrimethylammonium ion [35]. This E2
reaction was studied in alkoxide-alcohol solvent systems.

$$CH_3CH_2\underset{\underset{+NMe_3}{|}}{\overset{\overset{CH_3}{|}}{C}}HCHCH_3 \xrightarrow[\text{ROH}]{RO^-} CH_3CH=\underset{}{\overset{\overset{CH_3}{|}}{C}}HCHCH_3 + CH_3CH_2CH=\overset{\overset{CH_3}{|}}{C}CH_3$$

(Hofmann)   (Saytzeff)   (11.28)

On changing from a normal to a secondary to a tertiary
base-solvent system, the proportion of Hofmann product was
progressively increased. This was attributed [35] to increasing alkoxide base strength, with stronger bases forcing a more reactant-like nature of the transition state [30].
     A few elimination reactions are briefly described [36].
The Hofmann degradation is the final step in a scheme
usually applied to establish the structure of an amine,
especially alkaloids. First the amine is subjected to ex-
haustive methylation with methyl iodide, giving the quaternary ammonium iodide. Upon treatment with silver oxide
the quaternary ammonium hydroxide is obtained; this is
heated in alcoholic medium. β-Elimination occurs with
Hofmann orientation. The structure of the olefin is established by standard procedures, and the original amine
structure is inferred. A secondary cyclic amine requires
two cycles to eliminate the nitrogen from the final olefin,
as shown in Equation (11.29).
     A recent olefin synthesis, the Wittig reaction, uti-
lizes a suitable onium substrate (ammonium, sulfonium,
phosphonium) and a very strong base such as phenyl lithium.
The base abstracts a proton from the α' position, giving
an intermediate ylide, 27. In an intramolecular step the
carbanion removes the β proton. This is called the α'-β
mechanism.

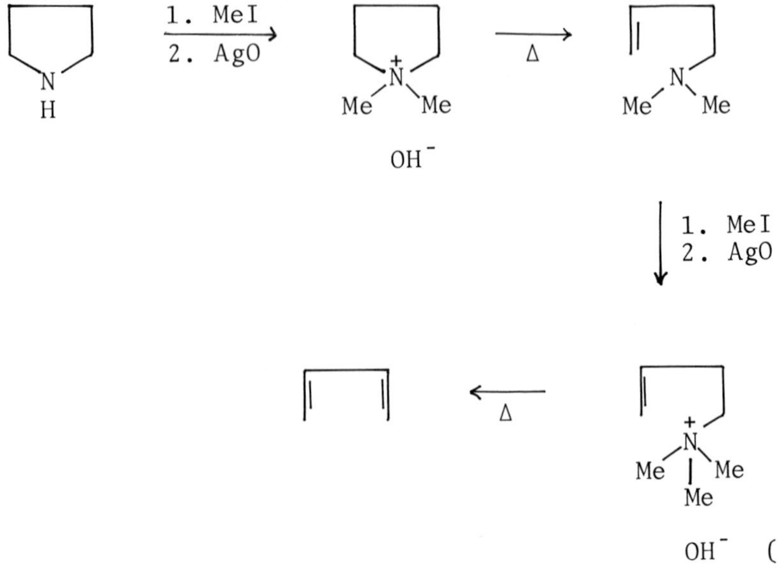

$$\text{(11.29)}$$

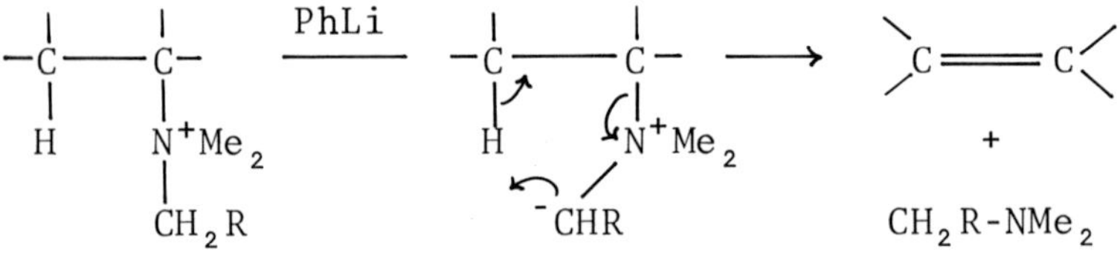

$$\underline{27} \qquad\qquad\qquad\qquad\qquad \text{(11.30)}$$

Syn-elimination occurs. The Wittig reaction is used synthetically to convert a carbonyl group to a substituted
methylene group by forming an ylide from the carbonyl and
(for example) methylene triphenylphosphorane. The sequence is [37]

$$Ph_3P=CH_2 + RCH \longrightarrow H-\underset{\underset{Ph_3P^+}{|}}{\overset{\overset{H}{|}}{C}}-\underset{\underset{O^-}{|}}{\overset{\overset{R}{|}}{C}}-H \longrightarrow CH_2=CHR + Ph_3P=O \quad \text{(11.31)}$$

β-Elimination is assisted by a carbonyl group adjacent
to the leaving proton, and elimination from Mannich bases

appears to be intramolecularly catalyzed by the leaving
amino group [38]. A two-step process is shown in Eq.
(11.32), but concerted cyclic transition states have also
been suggested. The reaction is reversible (the reverse
is an Ad_N2 reaction).

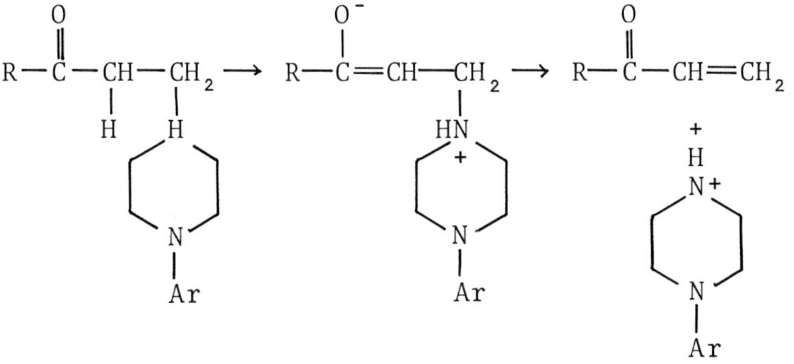

$$R-\overset{\overset{\displaystyle O}{\|}}{C}-\underset{\underset{\displaystyle H}{|}}{CH}-\underset{\underset{\displaystyle H}{|}}{CH_2} \longrightarrow R-\overset{\overset{\displaystyle O^-}{|}}{C}=CH-CH_2 \longrightarrow R-\overset{\overset{\displaystyle O}{\|}}{C}-CH=CH_2$$

(11.32)

The dehydrohalogenation of alkyl halides has been
mentioned several times in illustration of elimination
features. Saytzeff orientation is common. E2 is the
usual mechanism, though E1 can occur with secondary, tertiary, or α-aryl-substituted halides in ionizing solvents.
Dehydrohalogenation of vic-dihalides (28), gem-dihalides
(29), or halo-alkenes (30) by a strong base gives alkynes.

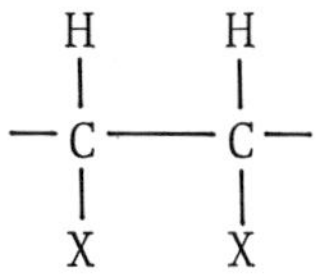

28 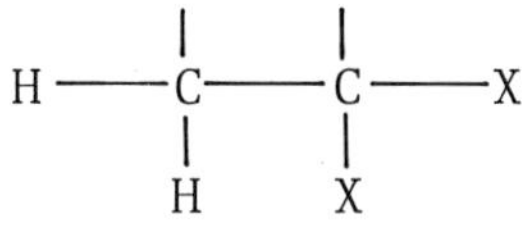   29 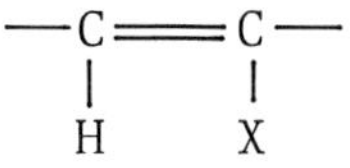   30

With the very strong base $NaNH_2$, the 1-alkyne R-C≡CH is
favored because it is more acidic than the 2-alkyne
R-C≡C-CH$_2$; the 2-alkyne is thermodynamically more stable,
however, so it is favored when weaker bases are used.

β-Elimination of halogen and another group can be promoted by metals. The debromination of vic-dibromides with
zinc is typical.

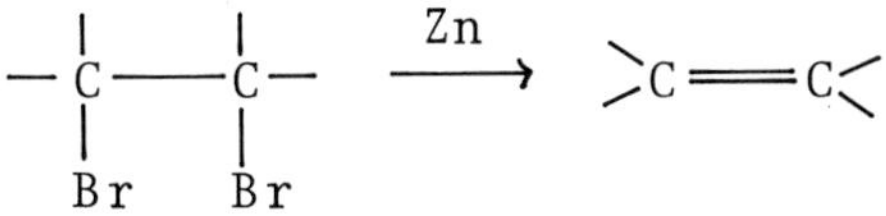

$$-\overset{|}{\underset{\underset{\displaystyle Br}{|}}{C}}-\overset{|}{\underset{\underset{\displaystyle Br}{|}}{C}}- \quad \overset{Zn}{\longrightarrow} \quad \overset{\displaystyle >}{}C=C\overset{\displaystyle <}{}$$

(11.33)

The reaction probably proceeds via a carbanion intermediate, because substitution and dehydrohalogenation do

not occur [39].

The dehydration of alcohols is the reverse of the addition of water to olefins (see Sec. 10.1). The reaction is acid catalyzed, and the mechanism can be inferred, according to the principle of microscopic reversibility, from the mechanism of olefin hydration. It is essentially an El reaction, the rate-determining step being $ROH_2^+ \rightarrow R^+ + H_2O$. Saytzeff orientation predominates, though this would be expected under the usual conditions of thermodynamic control. Certain special mechanisms have been identified; thus some acids, like sulfuric and phosphoric, can form esters with the alcohols, and the elimination occurs from the ester. A rate term second order in alcohol has been detected for the dehydration of 1,2-diphenyl-2-propanol (31) in acetonitrile [40]. This was interpreted as an elimination from the protonated ether 32, which is formed from the alcohol and its derived carbonium ion.

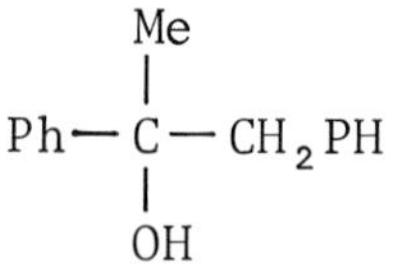

$$
\begin{array}{cc}
\overset{\displaystyle Me}{\underset{\displaystyle OH}{Ph-\overset{\textstyle |}{\underset{\textstyle |}{C}}-CH_2PH}} & \overset{\displaystyle Me \qquad Me}{\underset{\displaystyle Ph \qquad Ph}{PhCH_2-\overset{\textstyle |}{\underset{\textstyle |}{C}}-O-\overset{\textstyle |}{\underset{\textstyle |}{C}}-CH_2Ph}}
\end{array}
$$

$$\underline{31} \qquad\qquad\qquad\qquad \underline{32}$$

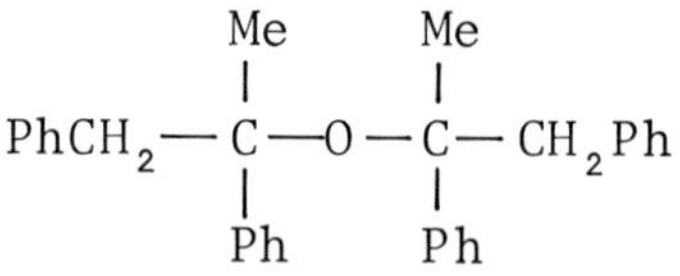

The dehydration of 9-fluorenylmethanol, Eq. (11.34), occurs by the Elcb mechanism, presumably because of the acidic proton and the poor leaving group [41].

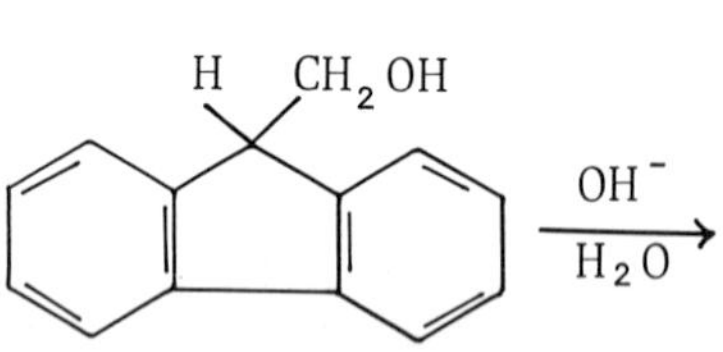

$$(11.34)$$

An iodine-catalyzed dehydration (the Hibbert reaction) appears to utilize an El pathway for tertiary and some secondary alcohols [42].

$$
\underset{\displaystyle Ph-CH-CH_3}{\overset{\displaystyle OH}{\overset{\textstyle |}{}}} \overset{I_2}{\longrightarrow} Ph-CH=CH_2 \qquad\qquad (11.35)
$$

Cis-trans isomerization of cinnamic acid, which is acid catalyzed in aqueous solution, appears to be a two-step hydration-dehydration sequence [43].

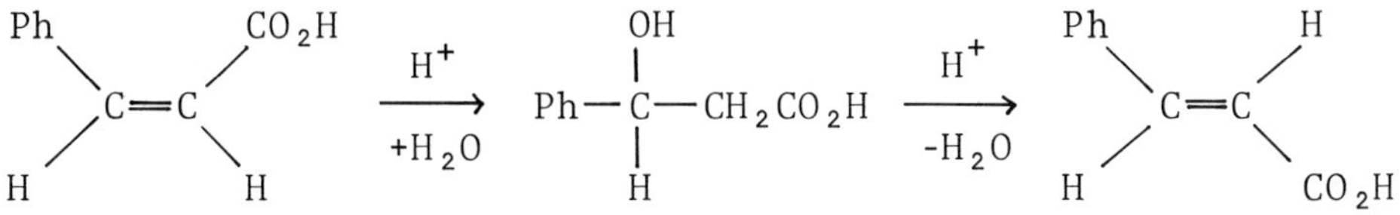

$$
\underset{H}{\overset{Ph}{\diagdown}}C=C\underset{H}{\overset{CO_2H}{\diagup}} \quad \xrightarrow[+H_2O]{H^+} \quad Ph-\underset{\underset{H}{|}}{\overset{\overset{OH}{|}}{C}}-CH_2CO_2H \quad \xrightarrow[-H_2O]{H^+} \quad \underset{H}{\overset{Ph}{\diagdown}}C=C\underset{CO_2H}{\overset{H}{\diagup}}
$$

$$(11.36)$$

An important series of carbonyl condensation reactions, followed by dehydrations, leads to olefins conjugated with carbonyl or carboxyl groups [37]. The preliminary additions are discussed in Chap. 12; they all involve attack by a nucleophilic carbon species on the electron-deficient carbonyl carbon. The aldol condensation, exemplified by the first step in Eq. (11.37), leads to α,β-unsaturated aldehydes and ketones.

$$
\underset{CH_3CH}{\overset{O}{\|}} + \underset{^-CH_2CH}{\overset{O}{\|}} \rightarrow \underset{CH_3CHCH_2CH}{\overset{O^-}{\underset{|}{}}\overset{O}{\|}} \xrightarrow{H^+} \underset{CH_3CH=CHCH}{\overset{O}{\|}} \qquad (11.37)
$$

The double bond preferentially moves into the site in conjugation with the carbonyl group. The Knoevenagel, Reformatsky, and Perkin reactions are similar in nature, differing primarily in the source of nucleophile.

A type of elimination quite different from heterolytic solution processes is pyrolytic elimination in the gas phase [44]. Some of these are synthetically valuable. The reactions occur principally through cyclic transition states (E_i mechanism); these are shown for esters (33), xanthates (34, the Chugaev reaction), and amine oxides (35, the Cope reaction).

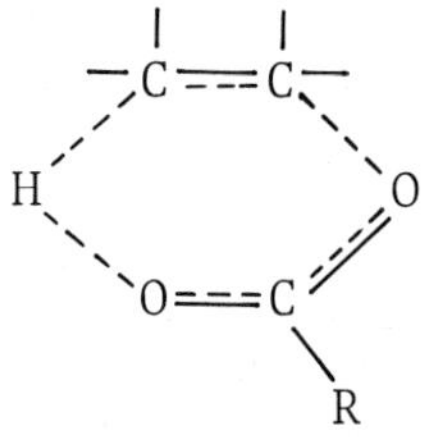

$$
\begin{array}{ccc}
\text{33} & \text{34} & \text{35}
\end{array}
$$

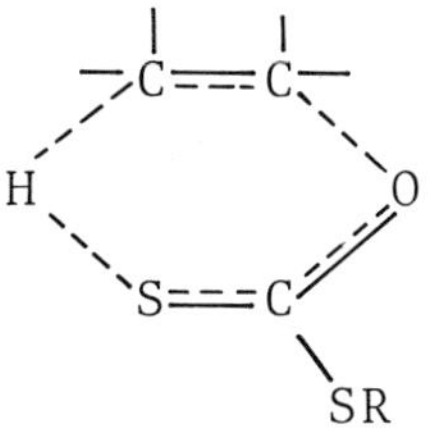

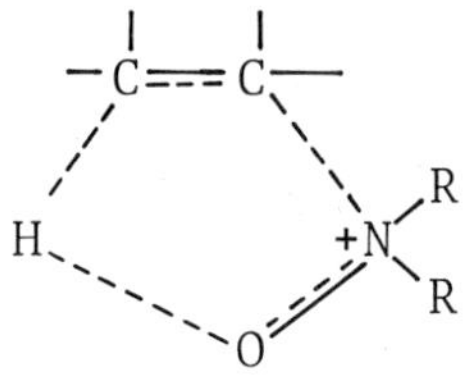

Syn-elimination occurs in these reactions. Although the transition states 33-35 imply fully concerted bond-making and bond-breaking steps, a range of transition states is available for these E_i reactions, as for the E2 reactions considered earlier.

11.2   Analytical Reactions

The Methods.   Few analyses are based on olefin-forming
β-eliminations.   Several dehydration reactions have been
utilized.   A method for propylene glycol (1,2-propanediol)
is interesting in its specificity [45].   This compound is
dehydrated by sulfuric acid to a mixture of allyl alcohol
and the enolic form of propionaldehyde.   This mixture of

$$CH_3CHCH_2 \xrightarrow[-H_2O]{H_2SO_4} CH_2=CHCH_2OH + CH_3CH=CHOH \qquad (11.38)$$
$$\quad\;\; |\;\; | $$
$$\quad OHOH$$

products gives a violet color with ninhydrin in the presence of sulfuric acid.   The specificity is partly a consequence of the ninhydrin reaction, which occurs only with
allyl alcohol and propionaldehyde; propylene glycol is
dehydrated to these compounds, whereas most other alcohols
and glycols are not.
     The aldopentoses (ribose, xylose, lyxose, and arabi-
nose) are converted by the action of mineral acids to
furfural.   The formal reaction is indicated in

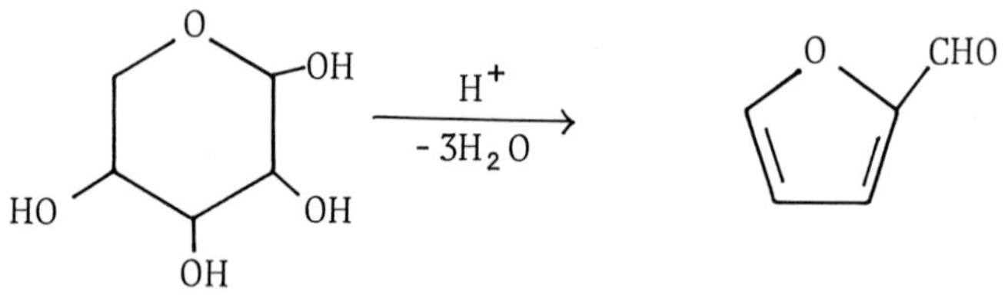

$$(11.39)$$

The ultraviolet absorption of the furfural chromophore
provides an assay method for the sugar.   Mixtures of ribose and 2-deoxy-D-ribose have been analyzed in this way,
since the latter compound produces a different chromophore [46].   A detailed study has been made by Garrett
and Dvorchik [47] of the kinetics and mechanisms of furfural production from the aldopentoses.   These sugars can
exist as equilibrium mixtures of open chain, furanose
(five-membered ring), and pyranose (six-membered ring)
forms.   Moreover, each cyclic form has two possible ano-
mers, and each pyranose anomer has two chair conforma-
tions.   It was found that the relative rates of furfural
production from the aldopentoses were ribose, 5.4; xylose,
2.2; lyxose, 2.2; arabinose, 1.0.   The results were interpreted on the assumption that the pyranose forms are the

reactive species, and that the different rates are consequences of stereochemical differences. The rate-determining step was considered to be the initial dehydration of
a trans, diaxial protonated hydroxy group and proton, with
some ring opening in the transition state of the concerted
process, as indicated for one possible sequence with
lyxose [47]:

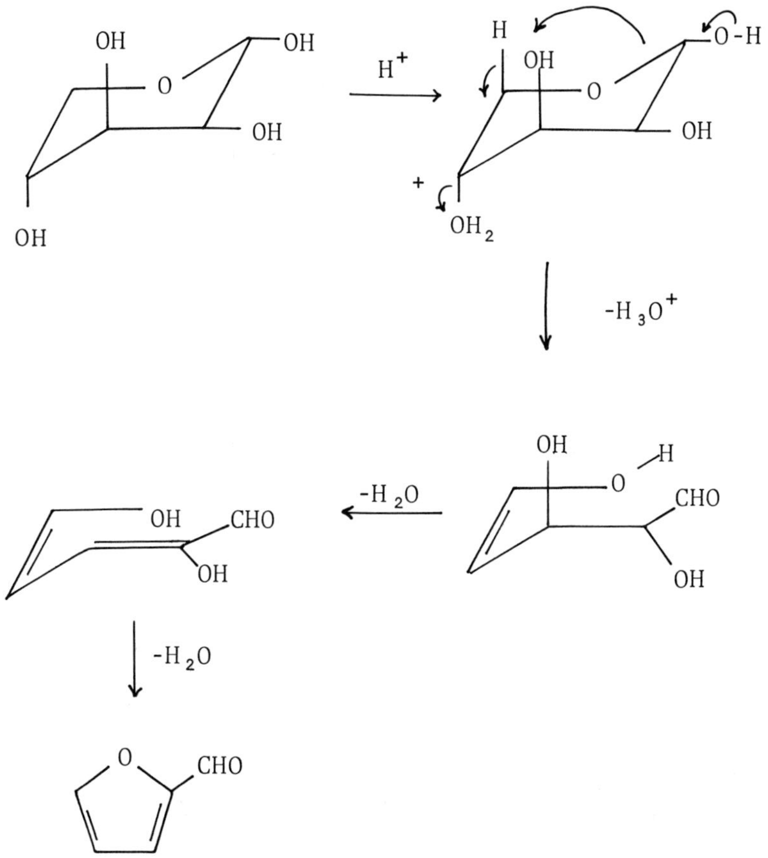

$$(11.40)$$

The ring cleavage in the transition state was postulated on the evidence of large positive entropies of activation. The suggested mechanism seems doubtful, however,
when considered in the light of other, related, elimination reactions. As noted earlier, acid-catalyzed dehydrations usually occur by the unimolecular (E1) mechanism,
with the ionization step being rate determining. The

energy of activation for all four sugars is 30.8 kcal/mole, which may be compared with 30.1 kcal/mole for the E1 dehydration of tert-butyl alcohol [48]. Large positive entropies of activation are typical of the unimolecular ionization pathway for ions such as onium ions [49]; a protonated hydroxy compound is such a species. It therefore seems that the aldopentose dehydration might be reinterpreted in E1 terms.

A method for secondary alcohols utilizes dehydration with an aluminosilicate catalyst in xylene; after reaction is complete the water produced is measured [50].

Some active methylene compounds have been determined by condensation with carbonyls followed by dehydration. Thus cyclopentadiene reacts with benzaldehyde under basic conditions to yield phenylfulvene [51].

$$(11.41)$$

The phenylfulvene is measured spectrophotometrically. Presumably this is an E1cb reaction. β-Hydroxy carbonyl compounds, or indeed nearly any alcohol carrying a strongly electronegative group beta to the hydroxy, should similarly be susceptible to dehydration. Such compounds are even dehydrated under pyridine-catalyzed acylation conditions [52].

Eliminations from compounds containing the leaving groups familiar in mechanistic and synthetic studies have not been analytically exploited. In a sense the Hofmann degradation scheme is an analytical tool for structure elucidation. α-Halo and β-halo carboxylic acids give quantitative yields of halide ions in aqueous solution, at least in part through a β-elimination [53].

As a preparative method for olefins, dehalogenation is not very useful because the 1,2-dihalo compounds are prepared from olefins. These compounds might be analyzed by dehalogenation, however. 1,2-Dichloro and 1,2-dibromo compounds can be identified by iodide-promoted dehalogenation [54]. The reaction is carried out in acetone solution; 1,2-dihalo compounds give a precipitate of the halide salt and they liberate free iodine. Siggia et al. [55] developed a method for oxyalkylene groups that is based on their conversion to 1,2-diiodo compounds upon treatment with hydriodic acid. The diiodo compound dehalogenates and the iodine released is titrated with thiosulfate. The reactions can be written as follows:

$$RO\text{-}CH_2CH_2\text{-}OR \xrightarrow{HI} I\text{-}CH_2CH_2\text{-}I \rightarrow CH_2{=}CH_2 + I_2 \qquad (11.42)$$

Polyalkylene glycols, 1,4-dioxane, and 1,2-glycols could be determined, but epoxides could not.

Pyrrolizidine alkaloids can be determined spectrophotometrically by a reaction sequence that formally includes a dehydration [56].

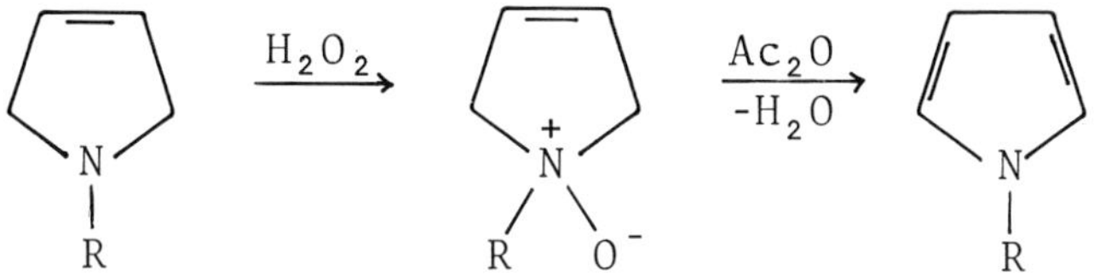

$$(11.43)$$

It has been suggested that the N-oxide is acetylated and that the elimination proceeds as in [57]

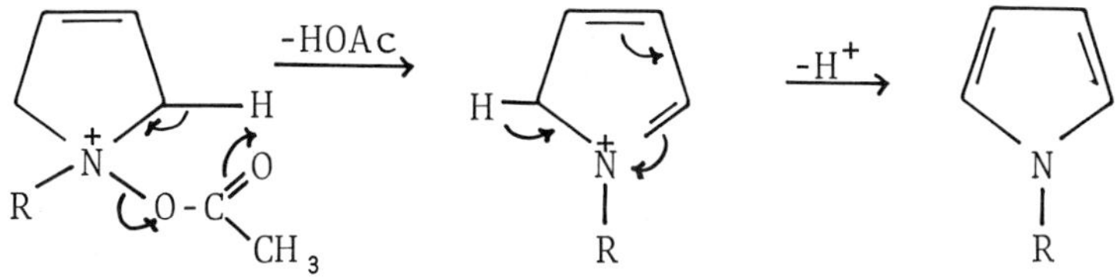

$$(11.44)$$

We now consider an ingenious and valuable end group analysis used for determining the base sequence of ribonucleic acids. A ribopolynucleotide chain, shown as 36, consists of ribonucleosides connected by repeating $C_3{\acute{}}\text{-}C_5{\acute{}}$ phosphodiester bonds. The sequencing procedure consists of three steps: (1) release of the terminal phosphate by the action of a phosphatase enzyme; (2) periodate oxidation of the resulting 1,2-glycol; (3) amine-catalyzed cleavage of the 5´-phosphate bond [58]. The released nucleoside is isolated and identified, and then the cycle is repeated on the nucleotide fragment. In this way the base components (R groups below) can be identified one by onc. The point of present interest is the nature of the final cleavage. Many experimental modifications have been suggested [59]. The role of the amine is not understood, but it appears that Schiff bases are formed by amine addition to the aldehyde groups [60]. These functions may act as intramolecular catalysts. That β-elimination occurs in the cleavage step was shown by experiments on model compounds [61]; note that the structure is condu-

cive to a β-elimination, with a good leaving group (a
phosphate ester) and an electron-withdrawing group adjacent to the potential double bond.

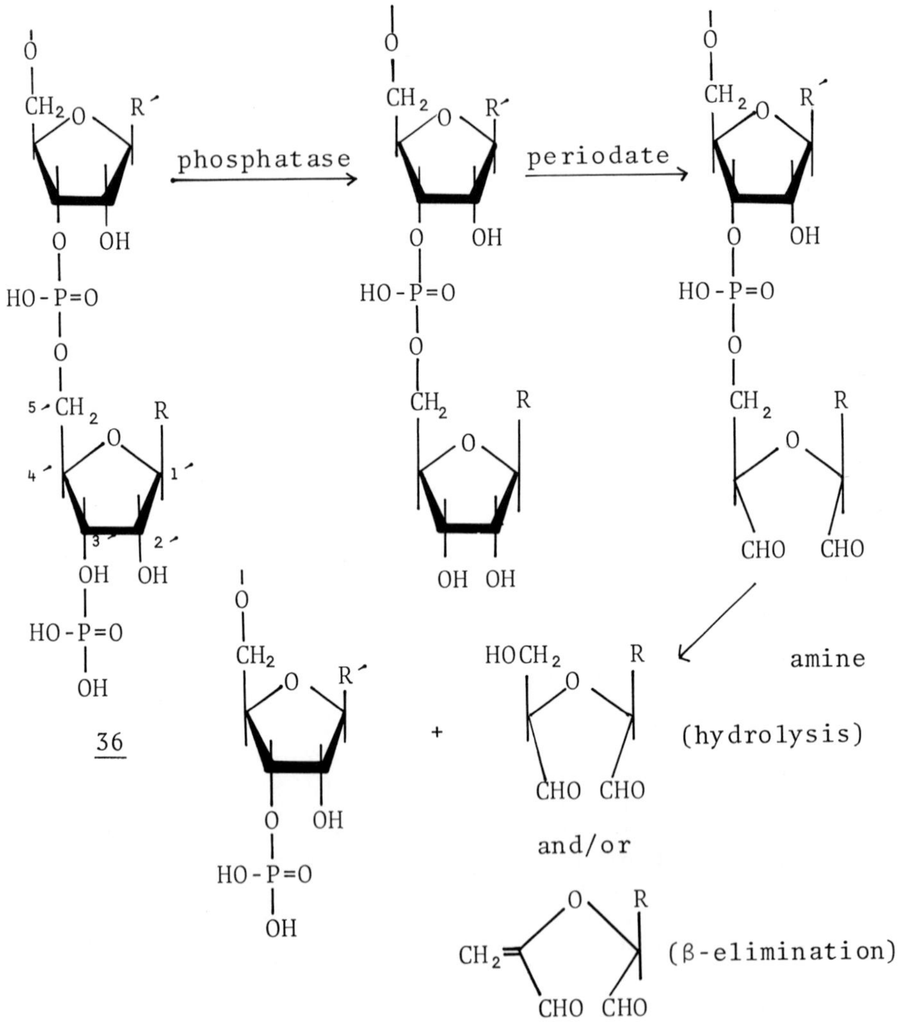

$$(11.45)$$

Comment. The main disadvantage of β-elimination in analytical practice is the possibility of competing substitution. If the reaction is followed by measuring the
leaving group, this disadvantage is overcome, though no
benefit has then been gained from the elimination. The
real desirability of β-eliminations is in their production
of olefins. The unsaturated function has very characteristic chemical and physical properties, and its generation
is worth accomplishing. Its spectral behavior, especially
when it can be placed in conjugation with another chromophoric group, is most important.
    It is likely that analytical procedures in which a
pronounced spectral shift occurs upon the addition of a
strong base involve a deprotonation, an elimination, or a
rearrangement.
    Some useful analytical exploitation probably could be
based on the carbonyl condensation reactions followed by
dehydration. These reactions give conjugated unsaturated
bonds suitable for spectroscopic measurement.
    Parker [62] has recommended β-elimination procedures
utilizing weak bases in polar aprotic solvents to accent-
uate the E2C-like transition state character, which resembles the product. These reactions can give better
control over product distributions and yields, and they
may be worth analytical attention.

Problems

1. Derive equations relating observed rate constants for
   olefin production to the rate constants for the individual steps of the carbanion mechanisms shown as Eqs.
   (11.12), (11.13), and (11.14).

2. Draw Newman projection formulas for the four stereoisomers of 3-phenyl-2-butyl tosylate [25].

3. From the data in Table 11.III calculate rate constants
   for elimination.

   Answer: see Ref. [33].

4. Postulate a concerted mechanism for elimination from
   a Mannich base involving a molecule of solvent.

5. Why is the 9-hydrogen of 9-fluorenylmethanol expected
   to be unusually acidic?

6. Account for the statement that, in the production of
   alkynes by dehydrohalogenation, strong bases prefer-

entially lead to 1-alkynes, whereas weaker bases favor production of 2-alkynes.

7. Develop an analytical procedure for the quantitative determination of a mixture containing $CH_3CH_2CH_2I$ and CH_3CHICH_2I.

References

1. β-Eliminations have been reviewed by C. K. Ingold, Structure and Mechanism in Organic Chemistry, Cornell Univ. Press, Ethaca, N. Y., 1953, Chap. VIII; E. S. Gould, Mechanism and Structure in Organic Chemistry, Holt, Rinehart and Winston, New York, 1959, Chap. 12; D. V. Banthorpe, Elimination Reactions, Volume 2 in Reaction Mechanisms in Organic Chemistry (E. D. Hughes, ed.), Elsevier, Amsterdam, 1963; W. H. Saunders, Jr., The Chemistry of Alkenes (S. Patai, ed.), Wiley (Interscience), New York, 1964; J. March, Advanced Organic Chemistry: Reactions, Mechanisms, and Structure, McGraw-Hill, New York, 1968, Chap. 17.
2. See C. K. Ingold, Ref. [1].
3. J. Hine, R. Wiesboeck, and O. B. Ramsay, J. Am. Chem. Soc., 83, 1222 (1961). The resulting olefin rapidly adds methanol.
4. Note that the reverse of the Michael addition (Chap. 10) is an E1cb reaction.
5. This is an oversimplification, for the departing group Y may influence the fate of the carbonium ion if the latter is very short-lived.
6. Recall from Chap. 9 that the constancy of this ratio is a mechanistic criterion of the S_N1 pathway in substitution reactions.
7. C. K. Ingold, Ref. [1], p. 426.
8. J. F. Bunnett, Angew. Chem., Intern. Ed., 1, 225 (1962) Structures 3 and 7 are actually intermediates rather than transition states, but are included to demonstrate the extremes in this spectrum.
9. J. F. Bunnett and E. Baciocchi, J. Org. Chem., 35, 76 (1970).
10. A. J. Parker, M. Ruane, G. Biale, and S. Winstein, Tetrahedron Letters, 1968, p. 2113.
11. See Sec. 4.3; nucleophilicity toward carbon is defined in terms of S_N2 reactivity in this treatment.
12. S. Winstein, D. Darwish, and N. J. Holness, J. Am. Chem. Soc., 78, 2915 (1956).
13. G. Biale, A. J. Parker, S. G. Smith, I. D. R. Stevens, and S. Winstein, J. Am. Chem. Soc., 92, 115 (1970).
14. E. C. F. Ko and A. J. Parker, J. Am. Chem. Soc., 90,

6447 (1968).

15. D. Eck and J. F. Bunnett, J. Am. Chem. Soc., 91, 3099 (1969); D. Cook and A. J. Parker, Tetrahedron Letters, 1969, p. 4901.
16. R. A. M. O'Ferrall, J. Chem. Soc., B1970, p. 274.
17. F. G. Bordwell, M. M. Vestling, and K. C. Yee, J. Am. Chem. Soc., 92, 5950 (1970).
18. F. G. Bordwell, K. C. Yee, and A. C. Knipe, J. Am. Chem. Soc., 5945 (1970). The β-elimination product undergoes a subsequent fast deprotonation.
19. E. S. Gould, Ref. [1], p. 492.
20. S. J. Cristol and E. F. Hoegger, J. Am. Chem. Soc., 79, 3438 (1957).
21. C. H. DePuy, R. D. Thurn, and G. F. Morris, J. Am. Chem. Soc., 84, 1314 (1962).
22. It has been suggested that syn-elimination has a greater requirement for carbanion character than does anti-elimination; C. H. DePuy, C. G. Naylor, and J. A. Beckman, J. Org. Chem., 35, 2750 (1970).
23. Bunnett and Baciocchi [9], however, prefer a conventional E2 transition state. A bridged halonium ion intermediate is favored by C. S. T. Lee, I. M. Mathai, and S. I. Miller, J. Am. Chem. Soc., 92, 4602 (1970).
24. G. S. Hammond and C. H. Collins, J. Am. Chem. Soc., 82, 4323 (1960).
25. D. J. Cram, J. Am. Chem. Soc., 74, 2137 (1952); D. J. Cram, Steric Effects in Organic Chemistry (M. S. Newman, ed.), Wiley, New York, 1956, Chap. 6.
26. Significant amounts of 2-phenyl-1-butene, 3-phenyl-1-butene, and acetoxy substitution products were also found.
27. Also spelled Zaitsev.
28. This work has been reviewed by C. K. Ingold, Ref. [1], pp. 427-448.
29. A simple statistical effect based on the numbers of hydrogens available for elimination does not solely account for Hofmann orientation.
30. Note that the inductive model is based on an initial state model of the transition state (relative acidity of hydrogens in the substrate), and the conjugative model is based on a final state model (stabilization of the double bond in the product).
31. H. C. Brown and I. Moritani, J. Am. Chem. Soc., 78, 2203 (1956), and preceding papers.
32. D. V. Banthorpe, Ref. [1], pp. 66-77.
33. P. D. Buckley, B. D. England, and D. J. McClennan, J. Chem. Soc., B1967, 98.
34. The observed effects are small and predictive ability is slight in these systems. Large effects have been predicted and observed, however, for reactions between

two neutral reactants or two charged reactants; see
C. K. Ingold, Ref. [1], pp. 455-460.

35. I. N. Feit and W. H. Saunders, Jr., J. Am. Chem. Soc.,
     92, 5615 (1970).

36. See also J. March, Ref. [1], pp. 753-780.

37. T. I. Crowell, The Chemistry of Alkenes (S. Patai,
     ed.), Wiley (Interscience), New York, 1964, Chap. 4.

38. P. Zuman, Advan. Phys. Org. Chem., 5, 1 (1967); J.
     A. Mollica, J. B. Smith, I. M. Nunes, and H. K. Govan,
     J. Pharm. Sci., 59, 1770 (1970).

39. D. V. Banthorpe, Ref. [1], p. 136.

40. I. Ho and J. G. Smith, Tetrahedron Letters, 1970,
     p. 3535.

41. R. A. M. O'Ferrall and S. Slae, J. Chem. Soc., B1970,
     p. 260.

42. A. J. Castro, J. Am. Chem. Soc., 72, 5311 (1950); D.
     V. Banthorpe, Ref. [1], p. 145.

43. D. S. Noyce, P. A. King, K. B. Kirby, and W. L. Reed,
     J. Am. Chem. Soc., 84, 1632 (1962).

44. A. Maccoll, The Chemistry of Alkenes (S. Patai, ed.)
     Wiley (Interscience), New York, 1964, Chap. 3.

45. L. R. Jones and J. A. Riddick, Anal. Chem., 29, 1214
     (1957).

46. E. R. Garrett, J. Blanch, and J. K. Seydel, J. Pharm.
     Sci., 56, 1560 (1967).

47. E. R. Garrett and B. H. Dvorchik, J. Pharm. Sci., 58,
     813 (1969).

48. I. Dostrovsky and F. S. Klein, J. Chem. Soc., 1955,
     p. 791.

49. C. K. Ingold, Ref. [1], p. 462.

50. A. N. Bashkirov, S. A. Lodzik, and V. V. Kamzolkin,
     Trudy Inst. Nefti, Akad. Nauk SSSR, 12, 297 (1958);
     N. D. Cheronis and T. S. Ma, Organic Functional Group
     Analysis by Micro and Semimicro Methods, Wiley (Inter-
     science), New York, 1964, p. 192.

51. K. Uhrig, E. Lynch, and H. C. Becker, Ind. Eng. Chem.,
     Anal. Ed., 18, 550 (1946). For related analyses see
     J. S. Powell, K. C. Edson, and E. L. Fisher, Anal.
     Chem., 20, 213 (1948); D. A. Skoog and H. D. DuBois,
     ibid., 21, 1528 (1949).

52. F. E. Critchfield, Organic Functional Group Analysis,
     Macmillan, New York, 1963, p. 91.

53. N. D. Cheronis and T. S. Ma, Ref. [50], pp. 460-461.
     For another possible analytical dehydrohalogenation
     see Ref. [177] in Chap. 10.

54. R. L. Shriner, R. C. Fuson, and D. Y. Curtin, The
     Systematic Identification of Organic Compounds, 5th
     ed., Wiley, New York, 1964, p. 170.

55. S. Siggia, A. C. Starke, Jr., J. J. Garis, Jr., and
     C. R. Stahl, Anal. Chem., 30, 115 (1958).

56. A. R. Mattocks, Anal. Chem., 39, 443 (1967); J. B. Bingley, ibid., 40, 1166 (1968).
57. R. Kreher and H. Pawelczyk, Angew. Chem. Intern. Ed., 3, 510 (1964).
58. P. R. Whitfield and R. Markham, Nature, 171, 1151 (1953).
59. H. C. Neu and L. A. Heppel, J. Biol. Chem., 239, 2927 (1964); J. X. Khym and M. Uziel, Biochemistry, 7, 422 (1968); H. L. Weith and P. T. Gilham, J. Am. Chem. Soc., 89, 5473 (1967); Science, 166, 1004 (1969); A. Steinschneider and H. Fraenkel-Conrat, Biochemistry, Biochemistry, 5, 2735 (1966).
60. J. X. Khym and W. E. Cohn, J. Biol. Chem., 236, PC9 (1961); A. Steinschneider and H. Fraenkel-Conrat, Biochemistry, 5, 2729 (1966).
61. D. M. Brown, M. Fried, and A. R. Todd, J. Chem. Soc., 1955, 2206.
62. A. J. Parker, Chem. Technol., 1971, p. 297.

Chapter 12. ADDITION TO CARBON-HETEROATOM MULTIPLE BONDS

12.1 Nature of the Reaction

Survey of the Reaction. Most of the basic heterolytic reaction types have been described in Chaps. 7-11. Chapters 12 and 13 consider reactions at double and triple bonds in which at least one of the multiply bonded atoms is not carbon. Most of the substrates thus encompassed contain the carbonyl group, $>C=O$. The chemical properties of carbonyl compounds depend markedly upon the nature of the remaining two bonds to the carbonyl carbon; classification by chemical behavior therefore leads to a familiar structural classification. Most of the reactions to be considered in Chaps. 12 and 13 are (or include) addition to the multiple bond. If the overall process conforms to Eq. (12.1), namely, acyl transfer or substitution on unsaturated carbon, the reaction is treated in Chap. 13 [1].

$$R-\overset{\overset{\displaystyle O}{\|}}{C}-X + Y \longrightarrow R-\overset{\overset{\displaystyle O}{\|}}{C}-Y + X \qquad (12.1)$$

In other words, if the carbonyl group remains intact in the overall process, the reaction is treated as an acyl transfer, whereas if the carbonyl group undergoes marked bonding changes in the overall process the reaction is considered here. With a few exceptions, therefore, Chap. 12 treats reactions of aldehydes and ketones, and Chap. 13 discusses reactions of carboxylic acid derivatives. It is common to apply the term "carbonyl compound" to only aldehydes and ketones.

Carbonyl additions occur with the formation of a bond from the carbonyl carbon to the nucleophilic fragment of an adding species. When the reagent possesses a single dissociable proton, the reaction can be described by

$$>C=O + HX \rightleftharpoons \;>C<\!\!{}^{OH}_{X} \qquad (12.2)$$

We refer to this as "simple addition." If the reagent has
two protons, a subsequent β-elimination can occur, as in

$$\text{>C=O} + H_2X \rightleftharpoons \begin{array}{c} \diagdown \diagup OH \\ C \\ \diagup \diagdown XH \end{array} \rightleftharpoons \text{>C=X} + H_2O \qquad (12.3)$$

This is called the addition-elimination route. These reactions are reversible, so their mechanisms can be studied
from either direction. Note that the reverse of Eq. (12.3)
is the hydrolysis of the C=X bond. Additions and their
retrogressions are discussed in this chapter.

Nucleophilic addition to the olefinic bond of α,β-
unsaturated carbonyl compounds, C=C-C=O, has been discussed
in Sec. 10.2. This conjugated system behaves very much
like the isolated carbonyl group, with the terminal carbon
being the site of nucleophilic addition.

Table 12.I shows a few of the important carbonyl addition reactions. In the presence of proton acids or Lewis
acids, hemiacetals are converted to acetals; this is a
nucleophilic substitution on saturated carbon, occurring
by the carbonium ion mechanism.

$$\begin{array}{ccccc} O & & OH & & OR' \\ \parallel & & | & H^+,R'OH & | \\ R-C-H + R'OH & \rightleftharpoons & R-C-H & \rightleftharpoons & R-C-H + H_2O \\ & & | & & | \\ & & OR' & & OR' \end{array}$$

Carbon nucleophiles can add to the carbonyl group;
several important synthetic reactions are based on this
addition. Aldol condensations, of which there are several
variants, involve reaction of an active methylene with the
carbonyl group. Since the electron-withdrawing carbonyl
activates adjacent protons, self-condensations can occur,
as shown in Eq. (12.5) for acetaldehyde.

$$\begin{array}{ccc} O & O & OH \\ \parallel & \parallel & | \\ CH_3-C-H + H-CH_2-C-H & \rightleftharpoons & CH_3-C-H \qquad (12.5) \\ & & | \\ & & CH_2CHO \end{array}$$

These additions may be followed by dehydrations, as described in Chap. 11, to give α,β-unsaturated carbonyl compounds. The Grignard reaction is a carbonyl addition by
a carbon nucleophile:

$$
\underset{\substack{\| \\ O}}{R'\text{-}C\text{-}R} + CH_3MgBr \rightleftharpoons \underset{\substack{| \\ CH_3}}{\overset{OMgBr}{\overset{|}{R'\text{-}C\text{-}R}}} \tag{12.6}
$$

The adduct is then hydrolyzed to the alcohol.

Table 12.I   Examples of Carbonyl Addition Reactions

Overall reaction	Product name
Simple additions	
$R'\text{-}\underset{\substack{\| \\ O}}{C}\text{-}R + H_2O \rightleftharpoons R'\text{-}\underset{\substack{| \\ OH}}{\overset{OH}{C}}\text{-}R$	hydrate
$R'\text{-}\underset{\substack{\| \\ O}}{C}\text{-}H(R) + R'OH \rightleftharpoons R'\text{-}\underset{\substack{| \\ OR'}}{\overset{OH}{C}}\text{-}H(R)$	hemiacetal (hemiketal)
$R'\text{-}\underset{\substack{\| \\ O}}{C}\text{-}R + HCN \rightleftharpoons R'\text{-}\underset{\substack{| \\ CN}}{\overset{OH}{C}}\text{-}R$	cyanohydrin
$R'\text{-}\underset{\substack{\| \\ O}}{C}\text{-}R + HSO_3^- \rightleftharpoons R'\text{-}\underset{\substack{| \\ SO_3^-}}{\overset{OH}{C}}\text{-}R$	hydroxylsulfonate
Addition-eliminations	
$R'\text{-}\underset{\substack{\| \\ O}}{C}\text{-}R + R''NH_2 \rightleftharpoons R'\text{-}\underset{\substack{\| \\ }}{\overset{N\text{-}R''}{C}}\text{-}R + H_2O$	imine (Schiff base)

$$R'-\overset{\displaystyle O}{\overset{\|}{C}}-R + NH_2OH \rightleftharpoons R'-\overset{\displaystyle N-OH}{\overset{\|}{C}}-R + H_2O \qquad \text{oxime}$$

$$R'-\overset{\displaystyle O}{\overset{\|}{C}}-R + NH_2NH_2 \rightleftharpoons R'-\overset{\displaystyle N-NH_2}{\overset{\|}{C}}-R + H_2O \qquad \text{hydrazone}$$

$$R'-\overset{\displaystyle O}{\overset{\|}{C}}-R + NH_2NHCONH_2 \rightleftharpoons R'-\overset{\displaystyle N-NHCONH_2}{\overset{\|}{C}}-R \quad + H_2O \quad \text{semicarbazone}$$

Carbonyl reductions can be formulated as additions. The conversion of a carbonyl to an alcohol with lithium aliminum hydride is initiated by addition of a hydride ion to the carbonyl carbon.

$$R'-\overset{\displaystyle O}{\overset{\|}{C}}-R + LiAlH_4 \rightleftharpoons R'-\underset{\displaystyle H}{\overset{\displaystyle OLiAlH_3}{\overset{|}{\underset{|}{C}}}}-R \xrightarrow{H^+} R'-\underset{\displaystyle H}{\overset{\displaystyle OH}{\overset{|}{\underset{|}{C}}}}-R \qquad (12.7)$$

It will be clear that all of these reactions are considered as Ad_N reactions, with the carbonyl carbon being the electron-deficient site, as suggested by the valence-bond representation 1 [2].

$$\overset{\displaystyle >}{C}{=}O \longleftrightarrow \overset{\displaystyle >}{C}{\overset{+}{-}}O^-$$

1

Besides their reactions at the carbonyl group itself, aldehydes and ketones may undergo reactions elsewhere in the molecule. These are influenced by the carbonyl group. The most characteristic of these reactions are consequent upon the phenomenon of keto-enol tautomerism, which is demonstrated below:

$$-CH_2\overset{\displaystyle O}{\overset{\|}{C}}- \rightleftharpoons -CH{=}\overset{\displaystyle OH}{\overset{|}{C}}-$$

This is a true equilibrium. If the methylene group is flanked by electron-withdrawing groups, as in β-dicarbonyl compounds, the enol form is stabilized. The halogenation of ketones proceeds through the enol form, with the enolization being rate determining [3].

$$CH_3\overset{\overset{\textstyle O}{\|}}{C}CH_3 \underset{}{\overset{OH^-}{\rightleftharpoons}} \overset{\overset{\textstyle OH}{|}}{CH_2}=C-CH_3 \xrightarrow{Br_2} BrCH_2\overset{\overset{\textstyle O}{\|}}{C}CH_3$$

In addition to these reactions of carbonyl compounds, several other types of unsaturation may be included here. It has already been noted that additions to imines (the C=N group) are related to carbonyl additions. The thiocarbonyl (C=S) and nitrile (C≡N) groups undergo nucleophilic addition; thus nitriles add water to give the tautomeric form of an amide (which may subsequently hydrolyze to the acid).

$$R-C{\equiv}N \overset{H_2O}{\rightleftharpoons} R-\overset{\overset{\textstyle OH}{|}}{C}=NH \rightleftharpoons R-\overset{\overset{\textstyle O}{\|}}{C}-NH_2 \qquad (12.8)$$

Nucleophilic additions to heterocumulenes are exemplified by Eq. (12.9), the addition of amines to isocyanates to give substituted ureas, and Eq. (12.10), the addition of amines to carbon disulfide, yielding dithiocarbamic acids.

$$R-N=C=O + R'NH_2 \rightarrow RNH-\overset{\overset{\textstyle O}{\|}}{C}-NHR \qquad (12.9)$$

$$S=C=S + RNH_2 \rightarrow RNH-\overset{\overset{\textstyle S}{\|}}{C}-SH \qquad (12.10)$$

These are exceptions to the classification scheme suggested at the beginning of this section, but it seems preferable to consider these as simple additions followed by tautomeric proton shifts.

Kinetics and Mechanisms. Because of the polarization of the C=O bond, orientation of addition is always clearcut, with the nucleophile going to the carbon and the electrophile to the oxygen. Two of the possible mechanisms can be illustrated with cyanohydrin formation; these differ in the order of attack by hydrogen ion and cyanide ion.

$$R-\overset{\overset{\textstyle O}{\|}}{C}-R + CN^- \rightleftharpoons R-\overset{\overset{\textstyle O^-}{|}}{\underset{\underset{\textstyle CN}{|}}{C}}-R \overset{H^+}{\rightleftharpoons} R-\overset{\overset{\textstyle OH}{|}}{\underset{\underset{\textstyle CN}{|}}{C}}-R \qquad (12.11)$$

$$\underset{R-C-R}{\overset{\displaystyle O}{\overset{\|}{}}} + H^+ \rightleftharpoons \underset{R-C-R}{\overset{\displaystyle {}^+OH}{\overset{\|}{}}} \overset{CN^-}{\rightleftharpoons} \underset{\overset{|}{CN}}{\underset{R-C-R}{\overset{\displaystyle OH}{\overset{|}{}}}} \tag{12.12}$$

Both of these possibilities were discussed by Lapworth in
1903 [4]. On the basis that the reaction velocity is decreased by acid but increased by alkali and by cyanide
salts, Lapworth selected Eq. (12.11) as the correct mechanism. This is an Ad_N reaction, the attack by cyanide
being the slow step. The reaction is overall second order
and is not significantly subject to general acid-base
catalysis [5].
    Although Eq. (12.11) describes cyanohydrin formation
satisfactorily, Eq. (12.12) gives an appropriate description of some other additions. Since the initial proton
transfer will usually be very fast relative to the nucleophilic attack, this mechanism is considered to be an acid-
catalyzed Ad_N process, and therefore not much different
from the uncatalyzed Ad_N mechanism. Jencks [6] suggests
that catalysis occurs where it is most needed; thus additions of strong nucleophiles, like CN^-, SO_3^{2-}, and amines,
are not subject to appreciable acid catalysis, whereas the
weaker nucleophiles H_2O, ROH, and amides add to carbonyl
groups with catalytic assistance by general acids and
bases. Specific base catalysis is characteristic, but
this is often just another way of saying that the attacking reagent must be in its conjugate base form. For example, thiols add to carbonyls by a specific base-catalyzed
route that is identical with the cyanohydrin mechanism,
Eq. (12.11) [7]. This addition is also catalyzed by general acids, the suggested transition state being shown as
2.

$$
\begin{array}{c}
B\delta^- \\
\diagup \\
H \\
\diagup \\
O \\
\| \\
CH_3\!\!-\!\!C\!\!-\!\!H \\
| \\
S\delta^+ \\
\diagup\;\diagdown \\
R\quad H
\end{array}
$$

2

In 2 a partial proton transfer from acid HB to the carbonyl
oxygen augments the polarization of the carbonyl group,

thus assisting the nucleophilic attack.

Carbonyl additions that proceed by the two-step addition-elimination route can exhibit complicated kinetic
behavior. One manifestation of this is the pH dependence
of rate constants for the overall reaction; the pH-rate
profile often reveals a maximum, with a well-developed
bell-shaped curve. Figure 12.1 shows such a curve for
the oximation of acetone [8]. Similar behavior in semicarbazone formation, which is general acid catalyzed, had
earlier been interpreted to mean that the rate increase
on the alkaline side of the pH curve was the result of
general acid catalysis, since at lower pH a greater proportion of any buffer will be in the catalytically active
acid form. The rate drop-off at still lower pH was ascribed to protonation of the attacking nucleophile [9].
On this basis it was possible to account reasonably well
for the location of the maximum in the pH-rate curve as
a function of the pK_a's of the catalyst acid and of semicarbazide [10]. Subsequent studies, primarily by Jencks
[6,8,11] have shown the inadequacy of this description,
and have replaced it with a mechanism that recognizes the
possibility of a change in rate-determining step as the
reaction conditions are changed. The two-step addition
of an amine to a carbonyl is shown in

$$\text{>C=O} + \text{RNH}_2 \rightleftharpoons \overset{OH}{\underset{NHR}{\overset{\diagdown \diagup}{\underset{\diagup \diagdown}{C}}}} \rightleftharpoons \text{>C=NR} + \text{H}_2\text{O} \qquad (12.13)$$

$$\underline{3}$$

The carbinolamine $\underline{3}$ is often called a tetrahedral intermediate, for obvious reasons. Its presence, which is not
in question, requires that the reaction occur in at least
two steps, either of which may be rate limiting.

When a carbonyl compound is treated with hydroxylamine
or other typical nucleophiles near neutral pH, the carbonyl
absorption intensity decreases very rapidly; this spectral
change is followed by a much slower absorption increase
that is due to the final imine-type product. This suggests
that, at such pH values, the initial addition is very rapid,
and the rate-determining step is the second step, dehydration of the carbinolamine. This reaction is catalyzed by
hydrogen ion and by general acids; it therefore accounts
for the rate increase as the pH is shifted from neutrality
to more acid solutions (see Fig. 12.1). This increase does
not continue indefinitely, however, because with the
lowered pH, a greater fraction of the nucleophile exists

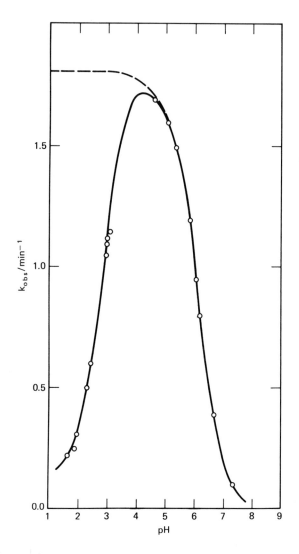

Fig. 12.1. pH-rate profile for the reaction of hydroxylamine with acetone in water at 25°. Dashed linc: rate
of the acid-catalyzed dehydration step; full line:
observed rate.

in its unreactive protonated form. These two effects,
acid-catalyzed dehydration and conversion to the non-
nucleophilic conjugate acid, compensate each other, so
that the rate of the second step finally reaches a

limiting value, independent of pH, at low pH (dashed line in Fig. 12.1).

However, as Fig. 12.1 shows, at very low pH the observed rate actually decreases. Since, as the preceding argument shows, rate-limiting dehydration should result in a pH-independent rate at low pH, this decreased rate must mean that the rate-determining step has changed. This is reasonable, for, at pH values well below the pK_a of the nucleophile, the decreasing proportion of amine in the unprotonated form will decrease the rate of the initial addition. At some pH, then, the rate of the addition step will fall below that of the dehydration step, and the observed rate curve will be lower than the predicted rate for the dehydration [12].

These qualitative arguments are reinforced by the quantitative kinetic analysis [13]. The oximation of a carbonyl compound is conveniently represented by Eq. (12.14); the dehydration step is essentially irreversible under the conditions of the kinetic measurements.

$$\text{>C=O} + \text{NH}_2\text{OH} \underset{\substack{k_{-1} \\ k_{-2}[\text{H}^+]}}{\overset{\substack{k_1 \\ k_2[\text{H}^+]}}{\rightleftharpoons}} \text{>C}\begin{smallmatrix}\text{OH}\\\text{NHOH}\end{smallmatrix} \xrightarrow{k_3[\text{H}^+]} \text{>C=NOH} + \text{H}_2\text{O} \quad (12.14)$$

Equation (12.14) expresses, besides the points already described, the observation of a small acid-catalyzed contribution to the first step. If the tetrahedral intermediate concentration never achieves a significant level relative to the carbonyl concentration (which may be assured for this system by using a low hydroxylamine concentration), the steady-state approximation may be applied to the intermediate. Letting [I] be the concentration of intermediate, this gives

$$[I] = \frac{(k_1 + k_2[\text{H}^+])[\text{carbonyl}][\text{NH}_2\text{OH}]}{k_{-1} + k_{-2}[\text{H}^+] + k_3[\text{H}^+]} \quad (12.15)$$

Since the reaction rate is given by $v = k_3[\text{H}^+][I]$ and the apparent first-order rate constant by $k_{obs} = v/[\text{carbonyl}]$, the resulting equation for k_{obs} is

$$k_{obs} = \frac{k_3[\text{H}^+](k_1 + k_2[\text{H}^+])[\text{NH}_2\text{OH}]}{k_{-1} + k_{-2}[\text{H}^+] + k_3[\text{H}^+]} \quad (12.16)$$

In this equation [NH$_2$OH] represents the concentration of hydroxylamine in the conjugate base form; that is, the equation does not automatically account for the protonation of hydroxylamine. Since its pK_a is known, however, the free hydroxylamine concentration can be calculated at each pH for any given total hydroxylamine concentration. By applying limiting conditions to Eq. (12.16), estimates of some of the rate constants can be obtained. For example, at high pH, Eq. (12.16) reduces to $k_{obs} = k_1 k_3 [H^+][NH_2OH]/k_{-1} = Kk_3 [H^+][NH_2OH]$, where K is the equilibrium constant for carbinolamine formation from reactants. Thus, from data at high pH, the quantity Kk_3 can be evaluated. This special case demonstrates that the rate on the alkaline side of the curve depends on the equilibrium concentration of the intermediate and upon its rate of dehydration, in agreement with the description given earlier. At low pH, the intermediate formation becomes rate determining because its dehydration is much faster than its return to reactants; that is, $k_3[H^+]>>k_{-1}$ and $k_3[H^+]>>k_{-2}[H^+]$, giving, from Eq. (12.16),

$$k_{obs} = (k_1 + k_2[H^+])[NH_2OH] \qquad (12.17)$$

A plot of $k_{obs}/[NH_2OH]$ against $[H^+]$, for low pH, therefore yields estimates of k_1 and k_2. Estimates obtained in this way can then be inserted into the full equation, (12.16), to develop a calculated pH-rate profile, as with the full curve in Fig. 12.1 [11].

The mechanism just described for oximation appears to be widely applicable for carbonyl additions: rate-determining attack of the nucleophile on the carbonyl compound at low pH, with a transition to rate-determining dehydration of the tetrahedral intermediate at higher pH. The transition occurs in the range pH 2-5 for many nucleophiles, with a maximum in the pH-rate profile being observed in oximation [8] (Fig. 12.1) and Schiff base formation from aniline and p-chlorobenzaldehyde [14]. The reactions of semicarbazide and thiosemicarbazide with substituted benzaldehydes also give pH-rate curves revealing the transition [15], but in these reactions the acid-catalyzed addition can make a greater contribution to the rate at low pH, because semicarbazide (pK_a 3.65) and thiosemicarbazide (pK_a 1.88) [15b] are weaker bases than are aniline (pK_a 4.60) and hydroxylamine (pK_a 5.97).

Base catalysis has been observed in these reactions at high pH. The oximation of benzaldehyde is catalyzed by hydroxide ion [16]; this must be catalysis of the dehydration, which is rate limiting in this pH region. Acetone oximation is similarly base catalyzed, and also exhibits a pH-independent reaction from pH 8.5 to 10 [16]. The pH-independent reaction was ascribed to general acid catalysis of dehydration by H_2O, and the base catalysis to

specific base catalysis [16]. This interpretation is
doubted by Sayer and Jencks [15b], who detected general
base catalysis of thiosemicarbazone formation. The following two kinetically equivalent mechanisms can be drawn for
the general base-catalyzed dehydration.

$$
\underset{\substack{|\\N-H\quad B\\|\\R}}{\overset{OH}{\underset{/}{\diagup}}C\diagdown}
\quad\overset{slow}{\rightleftharpoons}\quad
{>}C{=}NR + OH^- + BH^+ \quad\overset{fast}{\rightleftharpoons}\quad {>}C{=}NR + H_2O + B
$$

(12.18)

$$
\overset{OH}{\underset{NHR}{\diagup}}C\diagdown + B \quad\overset{fast}{\rightleftharpoons}\quad
\overset{OH}{\underset{\substack{N^-\\|\\R}}{\diagup}}C\diagdown \; H{-}B^+ \quad\overset{slow}{\rightleftharpoons}\quad {>}C{=}NR + H_2O + B
$$

(12.19)

Equation (12.18) is a general base catalysis of the dehydration, the slow step in the reverse reaction being attack by hydroxide ion on the imine, with assistance by
the conjugate acid of the base (details not shown). The
alternative, (12.19), involves a fast pre-equilibrium,
with the slow step being general acid-assisted loss of
hydroxide from the anion of the carbinolamine. The reverse of this is general base catalysis of water attack
on the amine.
    Sayer and Jencks prefer the mechanism of Eq. (12.18)
primarily on the basis, mentioned earlier, that catalysis
occurs where it is most needed; in this case it avoids
formation of the very unstable carbinolamine anion. The
reverse reaction also is chemically more reasonable when
formulated as in Eq. (12.18) [17].
    Nucleophilic catalysis of carbonyl addition-elimina-
tion reactions occurs with the intermediate formation of
a reactive imine. Aniline catalyzes semicarbazone formation in this way [18]. Equations (12.20) and (12.21) show
the course of the catalysis; $ArNH_2$ represents aniline and
RNH_2 is semicarbazide.

$$
{>}C{=}O + ArNH_2 \quad\overset{slow}{\longrightarrow}\quad {>}C{=}NAr + H_2O
$$

(12.20)

$$
{>}C{=}NAr + RNH_2 \quad\overset{fast}{\longrightarrow}\quad {>}C{=}NR + ArNH_2
$$

(12.21)

This remarkable catalysis occurs because (1) aniline is more reactive toward the carbonyl substrate than is semi-carbazide; (2) semicarbazide attacks the intermediate Schiff base more readily than it does the carbonyl; (3) the equilibrium constant for semicarbazone formation is greater than that for the Schiff base intermediate. The last condition means that the relative thermodynamic stabilities of the products bear an inverse relationship to the kinetic reactivities of the reactants. Condition (1) is not surprising, considering the basicities of the two amines. The second condition appears to be a consequence of the greater basicity of the Schiff base relative to the carbonyl group, so that at any given pH a greater proportion of the Schiff base exists in the protonated form, which is very susceptible to attack by semicarbazide [19].

The mutarotation of 2,3,4,6-tetramethyl-D-glucose is a carbonyl reaction whose study has influenced concepts of acid-base catalysis. This molecule can exist in an open-chain aldehyde form and as two stereoisomeric cyclic hemi-acetals. In the presence of suitable catalysts either of the hemiacetal forms will undergo conversion through the open-chain form to give an equilibrium mixture of the hemiacetals; this configurational change, which is accompanied by a change in optical activity, is called mutarota-tion. Swain and Brown [20] found that the mutarotation, in benzene solution, is catalyzed by a mixture of phenol and pyridine, presumably by a general acid-general base mechanism. When the acidic and basic groups were located in the same molecule, 2-hydroxypyridine, greatly enhanced catalytic activity was observed. This was attributed to a concerted general acid-base catalysis, as in

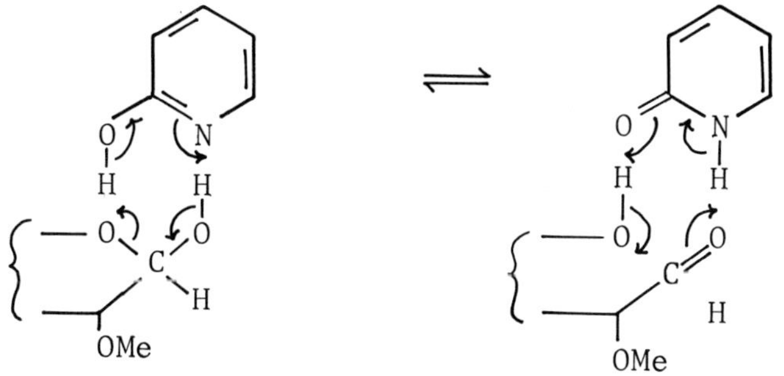

(12.22)

Rony [21] has proposed that the catalytic effectiveness of 2-hydroxypyridine in the mutarotation is a consequence of this molecule's ability to exist in two neutral tautomeric forms, 2-hydroxypyridine and 2-pyridone; thus the proton transfers occurring in the catalysis do not lead to an unstable ionic species. Benzoic acid is an effective mutarotation catalyst; this molecule also is capable of existence in two neutral tautomeric forms (indistinguishable in this case). Such tautomeric catalysis could be more important than classical concerted general acid-base mechanisms in reactions catalyzed by weakly acidic and basic species.

Reactions of Oxygen Nucleophiles. Some carbonyl additions are now reviewed. The addition of water, Eq. (12.23), which might seem to be the simplest of these, is still incompletely understood.

$$\begin{array}{c}R_1 \\ \diagdown \\ \diagup \\ R_2 \end{array} C{=}O \; + \; H_2O \; \rightleftharpoons \; \begin{array}{c} R_1 \\ \diagdown \quad \diagup \; OH \\ C \\ \diagup \quad \diagdown \; OH \\ R_2 \end{array} \qquad (12.23)$$

The extent of hydrate formation is governed by the polar and steric characteristics of the substituents on the carbonyl group, in ways that can be readily accounted for.

Table 12.II   Equilibrium Constants for Hydration of Carbonyls in Aqueous Solution[a]

Carbonyl compound	K_h at 25°	$-\Delta H°$/kcal-mole^{-1}
CCl_3CHO	2.8×10^4	14.0
CH_2O	2×10^3	
CH_3CHO	1.4	6
CH_3CH_2CHO[b]	0.87	5.4
$(CH_3)_2CHCHO$[b]	0.66	5.8
$(CH_3)_3CCHO$[b]	0.24	4.4
$(CH_3)_2CO$	0.002	

[a]From R. P. Bell, Advan. Phys. Org. Chem., 4, 1 (1966), except as noted.
[b]Y. Pocker and D. G. Dickerson, J. Phys. Chem., 73, 4005 (1969).

Table 12.II gives some hydration equilibrium constants, K_h, for Eq. (12.23) in aqueous solution [22].

$$K_h = \frac{[R_1R_2C(OH)_2]}{[R_1R_2CO]} \qquad (12.24)$$

Some carbonyl hydrates are isolable crystalline compounds; chloral hydrate, 4, and ninhydrin, 5, are well-known examples. These hydrates are stabilized by powerfully electro-

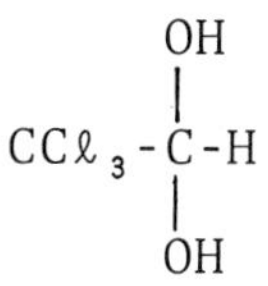

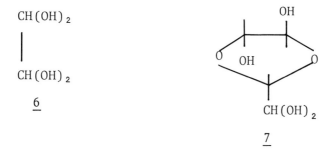

$$CCl_3-\overset{\displaystyle OH}{\underset{\displaystyle OH}{C}}-H$$

4

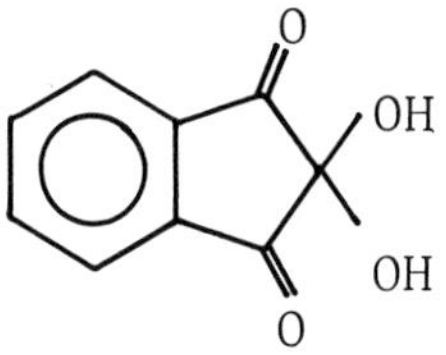

5

negative groups adjacent to the gem-diol function. It is clear, from the hydration constants shown, that a considerable fraction of an aldehyde solute may exist as the hydrate in aqueous solution [23]. Glyoxal, $(CHO)_2$, exists predominantly as the hydrated monomer 6, but a significant amount of the cyclic acetal dimer 7 is also found [24].

$$CH(OH)_2$$
$$|$$
$$CH(OH)_2$$

6

7

The kinetics of hydration and dehydration are rapid, but are measurable with ordinary techniques. These reactions are subject to general acid and general base catalysis, and they display good linearity in Brønsted plots over quite remarkable ranges of acid and base strength (up to 17 pK units [22]). The Brønsted coefficients are typically about 0.5. Postulated mechanisms must be consistent with these observations. Equation (12.25) shows a general acid catalyzed hydration,, the dehydration being specific acid-general base catalyzed [25]. Equation (12.26) is a kinetically indistinguishable alternative, the hydration being specific acid-general base catalyzed and the

$$\begin{array}{ll}\underset{H}{\overset{H}{\diagdown}}O{\cdots}C{=}O \quad H{-}A \rightleftharpoons \underset{H}{\overset{H}{\diagdown}}\overset{+}{O}{-}\overset{|}{\underset{|}{C}}{-}O{-}H \quad A^- & (12.25)\end{array}$$

$$\begin{array}{ll}\overset{-}{A}\ \ H \\ \quad \underset{H}{\overset{H}{\diagdown}}O{\cdots}C{=}\overset{+}{O}H \rightleftharpoons \overset{A{-}H}{\underset{H}{\overset{\diagup}{\diagdown}}}O{-}\overset{|}{\underset{|}{C}}{-}O{-}H & (12.26)\end{array}$$

dehydration general acid catalyzed. Although these mechanisms seem plausible, each of them requires the intervention of a high-energy unstable intermediate, the conjugate
acid of the gem-diol in Eq. (12.25) and the conjugate acid
of the carbonyl compound in (12.26). These will be present
in such low concentrations [26] that the rate constant for
either of these reactions would have to be larger than the
diffusion-controlled limit of about $10^{12}sec^{-1}$ in order for
the observed rate to be achieved. These mechanisms are
therefore improbable [27]. A concerted mechanism, which
avoids the formation of high-energy intermediates by utilizing a cyclic transition state incorporating one or more
molecules of water [27], seems to be consistent with the
data. Structure 8 shows such a transition state for the
general acid catalysis.

$$\begin{array}{ccc} \underset{R_2-C=O}{\overset{R_1}{|}} & & \underset{R_2-C{=\!=\!=}O}{\overset{R_1}{|}} \\ H{-}O\diagup\quad H & \text{or} & H{-}O\diagup\quad H \\ H{\diagdown}A\diagup & & H{\text{---}}A \end{array}$$

$$\underline{8}$$

Hydrogen peroxide adds to aldehydes with specific
acid-base and general acid-base catalysis, probably by
mechanisms similar to hydration reactions [28]. Even very
weak nucleophiles can add to the carbonyl group if aided
by acid catalysis; thus acetone and perchloric acid in
chloroform yield a liquid believed to be the gem-diper-
chlorate [29], $(CH_3)_2C(OC\ell O_3)_2$.
The hydrolysis of acetals ($\underline{9}$, R_2 = H), ketals, and
ortho esters ($\underline{9}$, R_2 = OR) has attracted much mechanistic
attention [30].

$$\underset{R_2}{\overset{R_1}{\diagdown}}C\underset{OR}{\overset{OR}{\diagup}}$$

$$\underline{9}$$

The overall reaction is shown as

$$\underset{R_2}{\overset{R_1}{\diagdown}}C\underset{\diagup OR}{\diagdown OR} + H_2O \rightleftharpoons \underset{R_2}{\overset{R_1}{\diagdown}}C{=}O + 2ROH \qquad (12.27)$$

The following observations apply to most compounds in these classes: (a) The reaction is always catalyzed by hydronium ion, sometimes by general acids, and never by bases; (b) linear free energy correlations give large negative ρ values; (c) the hydrolysis of acetals derived from optically active alcohols shows that the carbonyl carbon-oxygen bond is cleaved; (d) entropies of activation are positive or slightly negative [31]; (e) added nucleophiles change the product distribution without affecting the observed rate constant for the overall reaction.

All of these indications support an A1 mechanism (unimolecular reaction of the conjugate acid), closely analogous to the S_N1 mechanism. The pathway is shown in Eqs. (12.28)-(12.31).

$$\underset{R_2}{\overset{R_1}{\diagdown}}C\underset{\diagup OR}{\diagdown OR} + H^+ \rightleftharpoons \underset{R_2}{\overset{R_1}{\diagdown}}C\underset{\diagup OR}{\diagdown \overset{H}{\overset{+}{O}R}} \qquad (12.28)$$

$$\underset{R_2}{\overset{R_1}{\diagdown}}C\underset{\diagup OR}{\overset{\overset{H}{OR}}{\diagdown}}{}^+ \rightleftharpoons \underset{R_2}{\overset{R_1}{\diagdown}}C^+\diagdown OR + ROH \qquad (12.29)$$

$$\underset{R_2}{\overset{R_1}{\diagdown}}C^+\diagdown OR + H_2O \rightleftharpoons \underset{R_2}{\overset{R_1}{\diagdown}}C\underset{\diagup \underset{H}{\overset{+}{O}R}}{\diagdown OH} \qquad (12.30)$$

$$\underset{R_2}{\overset{R_1}{\diagdown}}C\underset{\diagup \underset{H}{\overset{+}{O}R}}{\diagdown OH} \rightleftharpoons \underset{R_2}{\overset{R_1}{\diagdown}}C{=}\overset{+}{O}H + ROH \qquad (12.31)$$

Equation (12.29) appears to be the rate-determining step for many of these reactions. Evidence exists, however, that with some cyclic acetals Eq. (12.31) is rate determining [32].

Much of the interest in acetal chemistry is related to carbohydrate reactions, for these polyhydroxyaldehydes exist predominantly as cyclic hemiacetals. The phenomenon of mutarotation was referred to earlier. Capon has reviewed the mechanisms of carbohydrate reactions [33]. The enzyme lysozyme catalyzes the hydrolysis of an acetal

linkage in polysaccharides, and the pH-rate profile suggests the involvement of two carboxylic acid groups as
catalysts of the reaction. Dunn and Bruice [34] have interpreted their kinetic results for the hydrolysis of the
model acetal 2-methoxymethoxybenzoic acid as an intramolecular general-acid catalysis by the undissociated
ortho carboxylic acid group, as in 10, which they picture
as an electrostatic facilitation by the o-carboxylate
anion of an A1 mechanism.

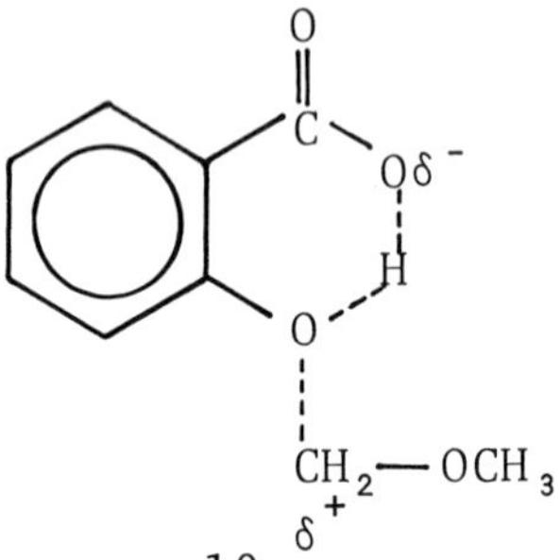

10

The hydrolysis of imines is the reverse of the addition
of amines to carbonyls, so the two-step scheme of Eq.
(12.32) applies.

$$\text{>C=NR} + H_2O \rightleftharpoons \begin{matrix} \nearrow \text{OH} \\ \text{>C} \\ \searrow \text{NHR} \end{matrix} \rightleftharpoons \text{>C=O} + RNH_2 \qquad (12.32)$$

The mechanism of imine formation is similar to the oximation mechanism discussed earlier, and it follows that the
mechanism of imine hydrolysis must be closely related.
The pH-rate curves for the hydrolysis of imines show several distinct segments. In the pH 0-4 range the rate increases with increasing pH; in the pH 8-14 range the rate
is independent of pH; and in the range 4-8 the transition
behavior depends upon the polar characteristics of the
amine. The explanation involves, as in the earlier discussion, a change in rate-determining step, complicated
by the appreciable basicity of the imine [35]. In the
alkaline region, where dehydration is rate determining for
imine formation, the addition of water must be rate determining in the reverse reaction. This could be either the
attack of water on neutral imine or the kinetic equivalent,
attack of hydroxide on the protonated imine. Since substituent effects on the rate in this region are small, the
second alternative is preferred.

$$\begin{array}{ccc}
& & \text{OH} \\
& & \diagdown \ \diagup \\
\diagup{\overset{+}{C}}{=}\text{NHR} + \text{OH}^- \underset{\longleftarrow}{\overset{\text{slow}}{\longrightarrow}} \ \ \ \diagdown \text{C} \diagup \ \ \underset{\longleftarrow}{\overset{\text{fast}}{\longrightarrow}} \ \ \diagup{C}{=}\text{O} + \text{RNH} & (12.33) \\
& & \diagup \ \diagdown \\
& & \text{NHR}
\end{array}$$

At somewhat lower pH, as the fraction of protonated imine increases (typical pK_a values are 6-7) and the concentration of hydroxide decreases, attack of water on the protonated imine becomes important.

$$\begin{array}{ccc}
& & \text{OH} \\
& & \diagdown \ \diagup \\
\diagup{\overset{+}{C}}{=}\text{NHR} + \text{H}_2\text{O} \underset{\longleftarrow}{\overset{\text{slow}}{\longrightarrow}} \ \ \diagdown \text{C} \diagup \ \ \underset{\longleftarrow}{\overset{\text{fast}}{\longrightarrow}} \ \diagup{C}{=}\text{O} + \text{RNH}_3^+ & \\
& & \diagup \ \diagdown \\
& & \underset{+}{\text{NH}_2\text{R}} \\
& & (12.34)
\end{array}$$

The rate becomes briefly pH independent when all of the imine is protonated, but with further decrease in pH the rate-determining step shifts to loss of amine from the carbinolamine (recall the rate-determining step in the reverse reaction at acid pH).

$$\begin{array}{ccc}
& \text{OH} & \text{O}^- \\
& \diagdown \ \diagup & \diagdown \ \diagup \\
\diagup{\overset{+}{C}}{=}\text{NHR} + \text{H}_2\text{O} \underset{\longleftarrow}{\overset{\text{fast}}{\longrightarrow}} \ \diagdown \text{C} \diagup \ \underset{\longleftarrow}{\overset{-\overset{+}{\text{H}}}{\longrightarrow}} \ \diagdown \text{C} \diagup & \\
& \diagup \ \diagdown & \diagup \ \diagdown \\
& \underset{+}{\text{NH}_2\text{R}} & \underset{+}{\text{NH}_2\text{R}}
\end{array}$$

$$\Big\Downarrow \text{slow}$$

$$\diagup{C}{=}\text{O} + \text{RNH}_2 \quad (12.35)$$

The hydrolysis of eneamines, which are α,β-unsaturated amines, proceeds through the tautomeric imine. Since both the amine and the imine can be protonated, several kinetic possibilities are created.

$$\begin{array}{ccc}
\text{R-NH} & & \overset{+}{\text{R-NH}_2} \\
\diagdown \ \ \diagup & \underset{-\text{H}^+}{\overset{+\text{H}^+}{\longleftrightarrow}} & \diagdown \ \ \diagup \\
\ \ \text{C}{=}\text{C} & & \ \ \text{C}{=}\text{C} \\
\diagup \ \ \diagdown & & \diagup \ \ \diagdown
\end{array}$$

$$\text{HA} \Big\Updownarrow \text{A}^-$$

$$\begin{array}{ccc}
\overset{+}{\text{R-NH}} & & \text{R-N} \\
\diagdown \ | & \underset{+\text{H}^+}{\overset{-\overset{+}{\text{H}}}{\longleftrightarrow}} & \diagdown \ | \\
\ \ \text{C}{-}\text{C}{-} & & \ \ \text{C}{-}\text{C}{-} \\
\diagup \ | & & \diagup \ | \\
\ \ \text{H} & & \ \ \text{H} \qquad (12.36)
\end{array}$$

The rate-determining step at very low pH is hydrolysis of
the protonated imine; at higher pH the formation of the
imine is the slow step [36].

Reactions of Nitrogen Nucleophiles. The mechanism of the
addition-elimination pathway for reaction of primary
amines, RNH_2, with carbonyl compounds has already been
discussed. The product, which contains the $>C=N-$ function,
is an imine, and several variants are distinguished by
special names, as shown in Table 12.I. Geometrical isomerism about the carbon-nitrogen double bond gives two
possible forms of the imine. This isomerism has been most
carefully studied with oximes [37]. The phenomenon, and
its nomenclature, are illustrated by 11 (syn-benzaldoxime),
12 (anti-benzaldoxime), and 13 (either syn-phenyltolyl-
ketoxime or anti-tolylphenylketoxime).

$$
\begin{array}{ccc}
\overset{\displaystyle N-OH}{\underset{\displaystyle C_6H_5-C-H}{\|}} & \overset{\displaystyle N-OH}{\underset{\displaystyle H-C-C_6H_5}{\|}} & \overset{\displaystyle N-OH}{\underset{\displaystyle CH_3C_6H_4-C-C_6H_5}{\|}} \\
11 & 12 & 13
\end{array}
$$

The reactions of carbonyl compounds with hydroxylamine,
semicarbazide, and hydrazine and its derivatives are so
characteristic that these nucleophiles are collectively
referred to as "carbonyl reagents." It is not surprising
that nucleophiles react more rapidly with an unhydrated
carbonyl compound than with its hydrate, and the rate of
dehydration can be rate limiting for the very reactive
nucleophiles [22]. It is similarly to be expected that
the open-chain aldehyde form of sugars will be more reactive toward carbonyl reagents than are the cyclic hemiacetal forms. The rate-determining step can be either the
ring opening or the subsequent reaction with the nucleophile, depending upon the reactants and the pH [33].
Table 12.III gives equilibrium constants for the initial addition of many nucleophiles to pyridine-4-carbox-
aldehyde and to p-chlorobenzaldehyde [Eq. (12.37)].

$$
\underset{R}{\overset{R}{{}}}\!\!>C=O + HX \;\overset{K}{\rightleftharpoons}\; \underset{H}{\overset{R}{{}}}\!\!>C\!\!<\!\!\underset{X}{\overset{OH}{{}}} \qquad (12.37)
$$

These constants were obtained from measurements of K_{obs},
which is defined by K_{obs} = [adduct]/[hydrated + unhydrated
aldehyde][nucleophile], and the hydration constant K_h
defined in Eq. (12.240 [38]. The K values clearly are not
correlated with nucleophile basicity. The high affinity
of the typical carbonyl reagents for the carbonyl carbon

is particularly striking. This equilibrium affinity for carbon is paralleled by the nucleophilic reactivity of these reagents in reactions at the carbon-oxygen double bond. Enhancement of nucleophilic reactivity (relative to basicity) in substances possessing unshared electron pairs on the atom adjacent to the nucleophilic atom has been called the alpha-effect [39]. It is probable that the α-effect is a manifestation of several factors, such as polarizability, steric effects, intramolecular catalytic modes, and resonance stabilization of transition states [40].

Table 12.III   Equilibrium Constants for Eq. (12.37) [38][a]

Nucleophile	K/M^{-1}	
	Pyridine-4-carboxaldehyde	p-Chlorobenzaldehyde
Methylamine	87	--
Ethylamine	43	--
n-Propylamine	39	--
Piperidine	--	0.45
Morpholine	32	0.40
Piperazine	32^{b}	0.43^{b}
Piperazine cation	17	0.22
Urea	1.3^{b}	--
Formamide	1.7	--
Hydroxylamine	1500	24
N-Methylhydroxylamine	--	6.6
Hydrazine	--	7.0^{b}
Semicarbazide	250	3.1
Methanol	0.50	--
Hydrogen peroxide	19.8^{b}	0.44^{b}
Water	0.023	
Hydrogen cyanide	--	300
Bisulfite	--	1.14×10^{4}

[a]In water at 25°.
[b]Corrected by statistical factor of 2.

The earlier mechanistic treatment [see Eqs. (12.14) and (12.16)] showed that at neutral pH the overall rate constant for product formation is given by the product of the equilibrium constant K and the rate constant for dehydration of the carbinolamine. Table 12.IV gives rate and equilibrium data for semicarbazone formation from

substituted benzaldehydes [41]. Equation (12.38) shows
the reaction (RNH$_2$ ≡ semicarbazide).

$$X\text{-}C_6H_4\text{-}CHO \ + \ RNH_2 \xrightleftharpoons{K} X\text{-}C_6H_4\text{-}CH(OH)NHR \ \xrightleftharpoons{k_2[H^+]}$$

$$X\text{-}C_6H_4\text{-}CH{=}NR \ + \ H_2O \qquad\qquad (12.38)$$

Table 12.IV   Rate and Equilibrium Constants for
Semicarbazone Formation [41][a] [Eq. (12.38)]

X	$K/\underline{M}^{-1}$	$10^{-7}k_2/\underline{M}^{-1}\text{-min}^{-1}$	$10^{-5}K^b_{overall}/\underline{M}^{-1}$
H	1.32	0.82	6.9
o-MeO	1.67	2.84	8.2
p-MeO	0.34	1.25	1.5
o-OH	0.33	3.10	4.5
p-OH	0.073	6.30	1.1
o-Me	1.05	1.42	7.0
p-Me	0.62	1.49	3.4
o-Cℓ	19.0	0.102	23.3
p-Cℓ	4.14	0.207	10.5
o-NO$_2$	27.1	0.019	75
p-NO$_2$	40.1	0.024	83

[a]At 25° in 25% ethanol.
[b]Equilibrium constant for semicarbazone formation.

For every substituent except nitro, the equilibrium constant K for adduct formation is larger for the ortho- than
for the para-substituted substrate. The dehydration rate
constants do not show this pattern. The relative overall
rates therefore reflect these differences in affinity in
the initial addition. These results were attributed to
greater resonance stabilization of p- than of o-substituted
benzaldehydes by electron-donating substituents [41].
     A carbonyl addition appears to be the first step in the
Mannich reaction, in which an aldehyde, an amine, and an
active hydrogen compound undergo condensation. The overall
reaction for formaldehyde, ammonia, and a methyl ketone is
as follows:

$$\underset{HCH}{\overset{O}{\|}} + NH_3 + \underset{CH_3CR}{\overset{O}{\|}} \rightarrow H_2NCH_2\underset{CH_2CR}{\overset{O}{\|}} \qquad (12.39)$$

As in Eq. (12.39), the usual product contains an amino group beta to a carbonyl; this compound is called a Mannich base. Mannich bases readily eliminate ammonia to give α,β-unsaturated carbonyl compounds.

The mechanism of the Mannich reaction is not fully established, but kinetic studies show that it is third order (first order in each reactant) at high and low pH; at intermediate pH third-order kinetics were not observed [42]. The initial step is postulated to be the equilibrium formation of the carbinolamine adduct from the amine and formaldehyde. In basic solution the carbanion derived from the active hydrogen compound might carry out a rate-determining S_N2 displacement on the carbinolamine, though hydroxide is a poor leaving group.

$$
\begin{array}{c}
\text{OH} \\
| \\
\text{H-C-H} \\
| \\
\text{NH}_2
\end{array}
+ \ ^{-}\text{CH}_2\overset{\displaystyle \text{O}}{\overset{\|}{\text{C}}}\text{R} \rightarrow \text{H}_2\text{NCH}_2\text{CH}_2\overset{\displaystyle \text{O}}{\overset{\|}{\text{C}}}\text{R} + \text{OH}^- \qquad (12.40)
$$

In acid solution the carbinolamine could give a carbonium ion, which reacts with the enol form of the active hydrogen compound.

$$
\begin{array}{c}
\text{OH} \\
| \\
\text{H-C-H} \\
| \\
\text{NH}_2
\end{array}
\overset{\text{H}^+}{\longrightarrow}
\begin{array}{c}
+ \\
\text{H-C-H} \\
| \\
\text{NH}_2
\end{array}
\xrightarrow[\text{-H}^+]{\ \overset{\text{OH}}{\overset{|}{\text{CH}_2=\text{CR}}}\ } \text{H}_2\text{NCH}_2\text{CH}_2\overset{\displaystyle \text{O}}{\overset{\|}{\text{C}}}\text{R} \qquad (12.41)
$$

Another multistep reaction based on carbinolamine formation is the reductive alkylation of amines. Equation (12.42) shows the probable course of the reaction [43].

$$
\overset{\displaystyle \text{O}}{\overset{\|}{\underset{/\backslash}{\text{C}}}} + \text{RNH}_2 \rightleftharpoons
\begin{array}{c}
\text{OH} \\
| \\
-\text{C}- \\
| \\
\text{NHR}
\end{array}
\xrightarrow{\text{hydrogenolysis}}
\begin{array}{c}
\text{H} \\
| \\
-\text{C}- \\
| \\
\text{NHR}
\end{array}
$$

$$
\searrow_{-\text{H}_2\text{O}} \qquad \nearrow \text{hydrogenation}
$$

$$
\begin{array}{c}
\diagdown\text{C}\diagup \\
\| \\
\text{NR}
\end{array} \qquad (12.42)
$$

In effect, a carbonyl group is converted to an amine. Ammonia and primary amines can take the two pathways shown, but secondary amines cannot form the imine. The reaction

is known by several names depending upon the reducing
agent used; a common method utilizes formic acid, the procedure then being called the Leuckart reaction.

Reactions of Carbon Nucleophiles. The mechanism of cyanohydrin formation was discussed earlier. Another simple
carbonyl addition by a carbon nucleophile is the synthetically valuable Grignard reaction and related reactions of
organometallic compounds. The Grignard reaction mechanism
is complicated by the nature of the Grignard reagent in
solution. It appears that the reagent engages in equilibria involving monomers and dimers as in [44]

$$(RMgX)_2 \rightleftharpoons 2RMgX \rightleftharpoons R_2Mg + MgX_2 \rightleftharpoons R_2Mg \cdot MgX_2 \quad (12.43)$$

The extent of formation of the various species depends
upon R, X, and the solvent. A cyclic mechanism for Grignard addition has been suggested in which the unsymmetrical
dimer $R_2Mg \cdot MgX_2$ is the effective reagent, as in 14 [45].

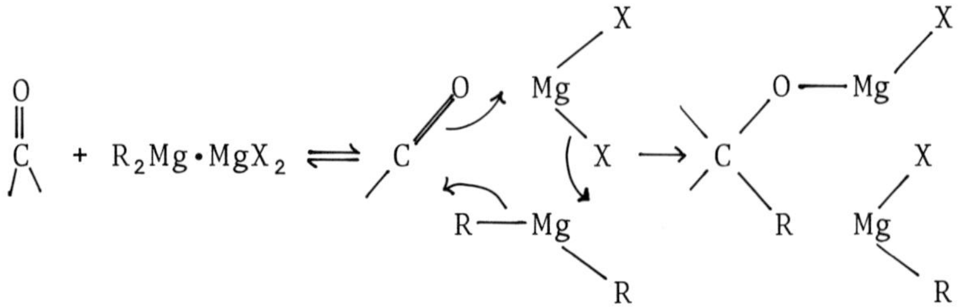

14   (12.44)

Ashby and Smith [44] have shown that the Grignard reagent
can exist as the monomer (especially at low concentrations)
in ether, and they suggest that attack can occur by the
monomer or the symmetrical dimer. The rate-determining
step is a displacement by the carbonyl oxygen of a solvating ether molecule on magnesium:

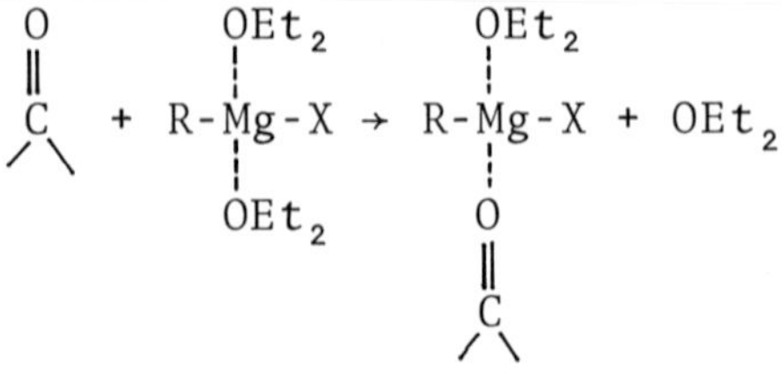

(12.45)

A second monomer reacts in a fast step with this associa-
tion complex:

$$\text{(12.46)}$$

This mechanism provides an explanation of the observation that the rate of a Grignard reaction decreases markedly after 50% of the "R" groups in the reagent have been consumed in the alkylation. The product of Eq. (12.46), $RMgOR'$, should be a "Grignard reagent" of much different reactivity than the reagent $RMgX$. Mechanism (12.44) does not account for this reactivity alteration.

Two important side reactions can occur in Grignard reactions with bulky groups on the reagent or the carbonyl. If the carbonyl compound possesses an alpha hydrogen, this can be transferred to the Grignard reagent. Upon hydrolysis the carbonyl compound is regenerated.

$$R'\text{-}\underset{\displaystyle}{\overset{O}{\overset{\|}{C}}}\text{-}\underset{\displaystyle}{\overset{H}{\overset{|}{C}}}\text{-} \; + \; RMgX \; \rightarrow \; R'\text{-}\underset{\displaystyle}{\overset{OMgX}{\overset{|}{C}}}\text{=}\underset{\displaystyle}{C}\text{-} \; + \; RH \qquad \text{(12.47)}$$

This reaction is called enolization. If the Grignard reagent has a beta hydrogen, reduction can occur, with the carbonyl group being converted, upon hydrolysis, to an alcohol.

$$\underset{/\backslash}{\overset{O}{\overset{\|}{C}}} \; + \; \text{-}\underset{\displaystyle}{\overset{H}{\overset{|}{C}}}\text{-C-MgX} \; \rightarrow \; \text{-}\underset{\displaystyle}{\overset{OMgX}{\overset{|}{C}}}\text{-} \; + \; {>}C{=}C{<} \qquad \text{(12.48)}$$

Organolithium compounds are utilized synthetically in much the same way as are Grignard reagents. They sometimes offer convenient routes to normal addition products when Grignard reagents give unacceptable levels of enolization or reduction.

The Reformatsky reaction, Eq. (12.49), is formally similar to the Grignard reaction, though an alkylzinc bromide is not the reagent.

$$\underset{/\backslash}{\overset{O}{\overset{\|}{C}}} \; + \; \text{-}\underset{\displaystyle}{\overset{Br}{\overset{|}{C}}}\text{-CO}_2\text{R} \; \xrightarrow{Zn} \; \text{-}\underset{\displaystyle}{\overset{OZnBr}{\overset{|}{\underset{\displaystyle}{\overset{C}{\underset{|}{\underset{\displaystyle}{\overset{C}{\underset{|}{CO_2R}}}}}}}} \qquad \text{(12.49)}$$

Upon hydrolysis the alcohol is produced. The halide compound is usually an α-halo ester, as shown.

Several important carbonyl addition reactions fit the
general scheme of

$$
\begin{array}{ccccc}
\text{O} & & \text{O}^- & \text{OH} & \\
\| & & | & | & \diagup \\
\text{C} & + \ \ ^-\text{C-} \ \ \to & \text{-C-} \ \to & \text{-C-} \ \to & \text{C} \\
\diagup \ \backslash & | & | & | & \| \\
\text{H} & & \text{-C-} & \text{-C-} & \text{C} \\
& & | & | & \diagup \ \backslash \\
& & \text{H} & \text{H} &
\end{array}
\qquad (12.50)
$$

An active hydrogen compound, upon treatment with base,
yields a carbanion, which attacks the carbon atom of a
carbonyl group. Dehydration may follow the initial addition if a hydrogen is present as shown, and especially if
the new double bond is in conjugation with other unsaturation. In the aldol condensation [46] the alpha-carbon of
an aldehyde or ketone provides the carbanion, which adds
to another aldehyde or ketone. Several possible combinations can be considered [47]. The reaction may be between
two molecules of the same aldehyde, often a very favorable
reaction. The self-condensation of propionaldehyde, Eq.
(12.51), illustrates this with the product drawn so as to
clarify the mode of addition.

$$
\begin{array}{c}
\text{OH} \\
|
\end{array}
$$
$$
2\text{CH}_3\text{CH}_2\text{CHO} \ \rightleftharpoons \ \text{CH}_3\text{CH}_2\text{-C-H} \qquad (12.51)
$$
$$
\begin{array}{c}
|\\
\text{CH}_3\text{-CH-CHO}
\end{array}
$$

The addition may be followed by dehydration, as noted
above, especially if acid is added. The product of the
simple addition (an aldol) may engage in other reactions,
since it is itself an aldehyde; thus it may undergo further
aldol condensation, and acetaldol, the product of self-
condensation of acetaldehyde, readily forms cyclic hemiacetals [48].

Mixed aldol condensation may occur with two different
aldehydes. The reaction mixture can give four aldols if
both aldehydes have α-hydrogens, but if one of the aldehydes lacks an α-hydrogen, good yields of a single product
may be obtained. Equation (12.52) exemplifies such a
reaction.

$$
\text{Ph-CHO} + \text{Ph-CH}_2\text{-CHO} \ \xrightarrow{\ \text{OMe}^-\ } \ \text{Ph-CH=C-CHO} + \text{H}_2\text{O} \qquad (12.52)
$$
$$
\begin{array}{c}
|\\
\text{Ph}
\end{array}
$$

Ketones are less effective than aldehydes in the aldol condensation, and the equilibrium favors the reactants. The self-addition of acetone, however, can be carried out with 71% yield of diacetone alcohol by suitably removing the adduct from the reaction mixture.

$$2CH_3CCH_3 \underset{base}{\rightleftharpoons} (CH_3)_2\overset{OH}{\underset{}{C}}-CH_2\overset{O}{\underset{}{C}}CH_3 \qquad (12.53)$$

Catalytic amounts of iodine lead to dehydration of this adduct to mesityl oxide, $(CH_3)_2C=CHCOCH_3$.

Self-condensation of methyl ethyl ketone gives 15, showing that the methyl group rather than the methylene carbon functioned as the carbanion nucleophile.

$$CH_3CH_2\overset{CH_3}{\underset{}{C}}=CHCOCH_2CH_3$$

<u>15</u>

Either acid or base can lead to aldol condensation of cyclic ketones, as in

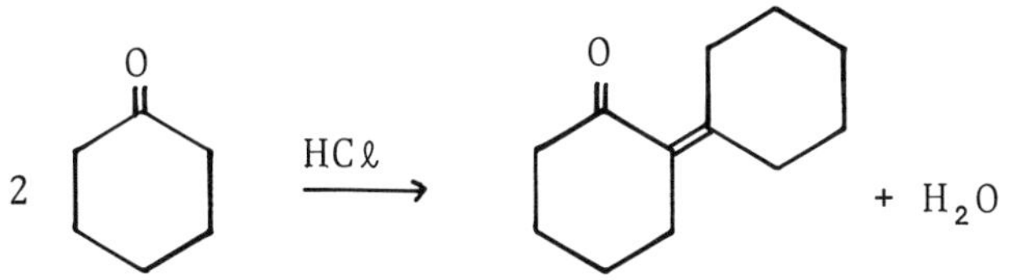

$$+ H_2O \qquad (12.54)$$

Condensations between an aldehyde and a ketone proceed with the ketone providing the carbon nucleophile and the aldehyde the carbonyl group. Yields are improved by using aromatic aldehydes, which cannot undergo self-condensation. Equations (12.55) and (12.56) show two of these condensations (Claisen-Schmidt condensation), the nature of the product being controlled by the ratio of reactant concentrations.

$$Ph-CHO + CH_3COCH_3 \rightarrow Ph-CH=CH-COCH_3 + H_2O \qquad (12.55)$$

$$2Ph-CHO + CH_3COCH_3 \rightarrow Ph-CH=CH-CO-CH=CH-Ph + 2H_2O \qquad (12.56)$$

Other active hydrogen compounds can take part in carbonyl additions. Common to all such compounds is one or two strongly electronegative groups adjacent to the incipient carbanion site. Esters will condense with aldehydes or ketones in the presence of very powerful bases.

The reaction is quite facile with diethyl succinate (the Stobbe condensation). One ester group is always cleaved, and elimination occurs.

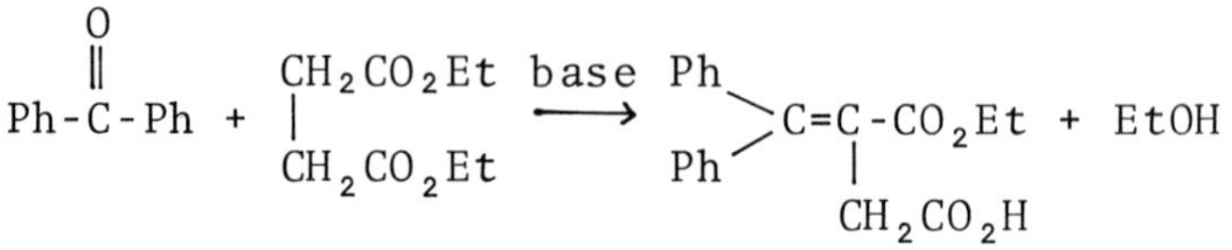

$$\underset{\text{Ph-C-Ph}}{\overset{\text{O}}{\parallel}} + \underset{\text{CH}_2\text{CO}_2\text{Et}}{\overset{\text{CH}_2\text{CO}_2\text{Et}}{\mid}} \xrightarrow{\text{base}} \underset{\text{Ph}}{\overset{\text{Ph}}{>}}\text{C}=\text{C}-\text{CO}_2\text{Et} + \text{EtOH} \quad (12.57)$$
$$\text{CH}_2\text{CO}_2\text{H}$$

An intermediate lactone, formed by intramolecular attack of the adduct oxyanion on an ester carbonyl, can account for the course of the reaction [49].

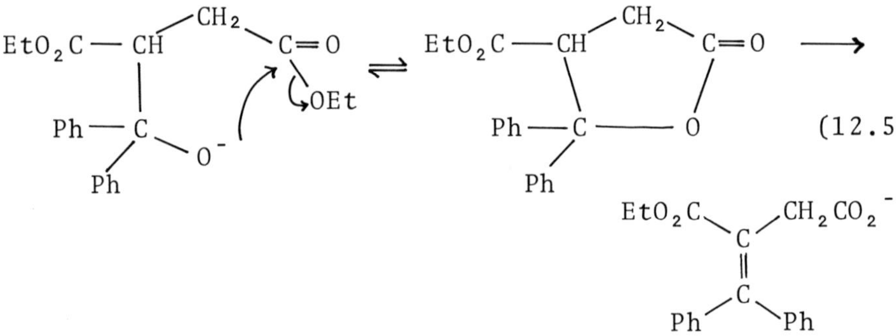

$$(12.58)$$

If the active hydrogen component of a carbonyl condensation reaction has the general structure $Z-CH_2-Z'$ or $Z-CHR-Z'$, where Z and Z' are electron-withdrawing groups, the reaction is called the Knoevenagel reaction [50]. The groups Z and Z' may be CHO, COR, COOR, CN, NO_2, etc. Dehydration usually occurs, and decarboxylation of an acid group is common. Some examples follow:

$$\text{Ph-CHO} + \text{CH}_2(\text{CO}_2\text{H})_2 \rightarrow \text{Ph-CH}=\text{CH-CO}_2\text{H} + \text{H}_2\text{O} + \text{CO}_2 \quad (12.59)$$

$$\text{Ph-CHO} + \underset{\text{CO}_2\text{Et}}{\overset{\text{CN}}{\text{H}_2\text{C}<}} \longrightarrow \text{Ph-CH}=\underset{\text{CO}_2\text{Et}}{\overset{\text{CN}}{\text{C}<}} + \text{H}_2\text{O} \quad (12.60)$$

$$\text{CH}_3\text{COCH}_3 + \underset{\text{CH}_3}{\overset{\text{CH}_3}{>}}\text{C} = \qquad + \text{H}_2\text{O} \quad (12.61)$$

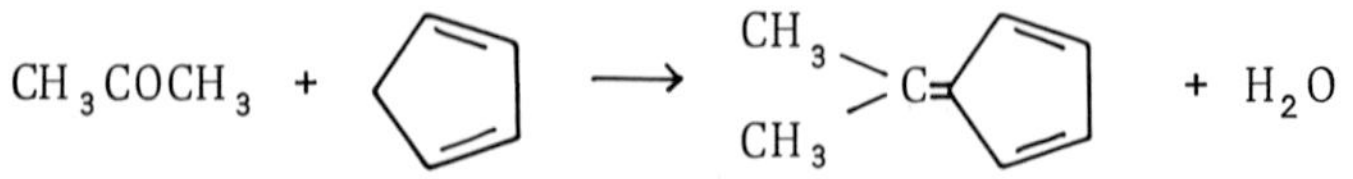

In the Perkin reaction an aromatic aldehyde is condensed with an acid anhydride; an alkali salt of the acid corresponding to the anhydride serves as the base. Cinnamic acid is prepared in this way.

$$\text{Ph-CHO} + \quad \begin{matrix} CH_3-C\overset{\displaystyle O}{\underset{\displaystyle O}{\diagdown}} \\ CH_3-C\underset{\displaystyle O}{\overset{\diagup}{\diagdown}} \end{matrix} \quad \xrightarrow{\text{NaOAc}} \quad \text{Ph-CH=CH-CO}_2\text{H} + CH_3CO_2H \qquad (12.62)$$

In this reaction it is the anhydride, not the acid anion, that adds to the aldehyde, since the salt can be replaced by another base. The interpretation of the following result,

$$\text{Ph-CHO} \quad \xrightarrow[\text{Ac}_2\text{O}]{\text{Ph-CH}_2\text{CO}_2\text{Na}} \quad \overset{\displaystyle \text{Ph}}{\underset{\displaystyle}{\text{Ph-CH=C-CO}_2\text{H}}} \qquad (12.63)$$

in which it appears that the salt has added, is that an acyl transfer equilibrium generates some phenylacetic anhydride (or, more probably, the mixed phenylacetic acetic anhydride):

$$\text{PhCH}_2\text{CO}_2{}^- + \text{Ac}_2\text{O} \rightleftharpoons \text{AcO}^- + \quad \begin{matrix} Ph-CH_2-C\overset{\displaystyle O}{\underset{\displaystyle O}{\diagdown}} \\ CH_3-C\underset{\displaystyle O}{\overset{\diagup}{\diagdown}} \end{matrix} \qquad (12.64)$$

The more reactive methylene, which is the one in the phenylacetic moiety, then adds to the carbonyl group.

The Wittig reaction belongs to this general type. This process, which was discussed in Chap. 11, involves the addition of a phosphorane to a carbonyl compound, giving, after elimination, an olefin.

$$\underset{\diagup\diagdown}{\overset{\displaystyle O}{\overset{\|}{C}}} + \text{Ph}_3\overset{+}{P}\text{-}\overset{-}{C}\text{H}_2 \rightarrow \underset{\displaystyle \underset{CH_2}{\overset{\|}{C}}}{\diagdown\diagup} + \text{Ph}_3\text{PO} \qquad (12.65)$$

Many of the reactions described here as carbonyl additions have variants in which the addition takes place at another type of multiple bond. Thus Grignard reagents add to nitriles, and nitriles can undergo the Thorpe reaction, which is similar to the aldol condensation:

$$2 \ \ -\overset{\displaystyle |}{\underset{\displaystyle}{\text{CH}}}\text{-CN} \xrightarrow{\text{OEt}^-} -\overset{\displaystyle |}{\underset{\displaystyle \underset{C\equiv N}{\overset{|}{\underset{|}{C}}}}{\text{CH}}}\text{-C=NH} \qquad (12.66)$$

12.2   Analytical Reactions

Most of the analytical examples of addition reactions to carbon-heteroatom multiple bonds involve, as in the preceding treatment of reactions and mechanisms, the carbonyl group. The determination of this functional group has been reviewed by many authors [51].

The subdivision of this section is according to the nucleophilic atom in the adding species. An analysis may involve an addition directly to the analytical sample compound, as in the determination of an aldehyde by oximation; or it may require a preliminary reaction to convert the sample compound into a form suitable for the addition reaction, as when an acetal is hydrolyzed to its parent aldehyde, which is then determined. So many analytical methods are based upon additions to carbon-heteroatom multiple bonds that this section gives a selection chosen to illustrate the important methods, some interesting applications, and the great variety of determinations based on this reaction type.

Oxygen and Sulfur Nucleophiles. We first consider reactions in which water is the source of nucleophilic oxygen. Carbonyl hydrate formation is not a useful analytical reaction. The hydrolysis of a ketal, however, serves as the basis of an ingenious method for drying samples [52] and determining water [53]. The reagent, 2,2-dimethoxypropane, undergoes acid-catalyzed hydrolysis according to

$$\begin{array}{c} CH_3 \\ \\ CH_3 \end{array} C \begin{array}{c} OCH_3 \\ \\ OCH_3 \end{array} + H_2O \xrightarrow{H^+} \begin{array}{c} CH_3 \\ \\ CH_3 \end{array} C=O + 2CH_3OH \qquad (12.67)$$

The acetone formed is determined by its absorption in the infrared to give a measure of the water consumed. By utilizing a more sensitive finish, such as a colorimetric measurement of the carbonyl compound, it might be possible to increase the sensitivity of this analysis.

Most analytical methods for acetals, ketals, and vinyl ethers are based upon their preliminary acid hydrolysis to the parent carbonyl compounds [Eqs. (12.27) and (12.68)], which are determined by the usual methods [54].

$$ROCH=CH_2 + H_2O \xrightarrow{H^+} CH_3CHO + ROH \qquad (12.68)$$

Compounds containing the functions $-O-CH_2-O-$, $-O-CH_2-S-$, $-N-CH_2-S-$, and $-N-CH_2-N-$ liberate formaldehyde upon acid hydrolysis. Acetonides, which are cyclic ketals of acetone and 1,2-diols, produce acetone when hydrolyzed [55].

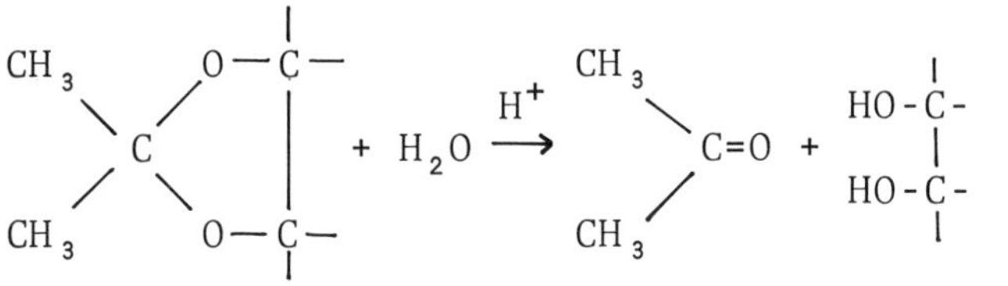

$$\underset{CH_3}{\overset{CH_3}{\diagdown}}C\underset{O-\overset{|}{C}-}{\overset{O-\overset{|}{C}-}{\diagup}} + H_2O \xrightarrow{H^+} \underset{CH_3}{\overset{CH_3}{\diagdown}}C=O + \begin{array}{c} HO-\overset{|}{C}- \\ | \\ HO-\overset{|}{C}- \\ | \end{array} \qquad (12.69)$$

In these methods it is the carbonyl compound that is sub-sequently measured. Thioacetals have been determined by hydrolysis [Eq. (12.70)], and bromometric titration of the thiol [56].

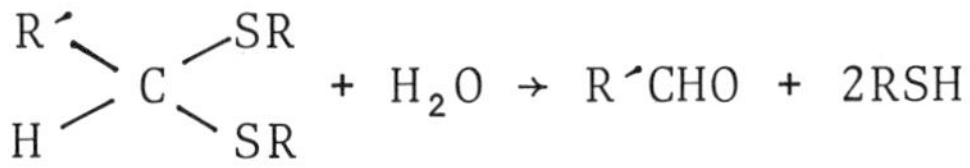

$$\underset{H}{\overset{R'}{\diagdown}}C\underset{SR}{\overset{SR}{\diagup}} + H_2O \rightarrow R'CHO + 2RSH \qquad (12.70)$$

$$2RSH + Br_2 \rightarrow RSSR + 2HBr \qquad (12.71)$$

Addition of water to a carbon-heteroatom multiple bond provides several analytical methods. Isocyanates and iso-thiocyanates have been determined by hydrolysis and deter-mination of carbon dioxide [Eq. (12.72)] [57] or amine [Eq. (12.73)] [58].

$$R-N=C=O + H_2O \xrightarrow{H^+} RNH_2 + CO_2 \qquad (12.72)$$

$$R-N=C=S + 2H_2O \xrightarrow{OH^-} RNH_2 + CO_2 + H_2S \qquad (12.73)$$

An analysis of the dithiocarbamate group first decomposes this function according to Eq. (12.74); the carbon disul-fide is then passed into methanolic potassium hydroxide, where potassium methyl xanthate is formed [Eq. (12.75)]. This is titrated with iodine [59].

$$\underset{R_2N-\overset{||}{C}-SH}{\overset{\overset{S}{||}}{}} \xrightarrow{H^+} R_2NH + CS_2 \qquad (12.74)$$

$$CS_2 + KOH + CH_3OH \rightarrow CH_3OCSSK + H_2O \qquad (12.75)$$

Alternatively the carbon disulfide may be collected in Vile's reagent, a mixture of copper acetate and diethyl-amine; cupric diethyldithiocarbamate, a yellow substance, is formed and is measured spectrophotometrically [60].

Schiff bases and cyclic acetals of benzaldehyde have been determined by hydrolysis to the aldehyde, which was then converted to the 2,4-dinitrophenylhydrazone for analysis [61]. Ketenes are hydrolyzed to the correspond-ing carboxylic acid, Eq. (12.76); the acid is titrated with alkali.

$$CH_2=C=O + H_2O \rightarrow CH_3COOH \qquad (12.76)$$

Alcohols as nucleophiles are involved in numerous analytical methods. The hemiketals formed from alcohols and hexafluoroacetone, Eq. (12.77), are suitable derivatives for the characterization of alcohols by ^{19}F nuclear magnetic resonance spectroscopy [62].

$$\begin{matrix} CF_3 \\ \end{matrix} \diagdown C=O + ROH \rightleftharpoons \begin{matrix} CF_3 \\ CF_3 \end{matrix} \diagup C \diagup \diagdown \begin{matrix} OH \\ OR \end{matrix} \qquad (12.77)$$

Since the adduct contains six fluorine atoms per molecule, the method is more sensitive than acylation to give the trifluoroacetyl ester. The ^{19}F chemical shift of the hemiketal, relative to the trifluoroacetone hydrate, is sensitive to the nature of the R group. Thiols and amines give similar behavior. The equilibrium constant for Eq. (12.78), which could be measured by ^{19}F NMR, was found to be 6.2 $\underline{M}^{-1}$ at 32° in acetone.

$$(CF_3)_2C=O + (CF_3)_2CHOH \rightleftharpoons (CF_3)_2C \diagup \diagdown \begin{matrix} OH \\ OCH(CF_3)_2 \end{matrix} \qquad (12.78)$$

The hemiketals of steroidal alcohols with hexafluoroacetone, as well as other hexahaloacetones, can be isolated in crystalline form, but they are unstable to gas-liquid chromatography [63]. The hemiketals were converted to the thermally stable methyl ketals for glc and combined glc-mass spectrometry [63].

$$(CX_3)_2C=O + ROH \rightarrow (CX_3)_2C \diagup \diagdown \begin{matrix} OH \\ OR \end{matrix} \xrightarrow{CH_2N_2} (CX_3)_2C \diagup \diagdown \begin{matrix} OCH_3 \\ OR \end{matrix} \qquad (12.79)$$

Crowell, Powell, and Varsel [64] have exploited the change in ultraviolet absorption when an aldehyde is converted to its dimethyl acetal for qualitative and quantitative analysis. The spectrum of the aldehyde is measured in methanol solution and in methanol acidified with sulfuric acid, the latter solution leading to rapid and extensive acetal formation for many aldehydes. Figure 12.2 shows these spectra for cinnamaldehyde and cinnamaldehyde dimethyl acetal. The pattern of spectral changes upon acetal formation is helpful for identification, and the change in absorption is a measure of aldehyde concentration. Some selectivity can thus be achieved if other absorbing components of a mixture do not undergo spectral alterations upon acidification. The fluorescence properties of aldehydes are also affected by acetal formation. The fluorescence resulting from the presence of a fluorophor other than the -CHO group is often enhanced by acetal formation, presumably because the quenching effect of the carbonyl group is eliminated in the acetal [65].

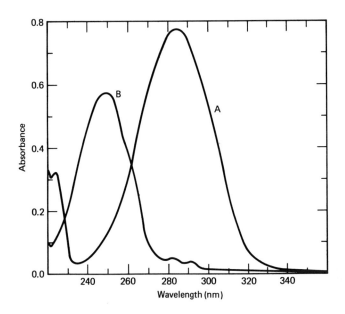

Fig. 12.2.   Spectrum of 2.03×10^{-5} $\underline{M}$ cinnamaldehyde in methanol (spectrum A) and in acidified methanol (spectrum B).

Alcohols add to aryl isocyanates, Eq. (12.80), to give urethanes (carbamates), which are crystalline derivatives suitable for characterization [66].

$$Ar-N=C=O + ROH \rightarrow Ar-NHCOOR \qquad (12.80)$$

Trichloroacetyl isocyanate, a very reactive reagent, gives derivatives useful for identification by nuclear magnetic resonance [67]. This reagent has been applied to the fractionation of alcohols in essential oils by chromatographic resolution of the initial addition compound or its hydrolysis product; the alcohol could also be ultimately regenerated [68].

$$CC\ell_3 \overset{\overset{\displaystyle O}{\|}}{C}-N=C=O \xrightarrow{ROH} CC\ell_3 \overset{\overset{\displaystyle O}{\|}}{C}-NHCOOR \xrightarrow{OH^-} H_2NCOOR \xrightarrow{OH^-} ROH \qquad (12.81)$$

In another use of the isocyanate addition, Pereira et al.
[69] reacted asymmetric secondary alcohols with R-(+)-1-
phenylethylisocyanate to give diastereoisomeric carbamates
that could be separated by gas chromatography.
    A few other oxygen nucleophiles give interesting reac-
tions. Nitriles are converted to amides by alkaline hydrogen peroxide; the reaction has been written according
to Eqs. (12.82) [70] or (12.83) [71].

$$RCN + 2H_2O_2 \xrightarrow{OH^-} RCONH_2 + H_2O + O_2 \qquad (12.82)$$

$$2RCN + 2H_2O_2 \xrightarrow{OH^-} 2RCONH_2 + O_2 \qquad (12.83)$$

α,β-Unsaturated nitriles may give the epoxide, as in Eq.
(12.84), but cinnamonitrile gives cinnamamide [70].

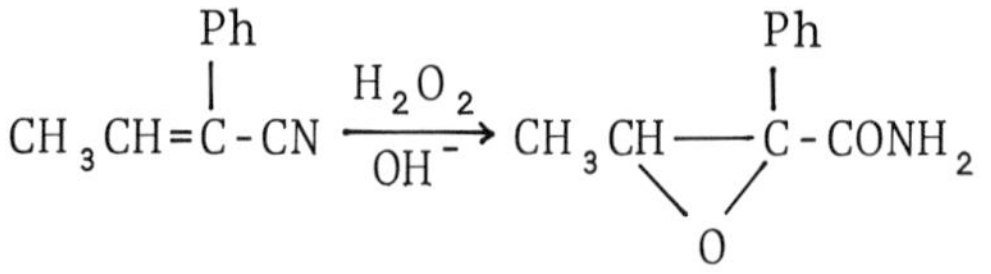

$$\underset{\substack{|\\CH_3CH=C-CN}}{\overset{Ph}{}} \xrightarrow[OH^-]{H_2O_2} CH_3CH\underset{\substack{\diagdown\diagup\\O}}{\overset{\substack{Ph\\|}}{-}}C-CONH_2 \qquad (12.84)$$

The yield of amide in this reaction is sensitive to the
solvent; Murray and Cloke preferred acetone. The method
has some synthetic utility. Whitehurst and Johnson [72]
applied the alkaline peroxide reaction to the determination
of nitriles. The amide produced was hydrolyzed to the
corresponding acid in situ, and the alkali consumed was
determined by titration with standard acid. A different
finish might extend the applicability of this method,
which as described suffers from numerous interferences.
    Sawicki, Stanley, and Hauser [73] describe a sensitive
color test for aromatic aldehydes in which the aldehyde is
treated with trifluoroacetic anhydride and fluoranthrene.
Sawicki [74] writes the reaction as follows:

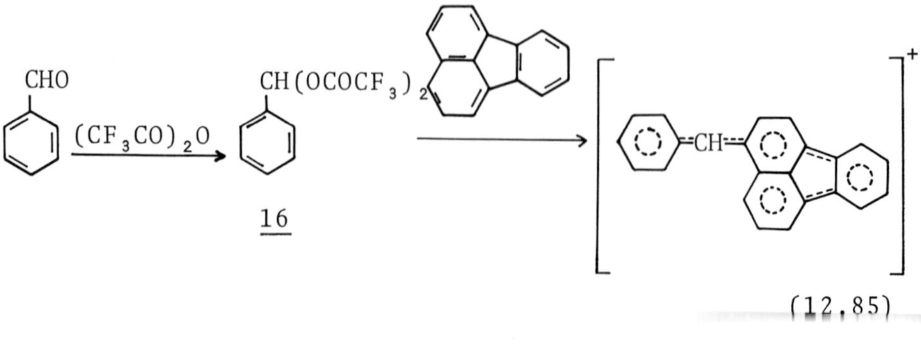

$$(12.85)$$

The fuchsin-sulfurous acid color test for aldehydes is based on the following reactions [75]. The dye p-fuchsin, 17, is decolorized by treatment with SO_2 to give the leuco C-sulfonic acid-N-sulfinic acid 18. This adds to an aldehyde, and 19 eliminates sulfurous acid producing the pink quinoid dye 20. Other triphenylmethane dyes undergo similar reactions.

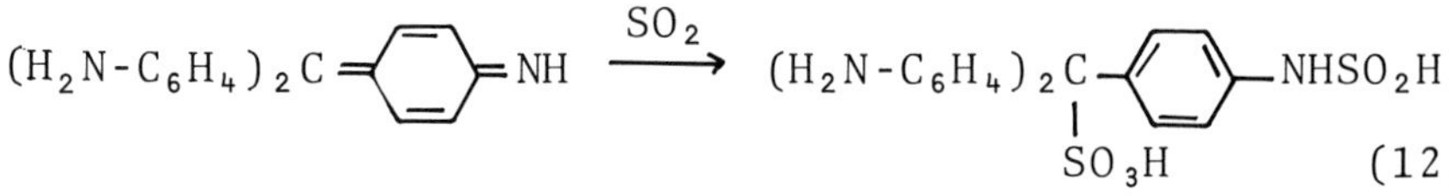

$$(H_2N-C_6H_4)_2C \!=\!\!\bigcirc\!\!= NH \xrightarrow{\ SO_2\ } (H_2N-C_6H_4)_2\underset{SO_3H}{\overset{|}{C}}\!\!-\!\!\bigcirc\!\!-NHSO_2H \qquad (12.86)$$

17   18

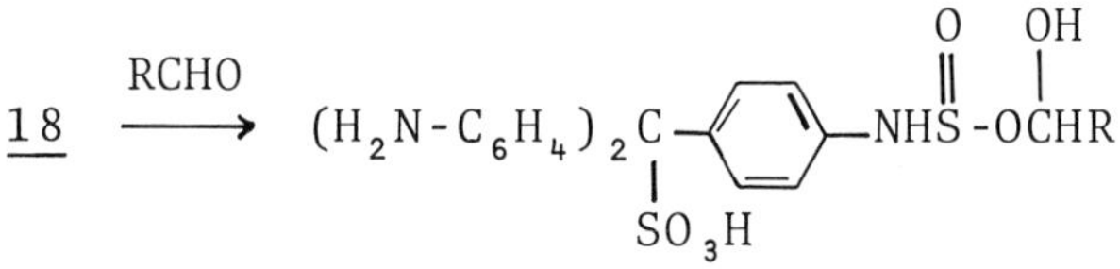

$$18 \xrightarrow{\ RCHO\ } (H_2N-C_6H_4)_2\underset{SO_3H}{\overset{|}{C}}\!\!-\!\!\bigcirc\!\!-NH\overset{O}{\overset{\|}{S}}\text{-}O\overset{OH}{\overset{|}{C}}HR \qquad (12.87)$$

19

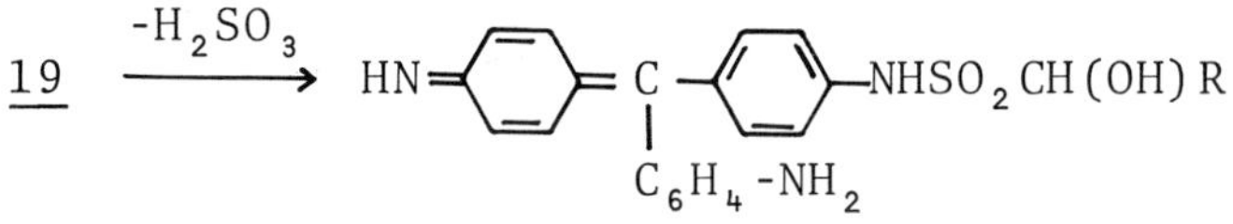

$$19 \xrightarrow{-H_2SO_3} HN\!=\!\!\bigcirc\!\!=\underset{C_6H_4-NH_2}{\overset{|}{C}}\!\!-\!\!\bigcirc\!\!-NHSO_2CH(OH)R \qquad (12.88)$$

20

The bisulfite addition reaction is the basis of a venerable method for aldehydes [76]. This simple addition can be formally written as

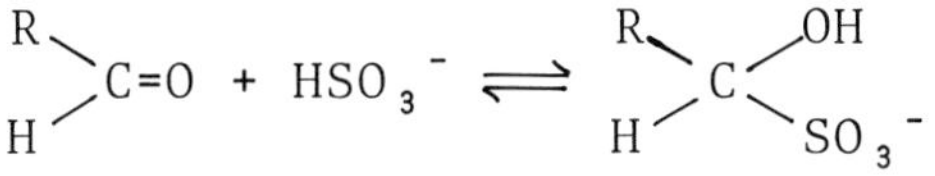

$$\underset{H}{\overset{R}{>}}C\!=\!O + HSO_3^- \rightleftharpoons \underset{H}{\overset{R}{>}}\underset{SO_3^-}{\overset{OH}{C}} \qquad (12.89)$$

The sulfite dianion, SO_3^{2-}, is the attacking nucleophile. As an analytical reaction bisulfite addition is not generally applicable to ketones because their equilibrium constants for adduct formation are not large enough. Kolthoff and Stenger [77] have given a detailed analysis of the analytical error introduced by incompleteness of the addition reaction (or, equivalently, "dissociation of the addition compound").

Several bisulfite procedures have been devised. In one of these the aldehyde is reacted with excess bisulfite, and the unreacted bisulfite is determined iodometrically. In another the sample is treated with excess sulfite and the

hydroxide produced according to Eq. (12.90) is titrated
with acid.

$$RCHO + SO_3^{2-} + H_2O \rightleftharpoons \begin{matrix} R \\ \diagdown \\ H \diagup \end{matrix} C \begin{matrix} \diagup OH \\ \diagdown SO_3^- \end{matrix} + OH^- \qquad (12.90)$$

Still another carries out the addition in excess bisulfite;
the unreacted bisulfite is oxidized with iodine, the addition compound is decomposed, and the bisulfite liberated
is titrated iodometrically [76]. Probably the most satisfactory procedure is that of Siggia and Maxcy [78], who
treat the aldehyde with excess sodium sulfite and standard
sulfuric acid in amounts such that $SO_3^{2-} > H_2SO_4 > RCHO$;
thus a sulfite-bisulfite mixture is generated.
    When many thiols are treated with o-phthaldialdehyde a
strong fluorescence is observed. The fluorescence intensity is pH dependent, as is its rate of development, both
the rate and intensity being low in acid and high above
neutrality. With glutathione, at least, p-phthaldialdehyde
does not develop a similar fluorescence [79]. Hemithioace-
tal formation has been suggested, but the nature of the
reaction is obscure. Simple removal of a quenching aldehyde group, analogous to the acetal formation procedure
discussed earlier [65], is a possibility, but a new fluorophor may be developed as a consequence of the ortho relationship of the two functional groups in a monohemithio-
acetal of o-phthaldialdehyde.

Nitrogen Nucleophiles. Imine formation is an extremely
important reaction in synthetic, analytical, and biochemical systems. The several variants of this reaction, including oximation and hydrazone formation, are considered
first, and then the addition of nitrogen nucleophiles to
other carbon-heteroatom multiple bonds is described.
    Oximes have some utility as carbonyl derivatives for
characterization, but they are usually lower melting than
the corresponding semicarbazones or 2,4-dinitrophenyl-
hydrazones. Sugars can be identified from their rates of
oximation at pH 6.50; the reaction is followed spectrophotometrically [80].
    Oximation is the method of choice for the general
determination of carbonyls on the macro and semi-micro
scales, but its utility is sharply controlled by the factors of rate, equilibrium, and analytical finish. Fritz
and Hammond [81], in their demonstration of the value of
a physical organic chemical approach to analytical development, selected oxime formation for detailed discussion
[82]. Schenk has subsequently treated the same reaction
and has reviewed the development of oximation methods [83],
so a brief treatment will suffice here.

The oximation reaction, Eq. (12.91), is not, as written, sufficient basis for the design of a satisfactory method.

$$\text{>C=O} + NH_2OH \rightleftharpoons \text{>C=NOH} + H_2O \qquad (12.91)$$

Most oximation methods for carbonyl determination utilize, as the finish, some measurement of the amount of hydroxyl-amine reacted. Unfortunately hydroxylamine is unstable in alkaline solution. This complication is one of the limitations on the method. Another is the position of equilibrium, which is pH dependent because both the hydroxylamine and the oxime are bases, but hydroxylamine is stronger than the oximes [84]. Oxime formation is therefore favored by an increase in pH. The rate of oximation is also a function of pH, with a bell-shaped pH-rate profile showing a maximum near pH 4-5, as illustrated for acetoxime formation in Fig. 12.1.

The several procedures that have been proposed can be viewed in terms of these conflicting pH requirements. A very simple method uses an aqueous solution of hydroxyl-amine hydrochloride, the net reaction being

$$\text{>C=O} + NH_3OH^+ \rightleftharpoons \text{>C=NOH} + H_3O^+ \qquad (12.92)$$

This has the advantage of simplicity and stability of the reagent, but the disadvantages of slowness of reaction and, especially, unfavorable equilibrium. Nevertheless, the method gives satisfactory results for most aldehydes and unhindered ketones. The finish is a titration with standard alkali of the strong acid liberated; since this acid is titrated in the presence of the hydroxylammonium ion, end points are not sharp, but good results are obtained by potentiometric titration [85] or by titration to a visual indicator comparison solution [86].

Bryant and Smith [87] carried out the oximation in ethanol containing pyridine, so the buffer system NH_3OH^+ + $C_6H_5N \rightleftharpoons NH_2OH + C_5H_5NH^+$ is set up. This establishes a "pH" in the alcoholic medium that provides practicable rates and equilibria of oxime formation. The analysis is completed by titration with alcoholic alkali, or, therefore, by titrating the pyridinium ion produced by the "liberated acid." Although pyridine and hydroxylamine are similar in base strength in water, apparently pyridine is, fortuitously, a weaker base in ethanol, thus permitting the titration of $C_5H_5NH^+$ in the presence NH_3OH^+. The end points are, however, not fully satisfactory. In 1956 Higuchi and Barnstein [88] introduced an ingenious modification by using hydroxylammonium acetate as the reagent [89] and glacial acetic acid as the solvent. In acetic acid, hydroxylammonium acetate is the conjugate base form

of the reagent. After oximation is complete, the excess
reagent is titrated potentiometrically with standard acetous perchloric acid. Oximes of dialkylketones are sufficiently basic to be titrated themselves. In this method
equilibrium limitations are unimportant, at least in part
because of the absence of water.

Subsequent proposals have followed the lead of Higuchi
and Barnstein in using a nonaqueous reaction medium, but
have achieved better end points. Closely related methods
use hydroxylammonium formate [91] as the reagent. Fritz
et al. [92] employ a mixture of hydroxylamine hydrochloride and 2-dimethylaminoethanol (chosen because it is a
tertiary amine, a stronger base than hydroxylamine, and
its hydrochloride is soluble in the nonaqueous solvent
system). After oximation the unreacted hydroxylamine is
titrated with standard acid.

Several hydrazines have been used as carbonyl reagents.
Phenylhydrazine, p-nitrophenylhydrazine, and 2,4-dinitro-
phenylhydrazine often yield crystalline hydrazones helpful
in characterizing aldehydes and ketones [93]; Eq. (12.93)
shows the reaction with the last of these.

$$\text{>C=O} + \text{H}_2\text{N-NH-}\underset{\text{NO}_2}{\bigcirc}\text{-NO}_2 \rightarrow \text{>C=N-NH-}\underset{\text{NO}_2}{\bigcirc}\text{-NO}_2 + \text{H}_2\text{O}$$

(12.93)

2,4-Dinitrophenylhydrazones are insoluble in acid solution,
and their precipitation ensures that the reaction goes to
completion. Because of this insolubility and the large
gravimetric factor, 2,4-dinitrophenylhydrazone formation
is the basis of a good gravimetric determination of carbonyls [94].

The best titrimetric carbonyl determination based on
hydrazone formation is that of Siggia and Stahl [95], which
uses unsymmetrical dimethylhydrazine, with potentiometric
back-titration of unreacted reagent with standard acid.
Ketones cannot be determined because their dimethylhydra-
zones are about as basic as the reagent; in fact, aromatic
aldehydes can be determined in the presence of ketones
(though aliphatic aldehydes cannot). Because the reaction
medium is alkaline, acetals, which are resistant to alkaline
hydrolysis, do not interfere [96].

Many spectrophotometric analyses are based on hydrazone
formation. The spectral properties of 2,4-dinitrophenyl-
hydrazones are determined primarily by the 2,4-dinitrophenyl
group, but sufficient variation is introduced by the carbonyl substituents that limited correlations of spectra with
structure are possible [97]. Treatment of a 2,4-dinitro-

phenylhydrazone with alkali leads to an intense red color
that has been attributed to a quinoidal electronic distribution in the anion:

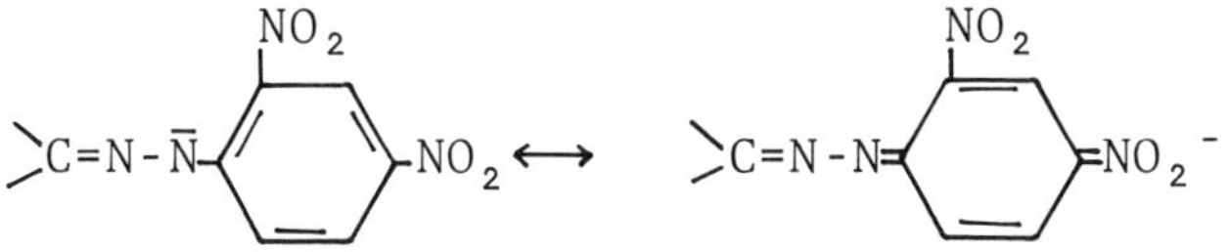

This gives a sensitive quantitative method for carbonyls
[98], because the molar absorptivities of 2,4-dinitrophe-
nylhydrazones in alkali are $2\text{-}3 \times 10^4$. Table 12.V lists
some of the hydrazines that have been employed in the
spectrophotometric determination of carbonyls (or of compounds, like alcohols, acetals, and epoxides, that upon
suitable treatment generate carbonyls). The table also
gives a few references to typical uses of these reagents.
MBTH and HBT are very sensitive reagents for aldehydes
because an oxidized molecule of reagent can carry out an
electrophilic substitution on the carbonyl carbon atom in
the initially formed imine, forming a highly conjugated
chromophore, as shown for MBTH [74].

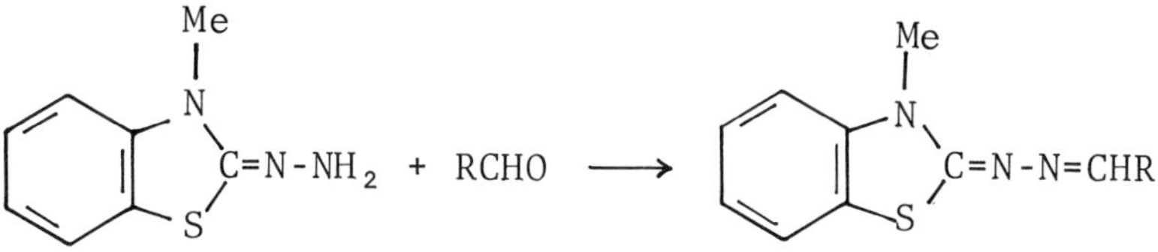

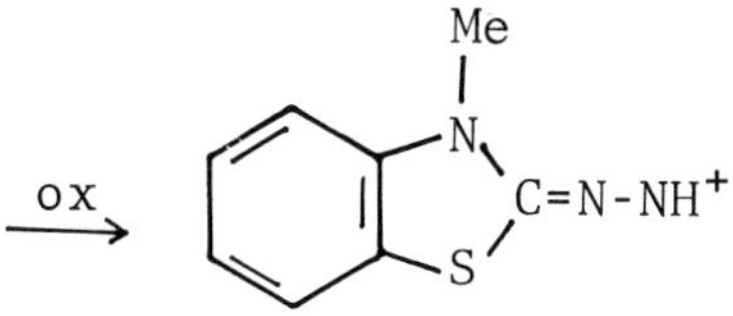

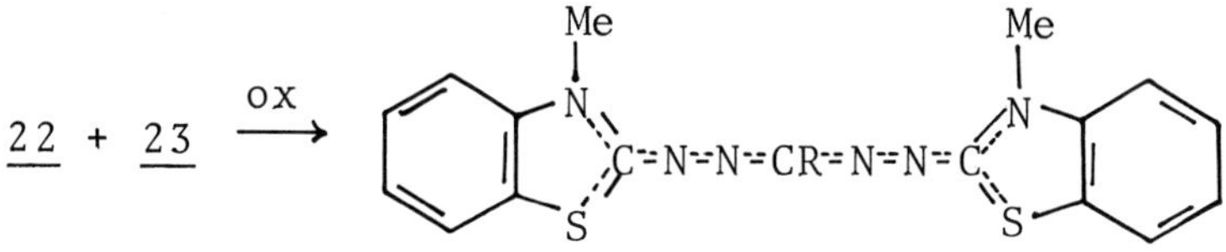

Table 12.V Hydrazines Used in Spectrophotometric
Determination of Carbonyls[a]

Hydrazine	Structure	Ref.
p-Nitrophenyl-hydrazine	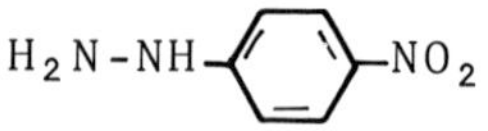 H_2N-NH— —NO_2	b
2,4-Dinitrophenyl-hydrazine	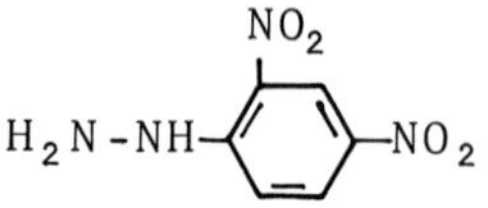 H_2N-NH— —NO_2 (with NO_2 above)	c,d
p,p -Dihydrazino-triphenylmethyl chloride	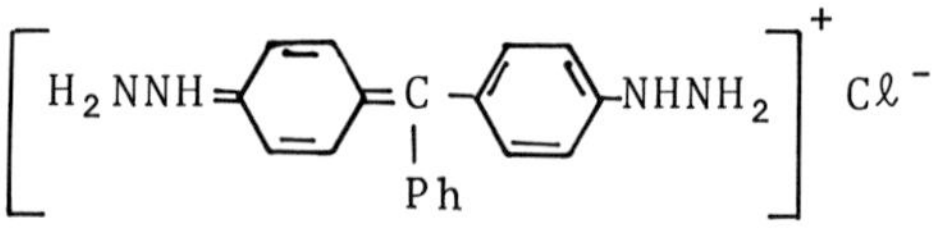 $[H_2NNH=$ =C$($ $)-NHNH_2]^+$ $C\ell^-$ (with Ph below C)	e
2-Diphenylacetyl-1,3-indandione 1-hydrazone	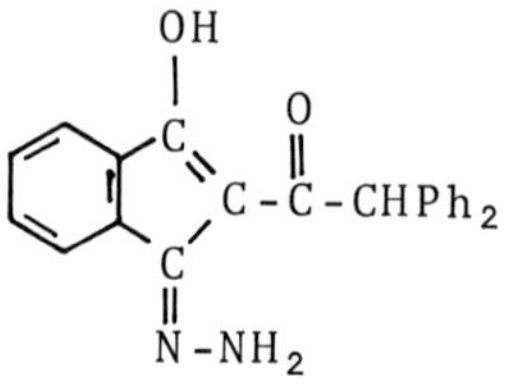 (structure with OH, C-C-CHPh$_2$, N-NH$_2$)	f,g
2-Hydrazino-benzo-thiazole (HBT)	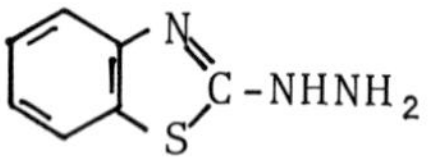 (benzothiazole) C-NHNH$_2$	h
3-Methyl-2-benzo-thiazolone hydrazone (MBTH)	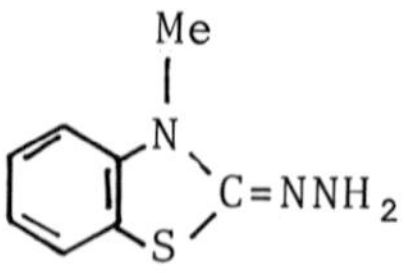 (benzothiazole with Me-N) C=NNH$_2$	h,i

[a]Dansyl hydrazine (1-dimethylaminonaphthalene-5-sulfonyl hydrazine) has been used for fluorometric detection of carbonyls: R. Chayen, R. Dvir, S. Gould, and A. Harell, Anal. Biochem., 42, 283 (1971).
[b]R. A. Gelman and J. R. Gilbertson, Anal. Biochem., 31, 463 (1969).

[c]F. E. Critchfield and J. A. Hutchinson, Anal. Chem., 32, 862 (1960).
[d]E. B. Sanders and J. Schubert, Anal. Chem., 43, 59 (1971).
[e]L. P. Kuhn and L. DeAngelis, J. Am. Chem. Soc., 71, 3084 (1949).
[f]R. A. Braun and W. A. Mosher, J. Am. Chem. Soc., 80, 3048 (1958).
[g]D. J. Pietrzyk and E. P. Chan, Anal. Chem., 42, 37 (1970).
[h]E. Sawicki, Ref. [74].
[i]F. W. Neumann, Anal. Chem., 41, 2077 (1969).

Extraction of carbonyl compounds from mixtures (such as essential oils or mixtures of steroids) can be accomplished with the aid of quaternary ammonium acetyl hydrazides, which form water-soluble hydrazones. The most widely used of these are the Girard-T(24) and Girard-P (25) reagents.

$$(CH_3)_3\overset{+}{N}CH_2CONHNH_2$$

24

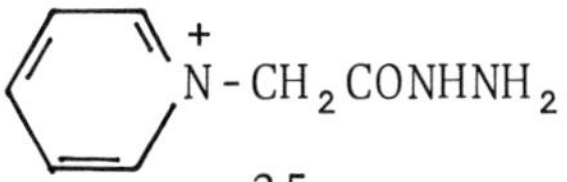

25

After recovery of the hydrazone, the original aldehyde or ketone can be regenerated by hydrolysis [99].
Some analyses (as in the MBTH and HBT methods) involve condensation of a carbonyl with a hydrazine followed by further steps. One of these complex reactions is the color test specific for aldehydes based on formation of triazolotetrazines as in [100]

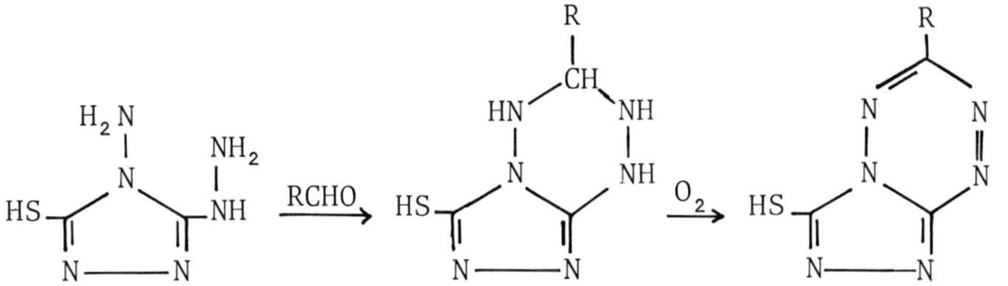

26   (12.94)

The final step is an air oxidation (dehydrogenation) [101].
Sugars react with phenylhydrazine giving, first, a phenylhydrazone by condensation with the carbonyl terminal of the free aldehyde form. Subsequently the adjacent hydroxy group is oxidized and a bis-phenylhydrazone, or osazone, is formed. Osazones are useful sugar derivatives.

The osazones prepared from 2,3-diketones and 2,4-dinitro-
phenylhydrazine can be hydrolyzed to regenerate the parent
carbonyls for glc detection [102].
    Schiff base (imine) formation can serve to analyze
either reactant, that is, the carbonyl compound or the
amine. Table 12.VI describes some amine analyses that
rely on imine formation. A particularly important condensation is that between formaldehyde and amines. In
aqueous solution formaldehyde exists mainly as its hydrate,
methylene glycol, $CH_2(OH)_2$. Reaction with a primary amine
can proceed in these two steps, giving successively a hy-
droxymethylamine and a di-(hydroxymethyl)amine:

$$RNH_2 + CH_2(OH)_2 \rightleftharpoons RNHCH_2OH + H_2O \qquad (12.95)$$

$$RNHCH_2OH + CH_2(OH)_2 \rightleftharpoons RN(CH_2OH)_2 + H_2O \qquad (12.96)$$

Polymeric condensation products also can form. Ammonia is
a unique amine, and reacts according to the equation
$6CH_2O + 4NH_3 \rightleftharpoons (CH_2)_6N_4$; the product, hexamethylenetetra-
amine (hexamine, urotropin), has structure 27.

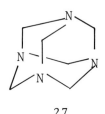

27

    Treatment of amino acids and proteins with formaldehyde
results in some complicated and interesting chemistry [103].
The formol titration of amino acids is based on Eqs.
(12.95) and (12.96). Amino acids cannot be accurately titrated with alkali in aqueous solution, in effect because
the solution is buffered by the basic amino group. In the
formol titration the carboxylic acid group of an amino acid
is titrated with alkali in an aqueous formaldehyde solution.
The primary amino group is transformed into the di-(hydroxymethyl)amino group, which is essentially nonbasic under the
titration conditions, and the carboxylic acid can be titrated in the conventional way [104].
    The well-known ninhydrin test for α-amino acids [106]
appears to be initiated by imine formation. Ninhydrin is
triketohydrindene hydrate, 28. Upon reaction with an
α-amino acid, carbon dioxide and an aldehyde are produced
[107], and a purple color appears that is probably due to
the presence of several condensation products. One of

these is diketohydrindylidenediketohydrindamine, 30, which
could arise from condensation of ninhydrin with diketohydrindamine, 29 [108], as shown in the following scheme.

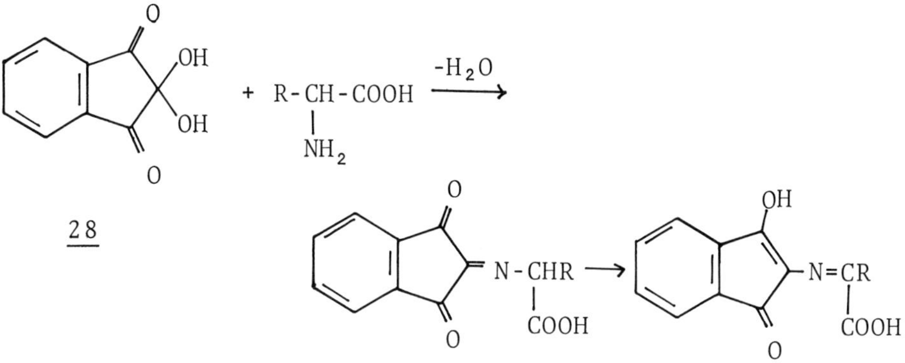

$$+ \text{ R-CH-COOH} \xrightarrow{-H_2O}$$

28

(12.97)

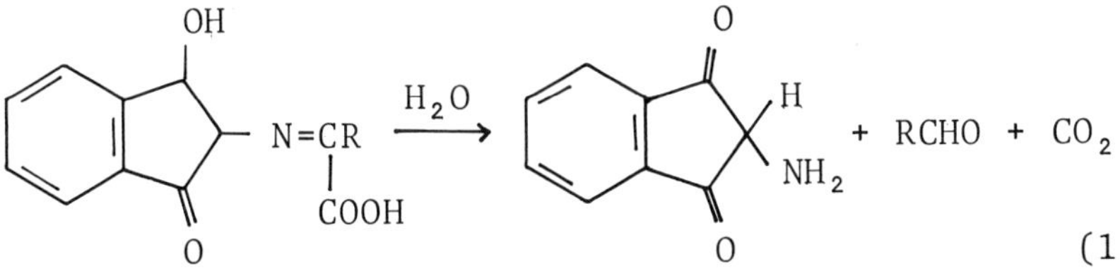

$$\xrightarrow{H_2O} \quad + \text{ RCHO } + CO_2$$

29

(12.98)

$$28 + 29 \xrightarrow{-2H_2O}$$

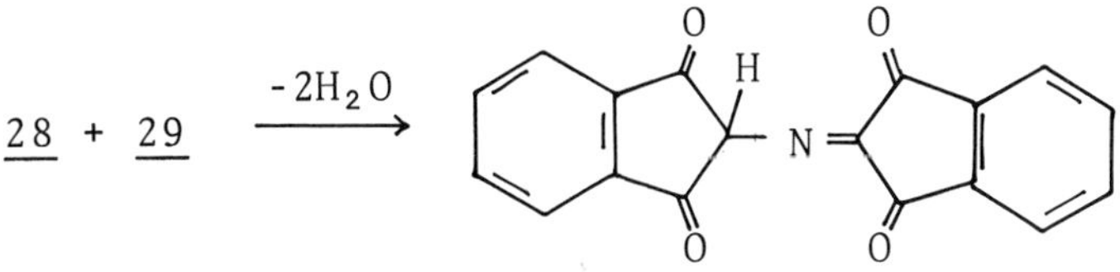

30   (12.99)

Table 12.VI Examples of Imine Formation Used in Amine Analysis

Carbonyl reagent	Applicability	Finish	Ref.
Salicylaldehyde	primary aliphatic amines	titration of excess reagent as acid	a
Salicylaldehyde	primary aliphatic amines	spectrophotometric measurement of imine	b
Salicylaldehyde	primary aliphatic amines	spectrophotometric after extraction of Cu^{2+}-imine complex	c
Salicylaldehyde	eliminates interference by prim. amines in determination of sec and tert amines	sec and tert amines titrated with acid	d,e
2-Ethylhexaldehyde	eliminates prim. amine interference	sec amines treated with CS_2, forming dithio-carbamates	f
2,4-Pentanedione	primary aliphatic amines	titration of excess reagent as the enol	g
p-Dimethylamino-cinnamaldehyde	biotin; cyclic ureides	spectrophotometric, of imine	h
Succinaldehyde	primary amines	pyrrole formed is con-densed with p-dimethyl-aminobenzaldehyde (Erlich's reagent)	i
Malonaldehyde	arginine in proteins	spectrophotometric, of pyrimidine formed	j
2-Ethylhexaldehyde	primary aliphatic amines	spectrophotometric titration	k

a J. B. Johnson and G. L. Funk, Anal. Chem., $\underline{28}$, 1977 (1956).

b A. J. Milun, Anal. Chem., $\underline{29}$, 1502 (1957).

c E. Critchfield and J. B. Johnson, Anal. Chem., $\underline{28}$, 436 (1956).

d F. O. Wagner, R. H. Brown, and E. D. Peters, J. Am. Chem. Soc., $\underline{69}$, 2611 (1947).

e C. Siggia, J. G. Hanna, and I. R. Kervenski, Anal. Chem., $\underline{22}$, 1295 (1950).

f S. E. Critchfield and J. B. Johnson, Anal. Chem., $\underline{28}$, 430 (1956).

g F. E. Critchfield and J. B. Johnson, Anal. Chem., $\underline{29}$, 1174 (1957).

h D. B. McCormick and J. A. Roth, Anal. Biochem., $\underline{34}$, 226 (1970).

i E. Sawicki and H. Johnson, Chemist-Analyst, $\underline{55}$, 101 (1966).

j T. P. King, Biochemistry, $\underline{5}$, 3454 (1966).

k Y. L. G. Liu and C. A. Reynolds, Anal. Chem., $\underline{34}$, 543 (1962).

Note that Eqs. (12.97) and (12.98) represent imine formation, tautomeric conversion into the imine of a different
carbonyl compound, and hydrolysis of this second imine.
   Diketohydrindamine can lose ammonia, forming reduced
ninhydrin, 31.

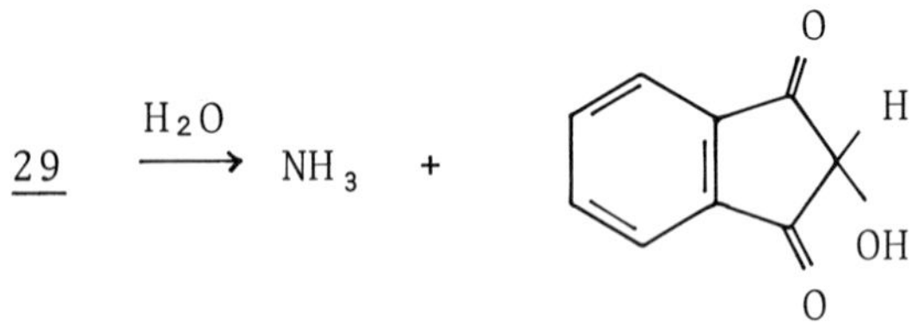

$$\underline{29} \xrightarrow{\text{H}_2\text{O}} \text{NH}_3 +$$

31   (12.100)

Condensation of ninhydrin with reduced ninhydrin gives 32,
which is another contributor to the color [109].

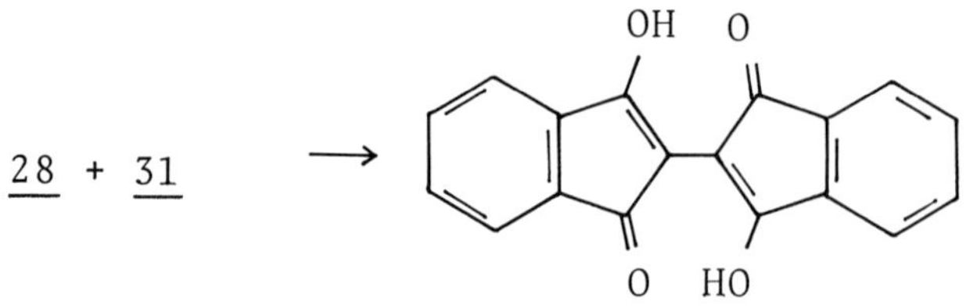

$$\underline{28} + \underline{31} \longrightarrow$$

32   (12.101)

   Ninhydrin reacts with propionaldehyde in concentrated
sulfuric acid to give a red-blue color [110]. The specificity of this reaction is remarkable; its basis is not
known.
   Several sensitive color tests for carbonyls involve
imine formation. o-Dianisidine gives colors with most
aldehydes but not with ketones [111].

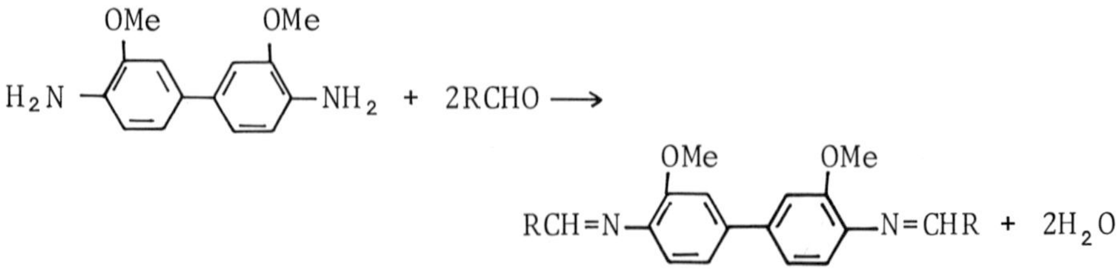

$$\text{H}_2\text{N} - \overset{\text{OMe}}{\bigodot} - \overset{\text{OMe}}{\bigodot} - \text{NH}_2 + 2\text{RCHO} \longrightarrow$$

$$\text{RCH=N} - \overset{\text{OMe}}{\bigodot} - \overset{\text{OMe}}{\bigodot} - \text{N=CHR} + 2\text{H}_2\text{O}$$

(12.107)

Furfural gives a purple color in the presence of aniline and an acid. The initial step is imine formation to give the anil 33.

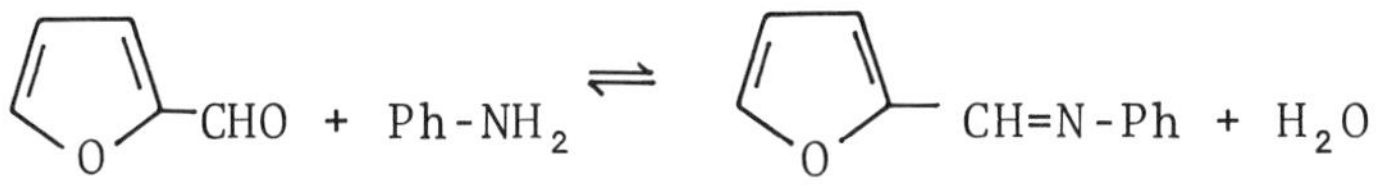

$$\text{furyl-CHO} + \text{Ph-NH}_2 \rightleftharpoons \text{furyl-CH=N-Ph} + \text{H}_2\text{O}$$

33   (12.102)

In aqueous medium the furan ring can hydrolyze and tautomerize, then condense with another aniline molecule to give the purple compound 34.

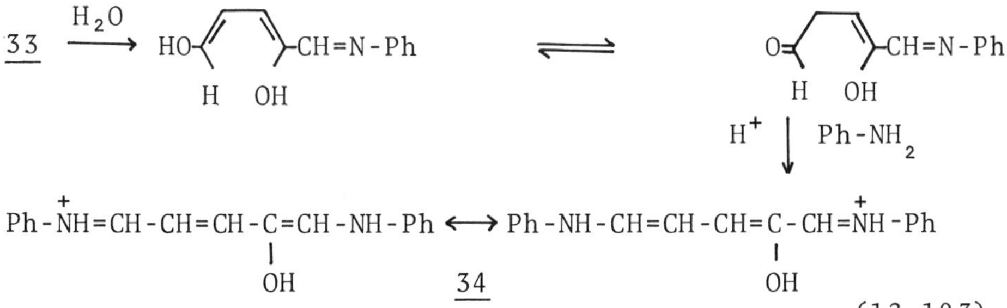

$$33 \xrightarrow{\text{H}_2\text{O}} \text{HO-}\langle\rangle\text{-CH=N-Ph} \rightleftharpoons \text{O=}\langle\rangle\text{-CH=N-Ph}$$

H   OH   H   OH

$$\text{H}^+ \downarrow \text{Ph-NH}_2$$

$$\text{Ph-}\overset{+}{\text{NH}}\text{=CH-CH=CH-C=CH-NH-Ph} \longleftrightarrow \text{Ph-NH-CH=CH-CH=C-CH=}\overset{+}{\text{NH}}\text{-Ph}$$

          | |
         OH   34   OH

(12.103)

An alternative route has been proposed in which 1,6-addition of HCl to the anil is followed by nucleophilic displacement of the allylic chloride by aniline [112]:

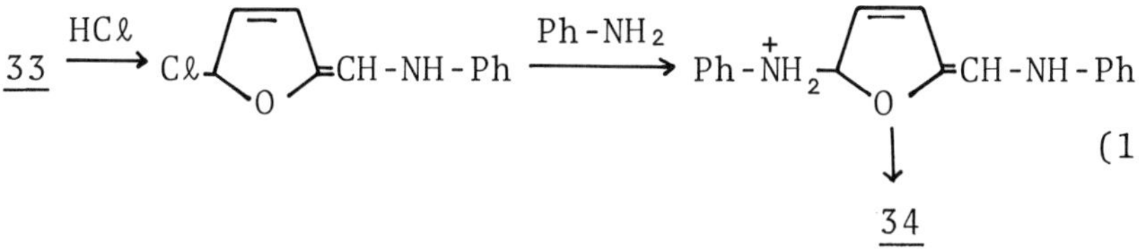

$$33 \xrightarrow{\text{HCl}} \text{Cl-}\langle\rangle\text{=CH-NH-Ph} \xrightarrow{\text{Ph-NH}_2} \text{Ph-}\overset{+}{\text{NH}}_2\text{-}\langle\rangle\text{=CH-NH-Ph}$$

(12.104)

$$\downarrow$$

34

A few analyses are mentioned in which nitrogen nucleophiles attack carbon-heteroatom double bonds other than the carbonyl group. A very sensitive test for secondary amines, even in the presence of excess primary or tertiary amine, utilizes a reagent of ammonia, nickel (II), and carbon disulfide [113]. A precipitate indicates a secondary amine. Apparently the dialkyldithiocarbamate formed

in the addition, $R_2NCSS^-NH_4^+$, gives an insoluble nickel
complex. Essentially the same chemistry occurs in a spectrophotometric quantitative method for secondary amines in
which the dithiocarbamic acid is complexed with copper
[114]. Critchfield and Johnson [115] determined primary
and secondary amines by reacting them with CS_2 and titrating the resulting dithiocarbamic acids with standard base.

Addition of an amine to an isocyanate RNCO or an isothiocyanate RNCS gives a substituted urea or thiourea,
respectively. Typical procedures for the determination of
isocyanates or isothiocyanates react these compounds with
excess amine and measure the unreacted amine by titration
with standard acid. n-Butylamine [116] and piperidine
[117] have been recommended. Polymeric isocyanates were
determined in the presence of halides in this way by using
dicyclohexylamine, which adds to the unsaturated molecule
much faster than it substitutes the halide [118].

A fluorescence assay of primary aliphatic amines uses
9-isothiocyanatoacridine, 35, as the reagent [119].

NCS

35

The thiourea derivative is transformed into a highly
fluorescent substance by treatment with ethoxide.

A technique for peptide sequence analysis begins with
addition of the N-terminal amino group to phenyl isothiocyanate. The terminal residue is hydrolyzed and cyclized,
ultimately, to the substituted phenylthiohydantoin corresponding to the N-terminal amino acid of the peptide. This
thiohydantoin is identified, and the procedure is repeated
with the next residue in the peptide. The method, which
is called the Edman degradation [120], is outlined on page
501. The phenylthiohydantoins can be identified by several
techniques, primarily chromatographic and spectroscopic
methods. Many variants have been introduced, such as an
insoluble isothiocyanate reagent to facilitate the stepwise sequencing procedure [121], and other isothiocyanates
chosen to aid in the final identification; thus methyl isothiocyanate gives thiohydantoins easily resolved by glc
[122], and 1-naphthyl isocyanate yields derivatives with
good chromatographic and fluorescent properties [123].

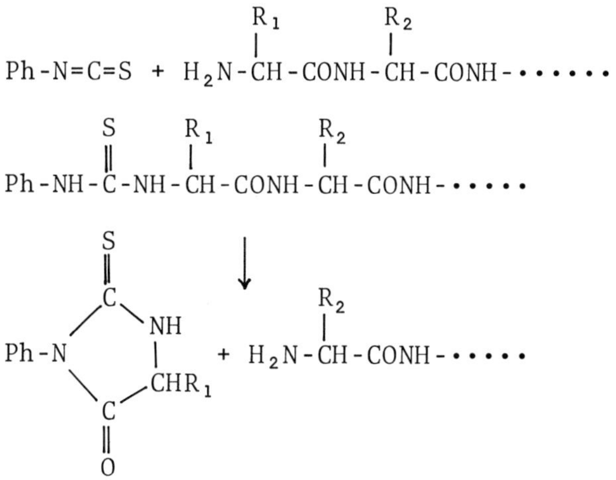

$$Ph-N=C=S \; + \; H_2N-\overset{\overset{\displaystyle R_1}{|}}{CH}-CONH-\overset{\overset{\displaystyle R_2}{|}}{CH}-CONH-\cdots\cdots$$

$$Ph-NH-\overset{\overset{\displaystyle S}{||}}{C}-NH-\overset{\overset{\displaystyle R_1}{|}}{CH}-CONH-\overset{\overset{\displaystyle R_2}{|}}{CH}-CONH-\cdots\cdots$$

Carbon Nucleophiles.

Some use has been made of condensation reactions with active methylene compounds in order to determine the active methylene component [124], but most applications of this reaction employ the active methylene compound as the reagent and a carbonyl compound as the sample. Many of these reactions are Knoevenagel reactions, or are similar to them, at least in their initial portions. Table 12.VII lists some reagents that function as carbon nucleophiles, with references to typical applications. Many of these condensations progress to the stage of 2:1 stoichiometry (carbon nucleophile:carbonyl). The formaldehyde-chromotropic acid product, for example, can be drawn as in 36 [125].

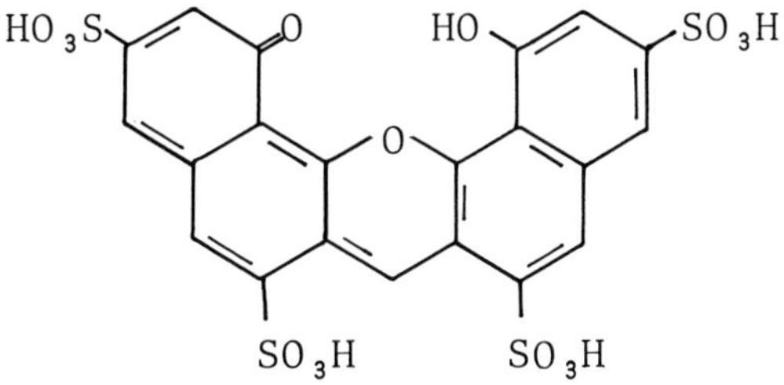

36

The specificity of this reagent for formaldehyde has been ascribed to the steric requirements of the sulfonic acid group.

The reaction of tetrahydro-6-oxo-3-pyridazine carboxylic acid, 37, with o-nitrobenzaldehyde to give a fluorescent product has been attributed to an aldol condensation [126].

Table 12.VII   Some Carbon Nucleophile Reagents
for Carbonyl Compounds

Name	Structure	Application	Ref.
5,5-Dimethylcyclo-hexane-1,3-dione (dimedon)		identification of aldehydes	a,b
1,8-Dihydroxynaph-thalene-3,6-disul-fonic acid (chro-motropic acid)		specific for formaldehyde	c
2-Amino-5-naphthol-7-sulfonic acid (J-acid)		specific for formaldehyde	c
2-Thiobarbituric acid		aldehydes	c,d
4-Nitrophenyl-acetonitrile	O_2N—⬡—CH_2-CN	quinones	e
Thiophene		test for 1,2-diketones (also for thiophene)	f
Indole		aldehydes	c
Quinaldine ethiodide		chloral	c
Resorcinol		glyoxylic acid	c
Malononitrile	$CH_2(CN)_2$	quinones	g

[a]W. Weinberger, Ind. Eng. Chem., Anal. Ed., 3, 365 (1931).
[b]J. H. Yoe and L. C. Reid, Ind. Eng. Chem., Anal. Ed., 13, 238 (1941)

CRef. [74].
dS. Futterman and L. D. Saslaw, J. Biol. Chem., 236, 1652
(1961).
eL. Legradi, Analyst, 95, 590 (1970).
fF. Feigl, Spot Tests in Organic Analysis, 6th ed., Elsevier, Amsterdam, 1960, pp. 220, 472.
gC. E. Gonter and J. J. Petty, Anal. Chem., 35, 663 (1963);
H. Junek and H. Hamboeck, Mikrochim. Acta, 1966, 522.

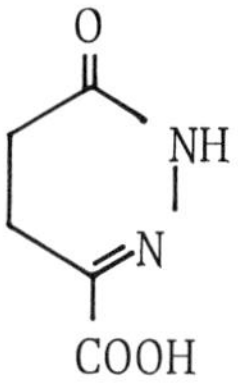

37

Diacetyl, $CH_3COCOCH_3$, reacts in alkali, presumably by aldol condensation, to yield a color suitable for analysis
[127].
     In the determination of hydroxy groups by acylation
with acetic anhydride, interference by aldehydes has been
noted. Although possible reasons for this interference
have been discussed [128], it is not yet understood.
Since phthalic anhydride is not subject to this phenomenon,
the most likely reason for the interference seems to be a
Perkin reaction; acetic anhydride possesses an active
methylene group, whereas aromatic anhydrides do not.
     The Komarowski reaction, in which a color is formed
when an alcohol is treated with sulfuric acid and an aromatic aldehyde, has attracted interest. The usual aldehydes are p-hydroxybenzaldehyde, salicylaldehyde, and p-dimethylaminobenzaldehyde. Primary alcohols are not effective. The course of the reaction is not agreed upon. One
suggestion is that the protonated alcohol loses water to
give a carbonium ion, this electrophile attacks the aldehyde oxygen, and then bond scission occurs to give a benzyl cation and a ketone, as shown below [129].

$$\overset{+OH_2}{R-CH-R} \rightleftharpoons \overset{+}{R-CH-R} + H_2O$$

$$\overset{+}{R-CH-R} + Ar-CHO \rightarrow \overset{+}{Ar-CH-OCHR_2}$$

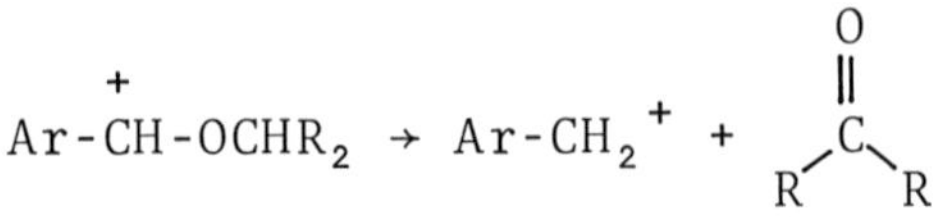

$$Ar\text{-}\overset{+}{C}H\text{-}OCHR_2 \rightarrow Ar\text{-}CH_2^+ + \underset{R \quad R}{\overset{\overset{\displaystyle O}{\displaystyle \|}}{C}}$$

The net result has been the oxidation of the alcohol to a ketone, which then may undergo aldol condensation with the aldehyde to give colored products. In agreement with this, dibenzalacetone, $(Ph\text{-}CH=CH\text{-})_2C=O$, is isolated from a Komarowski reaction of benzaldehyde and isopropyl alcohol [129]. This protonated ketone has spectral properties consistent with the color developed in the reaction [130]. Similarly, tristyrylcarbinol, $(Ph\text{-}CH=CH\text{-})_3C\text{-}OH$, is implicated as the chromophore in solutions of tert-butanol, benzaldehyde, and sulfuric acid [130]. On the other hand, it has been claimed that the color in the Komarowski reaction results from a species whose structure is independent of the alcohol tested [131]. The first step is thought to be acid-catalyzed dehydration of the alcohol to the olefin, to which the protonated aldehyde adds. Brieskorn and Otteneder [131] postulate a role for the bisulfate ion also. The sequence is shown below.

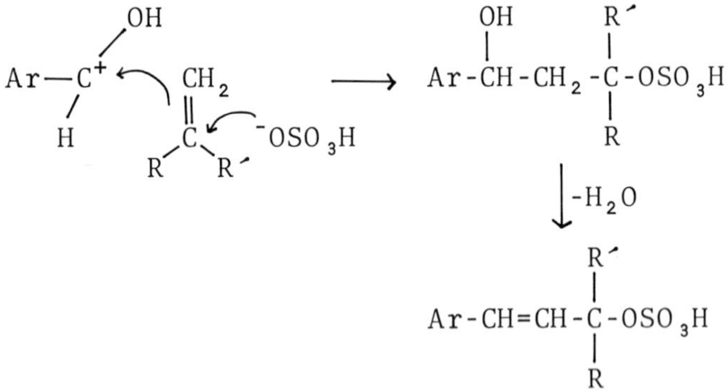

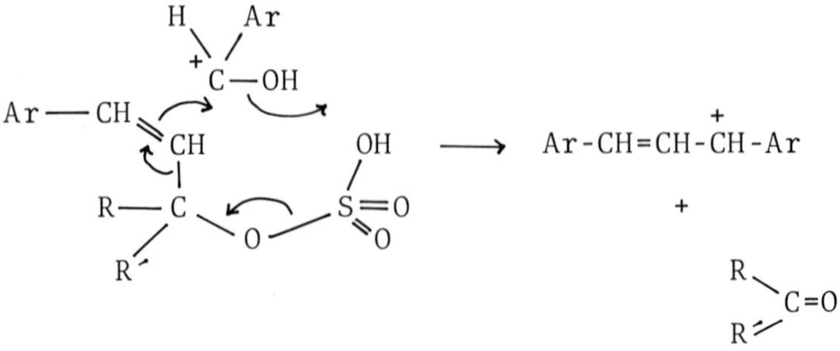

It is not unlikely, considering the rigorous conditions of
this reaction, that a mixture of colored products is formed
by several reactions.
     Addition of Grignard reagents has limited analytical
use because so many substances react with these reagents.
Any active hydrogen compound (alcohols, thiols, acids,
amines, water, etc.) will produce methane when treated with
the reagent CH_3MgX. Carbonyl compounds, however, add the
reagent and do not produce methane. The usual technique
is to add excess Grignard reagent to the carbonyl sample
and then to determine the unreacted reagent by treating it
with an active hydrogen compound and measuring the methane
evolved. Aldehydes and ketones can be determined in this
way, and so can esters, acid chlorides, nitriles, and iso-
nitriles [132].

Comment. Obviously addition reactions to carbon-heteroatom
multiple bonds have not been neglected by analytical chemists. Just as clearly, however, improvements continue to
be made and sought. Some possible modifications of benefit are suggested by this survey of analytical reactions.
Thus present titrimetric methods for carbonyls, especially
oximation, are chiefly based upon measurement of the unreacted reagent. Some of the difficulties of the oximation procedure follow from this practice, because free hydroxylamine is unstable. Direct measurement of the oxime
formed would be an attractive route. Spectrophotometric
measurement may be possible [80].
     Nucleophilic catalysis of carbonyl condensations, as
in the catalysis of oxime or semicarbazone formation by
aniline, occurs via a reactive Schiff base intermediate,
as discussed earlier in this chapter [18]. This phenomenon
might be exploited analytically.
     Considering the great synthetic utility of carbonyl
additions by carbon nucleophiles, further analytical development of these reactions may be fruitful. At present
most methods of this type utilize the carbon nucleophile
as the reagent and a carbonyl compound as the sample, often
with spectrophotometric or fluorometric detection of the
condensation product. The system could perhaps be reversed
to provide sensitive methods for carbon nucleophiles.
     The literature must already contain many observations
that are worth analytical exploitation. An excellent example of such analytical development, related to the subject of this chapter, is provided by recent work of Udenfriend and co-workers [133]. An earlier report [134] of a
fluorometric ninhydrin method specific for phenylalanine
led to an investigation showing that ninhydrin converts
phenylalanine to phenylacetaldehyde [see Eqs. (12.97) and
(12.98)]; a condensation product, which is highly fluores-

cent, then forms from one molecule each of phenylacetaldehyde, ninhydrin, and an added amine. By using phenylacetaldehyde and ninhydrin as the reagents, a fluorescence
method for amino acids was developed having 10-100 times
the sensitivity of the conventional colorimetric ninhydrin
method. Continued studies resulted in identification of
the fluorescent product [135] and the synthesis of a new
reagent to replace the ninhydrin procedure [136]. This
reagent (fluorescamine) reacts in milliseconds with primary
amines to produce a fluorophor, as little as 20 pmole of
amino acid being detectable.

Problems

1. The bell-shaped pH-rate curve in oxime formation has
    been ascribed to rate-determining dehydration on the
    alkaline side and rate-determining addition on the
    acid side. A kinetically equivalent interpretation
    is that dehydration is rate determining on the acid
    side and addition is rate determining on the alkaline
    side. What evidence rules out the second interpreta-
    tion?

    Answer: See W. P. Jencks, Progr. Phys. Org. Chem.,
    2, 63 (1964).

2. Formulate a chemically reasonable mechanism for simple
    addition of an amine to a carbonyl if the experimental
    rate equation is $v = k[\text{carbonyl compound}]\ [\text{RNH}_3^+]$.

3. In the oximation of pyruvate anion, the apparent first-
    order rate constant varies linearly with hydroxylamine
    concentration at very low concentration, but at very
    high concentration it is independent of $[\text{NH}_2\text{OH}]$.
    Account for this behavior.

    Answer: See W. P. Jencks, Progr. Phys. Org. Chem., 2,
    63 (1964).

4. Derive Eqs. (12.15) and (12.16).

5. Postulate a mechanism for general base catalyzed hy-
    drolysis of an imine that is consistent with the mech-
    anism of Eq. (12.18) for general base-catalyzed dehy-
    dration of a carbinolamine.

6. Rationalize the observation that aldehydes may be very
    extensively hydrated in aqueous solution, whereas ke-
    tones are very slightly hydrated.

Problems

7. For semicarbazone formation from substituted benzaldehydes these Hammett rho values were found by B. M.
   Anderson and W. P. Jencks, J. Am. Chem. Soc., $\underline{82}$,
   1773 (1960): at neutral pH, ρ for the equilibrium
   constant for tetrahedral intermediate formation is
   +1.81 and ρ for the acid-catalyzed dehydration rate
   is -1.74; at pH 1.75, ρ for the overall reaction is
   +0.91.
   (a) Account for the signs of the rho values.
   (b) Estimate ρ for the overall rate in neutral solution.

8. Define these constants in aqueous solution:
   $K_1 = [R_1R_2C(OH)O^-][H^+]/[R_1R_2C(OH)_2]$
   $K_2 = [R_1R_2C(OH)O^-]/[R_1R_2C(OH)_2][OH^-]$
   $K_3 = [R_1R_2C(OH)O^-][H^+]/R_1R_2CO]$
   $K_4 = [R_1R_2C(OH)O^-]/[R_1R_2CO][OH^-]$
   $K_d = [R_1R_2CO]/[R_1R_2C(OH)_2]$
   Find relationships among the constants.

   Answer: $K_2 = K_1/K_w$; $K_3 = K_1/K_d$; $K_4 = K_1/K_wK_d$

9. Postulate a cyclic transition state, analogous to $\underline{8}$,
   for the general base-catalyzed hydration of a carbonyl.

10. Why are acetals and ketals resistant to hydrolysis by
    alkali?

11. Derive an equation for K in Eq. (12.37) from the ex-
    perimental constant K_{obs} and the hydration constant
    K_h.

    Answer: $K = K_{obs}(1 + K_h)$

12. Using data of Table 12.IV, calculate equilibrium con-
    stants for the second step of Eq. (12.38) and rate
    constants for the overall reaction.

    Answer: $K_2 = K_{overall}/K$ and $k_{overall} = Kk_2$

13. Postulate a mechanism for the hydrolysis of primary
    or secondary aliphatic nitro compounds to aldehydes
    or ketones (the Nef reaction). In this reaction the
    salt of the nitro compound is hydrolyzed with acid:

    Answer: See M. F. Hawthorne, J. Am. Chem. Soc., $\underline{79}$,

2510 (1957).

14. Postulate cyclic transition states for the enoliza-
    tion and reduction side reactions of Grignard reagents
    with carbonyls.

15. Formulate reaction mechanisms for the hydrolysis of
    isocyanates and isothiocyanates, Eqs. (12.72) and
    (12.73).

16. Suggest a possible reaction pathway for the conver-
    sion of thiourea to urea by peroxide.

17. Propose a mechanism for Eq. (12.100), the conversion
    of diketohydrindamine to reduced ninhydrin.

18. Develop analytical procedures for the quantitative
    determination of each component in these mixtures:
    (a) Glycine, ethylamine, and aniline
    (b) Benzaldehyde and nitrobenzene [135]
    (c) Hydroxylamine and hydrazine.

References

1. Usually the C-X bond is broken in these reactions, but
   if X is a group rather than a single atom, a bond
   elsewhere in the group may be cleaved with the same
   apparent result (as in a mode of ester hydrolysis).
2. The Pauling electronegativities are, for carbon, 2.5,
   and oxygen, 3.5; L. Pauling, The Nature of the Chemical Bond, 3rd ed., Cornell Univ. Press, Ithaca, N. Y.,
   1960, p. 93. The polarization of the carbonyl group
   may be represented by $\overset{\delta+}{C}=\overset{\delta-}{O}$.
3. For discussions of keto-enol rates and equilibria see
   E. S. Gould, Mechanism and Structure in Organic Chemistry, Holt, Rinehart and Winston, New York, 1959,
   Chap. 10; R. P. Bell, The Proton in Chemistry, Cornell
   Univ. Press, Ithaca, N. Y., 1959, pp. 160-166. Carbonyl addition reactions have been reviewed by, among
   many, E. E. Royals, Advanced Organic Chemistry, Prentice-Hall, New York, 1954, Chap. 9; C. D. Gutsche,
   The Chemistry of Carbonyl Compounds, Prentice-Hall,
   Englewood Cliffs, N. J., 1967; J. March, Advanced
   Organic Chemistry: Reactions, Mechanisms, and Structure, McGraw-Hill, New York, 1968, Chap. 16.
4. A. Lapworth, J. Chem. Soc., 1903, 995. This is one
   of the earliest reaction mechanisms to be correctly
   formulated; it was, moreover, developed in a period
   when the concept of ionic organic species was not

fashionable.

5. W. J. Svirbely and J. F. Roth, J. Am. Chem. Soc., <u>75</u>, 3106 (1953).

6. W. P. Jencks, Progr. Phys. Org. Chem., <u>2</u>, 63 (1964); J. E. Reimann and W. P. Jencks, J. Am. Chem. Soc., <u>88</u>, 3973 (1966). A similar hypothesis has been applied to S_N reactions; see Ref. [45] of Chap. 9.

7. G. E. Lienhard and W. P. Jencks, J. Am. Chem. Soc., <u>88</u>, 3982 (1966).

8. W. P. Jencks, J. Am. Chem. Soc., <u>81</u>, 475 (1959).

9. J. B. Conant and P. D. Bartlett, J. Am. Chem. Soc., <u>54</u>, 2881 (1932); F. H. Westheimer, ibid., <u>56</u>, 1962 (1934).

10. L. P. Hammett, Physical Organic Chemistry, McGraw-Hill, New York, 1940, p. 331.

11. W. P. Jencks, Catalysis in Chemistry and Enzymology, McGraw-Hill, New York, 1969, Chap. 10.

12. pH-rate curves of this type should be corrected for general acid-base catalysis by extrapolating to zero buffer concentration. In Fig. 12.1, the alkaline side of the curve was studied in the absence of buffers, and the acid side is not subject to general acid-base catalysis.

13. W. P. Jencks, Ref. [11], p. 590.

14. E. H. Cordes and W. P. Jencks, J. Am. Chem. Soc., <u>84</u>, 832 (1962).

15. (a) E. H. Cordes and W. P. Jencks, J. Am. Chem. Soc., <u>84</u>, 4319 (1962); (b) J. M. Sayer and W. P. Jencks, ibid., <u>91</u>, 6353 (1969).

16. A. Williams and M. L. Bender, J. Am. Chem. Soc., <u>88</u>, 2508 (1966).

17. A great deal of effort has been expended in studies of the nature of acid-base catalysis in these reactions. One source of ambiguity in such studies is the difficulty of detecting general acid-base catalysis when the Brønsted slope is close to its extreme values of 0 or 1. Another is the problem of distinguishing between general acid catalysis by acid HA, and the kinetically equivalent (but mechanistically different) specific acid-general base catalysis by H^+ and A^-. Most of these problems concern relatively subtle differences in the timing of proton transfers. The subject can be followed further in W. P. Jencks, Ref. [11]; J. Am. Chem. Soc., <u>94</u>, 4731 (1972); Chem. Revs., <u>72</u>, 705 (1972).

18. E. H. Cordes and W. P. Jencks, J. Am. Chem. Soc., <u>84</u>, 826 (1962).

19. The Schiff base is roughly 10^7 more basic than the aldehyde; see E. M. Arnett, Progr. Phys. Org. Chem., <u>1</u>, 223 (1963).

20. C. G. Swain and J. F. Brown, Jr., J. Am. Chem. Soc., 74, 2534, 2538 (1952).
21. P. R. Rony, J. Am. Chem. Soc., 90, 2824 (1968); 91, 6090 (1969).
22. R. P. Bell, Advan. Phys. Org. Chem., 4, 1 (1966).
23. Aqueous solutions of acetaldehyde also contain the hemihydrate $CH_3CH(OH)OCH(OH)CH_3$; cf. G. Socrates, J. Org. Chem., 34, 2958 (1969).
24. E. B. Whipple, J. Am. Chem. Soc., 92, 7183 (1970). A cis-diol dimer is also present.
25. W. P. Jencks, Ref. [11], p. 211.
26. For example, the pK_a of benzaldehyde is -6.8; C. C. Greig and C. D. Johnson, J. Am. Chem. Soc., 90, 6453 (1968).
27. M. Eigen, Discussions Faraday Soc., 39, 7 (1965).
28. E. G. Sander and W. P. Jencks, J. Am. Chem. Soc., 90, 4377 (1968).
29. K. Baum, J. Am. Chem. Soc., 92, 2927 (1970).
30. E. H. Cordes, Progr. Phys. Org. Chem., 4, 1 (1967).
31. L. L. Schaleger and F. A. Long, Advan. Phys. Org. Chem., 1, 1 (1963).
32. M. S. Newman and R. E. Dickson, J. Am. Chem. Soc., 92, 6880 (1970).
33. B. Capon, Chem. Revs., 69, 407 (1969); see also E. A. Davidson, Carbohydrate Chemistry, Holt, Rinehart and Winston, New York, 1967.
34. B. M. Dunn and T. C. Bruice, J. Am. Chem. Soc., 92, 2410, 6589 (1970).
35. E. H. Cordes and W. P. Jencks, J. Am. Chem. Soc., 85, 2843 (1963); J. Hine, J. C. Craig, Jr., J. G. Underwood, II, and F. A. Via, ibid., 92, 5194 (1970).
36. J. C. Coward and T. C. Bruice, J. Am. Chem. Soc., 91, 5329 (1969); P. Y. Sollenberger and R. B. Martin, ibid., 92, 4261 (1970).
37. G. W. Wheland, Advanced Organic Chemistry, 3rd ed., Wiley, New York, 1960, pp. 437-450.
38. E. G. Sander and W. P. Jencks, J. Am. Chem. Soc., 90, 6154 (1968).
39. J. O. Edwards and R. G. Pearson, J. Am. Chem. Soc., 84, 16 (1962).
40. Alpha-effect nucleophilicity has been extensively studied in acyl transfer reactions, and is discussed further in Chap. 13.
41. R. Wolfenden and W. P. Jencks, J. Am. Chem. Soc., 83, 2763 (1961).
42. T. F. Cummings and J. R. Shelton, J. Org. Chem., 25, 419 (1960).
43. J. March, Advanced Organic Chemistry: Reactions, Mechanisms, and Structure, McGraw-Hill, New York, 1968, p. 668; O. H. Wheeler, The Chemistry of the

Carbonyl Group (S. Patai, ed.), Wiley (Interscience), New York, 1966, Chap. 11.

44. E. C. Ashby and M. B. Smith, J. Am. Chem. Soc., <u>86</u>, 4363 (1964).

45. J. Miller, G. Gregorion, and H. S. Mosher, J. Am. Chem. Soc., <u>83</u>, 3966 (1961).

46. A condensation is a reaction in which two or more molecules combine with the liberation of a molecule of water (or, by extension, other small molecule, such as HCℓ).

47. J. March, Ref. [3], pp. 694-695.

48. E. E. Royals, Ref. [3], p. 756.

49. W. S. Johnson and G. H. Daub, Org. Reactions, <u>6</u>, 1 (1951).

50. R. L. Reeves, The Chemistry of the Carbonyl Group (S. Patai, ed.), Wiley (Interscience), New York, 1966, Chap. 12.

51. J. Mitchell, Jr., Org. Anal., <u>1</u>, 243 (1953); S. Siggia, Quantitative Organic Analysis Via Functional Groups, 3rd ed., Wiley, New York, 1963, Chap. 2; N. D. Cheronis and T. S. Ma, Organic Functional Group Analysis by Micro and Semimicro Methods, Wiley (Interscience), New York, 1964, Chap. 6; F. E. Critchfield, Organic Functional Group Analysis, MacMillan, New York, 1963, Chap. 4; J. G. Hanna, The Chemistry of the Carbonyl Group (S. Patai, ed.), Wiley (Interscience), New York, 1966, Chap. 8; F. T. Weiss, Determination of Organic Compounds: Methods and Procedures, Wiley (Interscience), New York, 1970, Chap. 6; J. G. Hanna and S. Siggia, Treatise on Analytical Chemistry, Part II, Vol. 13 (I. M. Kolthoff and P. J. Elving, eds.), Wiley (Interscience), New York, 1966, pp. 131-213.

52. D. S. Erley, Anal. Chem., <u>29</u>, 1564 (1957).

53. F. E. Critchfield and E. T. Bishop, Anal. Chem., <u>33</u>, 1034 (1961).

54. J. Mitchell, Jr., Org. Anal., <u>1</u>, 309 (1953).

55. N. D. Cheronis and T. S. Ma, Ref. [51], p. 390.

56. B. Gauthier and J. Maillard, Ann. Pharm. Franc., <u>11</u>, 509 (1953).

57. N. D. Cheronis and T. S. Ma, Ref. [51], p. 300.

58. W. E. Kemp, Analyst, <u>64</u>, 648 (1939).

59. H. Roth and W. Beck, Mikrochim. Acta, 1957, 845; D. G. Clark, H. Baum, E. L. Stanley, and W. F. Hester, Anal. Chem., <u>23</u>, 1842 (1951).

60. W. K. Lowen, Anal. Chem., <u>23</u>, 1846 (1951).

61. M. Jurecek and K. Obruba, Chem. Listy, <u>52</u>, 2066 (1958); N. D. Cheronis and T. S. Ma, Ref. [51], p. 390.

62. G. R. Leader, Anal. Chem., <u>42</u>, 16 (1970).

63. G. A. Sarfaty and H. M. Fales, Anal. Chem., <u>42</u>, 288
     (1970). Instability of dimethylacetals of fatty
     aldehydes, $(MeO)_2CHCH_2R$, on some glc columns was
     noted by R. A. Stein and V. Slawson, J. Chromatogr.,
     <u>25</u>, 204 (1966); these adducts decompose to the parent
     aldehyde, to the corresponding methyl 1-alkenyl ether
     $MeOCH=CHR$, or both.
64. E. P. Crowell, W. A. Powell, and C. J. Varsel, Anal.
     Chem., <u>35</u>, 184 (1963).
65. E. P. Crowell and C. J. Varsel, Anal. Chem., <u>35</u>, 189
     (1963).
66. R. L. Shriner, R. C. Fuson, and D. Y. Curtin, The
     Systematic Identification of Organic Compounds, 5th
     ed., Wiley, New York, 1964, p. 241.
67. V. W. Goodlett, Anal. Chem., <u>37</u>, 431 (1965).
68. P. A. Hedin, R. C. Gueldner, and A. C. Thompson,
     Anal. Chem., <u>42</u>, 403 (1970).
69. W. Pereira, V. A. Bacon, W. Patton, B. Halpern, and
     G. E. Pollock, Anal. Letters, <u>3</u>, 23 (1970).
70. J. V. Murray and J. B. Cloke, J. Am. Chem. Soc., <u>56</u>,
     2749 (1934).
71. C. R. Noller, Org. Syn., Coll. Vol. <u>2</u>, 586 (1943).
     The mechanism has been studied by K. B. Wiberg, J.
     Am. Chem. Soc., <u>75</u>, 3961 (1953); <u>77</u>, 2519 (1955);
     and J. E. McIssac, R. E. Ball, and E. J. Behrman,
     J. Org. Chem., <u>36</u>, 3048 (1971).
72. D. H. Whitehurst and J. B. Johnson, Anal. Chem., <u>30</u>,
     1332 (1958).
73. E. Sawicki, T. W. Stanley, and T. R. Hauser, Chemist-
     Analyst, <u>47</u>, 31 (1958).
74. E. Sawicki, Proceedings, 1961 International Symposium
     on Microchemical Techniques (N. D. Cheronis, ed.),
     Wiley (Interscience), New York, 1962, pp. 59-106.
     This article is a valuable review of sensitive methods
     for aldehydes.
75. H. Wieland and G. Scheuing, Chem. Ber., <u>54</u>, 2527
     (1921).
76. For references to early papers see M. A. Joslyn and
     C. L. Comar, Ind. Eng. Chem., Anal. Ed., <u>10</u>, 364
     (1938); N. D. Cheronis and T. S. Ma, Ref. [51], pp.
     147-149.
77. I. M. Kolthoff and V. A. Stenger, Volumetric Analysis
     Vol. I, 2nd ed., Wiley (Interscience), New York,
     1942, pp. 213-222.
78. S. Siggia and W. Maxcy, Ind. Eng. Chem., Anal. Ed.,
     <u>19</u>, 1023 (1947). Siggia also applied the method to
     the determination of acetals and vinyl ethers by
     first subjecting them to acid hydrolysis; S. Siggia,
     ibid., <u>19</u>, 1025 (1947).
79. V. H. Cohn and J. Lyle, Anal. Biochem., <u>14</u>, 434 (1966)

T. L. McNeil and L. V. Beck, ibid., 22, 431 (1968);
P. C. Jocelyn and A. Kamminga, ibid., 37, 417 (1970).
Primary amides give cyclic products upon reaction
with o-phthaldialdehyde; R. D. Reynolds, D. L. Arend-
sen, D. F. Guanci, and R. F. Wickman, J. Org. Chem.,
35, 3940 (1970).

80. T. J. Mikkelson and J. R. Robinson, J. Pharm. Sci.,
57, 1180 (1968).

81. J. S. Fritz and G. S. Hammond, Quantitative Organic
Analysis, Wiley, New York, 1957, Chap. 2.

82. After Fritz and Hammond wrote, Jencks [8] clarified
the kinetics of oxime formation, as described in
Sec. 12.1.

83. G. H. Schenk, Organic Functional Group Analysis,
Pergamon, Oxford, 1968, Chap. 1

84. The pK_a of NH_3OH^+ is 6.0, whereas that for protonated
acetoxime is about 2.

85. S. Siggia, Quantitative Organic Analysis Via Functional Groups, 3rd ed., Wiley, New York, 1963, p. 74.

86. K. A. Connors, A Textbook of Pharmaceutical Analysis,
Wiley, New York, 1967, p. 437.

87. W. M. D. Bryant and D. M. Smith, J. Am. Chem. Soc.,
57, 57 (1935).

88. T. Higuchi and C. H. Barnstein, Anal. Chem., 28,
1022 (1956).

89. It should be possible to generate hydroxylammonium
acetate in situ by treating hydroxylamine hydrochloride in acetic acid with mercuric acetate [90].

90. C. W. Pifer and E. G. Wollish, Anal. Chem., 24, 300
(1952).

91. M. Pesez, Bull. Soc. Chim. France, 1957, p. 417; J.
E. Ruch, J. B. Johnson, and F. E. Critchfield, Anal.
Chem., 33, 1566 (1961).

92. J. S. Fritz, S. S. Yamamura, and E. C. Bradford,
Anal. Chem., 31, 260 (1959).

93. R. L. Shriner, R. C. Fuson, and D. Y. Curtin, Ref.
[66], pp. 126, 147.

94. H. A. Iddles and C. E. Jackson, Ind. Eng. Chem., Anal.
Ed., 6, 454 (1934).

95. S. Siggia and C. R. Stahl, Anal. Chem., 27, 1975
(1955).

96. Most carbonyl methods use acidic media, so acetals
are hydrolyzed to the parent aldehydes under the
assay conditions.

97. L. A. Jones, J. C. Holmes, and R. B. Seligman, Anal.
Chem., 28, 191 (1956); J. P. Phillips, J. Org. Chem.,
27, 1443 (1962); 29, 982 (1964).

98. G. R. Lappin and L. C. Clark, Anal. Chem., 23, 541
(1951).

99. O. W. Wheeler, Chem. Revs., 62, 205 (1962).

100. R. G. Dickenson and N. W. Jacobsen, Chem. Commun.,
      1970, 1719.
101. The reagent 26, which is 4-amino-3-hydrazino-5-
      mercapto-1,2,4-triazole, is formed from hydrazine
      and thiourea or isothiourea compounds. This reac-
      tion can be used to analyze for these compounds:
      R. G. Dickenson and N. W. Jacobsen, Anal. Chem.,
      41, 1324 (1969).
102. R. J. Ashworth, Anal. Chem., 42, 1445 (1970).
103. D. French and J. T. Edsall, Advan. Protein Chem.,
      2, 278 (1945).
104. Secondary amines have pK_a values 2.2-3.0 units high-
      er than their monohydroxymethyl adducts. The differ-
      ence must be much greater for primary amines [105].
105. R. G. Kallen and W. P. Jencks, J. Biol. Chem., 241,
      5864 (1966).
106. S. Ruhemann, J. Chem. Soc., 1910, p. 2025.
107. Amino acids have been determined by measuring the
      production of CO_2: D. D. van Slyke, R. T. Dillon,
      D. A. MacFadyer, and P. Hamilton, J. Biol. Chem.,
      141, 627 (1941); D. D. van Slyke, D. A. FacFadyer,
      and P. Hamilton, ibid., 141, 671 (1941).
108. S. Ruhemann, J. Chem. Soc., 1911, 1486.
109. R. Moubasher and M. Ibrahim, J. Chem. Soc., 1949,
      702. These authors postulate the further intermed-
      iate involvement of two molecules of amino acid in
      this conversion. D. J. McCaldin, Chem. Revs., 60,
      39 (1960) has given a modified version of these
      reactions; he postulates a concerted cyclic mecha-
      nism, involving decarboxylation, to replace the
      imine conversion of Eq. (12.97).
110. L. R. Jones and J. R. Riddick, Anal. Chem., 26, 1035
      (1954).
111. F. Feigl, Spot Tests in Organic Analysis, 6th ed.,
      Elsevier, Amsterdam, 1960, pp. 225-227.
112. W. M. Foley, Jr., G. E. Sanford, and H. McKennis,
      Jr., J. Am. Chem. Soc., 74, 5489 (1952).
113. F. R. Duke, Ind. Eng. Chem., Anal. Ed., 17, 196
      (1945).
114. G. R. Umbreit, Anal. Chem., 33, 1572 (1961). The
      same reaction is used to test for CS_2: N. Tischler,
      Ind. Eng. Chem., Anal. Ed., 4, 146 (1932).
115. F. E. Critchfield and J. B. Johnson, Ind. Eng. Chem.,
      Anal. Ed., 28, 430 (1956).
116. S. Siggia and J. G. Hanna, Anal. Chem., 20, 1084
      (1948); see also J. A. Vinson, ibid., 41, 1661 (1969).
117. R. Venkataraghavan and C. N. R. Rao, Chemist-Analyst,
      51, 48 (1962).
118. P. M. Beazley, Anal. Chem., 43, 148 (1971).
119. J. E. Sinsheimer, D. D. Hong, J. I. Stewart, M. L.

Fink, and J. H. Burckhalter, J. Pharm. Sci., $\underline{60}$, 141 (1971).

120. P. Edman, Acta Chem. Scand., $\underline{4}$, 277, 283 (1950); H. Fraenkel-Conrat, J. I. Harris, and A. L. Levy, Methods Biochem. Anal., $\underline{2}$, 359 (1955).

121. L. M. Dowling and G. R. Stark, Biochemistry, $\underline{8}$, 4728 (1969).

122. M. Waterfield and H. Haber, Biochemistry, $\underline{9}$, 832 (1970).

123. Z. Deyl, J. Chromatogr., $\underline{48}$, 231 (1970).

124. For some examples see K. Uhrig, E. Lynch, and H. C. Becker, Ind. Eng. Chem., Anal. Ed., $\underline{18}$, 550 (1946); J. S. Powell, K. C. Edson, and E. L. Fisher, Anal. Chem., $\underline{20}$, 213 (1948); S. Görög, ibid., $\underline{42}$, 560 (1970).

125. M. Kamel and R. Wizinger, Helv. Chim. Acta, $\underline{43}$, 594 (1960).

126. T. A. LaRue, Anal. Chem., $\underline{42}$, 541 (1970).

127. L. O'Daniel and L. B. Parsons, Oil and Soap, $\underline{20}$, 72 (1943).

128. S. Siggia, Quantitative Organic Analysis via Functional Groups, 3rd ed., Wiley, New York, 1963, p. 10. One postulated reaction is $RCHO + Ac_2O \rightarrow RCH(OAc)_2$; another is the self-aldol condensation $2\ R'CH_2CHO \rightarrow R'CH_2CH(OH)CH(R')CHO$. See also P. Baudet and C. Otten, Helv. Chim. Acta, $\underline{53}$, 1330 (1970).

129. F. R. Duke, Anal. Chem., $\underline{19}$, 661 (1947).

130. M. R. F. Ashworth and I. Venn, Anal. Chim. Acta, $\underline{49}$, 535 (1970).

131. C. H. Brieskorn and H. Otteneder, Chem. Ber., $\underline{103}$, 363 (1970).

132. K. G. Stone, Determination of Organic Compounds, McGraw-Hill, New York, 1956, pp. 101-109.

133. K. Samejima, W. Dairman, and S. Udenfriend, Anal. Biochem., $\underline{42}$, 222 (1971); K. Samejima, W. Dairman, J. Stone, and S. Udenfriend, ibid., $\underline{42}$, 237 (1971).

134. M. W. McCaman and E. Robins, J. Lab. Clin. Med., $\underline{59}$, 885 (1962).

135. M. Weigele, J. F. Blount, J. P. Tengi, R. C. Czaijkowski, and W. Leimgruber, J. Am. Chem. Soc., $\underline{94}$, 4052 (1972).

136. M. Weigele, S. L. DcBcrnardo, J. P. Tengi, and W. Leimgruber, J. Am. Chem. Soc., $\underline{94}$, 5927 (1972); S. Udenfriend, S. Stein, P. Böhlen, W. Dairman, W. Leimgruber, and M. Weigele, Science, $\underline{178}$, 871 (1972).

137. This is not a contrived problem, Benzaldehyde and nitrobenzene both have the odor of almond oil, and benzaldehyde is used to impart this odor. Nitrobenzene, which is highly toxic, is a potential adulterand of benzaldehyde.

Chapter 13.   ACYL TRANSFER

13.1   Nature of the Reaction

Survey of the Reaction.   In this chapter reactions of the
general form of Eq. (13.1) are reviewed.

$$R\text{-}\overset{\overset{\displaystyle O}{\|}}{C}\text{-}X + Y^- \longrightarrow R\text{-}\overset{\overset{\displaystyle O}{\|}}{C}\text{-}Y + X^- \tag{13.1}$$

The function RCO- is the acyl group, so these reactions
can be viewed as acyl transfers from X to Y; alternatively
they can be considered as nucleophilic substitutions on
carbonyl carbon.   Some closely analogous reactions can be
included, for example, reactions of phosphoric acid derivatives.   Primarily, however, this chapter deals with reactions of carboxylic acids and their derivatives.   Table
13.I shows some of these reaction types [1].
    Characteristic of substrates, RCOX, that undergo acyl
transfer reactions is possession of a good leaving group
X, so that the C-X bond can be cleaved and the integrity
of the R-C=O system maintained.   This behavior is in sharp
contrast with that of most aldehydes and ketones, as shown
in Chap. 12, for these compounds do not usually possess
good leaving groups.   When, however, a carbonyl compound
is substituted with powerful electron-withdrawing groups
so that the carbanion of the incipient leaving group is
stabilized, carbon-carbon bond cleavage can occur and such
aldehydes and ketones can be hydrolyzed, as shown below
[2,3].

$$CH_3\overset{\overset{\displaystyle O}{\|}}{C}\text{-}CH_2NO_2 + H_2O \rightarrow CH_3COOH + CH_3NO_2 \tag{13.2}$$

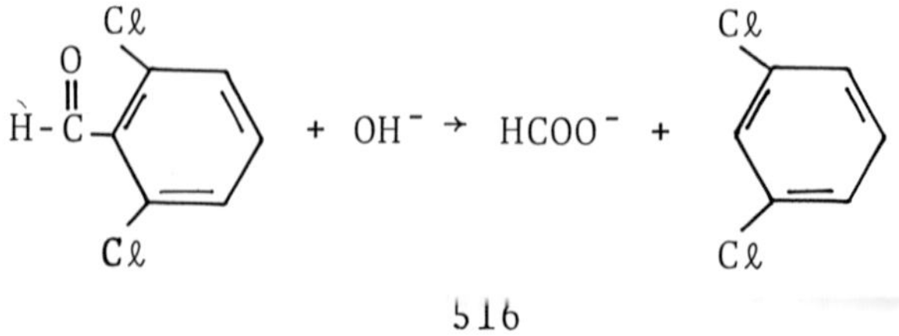

$$H\text{-}\overset{\overset{\displaystyle O}{\|}}{C}\text{-}\underset{C\ell}{\overset{C\ell}{\bigcirc}} + OH^- \rightarrow HCOO^- + \underset{C\ell}{\overset{C\ell}{\bigcirc}} \tag{13.3}$$

Table 13.I   Types of Acyl Transfer Reactions

Name	Reaction	Comment
Hydrolysis	RCOX + H$_2$O → RCOOH + HX	X = OR´, Cℓ, NHR´, OCOR´, SR´, etc.
Esterification	(1) RCOOH + R´OH → RCOOR´ + H$_2$O	
	(2) RCOX + R´OH → RCOOR´´+ HX	X = Cℓ, OCOR, etc.
Transesterification	(1) RCOOR´ + R´´OH → RCOOR´´ + R´OH	Alcoholysis
	(2) RCOOR´ + R´´COOH → R´´COOR´ + RCOOH	Acidolysis
Aminolysis	RCOX + R´NH$_2$ → RCONHR´ + HX	Amide formation
Claisen condensation	RCOOR´ + CH$_3$Z → RCOCH$_2$Z + R´OH	Z is electron-withdrawing

Some of the most important reactions to be discussed
here are those of carboxylic acids and esters with oxygen
nucleophiles. In these processes, which are exemplified
by ester hydrolysis and by esterification of acids, two
sites of bond cleavage are possible in the carboxyl compound, both leading to the same product; obviously it is
essential to know the type of bond fission in order to
construct a reaction mechanism. The two possibilities,
shown in structure 1, are denoted acyl-oxygen fission (Ac)
and alkyl-oxygen fission (Aℓ).

1

Several methods have been devised to establish the type of
bond fission. When an ester is hydrolyzed in water enriched in H$_2{}^{18}$O, incorporation of ^{18}O from the solvent will
be expected according to the pattern shown in Eq. (13.4).
Thus when methyl 2,4,6-trimethylbenzoate (methyl mesitoate)
was hydrolyzed (under alkaline conditions) in solvent containing added H$_2{}^{18}$O, the isolated mesitoic acid was found
to contain ^{18}O equivalent to the introduction of one ^{18}O
atom from the solvent; this result shows that acyl-oxygen

cleavage occurred [4]. In another method, neopentyl esters
were formed and then hydrolyzed, and neopentyl alcohol was
identified as the product [5]. If alkyl-oxygen cleavage
had occurred in either reaction, the neopentyl cation
$Me_3C-CH_2^+$ would have rearranged to the more stable
Me_2C-CH_2Me, and tert-amyl alcohol would have been obtained.
It follows that acyl-oxygen fission took place.

$$
\begin{array}{l}
\quad\quad\quad\quad\quad Ac \quad\quad\quad\quad\quad\quad\quad R-C-^{18}OH \ + \ R'-OH \\[4pt]
O \\
\| \\
R-C-O-R' \ + \ H_2^{18}O \\[8pt]
\quad\quad\quad\quad\quad A\ell \quad\quad\quad\quad\quad\quad R-C-OH \ + \ R-^{18}OH
\end{array}
\tag{13.4}
$$

      Discussion of acid and ester reaction mechanisms is
often based on a classification by Ingold that takes into
account the type of bond fission (Ac or Aℓ), the reaction
molecularity (1 or 2), and the ionic form of the substrate
(A for the conjugate acid $RC(OH)OR'^+$ and B for the conjugate base RCOOR) [6]. This classification is shown in
Table 13.II. Note that alkyl-oxygen fission constitutes
nothing more than a nucleophilic substitution on aliphatic
carbon, Eq. (13.5) [7,8], which was discussed in Chap. 9.

$$
RCH_2 \ \text{—}\ OCOR' \ + \ Y^- \longrightarrow RCH_2-Y \ + \ RCOO^- \tag{13.5}
$$
$$
A\ell
$$

The present chapter considers only reactions with acyl-
oxygen fission [9].

<u>The Tetrahedral Intermediate</u>. For the bimolecular reac-
tion, with Ac cleavage, of acids and their derivatives two
reasonable mechanisms have been suggested [6,10]. One of
these is a direct displacement analogous to the S_N2 mechanism of aliphatic nucleophilic substitution. The transition state for this hypothetical process is indicated by
<u>2</u>, but this is certainly an extreme form, for it does not
incorporate any electronic redistribution of the carbon-
oxygen double bond. The second mechanism is similar to
the addition-elimination pathway for carbonyl reactions;
it invokes the formation, by nucleophilic addition to the
carbonyl group, of a true <u>tetrahedral intermediate</u>, <u>3</u>.

Table 13.II   Classification of Ester Formation and
Hydrolysis [6]

		Type of fission	
		Acyl	Alkyl
Conjugate form	Acid	$A_{Ac}1$	$A_{A\ell}1$ (S_N1)
		$A_{Ac}2$	$A_{A\ell}2$ (unknown)
	Base	$B_{Ac}1$ (unknown)	$B_{A\ell}1$ (S_N1)
		$B_{Ac}2$	$B_{A\ell}2$ (S_N2)

$$Y^- + \underset{R}{\overset{O}{\underset{|}{\overset{\|}{C}}}}{-}X \rightleftharpoons \left[\overset{\delta^-}{Y} {-}{-}{-}\underset{R}{\overset{O}{\underset{|}{\overset{\|}{C}}}}{-}{-}{-}X^{\delta^-} \right] \rightleftharpoons Y{-}\underset{R}{\overset{O}{\underset{|}{\overset{\|}{C}}}} + X^- \qquad (13.6)$$

$$\underline{2}$$

$$Y^- + R{-}\overset{O}{\overset{\|}{C}}{-}X \rightleftharpoons R{-}\underset{Y}{\overset{O^-}{\underset{|}{\overset{|}{C}}}}{-}X \rightleftharpoons R{-}\overset{O}{\overset{\|}{C}}{-}Y + X^- \qquad (13.7)$$

$$\underline{3}$$

Bender, in 1951, provided the first diagnostic test of mechanism for these reactions [11].   The alkaline hydrolysis of several benzoate esters, a $B_{Ac}2$ reaction, was studied with esters enriched in ^{18}O at the carbonyl oxygen.   If the S_N2 mechanism, Eq. (13.8), is operative, no ^{18}O exchange with the solvent during hydrolysis would be expected, because the carbonyl oxygen does not engage in any reversible step.   The tetrahedral intermediate mechanism, however, provides a route for ^{18}O exchange, concurrent with hydrolysis, by the symmetrical partitioning shown in Eq. (13.9).

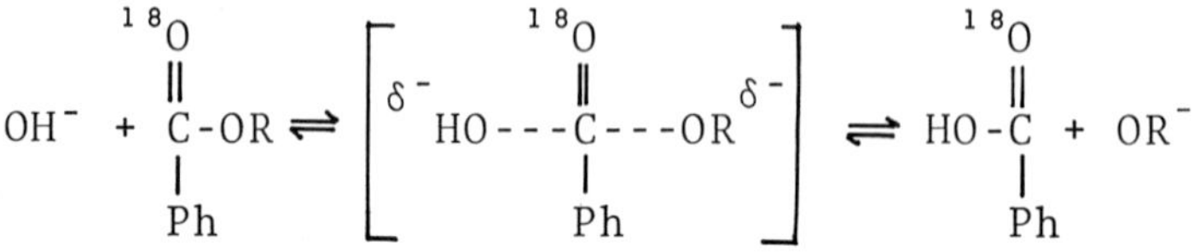

$$OH^- + \overset{\overset{{}^{18}O}{\|}}{\underset{\underset{Ph}{|}}{C}}-OR \rightleftharpoons \left[\overset{\delta^-}{HO}---\overset{\overset{{}^{18}O}{\|}}{\underset{\underset{Ph}{|}}{C}}---OR^{\delta^-}\right] \rightleftharpoons HO-\overset{\overset{{}^{18}O}{\|}}{\underset{\underset{Ph}{|}}{C}} + OR^- \qquad (13.8)$$

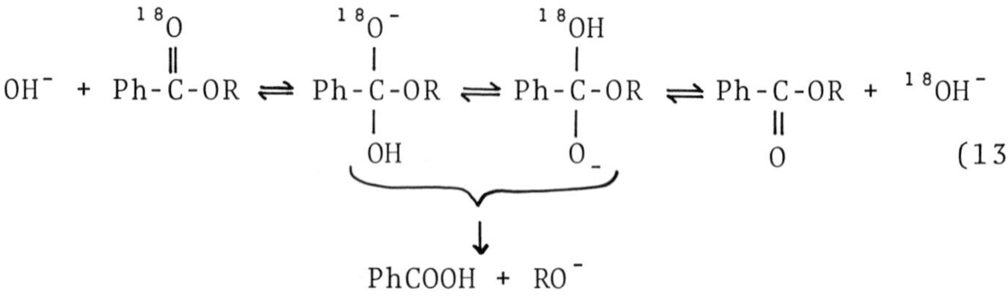

$$OH^- + Ph-\overset{\overset{{}^{18}O}{\|}}{C}-OR \rightleftharpoons Ph-\overset{\overset{{}^{18}O^-}{|}}{\underset{\underset{OH}{|}}{C}}-OR \rightleftharpoons Ph-\overset{\overset{{}^{18}OH}{|}}{\underset{\underset{O_-}{|}}{C}}-OR \rightleftharpoons Ph-\overset{\overset{}{}}{\underset{\underset{O}{\|}}{C}}-OR + {}^{18}OH^- \qquad (13.9)$$

$$\downarrow$$

$$PhCOOH + RO^-$$

When ethyl benzoate-carbonyl ${}^{18}O$ was subjected to alkaline
hydrolysis, samples of ester isolated during the progress
of the reaction were found to contain less ${}^{18}O$ than did
the initial ester. This demonstration of concurrent oxygen exchange and hydrolysis is strong (though not definitive) evidence for the tetrahedral intermediate mechanism.
    A quantitative kinetic analysis of oxygen exchange
studies gives valuable information about the kinetics of
the two-step mechanism. Writing an ester hydrolysis as
Eq. (13.10) and applying the steady-state approximation
to the tetrahedral intermediate:

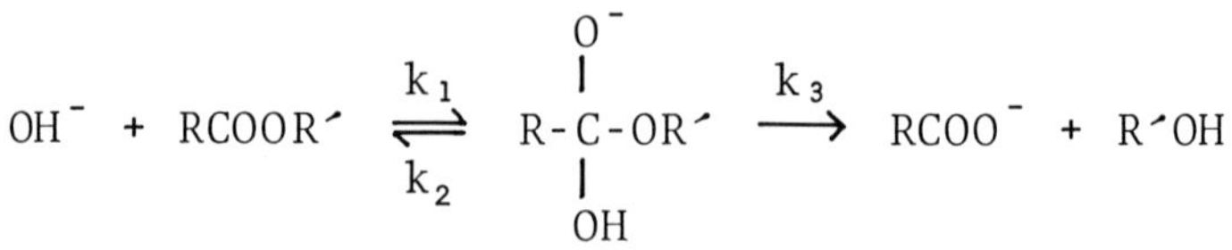

$$OH^- + RCOOR' \underset{k_2}{\overset{k_1}{\rightleftharpoons}} R-\overset{\overset{O^-}{|}}{\underset{\underset{OH}{|}}{C}}-OR' \xrightarrow{k_3} RCOO^- + R'OH \qquad (13.10)$$

$$[\text{Intermediate}] = \frac{k_1[RCOOR'][OH^-]}{k_2 + k_3} \qquad (13.11)$$

The rate of hydrolysis is given by $k_3[\text{Intermediate}]$, and
the rate of carbonyl oxygen exchange by $k_2[\text{Intermediate}]/2$;
then if k_h is the second-order rate constant for hydrolysis and k_e is the rate constant for exchange, the following relationships are obtained [12].

$$k_h = \frac{k_1}{(k_2/k_3) + 1} \qquad (13.12)$$

$$k_e = \frac{k_1 k_2}{2(k_2 + k_3)} \qquad (13.13)$$

$$\frac{k_2}{k_3} = 2 \frac{k_e}{k_h} \qquad (13.14)$$

$$k_1 = 2k_e + k_h \qquad (13.15)$$

The oxygen exchange experiments give the ratio k_e/k_h, and of course k_h is accessible from separate kinetic studies on the hydrolysis; with Eqs. (13.14) and (13.15) it is therefore possible to obtain k_1 and k_2/k_3.

Concurrent oxygen exchange and hydrolysis has been observed for several reactions, including the alkaline and acidic hydrolyses of some benzoate esters [4,11,13,14], and the alkaline hydrolysis of amides [15]. The values of k_h/k_e for the acidic and alkaline hydrolyses of ethyl benzoate differ by a factor of about 2 (k_h/k_e = 5.3 and 11.3 under acidic and basic conditions, respectively), whereas the k_h values under these two conditions differ by a factor of 10^4; it is therefore believed that a similar intermediate is formed in both cases [16].

Establishment of the occurrence of oxygen exchange is a difficult experimental problem, and it is estimated [7,14] that the technique cannot reliably estimate ratios of k_3/k_2 greater than about 100. Ratios greater than this are interpreted as indicating no detectable oxygen exchange. The absence of oxygen exchange is now thought to reflect merely quantitative differences in k_3/k_2; thus ethyl benzoate shows oxygen exchange whereas benzyl benzoate does not, but the tetrahedral intermediate mechanism, with different partition ratios, is believed to be operative for both esters.

Several kinds of kinetic evidence are consistent with a two-step mechanism, and therefore with the existence of an intermediate on the reaction path. Johnson [17] and Jencks [18] have discussed this evidence in detail. An effective result is the demonstration of a change in rate-determining step, which requires at least a two-step reaction. In Chap. 12 several examples of this behavior were discussed for additions to carbonyl compounds, reactions in which tetrahedral intermediates are sufficiently stable to be readily detected and sometimes even isolated. At the acyl level of oxidation, tetrahedral addition compounds can be directly observed only for very activated acyl groups or very powerful nucleophiles [19]. An example of a tetrahedral intermediate inferred from a change in rate-determining step is 4, the addition product of an amine with an imido ester. The reaction kinetics show a sharp bell-shaped pH profile that cannot be accounted for in terms of ionization of the reactants [20]. The kinetics were interpreted by proposing rate-determining formation of the tetrahedral

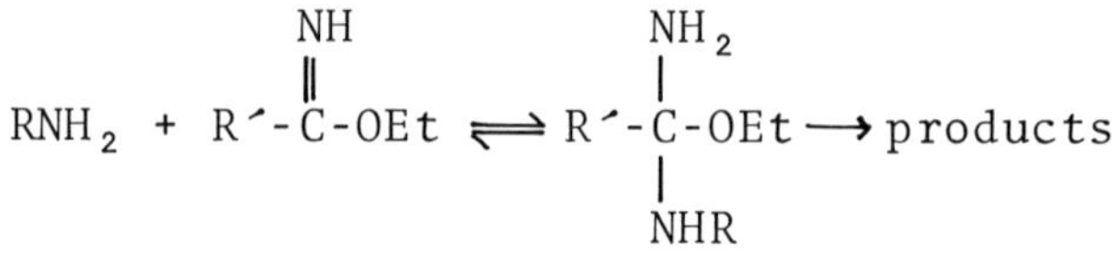

$$RNH_2 + R'-\overset{\overset{\displaystyle NH}{\|}}{C}-OEt \rightleftharpoons R'-\overset{\overset{\displaystyle NH_2}{|}}{\underset{\underset{\displaystyle NHR}{|}}{C}}-OEt \longrightarrow products \qquad (13.16)$$

$$\underline{4}$$

intermediate on the alkaline side of the pH-rate curve, and rate-determining decomposition of the intermediate on the acid side. One piece of evidence supporting a change in rate-determining step is that rate constants for kinetically identical rate terms differ by a factor of ten on the two sides of the curve. Another useful observation is the appearance of different products on the two sides of the pH-rate hump. The possibilities are shown in Eq. (13.17) for the reaction of ammonia with ethyl N-methylbenzimidate (ionic states not indicated).

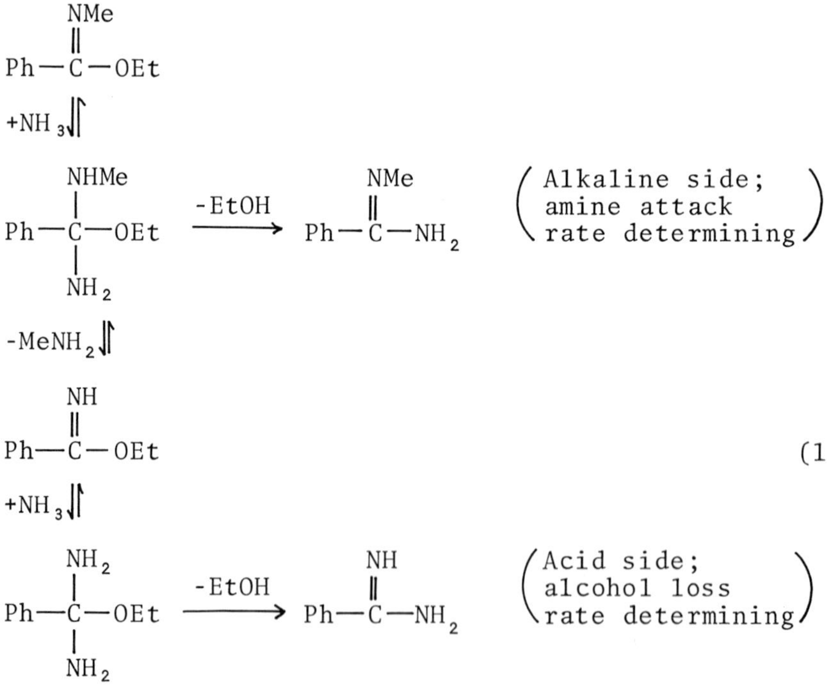

$$Ph-\overset{\overset{\displaystyle NMe}{\|}}{C}-OEt$$

$$+NH_3 \updownarrow$$

$$Ph-\overset{\overset{\displaystyle NHMe}{|}}{\underset{\underset{\displaystyle NH_2}{|}}{C}}-OEt \quad \xrightarrow{-EtOH} \quad Ph-\overset{\overset{\displaystyle NMe}{\|}}{C}-NH_2 \qquad \left(\begin{array}{l}\text{Alkaline side;}\\ \text{amine attack}\\ \text{rate determining}\end{array}\right)$$

$$-MeNH_2 \updownarrow$$

$$Ph-\overset{\overset{\displaystyle NH}{\|}}{C}-OEt \qquad\qquad\qquad\qquad\qquad\qquad (13.17)$$

$$+NH_3 \updownarrow$$

$$Ph-\overset{\overset{\displaystyle NH_2}{|}}{\underset{\underset{\displaystyle NH_2}{|}}{C}}-OEt \quad \xrightarrow{-EtOH} \quad Ph-\overset{\overset{\displaystyle NH}{\|}}{C}-NH_2 \qquad \left(\begin{array}{l}\text{Acid side;}\\ \text{alcohol loss}\\ \text{rate determining}\end{array}\right)$$

On the alkaline side of the pH-rate maximum the product is N-methylbenzamidine, showing that intermediate formation is rate determining, for as soon as it is formed it loses ethanol to give this product. On the acid side of the curve the product is benzamidine, which can form by pro-

equilibration of the first-formed intermediate with ammonia under conditions where decomposition of the intermediate is rate determining.

The bell-shaped pH-rate curve for the aminolysis of imido esters therefore is accounted for by postulating a change in rate-determining step with pH [21]. A different type of kinetic evidence for an intermediate is seen in the reaction of hydroxylamine with amides.

$$\text{NH}_2\text{OH} + \text{R-}\overset{\displaystyle O}{\overset{\|}{\text{C}}}\text{-NH}_2 \rightarrow \text{R-}\overset{\displaystyle O}{\overset{\|}{\text{C}}}\text{-NHOH} + \text{NH}_3 \qquad (13.18)$$

Again a sharp bell-shaped pH-rate profile is obtained. The reaction is catalyzed by the general acid hydroxyl-ammonium ion, and plots of apparent second-order rate constants against total hydroxylamine concentration show marked negative curvature [22]. This curvature can be accounted for as a consequence of a change in rate-determining step with change in catalyst concentration, the formation and decomposition of the tetrahedral intermediate being general acid catalyzed.

It is now widely accepted [23] that bimolecular reactions at carboxyl carbon, like those at the carbonyl group, generally proceed through tetrahedral intermediates formed by nucleophilic addition at the carbon atom. When the partition ratio k_2/k_3 [see Eqs. (13.10) and (13.12)] is smaller than unity, the rate is controlled by the addition (k_1) step; this is typical of ester hydrolyses. When k_2/k_3 exceeds unity, decomposition of the tetrahedral intermediate becomes rate limiting; this can happen when the leaving group is very poor or the nucleophile is very good. The solvent may play an important role in influencing kinetic behavior. For example, Menger and Smith [24] found an extremely high sensitivity to substituent variations in the leaving groups of esters subjected to aminolysis in aprotic solvents. They attributed this to rate-determining decomposition of the tetrahedral intermediate.

General Base and Nucleophilic Catalysis. Reactions of carboxylic acid derivatives are often catalyzed by basic (nucleophilic) reagents, and we have already seen that these substrates are susceptible to direct nucleophilic attack. It is therefore of interest to ascertain the manner in which basic reagents exert their catalytic effects. They may function as nucleophiles by attacking carbon, or they may serve as bases, that is, by attacking hydrogen. In either case the reagent brings an unshared electron pair to an electron-deficient site. Of course the reaction paths of general base catalysis and of

nucleophilic catalysis will be vastly different, but the
similarities of the two catalytic modes are worth considering. These can be seen by comparing Eqs. (13.19) and
(13.20), which are generalized representations of nucleophilic and general base reactions, with N: representing
the reagent.

$$\text{\underline{Nucleophilic}} \quad N: \widehat{} \overset{|}{\underset{|}{C}}{=}\overset{\frown}{O} \longrightarrow {}^{+}N{-}\overset{|}{\underset{|}{C}}{-}O^{-} \qquad (13.19)$$

$$\text{\underline{General base}} \quad N: \widehat{} H{\overset{\frown}{-}}Z \longrightarrow {}^{+}N{-}H + Z^{-} \qquad (13.20)$$

It is not surprising that conventional basicity, as
measured by pK_a values, is a good model property for
general base catalysis, because of the probable similarity
between the transition state for Eq. (13.20) and the model
reaction product. This comparison is usually made by means
of Brønsted plots of log $k_{catalytic}$ against pK_a of the
catalyst, as discussed in Sec. 4.3. Similar plots can be
made for nucleophilic catalyses, and these often yield good
correlations, revealing some similarity of the catalytic
process with the model process.

Both intermolecular and intramolecular examples of
general base and nucleophilic catalysis are known. We
first consider intermolecular reactions. The simplest
kinetic manifestation of a reaction catalyzed by a general
base or nucleophilic reagent is a rate term of the form
k[substrate][reagent]. By itself this cannot identify the
type of reaction, and other evidence must be applied. In
the $B_{Ac}2$ hydrolysis of an ester, Eq. (13.21), the hydroxide
ion must function by nucleophilic attack on the carboxyl
carbon.

$$\overset{O}{\overset{\|}{R{-}C{-}OR'}} + OH^{-} \rightarrow \overset{O}{\overset{\|}{R{-}C{-}OH}} + R'O^{-} \qquad (13.21)$$

Since the total alkalinity is depleted during the course
of the hydrolysis by the subsequent equilibrium RCOOH +
$R'O^{-} \rightleftharpoons RCOO^{-} + R'OH$, this is sometimes called a hydroxide
ion-promoted (rather than catalyzed) reaction. True nucleo-
philic catalysis of an ester hydrolysis is illustrated by
the imidazole-catalyzed hydrolysis of p-nitrophenyl acetate [25]. In contrast with this behavior is the imidazole-
catalyzed hydrolysis of ethyl acetate, which occurs by a
general base mechanism [26]. This may happen as shown in
Eq. (13.23), in which the general base imidazole increases
the nucleophilicity of a water molecule by partial abstraction of a proton in the transition state leading to the
tetrahedral intermediate.

$$HN\diagdown N + CH_3\overset{O}{\overset{\|}{C}}-O-\text{(ring)}-NO_2 \rightleftharpoons HO-\text{(ring)}-NO_2 + CH_3\overset{O}{\overset{\|}{C}}-N\diagdown N$$

$$\downarrow H_2O$$

$$CH_3COOH + HN\diagdown N$$

$$(13.22)$$

$$CH_3 - \overset{O}{\overset{\|}{C}} - OEt \rightleftharpoons CH_3 - \overset{O^-}{\underset{OH}{\overset{|}{C}}} - OEt \rightarrow products$$

$$HN\diagdown N \cdots H \diagdown H \qquad HN\diagdown NH \qquad (13.23)$$

The reactions shown in Eqs. (13.22) and (13.23) will each
have a rate equation of the form $v = k[\text{ester}][\text{imidazole}]$.
Some esters have a rate equation with another term that
is second order in imidazole, $k'[\text{ester}][\text{imidazole}]^2$. This
term is interpreted by combining the ideas of Eqs. (13.22)
and (13.23); that is, it reflects a general base-catalyzed
nucleophilic catalysis, with one molecule of imidazole
serving as the nucleophile and a second molecule of imidazole as the general base [27].
    Since the form of the rate equation does not distin-
guish between nucleophilic and general base catalysis,
other criteria must be applied to make assignments of the
type given above for imidazole catalysis of ester hydrolyses. Several criteria have been devised, and brief descriptions are given in Table 13.III. Johnson has given many
examples of their application [17]. Some of the methods
seek evidence of the intermediate that must be formed in a
nucleophilic catalysis; in Eq. (13.22), for example, this
intermediate is acetylimidazole, and catalysis is observed
because acetylimidazole is more labile than is p-nitrophenyl
acetate. In this reaction it is possible to detect the
intermediate spectrophotometrically [25a], and the time
course of its concentration can be correlated with the
kinetics of ester loss and p-nitrophenol production [25c].

Table 13.III Methods for Differentiating between General
Base and Nucleophilic Catalysis

Method	Criterion
Product analysis	Product identity or distribution attributable only to reaction via intermediate formed by nucleophilic attack
Trapping of intermediate	Product identity attributable only to diversion of intermediate by reaction with added trapping nucleophile
Physical detection of intermediate	Change of physical property during reaction attributable to intermediate formed by nucleophilic attack
Demonstration of reversible nucleophilic reaction	Inhibition of catalysis by added product of reaction
Solvent D_2O kinetic isotope effect	$k_H/k_D > 2$ for general base, 0.8-1.9 for nucleophilic catalysis (usually)
Relative reactivity of hindered and unhindered bases	k(hindered)/k(unhindered) same order for bases of same pK_a if general base catalysis, but this ratio very small for nucleophilic catalysis
Relative reactivity of imidazole and mono-hydrogen phosphate	$k(imidazole)/k(HPO_4^{2-})$ is order of unity for general base catalysis, but about 10^3 for nucleophilic catalysis
Relative reactivity of hydroxide and imidazole	k(hydroxide)/k(imidazole) is 10^5-10^6 for general base, but 10-10^3 for nucleophilic catalysis

[a]M. L. Bender and M. C. Neveu, J. Am. Chem. Soc., 80, 5388 (1958).
[b]D. G. Oakenfull, T. Riley, and V. Gold, Chem. Commun., 1966, 385.
[c]A. R. Fersht and W. P. Jencks, J. Am. Chem. Soc., 91, 2125 (1969).
[d]W. P. Jencks, F. Barley, R. Barnett, and M. Gilchrist, J. Am. Chem. Soc., 88, 4464 (1964).

Example	Ref.
Acetate-^{18}O-catalyzed hydrolysis of 2,4-dinitrophenyl benzoate gives ^{18}O-benzoic acid	a
Acetanilid formation in acetate-catalyzed hydrolysis of aryl acetates in presence of aniline	b
Spectrophotometric observation of acetylpyridinium ion in pyridine-catalyzed hydrolysis of acetic anhydride	c
Inhibition of nucleophilic component of acetate-catalyzed acetylimidazole hydrolysis by added imidazole	d
k_H/k_D = 3.1 for imidazole-catalyzed hydrolysis of ethyl trifluorothiolacetate	e
2,6-Lutidine is much less effective than pyridine in catalysis of acetic anhydride hydrolysis	f
This ratio is 0.25 for ethyl acetate hydrolysis, and 4.7 × 10^3 for p-nitrophenyl acetate hydrolysis	g
This ratio is 9.1 × 10^5 for ethyl acetate hydrolysis, and 160 for p-nitrophenyl acetate hydrolysis	g

[e]L. R. Fedor and T. C. Bruice, J. Am. Chem. Soc., 87, 4138 (1965).
[f]A. R. Butler and V. Gold, J. Chem. Soc., 1961, 4362.
[g]J. F. Kirsch and W. P. Jencks, J. Am. Chem. Soc., 86, 837 (1964).

Other methods for discriminating between general base and nucleophilic reactions are kinetic in nature. The ratio of catalytic rate constants for a substrate with imidazole and with monohydrogen phosphate is an indicator of mechanism, because these two reagents have about the same basicity. If they both operate by a general base mechanism, the rate ratio should therefore be about unity. If, on the other hand, they both operate by nucleophilic attack (or if imidazole does), the rate of imidazole catalysis will be much larger than that of phosphate catalysis, because imidazole is a much better nucleophile. The hydroxide/imidazole ratio tells if the imidazole reaction is general base or nucleophilic, because hydroxide always attacks as a nucleophile. Since imidazole is a better nucleophile that it is a base, the ratio will be high for a general base and low for a nucleophilic reaction.

Having introduced the concepts of general base and nucleophilic catalysis and briefly treated their experimental recognition, we can next consider the features of catalysis and its effectiveness. The type of catalysis depends upon the structures of the substrate and the nucleophile. Several generalizations of experimental observations can be made, as in the following items.

1. Alcohol-activated esters (esters with good leaving groups, like phenyl esters) are susceptible to nucleophilic catalysis, whereas acyl-activated esters (esters with poor leaving groups, like alkyl esters) are not, and show detectable general base catalysis [28]. A mixture of the two can occur, so that $k_{obs} = (k_{gb} + k_n)[nucleophile]$, and the proportion of the two catalytic contributions depends upon the leaving group [29].

2. The slope β of a Brønsted plot of log (catalytic rate constant) against pK_a of the catalyst is about 0.8 for nucleophilic attack and about 0.5 for general base catalysis [28]. For nucleophilic attack it is common to obtain separate and nearly parallel correlation lines for different classes of nucleophiles (such as pyridines, imidazoles, and oxygen anions) [30].

3. β for variation in basicity of the attacking nucleophile is roughly equal to $-\beta$ for variation in the basicity of the leaving group for nucleophilic reactions on alkyl and aryl acetates and on acetylpyridinium ions [31].

4. In order for general base catalysis of ester hydrolysis by imidazole to be detectable, the leaving group
should be several pK units more basic than 7 (the pK_a of
imidazole); otherwise nucleophilic attack will be favored
[32].

5. In nucleophilic reactions of nucleophiles possessing an atom with an unshared electron pair adjacent to the
attacking atom (as in hydroxylamine, hydrazine, and hydrogen peroxide), the nucleophilic reactivity is often much
greater than would be anticipated on the basis of pK_a or
Brønsted plots. This anomalously high nucleophilicity is
called the α-effect [33].

Some of these points are illustrated in Fig. 13.1,
which is a Brønsted-type plot of k_n, the second-order rate
constant for nucleophilic attack on p-nitrophenyl acetate
(a favorite substrate for studies of this type), against
pK_a of the conjugate acid of the nucleophile. This figure
shows the typical dispersion into separate lines for nucleophiles of different types [34]. The β values of 0.8
and 0.95 demonstrate a high sensitivity of nucleophilicity
to basicity. For the oxyanions of high basicity this dependence decreases sharply. The reactivity of hydroxide
is significantly less than its basicity would predict, a
common observation. Several α-effect nucleophiles are included to show their high nucleophilicity. Figure 13.1
also reveals the basis of the tests for mechanism using
the relative rates k(imidazole)/k(phosphate) and k(hydroxide)/k(imidazole) (Table 13.III). Obviously other pairs
of nucleophiles could also be used for such tests.

Within the class of nucleophilic reactions at the acyl
group, dependence of reactivity on basicity, as expressed
by β with respect to the nucleophile and the leaving group,
reveals patterns that Fersht and Jencks [31] used to assign
these reactions to three types. Table 13.IV shows this
classification, which is not intended to exclude examples
intermediate to the three main classes. The key observation in interpreting the grouping is that β for the nucleophiles (determined from plots like Fig. 13.1) is roughly
equal to -β for the leaving groups (determined from a
similar plot for reactions of a single nucleophile with a
series of substrates of varied leaving group). This
equivalence is inferred to mean that bond formation (between the nucleophile and the acyl carbon) and bond cleavage (between the acyl carbon and the leaving group) have
proceeded to about the same extent in the transition state
[31]. Transition states for the three classes are pictured below, where Y is the nucleophile and X is the leaving group. In terms of the tetrahedral intermediate pathway discussed earlier, the class I transition state represents rate-determining formation of the tetrahedral inter

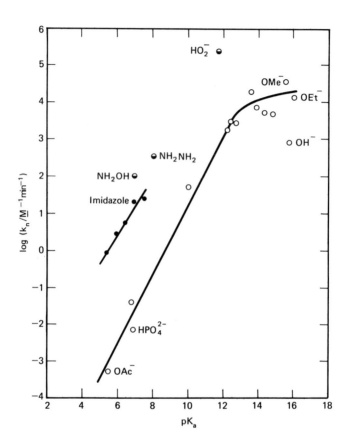

Fig. 13.1. Brønsted-type plot for nucleophilic reactions
of p-nitrophenyl acetate. Key: ●, simple imidazoles, in
28.5% ethanol at 30°, β = 0.80 [data from T. C. Bruice
and G. L. Schmir, J. Am. Chem. Soc., 80, 148 (1958)];   ,
oxygen anions, in water at 25°, β = 0.95 for linear portion [data from W. P. Jencks and J. Carriuolo, J. Am. Chem.
Soc., 82, 1778 (1960), and W. P. Jencks and M. Gilchrist,
ibid., 84, 2910 (1962)]; ◖, alpha-effect nucleophiles.
Several of the nucleophiles are identified.

Class I

$$Y\text{-}\text{-}\text{-}\overset{\displaystyle O}{\underset{\displaystyle R}{\overset{\|}{C}}}\text{—}X$$

Class II

$$Y\text{-}\text{-}\text{-}\overset{\displaystyle O}{\underset{\displaystyle R}{\overset{\|}{C}}}\text{-}\text{-}\text{-}X$$

Class III

$$Y\text{—}\overset{\displaystyle O}{\underset{\displaystyle R}{\overset{\|}{C}}}\text{-}\text{-}\text{-}X$$

mediate, and the class III transition state is for rate-determining decomposition of the tetrahedral intermediate. The class II transition state is explicable if the tetrahedral intermediate is extremely unstable. It is actually easier to describe these transition states in terms of the direct displacement (S_N2) mechanism [31].

Concepts of nucleophilic and general base catalysis have been strongly influenced by studies of the catalytic effects of imidazole [35]. These were stimulated by the possibility that proteolytic enzymes might exert their catalytic action through nucleophilic catalysis by the imidazolyl group of a histidine residue in the active site. It is now known that an acyl-enzyme intermediate is formed, but it is the hydroxy group of serine, rather than the imidazole of histidine, that is acylated. The model studies with imidazole have nevertheless revealed valuable mechanistic information about acyl-transfer catalytic processes.

The general rate equation for the imidazole-catalyzed hydrolysis of any ester can be written as Eq. (13.24), where Im represents neutral imidazole [36].

$$v = \{(k_n + k_{gb})[Im] + k_{ngb}[Im]^2 + k'[Im][OH^-]\}[ester]$$

$$(13.24)$$

Imidazole undergoes the acid-base equilibria of Eq. (13.25), so the rate term $k'[Im][OH^-]$ is kinetically equivalent to the term $k''[Im^-]$.

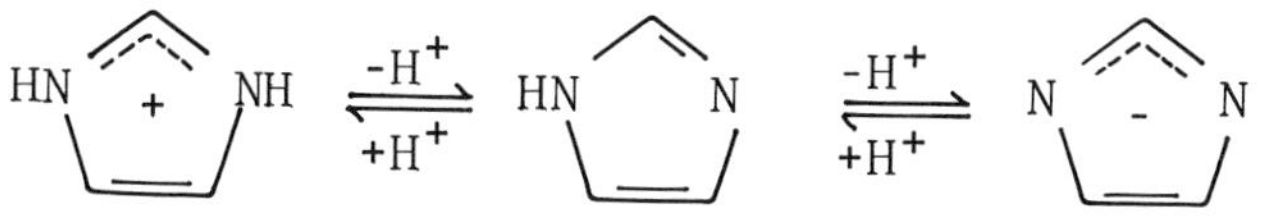

$$(13.25)$$

Equations (13.22) and (13.23) represent the k_n and k_{gb} terms, respectively. Detailed studies of the nucleophilic catalysis [25,35,37] have shown that Eq. (13.26) describes the course of the reaction. The N-acylimidazolium ion is in acid-base equilibrium ($pK_a = 3.6$ for $R=CH_3$) with N-acylimidazole, but it is the protonated form that undergoes hydrolysis. This was shown by experiments with N-methyl-N-acetyl-imidazolium ion, 5, which serves as a model for the N-acetylimidazolium ion 6. The rate equation for reactions of 5 with nucleophilic reagents HY will be of the form $v_5 = k_Y^5[AcImMe^+][Y^-]$, since 5 is constrained

Table 13.IV Classification of Nucleophilic
Acyl-Transfer Reactions Reactions [31]

Class	β (nucleophile) and -β (leaving group)	Characteristics
I	0-0.4	Strongly basic nucleophiles. Asymmetrical "early" transition state with little bond formation or bond cleavage
II	0.7-1.0	Intermediate basicity of nucleophiles or similar basicity of nucleophile and leaving group. Symmetrical transition state with moderate bond formation and bond cleavage
III	1.2-1.6	Weakly basic nucleophiles and poor leaving groups. Asymmetrical "late" transition state with extensive bond formation and bond cleavage

$$ R-\overset{O}{\overset{\|}{C}}-OR' + N\text{—}NH \rightleftharpoons R-\overset{O}{\overset{\|}{C}}-N^{+}NH + R'OH \qquad (13.26) $$

$$ \downarrow H_2O $$

$$ R-\overset{O}{\overset{\|}{C}}OH + N\text{—}NH $$

$$ CH_3-\overset{O}{\overset{\|}{C}}-N^{+}N-CH_3 $$

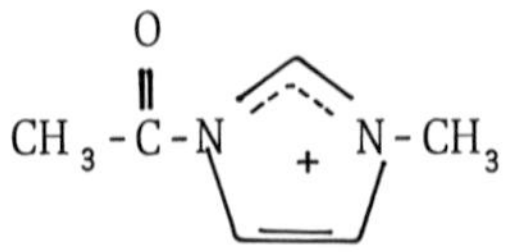

5

$$ CH_3-\overset{O}{\overset{\|}{C}}-N^{+} N-H $$

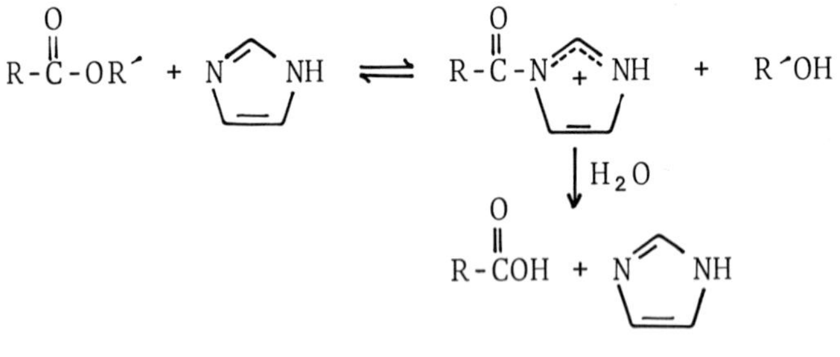

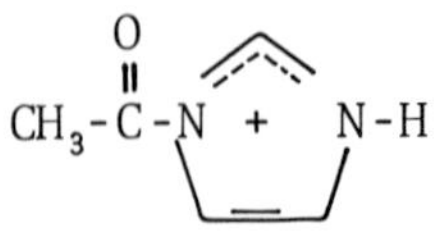

6

Nucleophile type	Substrate type
Basic alkoxides and OH$^-$	Most esters
Basic amines	Reactive esters (leaving group $pK < 5$)
Basic amines and alkoxides	Acetyl-$\overset{+}{N}\leqslant$
Oxyanions	Esters (similar pK of attacking and leaving groups)
Most amines	Phenyl esters
Most amines	Acetyl-$\overset{+}{N}\leqslant$
Oxyanions (low and intermediate basicity)	Acetyl-$\overset{+}{N}\leqslant$
Oxyanions (weakly basic)	Alkyl esters
Most amines	Alkyl esters

to react in the charged form, and Y^- is more nucleophilic than HY. The rate equation for the same reactions of 6 could take the kinetically equivalent forms $v_6 = k_{HY}^6[\text{AcIm}][\text{HY}] = k_Y^6[\text{AcImH}^+][Y^-]$, where k_Y^6 is related to k_{HY}^6 by the ratio of acid dissociation constants for AcImH$^+$ and HY. Table 13.V gives the second-order rate constants k_Y^5, k_{HY}^6, and k_Y^6, as defined by these rate equations [37b]. The very close correspondence of k_Y^5 and k_Y^6 shows that the protonated form of acetylimidazole is the reactive form.

The requirements for nucleophilic catalysis are rather special; they can be summarized as follows [38]:

1. The catalyst must be more nucleophilic than is the final acyl acceptor (for example, water in a hydrolysis reaction).
2. The intermediate formed by nucleophilic attack of the catalyst on the substrate must be more susceptible to attack by the final acyl acceptor than is the original substrate.
3. The final product must be thermodynamically more stable than the intermediate.

Table 13.V Rate Constants for Nucleophilic Reactions with N-Methylacetylimidazole (5) and Acetylimidazole (6) [25b]

Nucleophile, Y^-	Second-order rate constant in listed rate term[a]		
	k_Y^5 [AcImMe$^+$][Y$^-$]	k_Y^6 [AcImH$^+$][Y$^-$]	k_{HY}^6 [AcIm][HY]
H$_2$O	0.051[b]	0.051[b]	
CH$_3$COO$^-$	17	19	1.3
Succinate^{2-}	42	78	1.4
HPO$_4^{2-}$	2,230	2,100	1.2
NH$_3$	62,000	81,000	0.20

[a]At 25°; units are $\underline{M}^{-1}$-min^{-1}.
[b]Calculated using [$\underline{H}_2O$] = 55 $\underline{M}$.

These features are found in reactions of imidazole
with esters possessing good leaving groups. Adopting the
tetrahedral intermediate description, nucleophilic catalysis of ester hydrolysis by imidazole can be written as in

$$\text{R-C-OR}' + N\!\!\diagdown\!\!NH \underset{k_2}{\overset{k_1}{\rightleftharpoons}} \text{R-C-OR}' \underset{k_{-3}}{\overset{k_3}{\rightleftharpoons}} \text{R-C-N}\diagup\!\!\diagdown\!\!NH + \text{R}'\text{O}^-$$

$$\downarrow \text{H}_2\text{O}$$
$$\text{products} \qquad (13.27)$$

As shown in Fig. 13.1, imidazoles have very high nucleophilic reactivity for their basicity; this is partly because of the ready accessibility of the tertiary nitrogen
atom, which is not sterically protected by substituents as
it would be in aliphatic amines. The acetylimidazole
formed in the reaction is highly reactive, because it can
be protonated on a nitrogen atom other than the one bonded
to the acyl carbon; in this respect acylimidazoles are not
typical amides. This protonation provides a very good
leaving group. An additional factor contributing to the
reactivity of acylimidazoles is the aromatic character of
the imidazole ring, which limits participation of the
nitrogen atom in resonance delocalization over the carboxyl
group.
      The catalytic mechanism in imidazole-catalyzed hydro-
lysis of esters is determined by the structure of the ester. In terms of rate equation (13.24), this generalization can be made [37c]: For esters with very good leaving
groups, only the simple nucleophilic catalysis (k_n term)
is important; p-nitrophenyl acetate is a good example.
As the leaving group is made poorer (as in phenyl acetate),
assistance of the nucleophilic catalysis can be detected,
and the k_{ngb} and k' rate terms of Eq. (13.24) make a contribution. With still worse leaving groups (ethyl acetate), nucleophilic catalysis by any route is not detectable, and only the small k_{gb} term is seen. A finer analysis has been made by Kirsch and Jencks [32] of the k_n and
k_{gb} contributions as a function of ester structure.
Figure 13.2 is a plot of the logarithm of the second-order
rate constant for imidazole-catalyzed hydrolysis against
logarithm of the second-order rate constant for hydroxide
ion-catalyzed hydrolysis. The rate of alkaline hydrolysis
is used as a structural parameter in order to provide an
empirical measure with a common scale for variations in
both portions of the ester. The data of Fig. 13.2 fall in

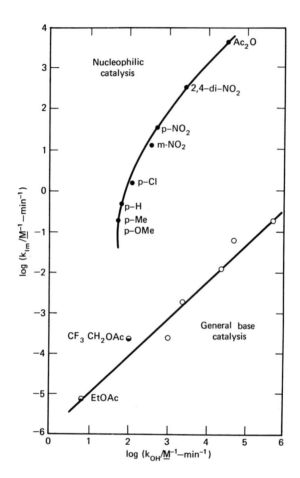

Fig. 13.2.   Rates of imidazole-catalyzed ester hydrolysis
as a function of rates of alkaline hydrolysis.   Key: ●,
nucleophilic reactions of acetate esters; ◐, general base
reactions of acetate esters; ○, general base reactions of
methyl and ethyl esters (data from Ref. [32]).

three regions: (1) the most reactive substrates (acetic
anhydride, 2,4-dinitrophenyl acetate, p-nitrophenyl acetate, m-nitrophenyl acetate) exhibit nucleophilic catalysis and are linearly correlated; (2) phenyl acetates with
poorer leaving groups ($pK_a > 10$) undergo nucleophilic
catalysis but the correlation line has negative curvature;
(3) esters with very poor leaving groups, such as alkyl
esters, show only general base catalysis, which is several
orders weaker than the nucleophilic catalysis by imidazole.

Possibly all of the acetate esters could be placed on a
single curve, with nucleophilic catalysis at the upper
end, general base catalysis at the lower end, and a transition between the two extremes. The curvature seen for
phenyl esters with poor leaving groups can be accounted
for in this way [32]. Consider Eq. (13.27). With a very
good leaving group, $k_3 > k_2$ and the tetrahedral intermediate will pass on to acetylimidazole more readily than
it reverts to ester; thus in the first region described,
formation of the tetrahedral intermediate is rate determining. As the leaving group is made poorer, k_3 becomes
smaller relative to k_2, the decomposition of the tetrahedral intermediate becomes rate determining, and the rate
becomes more sensitive to the nature of the leaving group;
this is seen in the curved portion of the upper curve.
With extremely poor leaving groups, no nucleophilic catalysis is observed.
     Nucleophilic reactivity has been discussed in several
places in this book (see Secs. 4.3 and 9.1). Many factors
can contribute to nucleophilicity, and the use of basicity
as a model property has been described in this section.
Plots like Fig. 13.1 show both the success and the failure
of basicity in correlating with nucleophilicity. Clearly
a correlation exists, with a high degree of dependence as
shown by high β values; but dispersion into separate lines
for nucleophilces of different classes indicates that some
factor other than basicity is also operative. Moreover it
is possible to see the effect on nucleophilic reactivity
of steric hindrance near the nucleophilic atom (see Fig.
4.3) so that this feature can be added to basicity as a
contributor to the net effect. The relative nucleophilicity of pyridine (7) and 2,6-lutidine (8) is a consequence
of the relative inaccessibility of the nitrogen atom of 8.

7   8

     That basicity need not be a good model for nucleophilic
reactivity is shown by its failure to correlate reactivity
in S_N2 displacements at saturated carbon. Even in reactions at acyl carbon, our present concern, additional factors must be present. This is most dramatically shown by
the enhanced reactivity (relative to their basicity) of α-
effect nucleophiles, which possess an unshared electron pair
on the atom adjacent to the attacking atom. The α-effect

has been extensively discussed [33,39]. It is probable
that the term collects nucleophiles that owe their high
reactivity to different factors into a single class, but
it has some utility in calling attention to the common
structural feature. Examples of the α-effect are seen in
Table 12.III and Fig. 13.1. Some α-effect nucleophiles
for reactions at acyl carbon are hydroxylamine, hydrazine,
hydrogen peroxide anion, and hypochlorite anion.
    The role of the α-electron pair in α-effect nucleo-
philes is sometimes ascribed to relief of the electron
deficiency at the attacking atom in the transition state.
For the presumed transition state in the attack of the
α-effect nucleophile hydrazine on a carboxylic acid derivative, to give a tetrahedral intermediate, 9 shows the
developing positive charge on the attacking atom.

$$H_2N \longrightarrow \underset{\underset{H}{|}}{\overset{\overset{H}{|}}{N}} \overset{\delta+}{\text{----}} \underset{\underset{X}{|}}{\overset{\overset{R}{|}}{C}} \text{===} O^{\delta-}$$

<u>9</u>

The hypothesis that electron-pair donation from the α-atom
will stabilize this transition state leads to the difficulty that the attacking atom must carry more bonds than
conventional valence bond symbolism admits. Despite this
problem, the general idea is expressed by 10 and its relationship to 9 by resonance. It is possible that transition state stabilization can be obtained in this way by
re-hybridization of the entire molecule [39b]. Klopman
et al. [39d] suggest that the α-effect arises from orbital

$$H_2N \overset{\delta+}{\text{====}} \underset{\underset{H}{|}}{\overset{\overset{H}{|}}{N}} \text{----} \underset{\underset{X}{|}}{\overset{\overset{R}{|}}{C}} \text{===} O^{\delta-}$$

<u>10</u>

splitting that raises the energy of the highest occupied
orbital of the nucleophile, thus destabilizing the initial
state relative to the transition state; on this basis the
α-effect will be observed mainly in those reactions in
which nucleophilic attack is the rate-determining step.
    Some examples of α-effect nucleophilicity may be the
result of concurrent general acid (or Lewis acid) assistance to the nucleophilic attack. The high nucleophilic

reactivity of hypochlorite may be rationalized in this
way, as suggested by 11.

$$\text{Cl}-\!\!-\!\!-\text{O}$$
$$-\text{O} \quad \overset{\text{O}}{\underset{\displaystyle R}{\overset{\|}{C}}}\!\!-\text{X}$$

11

Hydroxylamine, NH_2OH, is a remarkable α-effect nucleophile
that may function with internal general acid assistance.
Upon reaction with activated acyl groups (anhydrides, esters) a substantial fraction of the acylated hydroxylamine
is found to be O-acylhydroxylamine [40]. Jencks proposes
that the nucleophilic agent in the attack leading to O-
acylhydroxylamine is the zwitterionic form of hydroxylamine, $^{+}NH_3O^{-}$. This species combines in one molecule a
good nucleophile and a general acid. A concerted mechanism is favored, with the oxygen anion attacking the carboxyl carbon while the ammonium function polarizes the
carboxyl group by general acid interaction with the carboxyl oxygen, as in 12.

12

We now briefly consider intramolecular catalysis
("neighboring group participation") in acyl transfer reactions. Hydrolysis is the most thoroughly studied of these
reactions. Intramolecular catalysis requires the presence
of two suitably positioned functional groups, the catalytic
function and the substrate function, in the same molecule.
If intramolecular catalysis occurs, it will of course take
place concurrently with the normal intermolecular modes of
catalysis. The demonstration of intramolecular catalysis
requires that the total reaction rate be greater than that
expected for only the intermolecular reaction. This may
be a difficult experimental problem, because an estimate
of the pure intermolecular rate is not directly accessible
and is usually obtained from experiments on closely related
compounds. A favorite stratagem is to compare the reac-

tivities of ortho- and para-substituted aromatic substrates such as 13 and 14; the electronic (inductive and resonance) effects of the substituent should be about the same in the two cases, but only the ortho-substituted compound can undergo intramolecular catalysis.

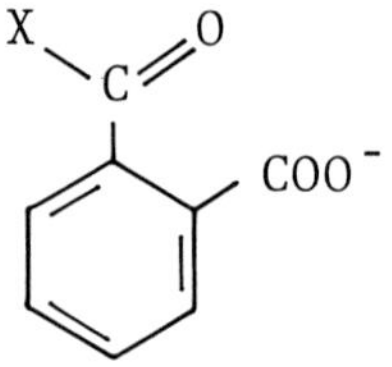

13 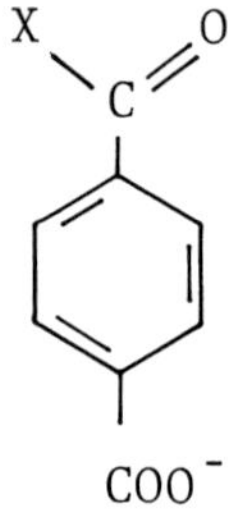   14

Another comparison is between o-hydroxy and o-methoxy compounds, as in 15 and 16; the electronic and steric effects will be (very roughly) similar, but only the hydroxy group is capable of intramolecular catalysis.

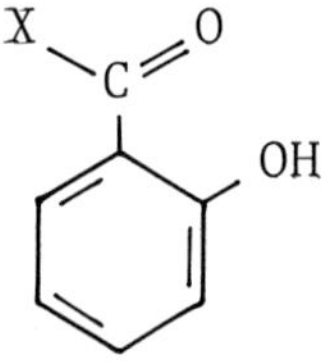

15 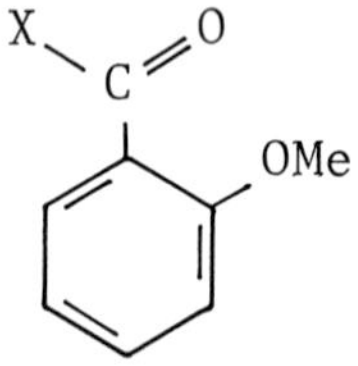   16

pH-rate profiles are valuable in identifying rate terms that may reflect intramolecular catalysis; a classic example, for the hydrolysis of aspirin, 17, is shown in Fig. 5.2, which also gives the profile for the corresponding

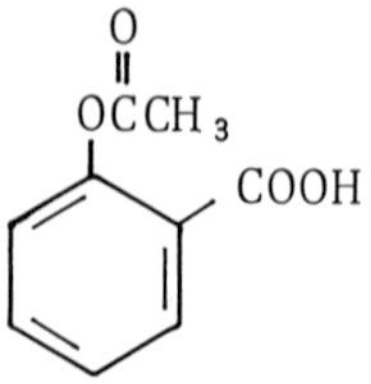

17

methyl ester. Analysis of the aspirin curve reveals that a rate term of the form k[aspirin anion] accounts for the pH-independent plateau region. This is consistent with an intramolecular catalysis of the ester hydrolysis by the carboxylate ion, and this is the accepted interpretation. The rate term is, however, also consistent with intermolecular attack by hydroxide on aspirin acid, because of the kinetic equivalence of the terms k[aspirin anion] and k´[aspirin acid][OH⁻].

The mechanisms available to intramolecular catalytic routes are the same as those of intermolecular reactions. Consider a generalized substrate containing an acyl function -COX and an ionizable catalytic function -YH.

$$
\begin{array}{ccc}
\ce{X-C=O}\\
\text{(ring with YH)}
\end{array}
\rightleftharpoons \text{H}^+ +
\begin{array}{ccc}
\ce{X-C=O}\\
\text{(ring with Y}^-)
\end{array}
\qquad (13.28)
$$

Three kinetically equivalent rate terms involving intramolecular participation are shown in Table 13.VI with representations of appropriate transition states (mechanisms). Differentiation among these possibilities can be difficult. Many examples of intramolecular general base and nucleophilic catalysis are known, and a few of these are collected in Table 13.VII. Bruice and Benkovic give a thorough treatment of intramolecularly catalyzed acyl transfers [41].

As with intermolecular catalysis, some general points can be made about the occurrence and mechanism of intramolecular catalysis:

1. Intramolecular catalysis is favored by the formation of five- or six-membered cyclic transition states. Structural factors that render the initial state conformation more like the cyclic transition state will enhance the intramolecular catalysis. Some of these factors are demonstrated in Table 13.VIII [42]. The glutarate possesses two methylene-methylene single bonds, rotation about which can oppose the juxtaposition of ester and carboxyl groups required for intramolecular catalysis. In the β,β-dimethylglutarate, steric compression by the gem-dimethyl groups reduces the conformational freedom, and this is reflected in a higher rate. The succinate possesses one bond fewer than the glutarate, leading to restricted rotamer distribution and a rate increase. Further "freezing" of the two groups in opposition, as with the maleate and the 3,6-

endoxo-Δ^4-tetrahydrophthalate, gives large rate enhance-
ments.
    2.   Intramolecular nucleophilic attack is possible
only if the nucleophilic group can approach the acyl group
roughly perpendicularly to the plane of the acyl group
[43].

Table 12.VI Kinetically Equivalent Intramolecular
Catalysis Mechanisms

Transition state	Mechanism	Rate term form
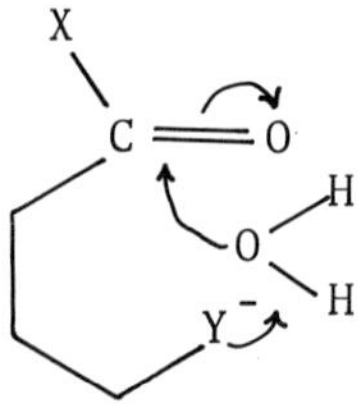	intramolecular general base	[anion][H_2O]
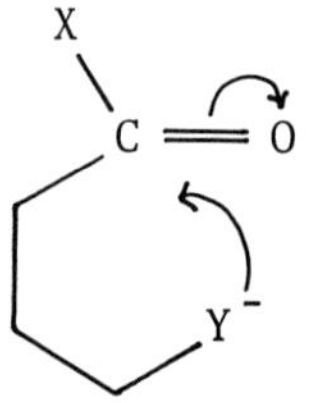	intramolecular nucleophilic	[anion]
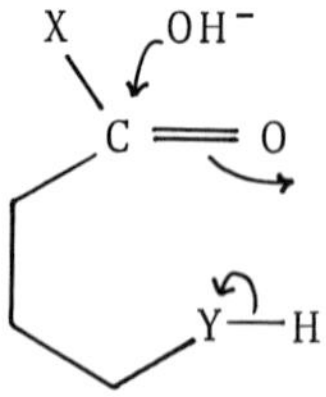	intramolecular general acid- intermolecular nucleophilic	[acid][OH^-]

    The concept is shown in 18. This orientation maximizes
orbital overlap between the incoming nucleophile and the
π-orbital of the carbon-oxygen bond. The very facile
hydrolysis of phthalamic acid can be accounted for by
intramolecular nucleophilic catalysis with perpendicular
attack [44]. A molecular model shows that the carboxamide
and carboxylic acid groups cannot both lie in the plane of

the aromatic ring. If the acid group is in the plane of
the ring, the amide group is constrained, sterically, to
be about perpendicular to it; thus their mutual orientation

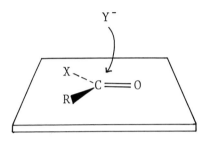

<u>18</u>

is suitable for perpendicular attack, as in <u>19</u>, and the
hydrolysis (via the intermediate anhydride) is rapid.

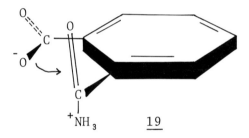

<u>19</u>

This interpretation is strengthened by the behavior of
o-carboxyphthalimide, <u>20</u>, in which the imide structure
requires that the carboximide group lie in the plane of
the ring.

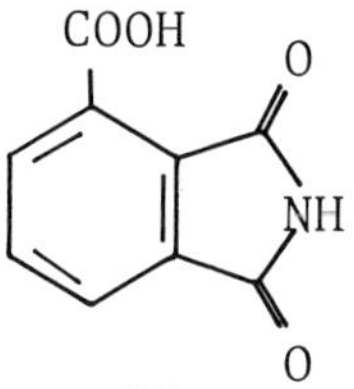

COOH

NH

<u>20</u>

Table 13.VII Examples of Intramolecularly Catalyzed
Hydrolysis Reactions

Substrate	Catalytic function

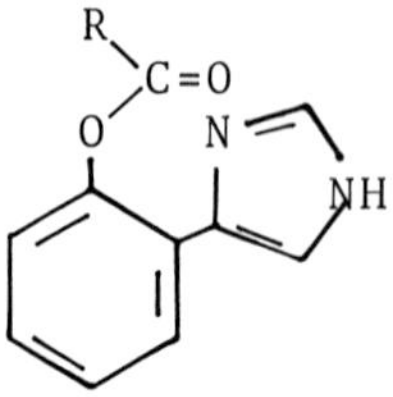

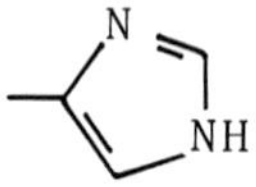

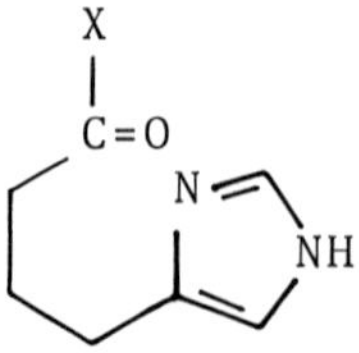

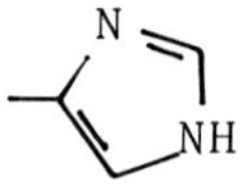

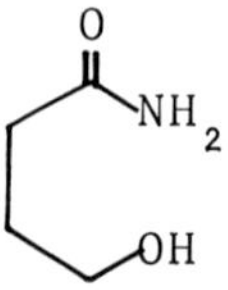

R–C=O, O–N=... NH (imidazole-substituted aryl ester)

N=...NH (imidazole)

X–C=O with imidazole chain

N=...NH (imidazole)

O=C–NH₂ with OH chain

$-OH, \ -O^-$

CONH₂, OH (salicylamide)

$-O^-$

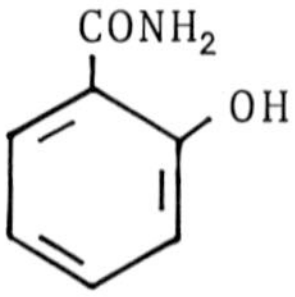

CO₂Me, CHO

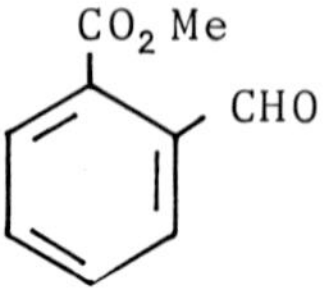

$$-\underset{\underset{O^-}{|}}{C}H-OH$$

Mechanism	Comment	Ref.
nucleophilic	R = CH$_3$ or aryl	a,b
nucleophilic	X = OAr or S-n-Pr	c,d
nucleophilic	intramolecular general acid-nucleophilic catalysis by -OH group	e
general base		f
nucleophilic	intramolecular attack by the anion of the hydrated formyl group	g

545

Table 13.VII continued

Substrate	Catalytic function

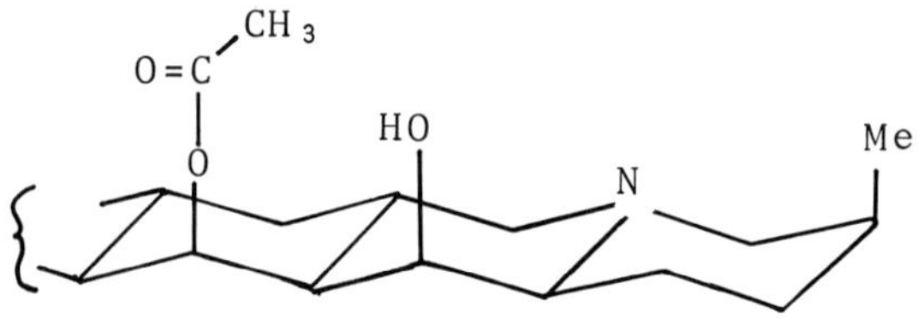

(chemical structure with $O=C$, CH_3, O, HO, N, Me) — $-OH$ and $-\overset{\mid}{N}-$

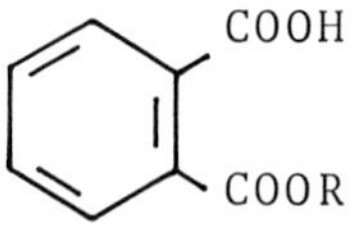

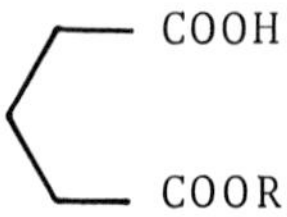

(benzene ring with COOH and COOR) — $-COOH$ or $-COO^-$

(chain with COOH and COOR) — $-COO^-$

(benzene ring with COOH and $OCOCH_3$) — $-COO^-$

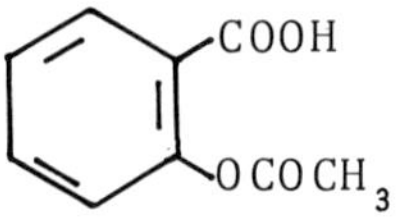

[a] G. L. Schir and T. C. Bruice, J. Am. Chem. Soc., 80, 1173 (1958).

[b] U. K. Pandit and T. C. Bruice, J. Am. Chem. Soc., 82, 3386 (1960).

[c] T. C. Bruice and J. M. Sturtevant, J. Am. Chem. Soc., 81, 2860 (1959).

[d] T. C. Bruice, J. Am. Chem. Soc., 81, 5444 (1959).

[e] T. C. Bruice and F.-H. Marquardt, J. Am. Chem. Soc., 84, 365 (1962).

[f] T. C. Bruice and D. W. Tanner, J. Org. Chem., 30, 1668 (1965).

[g] M. L. Bender, J. A. Reinstein, M. S. Silver, and R. Mikulak, J. Am. Chem. Soc., 87, 4545 (1965).

[h] S. M. Kupchan, S. P. Eriksen, and Y. T. Shen, J. Am. Chem. Soc., 85, 350 (1963).

[i] J. W. Thanassi and T. C. Bruice, J. Am. Chem. Soc., 88, 747 (1966).

Mechanism	Comment	Ref.
general acid-general base	intramolecular general acid (by hydroxy)-general base (by amino) catalysis of inter-molecular nucleophilic attack	h
nucleophilic	catalysis by -COOH if pK_a of ROH > 13.5; catalysis by -COO$^-$ if pK_a of ROH < 13.5	i
nucleophilic	R = aryl	j
general base		k

[j] E. Gaetjens and H. Morawetz, J. Am. Chem. Soc., **82**, 5328 (1960).
[k] A. R. Fersht and A. J. Kirby, J. Am. Chem. Soc., **89**, 4857 (1967).

Table 13.VIII   Relative Rates of Intramolecular Catalysis
of Ester Hydrolysis by Carboxylate Groups[a]

Ester[b]	Relative rate[c]
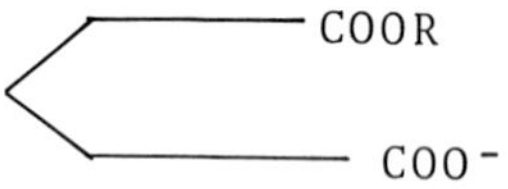	1.0
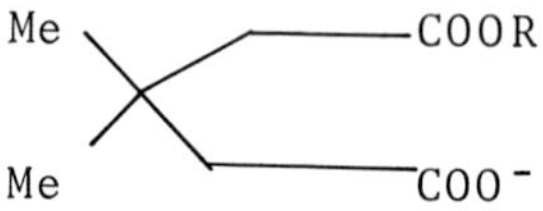	19.3
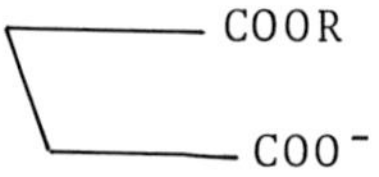	232
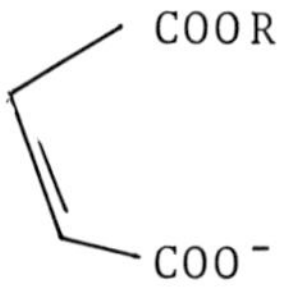	10,300
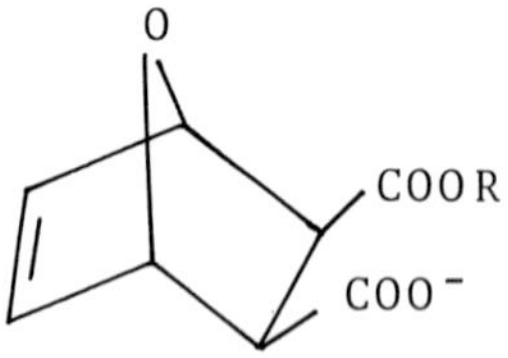	53,100

[a]From Ref. [42].
[b]R = p-bromophenyl or p-methoxyphenyl.
[c]In 50% dioxane at 30°.

Perpendicular attack by the carboxylate on the imide is
impossible in this compound, with the result that intramolecular nucleophilic catalysis is not observed; instead
an intramolecular general base or general acid catalysis
occurs [45].

3. Whether an intramolecular catalysis proceeds by
nucleophilic attack or a general base mechanism is largely
determined by the relative tenacities of the nucleophilic
group and the leaving group. For a given nucleophile, the
better the leaving group the greater the tendency toward
nucleophilic catalysis. Fersht and Kirby [46] argue that
the borderline between nucleophilic and general base intramolecular catalysis depends upon whether the leaving
group remains attached to the nucleophilic group. The
point can be made by comparing the hydrolyses of an acylsalicylate and a phthalate monoester. Equation (13.29)
shows intramolecular nucleophilic catalysis by an acylsalicylate, giving a mixed anhydride via a tetrahedral
intermediate.

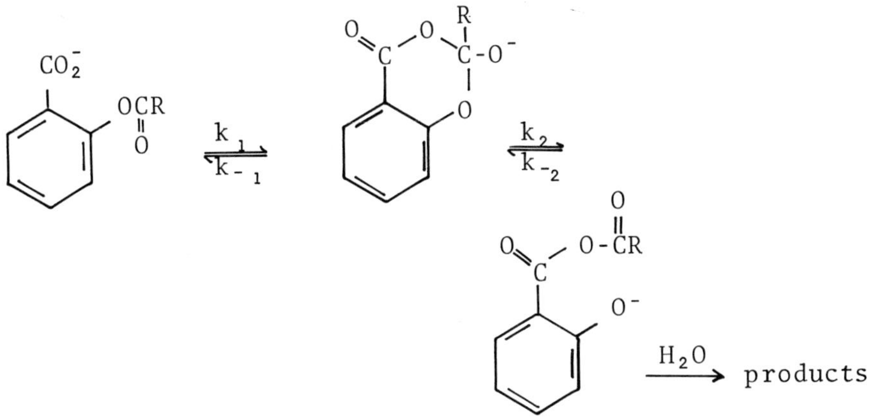

$$(13.29)$$

In this presumed reaction (formation of the mixed anhydride) the leaving group remains attached to the molecule.
This means that the reverse reaction (i.e., the k_{-2} step)
is also an intramolecular nucleophilic attack, and this
circumstance tends to cancel the advantage gained in the
k_1 step. In the case of a phthalate ester, shown in Eq.
(13.30), the leaving group is lost. For this reaction
the k_{-2} step is an intermolecular nucleophilic attack,
which is entropically less favorable than the intramolecular k_1 step. It is therefore expected that nucleophilic
catalysis is relatively more favorable in such a case than
it would be for Eq. (13.29) or for an intermolecular reaction. Fersht and Kirby [46] estimate that, for reactions
of oxyanions with esters, nucleophilic catalysis can occur
in systems like Eq. (13.30) (where the leaving group is
lost) when the leaving group is up to 6.5 pK units more

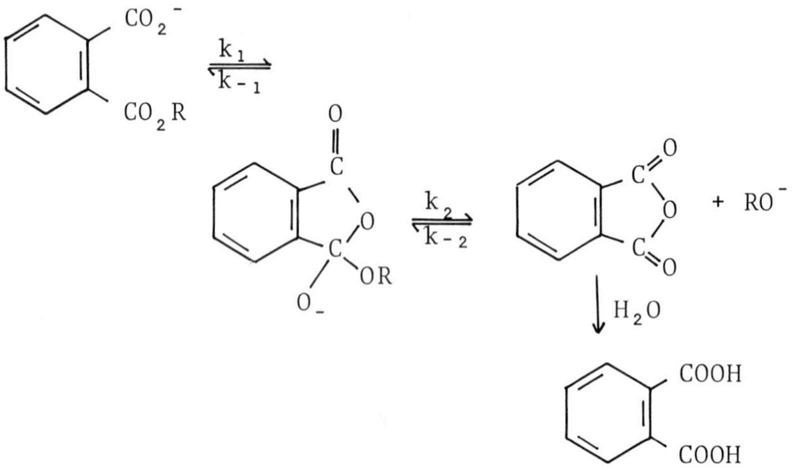

$$(13.30)$$

basic than the nucleophile, but that the borderline between nucleophilic and general base catalysis in systems
where the leaving group is not lost [as in Eq. (13.29)]
occurs when the pK's of the leaving group and the nucleophile are about equal.

   The efficiency of intramolecular catalysis relative to
intermolecular catalysis is difficult to assess.   One technique, already described as a method to detect intramolecular catalysis, is to make a comparison with an analog incapable of intramolecular catalysis.   Thus p-nitrophenyl
5-nitrosalicylate, 21, hydrolyzes about 2500 times faster
than p-nitrophenyl 2-methoxy-5-nitrobenzoate, 22 [47].
This type of comparison is between intramolecular and

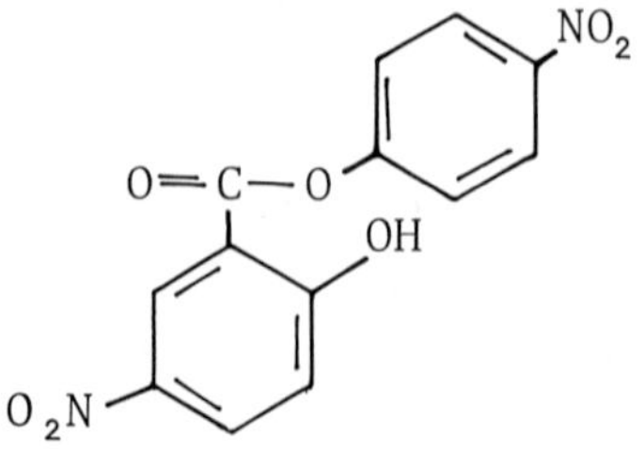

21

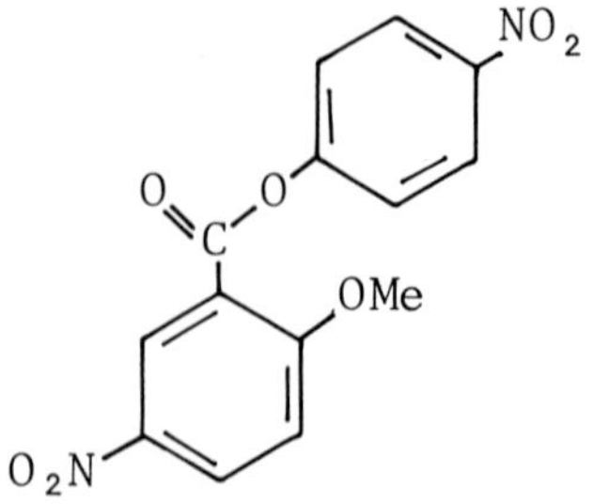

22

intermolecular reactions of somewhat different character,
and another approach is to compare the intramolecular reaction with an intermolecular reaction in which the reacting
groups closely resemble those in the intramolecular reaction. This method has its own disadvantage, because the
intermolecular reaction will usually be overall second
order, in accordance with the rate equation

$$v = k_2 [R-X][Y] \qquad (13.31)$$

whereas the analogous intramolecular reaction will be
first order,

$$v = k_1 [X-R-Y] \qquad (13.32)$$

First-order and second-order rate constants have different
dimensions and cannot be directly compared, so the following interpretation is made [48]. The ratio k_1/k_2 has the
units of moles/liter and is the molar concentration of Y
in Eq. (13.31) that would be required for the intermolecular reaction to proceed (under pseudo-first-order conditions) as fast as the intramolecular reaction. Jencks
[49] has pointed out that as a quantitative comparison
this method is arbitrary in its use of the molar concentration scale, and the unitary standard state (see Sec.
2.2) might be preferable. At any rate, comparisons made
in this way indicate that intramolecular catalysis can be
far too efficient to be accounted for simply by an increase
in "local concentration" of the nucleophile [50]. For example, the nucleophilic catalysis of phenyl acetate hydrolysis by tertiary amines has been studied as an intermolecular reaction [Eq. (13.33)] and an intramolecular reaction
[Eq. (13.34)] [51].

$$CH_3COOPh + NMe_3 \xrightarrow{-PhO^-} CH_3\overset{\displaystyle O}{\overset{\displaystyle \|}{C}}-NMe_3^+ \xrightarrow{OH^-} CH_3COOH + NMe_3 \quad (13.33)$$

$$\underset{\displaystyle \text{COOPh}}{} \quad \underset{\displaystyle \text{N-Me}}{\overset{\displaystyle Me}{|}} \xrightarrow{-PhO^-} \overset{\displaystyle O}{\overset{\displaystyle \|}{C}}-\underset{\displaystyle \text{N-Me}}{\overset{\displaystyle Me}{|}}_+ \xrightarrow{OH^-} HOOC-(CH_2)_3-NMe_2 \quad (13.34$$

The ratio k(intramol.)/k(intermol.) is 1260 $\underline{M}$ for these
reactions, clearly showing that the intramolecular catalysis is not merely the result of a proximity effect, because it is not possible to achieve such a concentration
[52].
     Storm and Koshland have observed very large intra-

molecular/intermolecular ratios for some lactonization
reactions, which they attribute to fine control of the
mutual orientation of reacting orbitals within the intramolecular system; they call this concept orbital steering
[53]. Even larger rate accelerations have been achieved
by stereopopulation control, which is a narrowing of the
distribution of conformational populations, with the fortuitous elimination of unproductive conformers. Some examples are shown in Table 13.VIII. A dramatic illustration is the relative rate of intramolecular acid-catalyzed
lactonization of 23 and 24 [54].

23   24

Compound 24 reacts 10^{11} times faster than 23. This level
of enhancement is comparable with that achieved by enzymes.

Acid Catalysis and Acylium Ions. The hydrogen ion-catalyzed hydrolysis and formation of esters have been carefully studied [55]. These reactions are the reverse of
each other, so if the mechanism is known for one of them.
that for the reverse reaction can be inferred by means of
the principle of microscopic reversibility. For most
esters in acidic aqueous solution the evidence is consistent with the $A_{Ac}2$ pathway, the reaction rates being
first order in substrate and in hydrogen ion. Concurrent
oxygen exchange and hydrolysis implicates a tetrahedral
intermediate, and it was earlier noted that this intermediate is thought to be the same one as occurs in alkaline hydrolysis [16]. The mechanism can be thought of as
a nucleophilic attack (by water or an alcohol) on the
protonated substrate, which will be more susceptible to
attack than is the unprotonated form. A possible mechanism is shown on the next page. This can be read in either
direction. The protonation equilibria, Eqs. (13.35) and
(13.37), should be very fast, and the slow steps are expected to be the attack of the nucleophile on the protonated substrate. Letting S represent unprotonated substrate (either acid or ester), the rate equation can be

$$R\text{-}\overset{\displaystyle O}{\overset{\|}{C}}\text{-}OR' + H^+ \rightleftharpoons R\text{-}\overset{\displaystyle {}^+OH}{\overset{\|}{C}}\text{-}OR' \qquad (13.35)$$

$$R\text{-}\overset{\displaystyle {}^+OH}{\overset{\|}{C}}\text{-}OR' + H_2O \underset{\text{slow}}{\rightleftharpoons} \underset{+OH_2}{\overset{OH}{\underset{|}{\overset{|}{R\text{-}C\text{-}OR'}}}} \rightleftharpoons \underset{OH}{\overset{OH}{\underset{|}{\overset{|}{R\text{-}\overset{+}{C}\text{-}OHR'}}}} \underset{\text{slow}}{\rightleftharpoons} R\text{-}\overset{\displaystyle {}^+OH}{\overset{\|}{C}}\text{-}OH + R'OH \qquad (13.36)$$

$$R\text{-}\overset{\displaystyle {}^+OH}{\overset{\|}{C}}\text{-}OH \rightleftharpoons R\text{-}\overset{\displaystyle O}{\overset{\|}{C}}\text{-}OH + H^+ \qquad (13.37)$$

written $v = k_H[S][H^+]$ or $k_H'[SH^+]$, where the second form expresses the mechanistic idea of Eq. (13.36). These kinetically equivalent terms are related by the acid dissociation constant $K_a = [H^+][S]/[SH^+]$, such that $k_H' = k_H K_a$. Typical values of k_H for the hydrolysis or formation of aliphatic esters are about 10^{-4} $M^{-1}\text{-sec}^{-1}$, and pK_a values for acids [56] and esters [57] are about -6 to -8. This means that k_H' must be 10^2-10^4 sec^{-1}. Though large, these values seem reasonable [58].

Studies on ester hydrolysis in strongly acid solutions, where the activity of water can be varied, indicate that two molecules of water are included in the transition state [57,59]. The second water molecule may function as a general base to increase the nucleophilicity of the attacking water molecule, as in 25 [59]. A cyclic transition state, 26, has also been proposed [57].

25   26

Amides are hydrolyzed in acid solution with overall second-order kinetics. The rate increases with acid concentration, finally reaching a maximum and then slowly decreasing with high concentrations of acid. This behavior has been attributed to two effects: (1) The rate increase is a consequence of acid catalysis via pre-protonation of the amide (exactly as described earlier for ester hydrolysis); since amides are relatively basic, protonation becomes essentially complete in the region

2.5-5 M acid. (2) The subsequent rate decrease is caused
by the decrease in activity of the water [7].

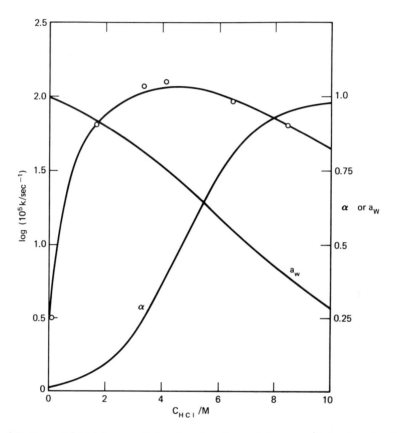

Fig. 13.3. Kinetics of hydrolysis of benzamide in hydrochloric acid solutions at 100.4° (experimental points and
left ordinate). Right ordinate: variation of water activity (a_w) and fraction of benzamide protonated (α), both
at 25°, in hydrochloric acid solutions.

Figure 13.3 shows data for the hydrolysis of benzamide
in hydrochloric acid solutions [60]. Bunton et al. [60]
have applied the following treatment. Let A and AH^+ represent unprotonated and protonated amide, respectively. Then
the assumed rate equation is

$$v = (k[A][H^+] + k'[AH^+])a_w \qquad (13.38)$$

where a_w is the water activity. It might seem that these
two rate terms are kinetically equivalent, and they would

be in dilute aqueous solution where the extent of protonation is determined by pH. In concentrated acid solutions,
however, the fraction of amide in the protonated form, α,
is determined by an acidity function other than pH, as
discussed in Sec. 4.4. Let $A_T = [A] + [AH^+]$; then $\alpha =
[AH^+]/A_T$ and $(1-\alpha) = [A]/A_T$. The experimental rate equa-
tion is $v = k_{obs}A_T$. Combining these relations,

$$k_{obs} = \{k(1-\alpha)[H^+] + k'\alpha\}a_w \qquad (13.39)$$

Figure 13.3 shows the variation of a_w with hydrochloric
acid concentration [61] and of α for benzamide over the
same range in acid concentrations; α was calculated from
the pK_a value -1.74 and the amide acidity function H_A.
The line correlating the experimental rate data in Fig.
13.3 was calculated with Eq. (13.39) and the values [60]
$k = 3.75 \times 10^{-4}$ M^{-1}-sec^{-1} and $k' = 16 \times 10^{-4}sec^{-1}$. The
fit is quite satisfactory.

Bunton, O'Connor, and Turney [60] have proposed two
reaction pathways on the basis of the preceding kinetic
analysis. These are shown as 27 (the k term) and 28 (the
k' term).

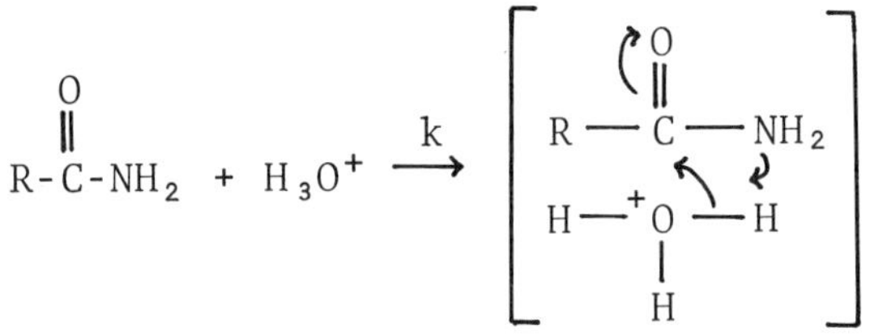

$$\begin{array}{c} O \\ \parallel \\ R\text{-}C\text{-}NH_2 \end{array} + H_3O^+ \xrightarrow{k} \left[\begin{array}{c} O \\ \parallel \\ R - C - NH_2 \\ \\ H - {}^+O - H \\ \mid \\ H \end{array} \right] \qquad (13.40)$$

<u>27</u>

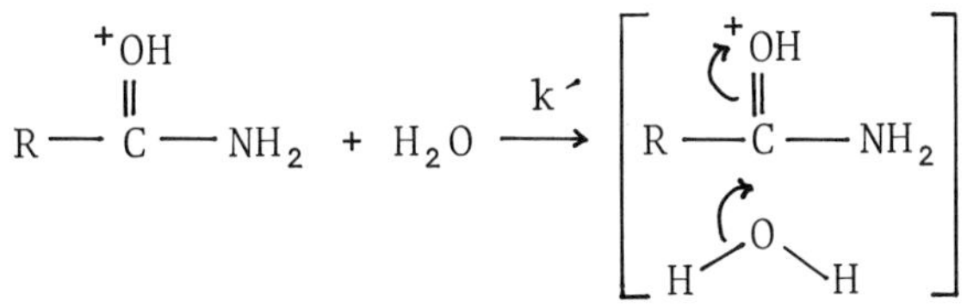

$$\begin{array}{c} {}^+OH \\ \parallel \\ R - C - NH_2 \end{array} + H_2O \xrightarrow{k'} \left[\begin{array}{c} {}^+OH \\ \parallel \\ R - C - NH_2 \\ \\ H \diagup {}^O \diagdown H \end{array} \right] \qquad (13.41)$$

<u>28</u>

Since acid catalysis of acyl transfer reactions seems
usually to occur via protonation of the substrate with enhancement of the electrophilicity of the acyl carbon, it
might be expected that metal ions, which are Lewis acids,
might catalyze these reactions. Such catalysis has been
observed for numerous systems, but it appears to be impor-

tant only when the metal ion forms a chelate structure with
the substrate. Thus Cu(II) catalyzes the hydrolysis of
amino acid esters in the presence of glycine, and the
catalysis was postulated to take place within a copper-
glycine-ester complex, 29 [62].

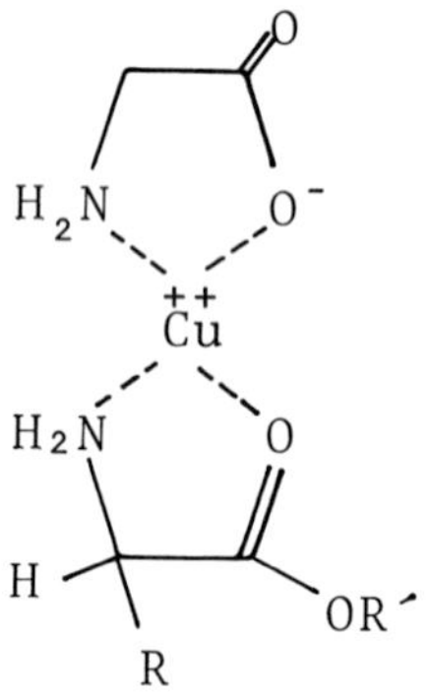

29

The hydrolysis of glycine esters is catalyzed by cobalt
(III), and it was shown that the chelate form 30 hydrolyzes much more rapidly than does the nonchelated complex
31 [63].

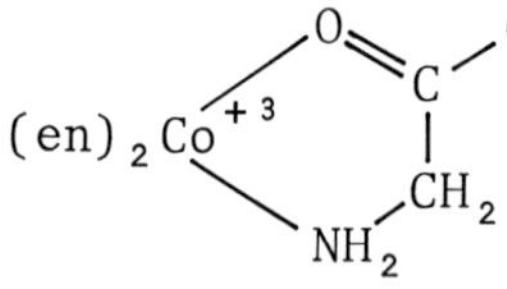

30 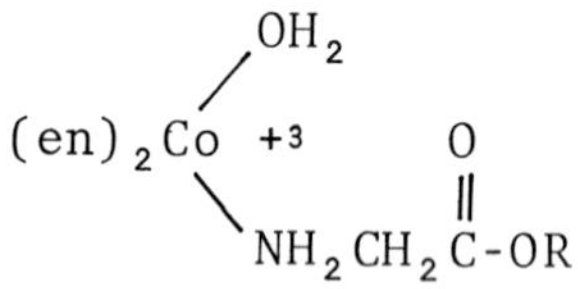   31

It is probably not a chelate structure itself that is required, but rather a strong coordination to the carbonyl
oxygen, and chelation promotes this interaction. Metal
binding to the ether oxygen of an ester could also activate
the acyl group toward nucleophilic attack. A possible
example is the hydrolysis of 8-acetoxyquinoline 5-sulfate
catalyzed by zinc pyridinecarboxaldoxime anion [64]. This
catalysis is postulated to occur within a mixed complex,
perhaps as in 32, with the zinc serving largely to bring
together the catalyst and the acyl group, but some enhancement of reactivity may occur by the electrophilic metal
ion.

Although most esters undergo $A_{Ac}2$ acid hydrolysis,
as described earlier, the other mechanisms ($A_{Ac}1$ and $A_{Al}1$)

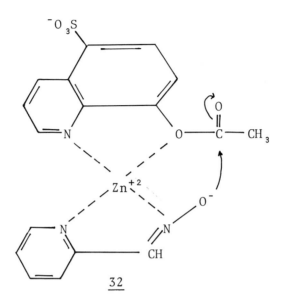

32

can be observed with appropriate substrates or conditions.
The $A_{Al}1$ reaction, it will be recalled, is an S_N1 reaction
with an ester leaving group; it is promoted by an alkyl
structure favoring carbonium ion stability. Esters of
tertiary alcohols, such as tert-butyl acetate, hydrolyze
by the $A_{Al}1$ route.

$$CH_3\overset{O}{\overset{\|}{C}}\text{-OCMe}_3 + H^+ \rightarrow CH_3COOH + Me_3C^+ \qquad (13.42)$$

The carbonium ion reacts rapidly with water yielding the
alcohol.
      More interesting in our present concern with acyl trans-
fer is the $A_{Ac}1$ hydrolysis of esters. This general reaction
occurs as shown in Eqs. (13.43) to (13.46), which, being
read in the reverse direction, show also the $A_{Ac}1$ esterification of an acid.

$$R\text{-}\overset{O}{\overset{\|}{C}}\text{-OR}' + H^+ \rightleftharpoons R\text{-}\overset{+OH}{\overset{\|}{C}}\text{-OR}' \qquad (13.43)$$

$$R\text{-}\overset{+OH}{\overset{\|}{C}}\text{-OR}' \rightleftharpoons R\text{-}\overset{+}{C}\text{=}O + R'OH \qquad (13.44)$$

$$\overset{+}{R-C}=O + H_2O \rightleftharpoons R-\overset{\overset{+OH}{\|}}{C}-OH \tag{13.45}$$

$$R-\overset{\overset{+OH}{\|}}{C}-OH \rightleftharpoons R-\overset{\overset{O}{\|}}{C}-OH + H^+ \tag{13.46}$$

The intermediate $R-\overset{+}{C}=O$ is called an <u>acylium ion</u>. Acyl groups that stabilize acylium ions by their electronic or steric features will promote $A_{Ac}1$ reaction. Thus methyl mesitoate (methyl 2,4,6-trimethylbenzoate) hydrolyzes in sulfuric acid solutions by the $A_{Ac}1$ mechanism [65]. The absence of <u>carbonyl-^{18}O</u> exchange and the observation of positive entropies of activation are consistent with this assignment [66]. Methyl mesitoate can be prepared by pouring a solution of mesitoic acid in 100% sulfuric acid into methanol [67]; this method is based on the pre-formation of the acylium ion according to

$$RCOOH + 2H_2SO_4 \rightleftharpoons RCO^+ + H_3O^+ + 2HSO_4^- \tag{13.47}$$

as revealed by cryoscopic measurements [68].

Acylium ion formation is not limited to special structures, but can be achieved with many substrates in highly acidic solutions [69]. Nuclear magnetic resonance spectroscopy has been very useful in studying these species. The positions of the protonation and dehydration equilibria, Eqs. (13.48) and (13.49), can be established by NMR.

$$RCOOH + H^+ \rightleftharpoons RCOOH_2^+ \tag{13.48}$$

$$RCOOH_2^+ \rightleftharpoons RCO^+ + H_2O \tag{13.49}$$

Acetic acid, for example, exhibits the equalities $[CH_3COOH] = [CH_3COOH_2^+]$, in 77% H_2SO_4, and $[CH_3COOH_2^+] = [CH_3CO^+]$ in 15% SO_3 in H_2SO_4 [70]. The transformation from > 90% $RCOOH_2^+$ to > 90% RCO^+ occurs within a narrow (~4%) range of SO_3 concentration. A general method of esterification has been suggested in which a solution of carboxylic acid in 25% SO_3 in H_2SO_4 is poured into an alcohol [70].

If acylium ions can exist as relatively stable species in strongly acid solution, then the unimolecular $A_{Ac}1$ mechanism should become competitive with the bimolecular $A_{Ac}2$ pathway in ester hydrolysis as the reaction medium is made more acidic. This behavior is in fact observed. Figure 13.4 summarizes some of the results of Yates and McClelland [59] on the acid hydrolysis of acetate esters. Three classes of behavior are seen. Primary acetates hydrolyze by the $A_{Ac}2$ mechanism, the rate passing through a maximum

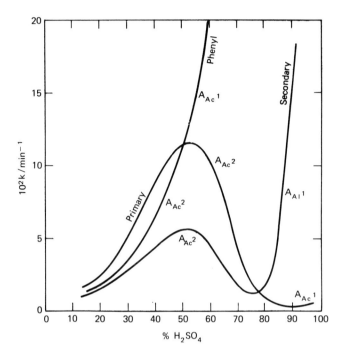

Fig. 13.4. Schematic representation of typical dependences of rate and mechanism of hydrolysis in aqueous sulfuric acid for primary, secondary, and phenyl acetates [59] (at 25°).

as the water activity decreases. A very low minimum rate is followed by a slight increase above 90% H_2SO_4 due to $A_{Ac}1$ reaction. Secondary acetates behave much the same way at moderate acidity, hydrolyzing by the $A_{Ac}2$ route, but a very sharp increase in rate, ascribable to the incursion of $A_{Al}1$ hydrolysis, occurs at high acidity; benzyl acetates behave similarly. Phenyl acetates do not show a rate maximum, but appear to undergo a mechanism change from $A_{Ac}2$ to $A_{Ac}1$ at moderate acidity. These mechanism assignments were based on the number of water molecules involved in the rate-determining step (2 for $A_{Ac}2$, 0 for $A_{Ac}1$), as determined from acidity function correlations [58].

The reactions of acyl halides present a complicated picture. Some authors have treated these in terms of a tetrahedral intermediate mechanism, whereas others interpret them as alkyl halide substitutions, that is, as S_N1 or S_N2 reactions. The S_N1 (acylium ion or ionization)

mechanism is favored by ionizing media and the absence of
strong nucleophiles. It is written as Eq. (13.50) for
the hydrolysis of an acid halide,

$$RCOX \underset{-X^-}{\overset{}{\rightleftharpoons}} \overset{+}{RC=O} \xrightarrow{H_2O} RCOOH + H^+ \tag{13.50}$$

with the formation of the acylium ion being rate determining. Only two acid halide hydrolyses have been found to
be acid catalyzed; these are the hydrolyses of benzoyl
fluoride [71] and mesitoyl chloride [72]. By analogy with
the known acid catalysis of benzyl fluoride hydrolysis, it
is postulated that catalysis of benzoyl fluoride takes
place by electrophilic assistance to removal of the fluoride ion [71]. Fluoride is a particularly strong hydrogen
bond acceptor. Mesitoyl chloride is sterically hindered
and therefore especially susceptible to acylium ion formation; its acid-catalyzed hydrolysis was interpreted, as
with benzoyl fluoride, as assistance to the departure of
the leaving group in an S_N1 reaction [72]. Mercuric ion
catalyzes the hydrolysis of benzoyl chloride, probably by
the same mechanism [73].

Hudson and Moss [74] favored a hydrated acylium ion
mechanism in accounting for mixed solvent studies on acetyl
chloride hydrolysis. This combines a tetrahedral intermediate with a carbonium ion intermediate mechanism.

$$\underset{\substack{\| \\ }}{\overset{O}{R-C-X}} \overset{H_2O}{\rightleftharpoons} \underset{\substack{| \\ OH}}{\overset{OH}{R-C-X}} \overset{-X^-}{\rightleftharpoons} \underset{\substack{| \\ OH}}{\overset{OH}{R-C+}} \rightarrow RCOOH + H^+ \tag{13.51}$$

Their kinetic data showed carbonium ion character in the
transition state (m = 0.81 in a correlation with $\underline{Y}$ values)
and markedly different rates in dioxane-water and acetone-
water mixtures of equal $\underline{Y}$ [75].

Bender and Chen [72] proposed rate-determining tetrahedral intermediate formation in base-catalyzed acid chloride hydrolysis. The reactions of acid chlorides with phenols were suggested by Briody and Satchell [76] to be simple S_N2 displacements, on the basis of salt and structural
effects. Sneen and Larsen [77] interpreted some benzoyl
chloride reactions with their concept of the ion-pair
intermediate common to all S_N reactions. It appears that
several mechanisms are accessible to acyl halides, and
their experimental differentiation is difficult. Because
their leaving groups are good, acyl halide reactions
probably are more akin to those of alkyl halides than of

most carboxylic acid derivatives.

An interesting acylium ion system is the mixture acetic acid-acetic anhydride-perchloric acid [78]. Both
acetic acid and acetic anhydride are dissociating solvents
according to the equilibria HOAc $\rightleftharpoons$ H$^+$ + OAc$^-$ and Ac$_2$O $\rightleftharpoons$
Ac$^+$ + OAc$^-$, respectively, where Ac$^+$ is the acetylium ion
CH$_3$CO$^+$. A mixture of perchloric acid in acetic acid,
though a very strong acid [79], does not act as an acetylating agent. When acetic anhydride is added to this solution, immediate O- or C-acetylation can take place [80].
When acetyl chloride is added to silver perchlorate, silver chloride precipitates and the remaining solution is a
powerful acetylating agent [81]. These observations can
be accounted for by assuming the presence of the acetylium
ion, which arises in the acetic acid-acetic anhydride-
perchloric acid system from the following equilibria [82]:

$$H^+C\ell O_4^- + Ac_2O \rightleftharpoons Ac_2OH^+C\ell O_4^- \qquad (13.52)$$

$$Ac_2OH^+C\ell O_4^- \rightleftharpoons Ac^+C\ell O_4^- + HOAc \qquad (13.53)$$

or, alternatively, the solvated proton H$_2$OAc$^+$ may transfer
a proton to Ac$_2$O. Acetyl perchlorate, Ac$^+$CℓO$_4^-$, is also
formed from the reaction of acetyl chloride and silver
perchlorate. Adding acetic acid to solutions of acetyl
perchlorate in nitromethane resulted in decreased yields
of acetylated products, indicating that Ac$^+$ is being converted to Ac$_2$OH$^+$ according to Eq. (13.53), and that Ac$^+$
is a more powerful acetylating agent than is Ac$_2$OH$^+$. These
systems are, in fact, similar to those employed in Friedel-
Crafts acylations (Chap. 7), and these acetylium ion solutions are capable of carrying out such electrophilic aromatic substitutions [81].

Structure and Reactivity. Esterification and ester hydrolysis have provided some of the most useful data leading
to correlations between structure and chemical reactivity.
Linear free energy relationships, especially the Hammett
and Taft treatments, have been developed and tested largely
with carboxylic acid derivative reactions. These relationships are described in Sec. 4.2, where some examples are
given of structure-reactivity correlations for ester reactions. Carboxylic acid derivatives are attractive substrates for such investigations because their reactions
are usually so "clean," in the sense that side reactions
or subsequent reactions seldom occur.

In attempting to establish relationships between chemical structure and reactivity, it is clear that relative
rates can be simply interpreted only if the reactions being
compared proceed by the same mechanism, and, further, that

the rate terms being compared have the same form. Let us
consider the bimolecular acid-catalyzed esterification of
aliphatic carboxylic acids. The dissociation constants of
these acids are all of the order 10^{-5}, the small variations
presumably being caused by minor differences in polar effects. The variations in esterification rates for these
acid are quite large, however, so that polar effects are
not responsible. Steric effects are therefore implicated
[84]. Table 13.IX gives second-order rate constants for
the acid-catalyzed esterification of some aliphatic acids
in methanol. Newman has drawn attention to the conformational role of the acyl group in limiting access to the
carboxyl carbon. He represents maximal steric hindrance to
attack as arising from a coiled conformation, shown for
n-butyric acid in $\underline{33}$.

Table 13.IX   Kinetics of Acid-Catalyzed Esterification
of Aliphatic Acids[a]

Acid	$10^3 k / \underline{M}^{-1}\text{-sec}^{-1}$[b]	$\dfrac{k_{RCOOH}}{k_{CH_3COOH}}$	Six-number[c]
CH_3COOH	132	1.00	0
CH_3CH_2COOH	111	0.84	0
$CH_3CH_2CH_2COOH$	65.2	0.495	3
$(CH_3)_2CHCOOH$	44.0	0.333	0
$(CH_3)_3CCOOH$	4.93	0.037	0
$(CH_3)_2CHCH_2COOH$	15.4	0.117	6
$(CH_3)_3CCH_2COOH$	3.09	0.023	9
$(CH_3)_2CHCH_2CH_2COOH$	63.4	0.48	3
$(CH_3)_3CCH_2CH_2COOH$	61.1	0.46	3
$CH_3CH_2CH(CH_3)COOH$	13.1	0.099	3
$(CH_3)_3CCH(CH_3)COOH$	0.0817	0.00062	9
$(CH_3)_3CC(CH_3)_2COOH$	0.0170	0.00013	9
$(CH_3)_3CCH_2CH(CH_3)COOH$	2.03	0.0015	3
$(CH_3)_3CHCH(C_2H_5)COOH$	0.0780	0.00059	9
$(CH_3CH_2)_3CCOOH$	0.0214	0.00016	9
$[(CH_3)_2CH]_2CHCOOH$	nil		12
$(n-C_4H_9)_2CHCOOH$	1.10	0.0083	6
$(CH_3)_3CCH(C_2H_5)COOH$	nil		12

[a]From Ref. [84].
[b]At 40° in 0.005 $\underline{M}$ HCl in methanol.
[c]See text.

$$\begin{array}{c}
\text{OH} \\
\diagdown \\
\underset{2}{\text{C}} = = \underset{1}{\text{O}} \\
\diagup \qquad \qquad \\
\underset{3}{\text{C}} \qquad \qquad \qquad \text{H}_6 \\
\diagdown \qquad \qquad \diagup \\
\underset{4}{\text{C}} \text{---} \underset{5}{\text{C}} \text{---H}_6 \\
| \\
\text{H}_6
\end{array}$$

<u>33</u>

Maximum interference arises from atoms in the six-position, numbering from the carbonyl oxygen. Thus n-butyric acid, which has three atoms in the six-position, is said to have a "six-number" of 3. The higher the six-number, the greater the possible steric resistance to ester formation [84,85]. This is called Newman's rule-of-six. The rates in Table 13.IX show that this rule is severely limited as a quantitative tool, but a plot of relative rates k_{RCOOH}/k_{CH_3COOH} against six-numbers does reveal that a smooth envelope of maximum rates is developed; individual acids may react much slower than this, but the rule gives a rough guide for estimated maximum esterification rates.

Taft developed an empirical quantitative measure of the steric effect in these reactions in his separation of polar and steric factors in aliphatic reactions [86]. The basic assumption, as in Newman's treatment, is that acid-catalyzed reactions are not markedly susceptible to polar effects, so that relative rate differences are ascribed to steric effects. A steric substituent constant E_s is defined by

$$E_s = \log \frac{k_{RCOOR'}}{k_{CH_3COOR'}} \qquad (13.54)$$

Thus E_s expresses the steric effect of the RCO- group relative to the CH_3CO- (i.e., acetyl) group. A set of E_s values has also been determined for ortho-substituted benzoates.

The effects of structure on reactivity in the alkaline hydrolysis of esters was dealt with in Sec. 4.2, where some linear free energy correlations were obtained. The interpretation of rho values was discussed, and it was noted that one of the factors that determine rho is the susceptibility of the reaction to electronic effects. A positive rho value indicates that reaction is facilitated by low electron density at the reaction site; thus $\rho = +2.27$ for the alkaline hydrolysis of substituted ethyl benzoates. Other examples are given in Table 4.III; see also the rate data for aliphatic ester hydrolysis in Table 4.IX and the associated discussion. Euranto [56] has summarized structural effects on ester hydrolysis rates.

We are now in a position to analyze these observed structural effects in terms of mechanism. If, as now seems very likely, $B_{Ac}2$ ester hydrolysis occurs via the tetrahedral intermediate mechanism, Eq. (13.55), then structural changes in R and R´ can be discussed in terms of their separate effects on k_1 and k_2/k_3, and the resultant effect on the observed hydrolysis rate constant k_h [7].

$$OH^- + R-\overset{\overset{\text{O}}{\|}}{C}-OR´ \underset{k_2}{\overset{k_1}{\rightleftharpoons}} R-\overset{\overset{\text{O}^-}{|}}{\underset{\underset{\text{OH}}{|}}{C}}-OR´ \overset{k_3}{\longrightarrow} RCOO^- + R´OH \qquad (13.55)$$

$$k_h = \frac{k_1}{(k_2/k_3) + 1} \qquad (13.56)$$

The simplest prediction to make is that a structural change in R will result in a change in k_1 but will not affect the partition ratio k_2/k_3. Then linear free energy relationships in k_h are expected, because k_h (or rather changes in k_h) is determined only by (changes in) k_1. This behavior is commonly observed. Electron withdrawal in R increases k_1 and therefore k_h; the mechanistic interpretation is that the electrophilicity of the acyl carbon is increased by electron withdrawal into R, thus facilitating attack by the nucleophile. This prediction is an oversimplification, of course, and in a series of methyl benzoates substantial variation was found in the ratio k_2/k_3; for each ester, however, k_2/k_3 was considerably smaller than unity, so $k_h \approx k_1$ and even large changes in k_2/k_3 had little influence on k_h [87].

Structural changes in R´ are less easy to analyze. Increased electron withdrawal by R´ should increase k_1, though the inductive effect should be less pronounced than that of the same structural change in R. Increased electron withdrawal in R´ will make RO⁻ a better leaving group (i.e., a more stable anion, or weaker base), increasing k_3 and therefore decreasing k_2/k_3. Both factors, the increase in k_1 and decrease in k_2/k_3, will increase k_h. A difficulty in these arguments is their neglect of steric and resonance effects. By comparing closely related substrates, these effects may be held more or less constant, but with marked changes in leaving groups they may dominate. The

acid chloride > acid anhydride > ester > amide

reactivity sequence can be rationalized by noting that
resonance interaction between the acyl group and the leaving group, as in 34, changes in the order amide > ester >
anhydride > chloride.

$$
\begin{array}{ccc}
\quad O & & O^- \\
\quad \| & & | \\
R\text{-}C\text{-}X & \longleftrightarrow & R\text{-}C\text{=}X^+
\end{array}
$$

34

These initial state effects will make k_1 smaller for amides
than for esters, etc.

The Hammett and Taft correlations have been most successful in dealing with acyl group variations. Alterations
in the alkyl group of esters are less easily handled, and
Sec. 4.2 describes attempts to use the pK_a of the leaving
group (alcohol or phenol) as a model property for correlating alkaline hydrolysis reactivity. Robinson and his co-
workers [88] have made a recent extensive study to test
the validity of this model. Table 13.X gives rate constants for the alkaline hydrolysis of substituted benzoate
esters with leaving group variations. As expected, acyl
group variations (alkyl group held constant) give linear
Hammett plots. Variation in the alkyl portion of aryl
benzoates (acyl group held constant) did not give linear
correlations with either σ or σ⁻; however, a linear combination of these two substituent parameters gave satisfactory correlations [89]. The most interesting aspect
of these data is seen by presenting them as in Fig. 13.5,
which is a plot of the logarithm of the alkaline hydrolysis rate constants against pK_a of the parent alcohol or
phenol for two acyl group series [87]. The two notable
features are (1) the curvature; (2) the dispersion into
separate lines for aliphatic and aryl esters of the same
acid. These phenomena are not, apparently, caused by
changes in mechanism or rate-determining step, but are
ascribed to differences in resonance and solvation interactions between the rate and equilibrium (model) processes.
Although it is common to think of aromatic esters as being
more reactive than aliphatic esters, Fig. 13.5 shows that
if two such esters of the same acid were available with
identical pK_a's of the alcohol and phenol, the aliphatic
ester would hydrolyze faster than the aryl ester. The
principal contributor to this difference is enhanced resonance stabilization of the initial state in the aromatic
ester; compare 35 and 36.

Reactions of Carboxylic Acid Derivatives. We now survey
a few acyl transfer reactions of carboxylic acids

Table 13.X   Kinetics of Alkaline Hydrolysis of Some Benzoate Esters[a,b]

Esters	$10^4 k_{OH}/\underline{M}^{-1}\text{-sec}^{-1}$	Esters	$10^4 k_{OH}/\underline{M}^{-1}\text{-sec}^{-1}$
p-Toluates		p-Fluorobenzoates	
Methyl	2.65	Methyl	12.1
Ethyl	0.879	Ethyl	4.05
n-Propyl	0.671	n-Propyl	2.65
Isopropyl	0.168	Isopropyl	0.623
n-Butyl	0.568	n-Butyl	2.44
Isobutyl	0.449	Isobutyl	1.90
Phenyl	15.1	Phenyl	67.3
p-Chlorophenyl	33.7	p-Chlorophenyl	115
p-Nitrophenyl	232	p-Nitrophenyl	772
p-Fluorophenyl	26.2	p-Fluorophenyl	89.3
p-Cyanophenyl	158	p-Cyanophenyl	552
p-Tolyl	10.5	p-Tolyl	35.6
Benzoates		p-Chlorobenzoates	
Methyl	6.08	Methyl	19.1
Ethyl	1.98	Ethyl	6.51
n-Propyl	1.67	n-Propyl	5.11
Isopropyl	0.319	Isopropyl	1.21
n-Butyl	1.41	n-Butyl	3.49
Isobutyl	1.18	Isobutyl	3.36
Phenyl	33.6	Phenyl	103
p-Chlorophenyl	77.9	p-Chlorophenyl	235
p-Nitrophenyl	536	p-Nitrophenyl	1460
p-Fluorophenyl	56.5	p-Fluorophenyl	191
p-Cyanophenyl	414	p-Cyanophenyl	1070
p-Tolyl	22.5	p-Tolyl	71.9
m-Cyanophenyl	226	2,4-Dinitrophenyl	7430
Monochloroethyl	12.4		
Dichloroethyl	31.9		
Trichloroethyl	57.8	p-Nitrobenzoates	
2,4-Dinitrophenyl	3580		
		Methyl	276
		Ethyl	98.8
p-Cyanobenzoates		n-Propyl	76.0
		Isopropyl	19.6
Methyl	195	n-Butyl	63.4
Ethyl	79.7	Isobutyl	60.0

Table 13.X continued

Esters	$10^4 k_{OH}/\underline{M}^{-1}\text{-sec}^{-1}$	Esters	$10^4 k_{OH}/\underline{M}^{-1}\text{-sec}^{-1}$
n-Propyl	52.6	Phenyl	1,140
Isopropyl	12.6	p-Chlorophenyl	2,570
n-Butyl	41.9	p-Nitrophenyl	20,900
Isobutyl	39.4	p-Fluorophenyl	2,000
Phenyl	736	p-Cyanophenyl	10,800
p-Chlorophenyl	1,660	p-Tolyl	752
p-Nitrophenyl	12,700	m-Cyanophenyl	8,450
p-Fluorophenyl	1,310	Monochloroethyl	504
p-Cyanophenyl	7,630	Dichloroethyl	1,816
p-Tolyl	518	Trichloroethyl	3,220
2,4-Dinitrophenyl	100,600		

[a]From Ref. [88].
[b]In 50% (v/v) acetonitrile-0.02 $\underline{M}$ phosphate buffer at 25°.

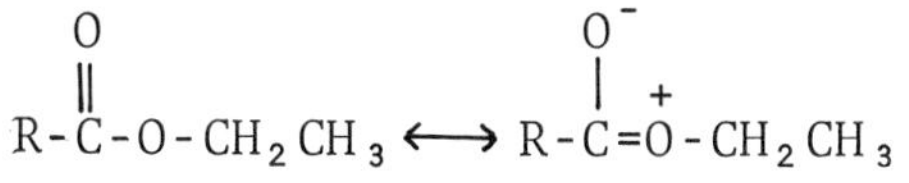

35

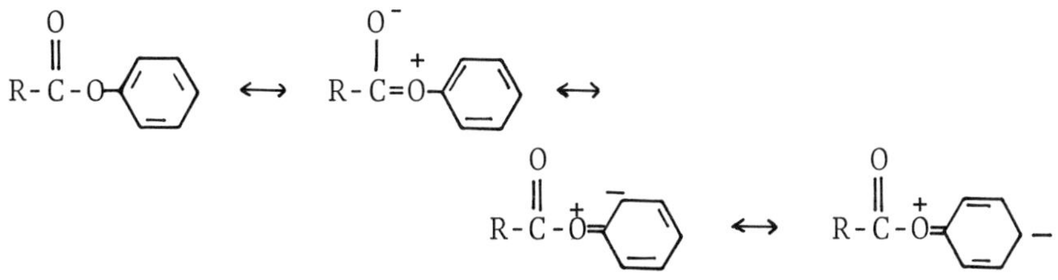

36

and their derivatives. Acyl transfer to (that is, acylation of) oxygen and nitrogen is considered first, and then
carbon acylation. Since solvolysis reactions, especially
ester hydrolyses, were extensively referred to earlier in
this chapter in illustration of mechanistic concepts, this
class of acylations is not given further description here.
    The most general synthetic procedures for carboxylic
acid esters and amides utilize the very reactive acid

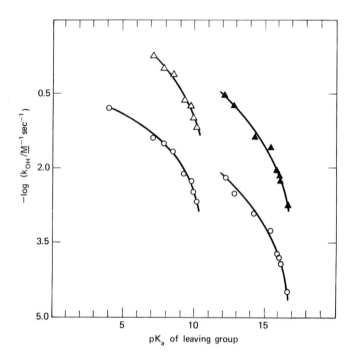

Fig. 13.5. Kinetics of alkaline hydrolysis of benzoates (circles) and p-nitrobenzoates (triangles) against pK_a of the parent phenols (open symbols) and alcohols (filled symbols) (data from Table 13.X and Ref. [88]).

chlorides and anhydrides according to these equations:

$$RCOC\ell + R'OH \rightarrow RCOOR' + HC\ell \qquad (13.57)$$

$$RCOC\ell + R'NH_2 \rightarrow RCONHR' + HC\ell \qquad (13.58)$$

$$(RCO)_2O + R'OH \rightarrow RCOOR' + RCOOH \qquad (13.59)$$

$$(RCO)_2O + R'NH_2 \rightarrow RCONHR' + RCOOH \qquad (13.60)$$

Pyridine is widely used as a catalyst (by a nucleophilic mechanism) [90]. A simple preparation of aromatic acetates adds acetic anhydride to a solution of the phenol in sodium hydroxide [91]. The anhydride, of course, is susceptible to rapid hydrolysis in this medium, but one gains more than he loses, because the phenol is present in its conjugate base form, which is a strong nucleophile, and high yields of ester are obtained. Since these esters are poorly soluble in water, they separate out on forming and

are not themselves rapidly hydrolyzed in the basic solution. Two recent synthetic advances, among many, in catalytic acylations are the introduction of p-dimethylaminopyridine, 37, which is said to be far superior to pyridine
in catalyzing anhydride acylations [92],

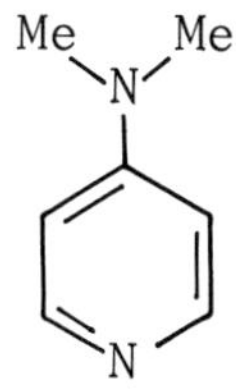

37

and a boron trifluoride-etherate reagent to esterify acids
under mild conditions [93].

$$RCOOH + R'OH \xrightarrow{BF_3} RCOOR' + H_2O \qquad (13.61)$$

The BF_3 may function as a Lewis acid catalyst, polarizing
the acyl group much as in the case of proton catalysis.
    The classical acylating agents of high reactivity,
anhydrides and acid chlorides, are apt to be rather unselective in their action. Other methods of activating
the acyl group have therefore been sought, primarily for
synthetic use. Woodward's reagent, N-ethyl-5-phenyliso-
xazolium-3'-sulfonate, 38, reacts with carboxylate anions
to give activated acyl compounds.

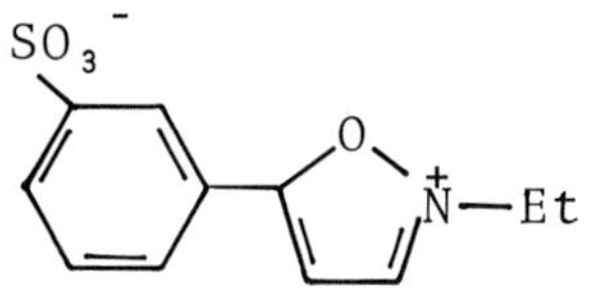

38

This reagent can be used to synthesize peptides according
to Equation (13.62) [94].
    Staab has utilized the high reactivity of certain
heterocyclic amides (azolides), of which N-acetylimidazole
is a familiar example from the earlier discussion, to per-
form acyl transfers [95]. A carboxylic acid is activated

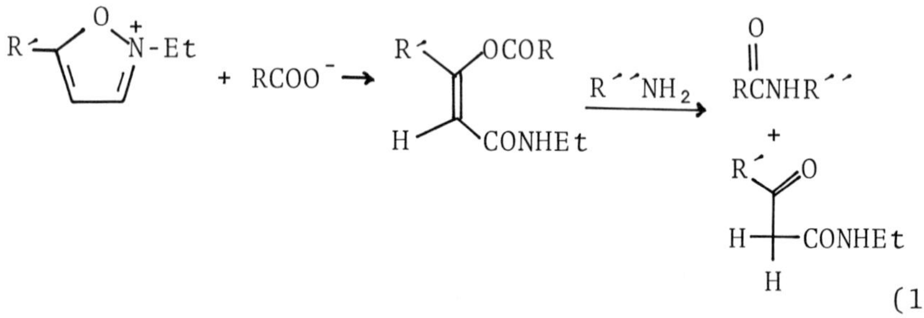

$$(13.62)$$

by treatment with N,N´-carbonyldiimidazole, 39, yielding
the N-acylimidazole; this will acylate numerous nucleophiles, allowing amides, esters, anhydrides, and other
derivatives to be synthesized.

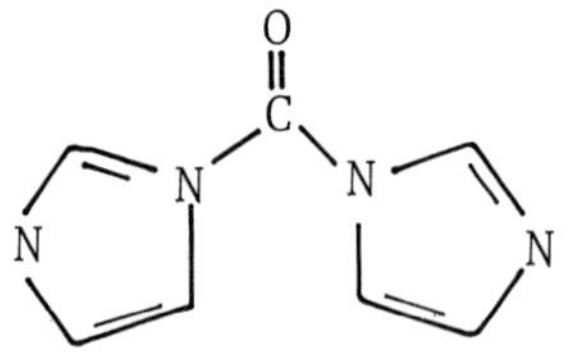

39

Letting ImH represent imidazole, the reaction sequence is
as shown below, where amide formation is illustrated.

$$RCOOH + Im\text{-}CO\text{-}Im \rightarrow RCOIm + ImH + CO_2 \qquad (13.63)$$

$$RCOIm + R´NH_2 \rightarrow RCONHR´ + ImH \qquad (13.64)$$

These reactions are carried out at room temperature in
nonhydroxylic solvents.

The acetylating agent 40, which is 3-acetyl-1,5,5-
trimethylhydantoin, is probably chemically similar to the
azolides.

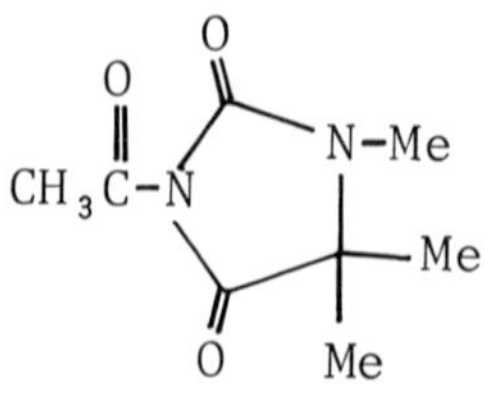

40

This reagent displays much greater reactivity toward phenols than toward alcohols in competitive acetylation experiments [96].

The development of selective acylating agents has been stimulated by attempts to synthesize peptides, especially peptides of natural origin. In order to form the peptide bond, the carboxylic acid group of one amino acid must be activated, and the amino group of the same amino acid must be "blocked" so that self-condensations do not occur. This N-blocked, C-activated amino acid is then reacted with the desired second amino acid to form an N-blocked dipeptide. Now before this fragment can be reacted with another N-blocked, C-activated amino acid, its amino group must be "de-blocked." Peptide synthesis therefore requires blocking and activating groups, with the further requirement that the blocking groups be easily removed under conditions that cause no racemization or peptide hydrolysis. These problems have been solved, and the first chemical synthesis of an enzyme was accomplished in 1969 by two groups [97].

The sequence of steps described above is shown schematically in the following equations, where we wish to form the tripeptide $\underline{41}$.

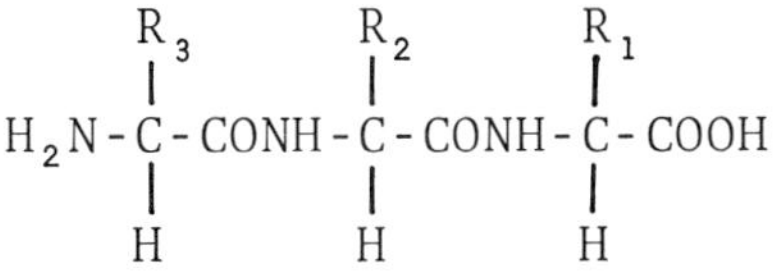

$$
\begin{array}{ccccc}
R_3 & & R_2 & & R_1 \\
| & & | & & | \\
H_2N-C-CONH- & C-CONH- & C-COOH \\
| & & | & & | \\
H & & H & & H
\end{array}
$$

$$\underline{41}$$

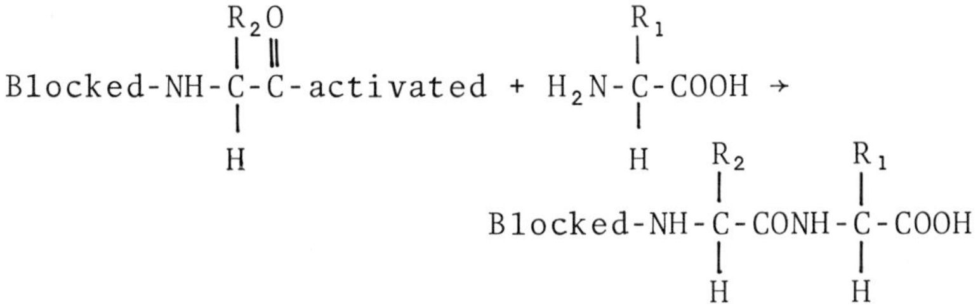

$$
\begin{array}{ccc}
R_2\,O & & R_1 \\
| \; \| & & | \\
\text{Blocked-NH-C-C-activated} & + & H_2N-C-COOH \rightarrow \\
| & & | \\
H & & H
\end{array}
$$

$$
\begin{array}{cccc}
& & R_2 & R_1 \\
& & | & | \\
& \text{Blocked-NH-C-CONH-C-COOH} & & (13.65) \\
& & | & | \\
& & H & H
\end{array}
$$

$$
\begin{array}{ccc}
R_2 \quad R_1 & & R_2 \quad R_1 \\
| \quad\; | & \text{de-blocking} & | \quad\; | \\
\text{Blocked-NH-C-CONH-C-COOH} & \xrightarrow{\hspace{2cm}} & H_2N-C-CONH-C-COOH \\
| \quad\; | & & | \quad\; | \\
H \quad H & & H \quad H
\end{array}
$$

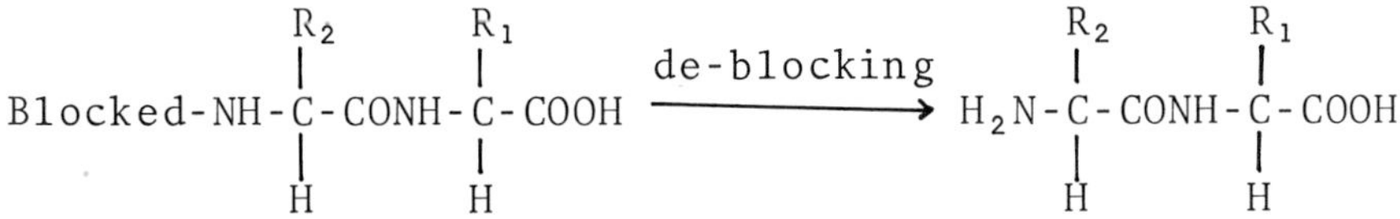

$$(13.66)$$

$$\underset{\substack{|\\H}}{\overset{\substack{R_3\;\;O\\|\;\;\parallel}}{Blocked\text{-}NH\text{-}C\!\!-\!\!C\text{-}activated}} + \underset{\substack{|\quad\;\;|\\H\quad\;\;H}}{\overset{\substack{R_2\quad R_1\\|\quad\;\;|}}{H_2N\text{-}C\text{-}CONH\text{-}C\text{-}COOH}} \longrightarrow$$

$$\underset{\substack{|\quad\quad\;\;|\quad\quad\;\;|\\H\quad\quad\;H\quad\quad\;H}}{\overset{\substack{R_3\quad\quad R_2\quad\quad R_1\\|\quad\quad\;\;|\quad\quad\;\;|}}{Blocked\text{-}NH\text{-}C\text{-}CONH\text{-}C\text{-}CONH\text{-}C\text{-}COOH}} \qquad (13.67)$$

$$\underset{\substack{|\quad\quad\;\;|\quad\quad\;\;|\\H\quad\quad\;H\quad\quad\;H}}{\overset{\substack{R_3\quad\quad R_2\quad\quad R_1\\|\quad\quad\;\;|\quad\quad\;\;|}}{Blocked\text{-}NH\text{-}C\text{-}CONH\text{-}C\text{-}CONH\text{-}C\text{-}COOH}} \xrightarrow{\text{de-blocking}} \underline{41} \qquad (13.68)$$

Several amino acids contain functional groups on their
side chains (R_1, R_2, R_3) that must also be protected by
blocking [98], and the terminal carboxy group may also be
protected.

The carbobenzyloxy group is a widely employed amino
blocking group [99], and its use is typical of many others.
Carbobenzyloxy amino acids are prepared by reacting alkaline aqueous solutions of amino acids with carbobenzyloxy
chloride (benzyl chloroformate).

$$\langle\!\!\!\bigcirc\!\!\!\rangle\text{-}CH_2OCOC\ell + H_2N\text{-}CHR\text{-}COO^- \rightarrow$$

$$\langle\!\!\!\bigcirc\!\!\!\rangle\text{-}CH_2OCONH\text{-}CHR\text{-}COO^- + HC\ell$$

$$(13.69)$$

The carbobenzyloxy group can be removed by catalytic hydrogenation, as well as by HBr in acetic acid and by sodium
in liquid ammonia. Other amino blocking groups are carbo-
tert-butyloxy (tert-butyl-oxycarbonyl, BOC), tosyl, trityl,
formyl, and phthaloyl. The BOC group is introduced as
tert-butyl azidoformate, which is obtained by treatment of
BOC-hydrazide with nitrous acid:

$$Me_3COCONHNH_2 \xrightarrow{HNO_2} Me_3COCON_3 \xrightarrow{H_2NCHRCOOH} Me_3COCONHCHRCOOH$$

$$(13.70)$$

Mildly acid conditions remove the BOC group.
Activation of the carboxy group can be achieved in many

ways. Conversion to a reactive ester, such as a p-nitrophenyl ester, is effective. The use of isoxazolium salts
has already been described [see Eq. (13.62)]. Several
derivatives that can be broadly classed as O-acylhydroxylamines are useful as activated amino acids; three of these
are the acylated N-hydroxyphthalimide 42, N-hydroxysuccin-
imide 43, and N-hydroxypiperidine 44.

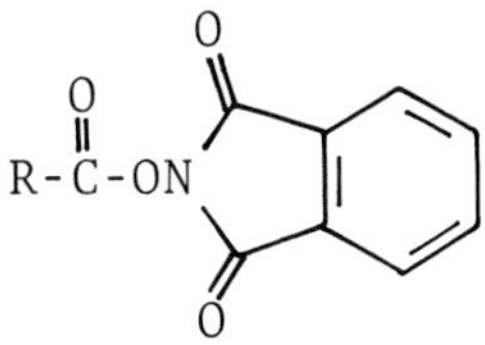

42

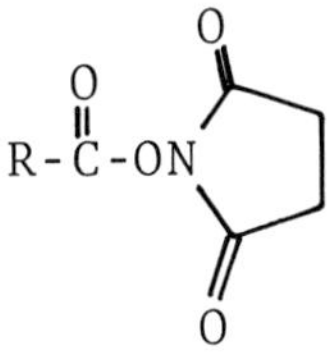

43

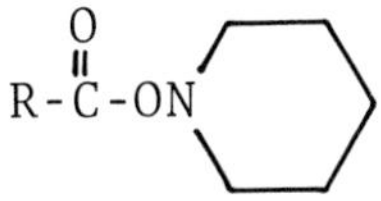

44

N,N´-dicyclohexylcarbodiimide, 45, acts as a condensing
agent by activating the carboxy group, either by forming
an intermediate anhydride or an O-acylurea [100].

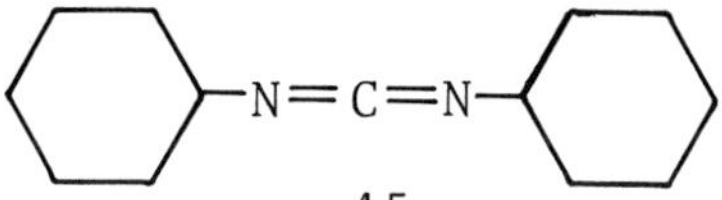

45

Amino protection and carboxy activation can be achieved
in a single step by formation of the N-carboxyanhydride
(Leuchs' anhydride) as in [101]

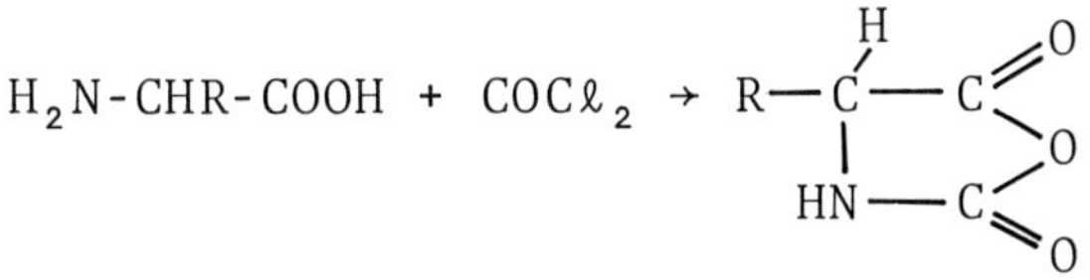

$$H_2N\text{-}CHR\text{-}COOH \; + \; COC\ell_2 \; \rightarrow \; R\text{-}\overset{\overset{\displaystyle H}{|}}{\underset{\underset{\displaystyle HN}{|}}{C}}\text{-}C \qquad\qquad (13.71)$$

The N-carboxyanhydride is coupled with an amino acid at 0°
in pH 10 aqueous solution, and de-blocking (which is a decarboxylation) is achieved by bringing the pH to 5.
   Next let us consider a reaction of no synthetic value
(at least in the laboratory), but of much mechanistic and
some biochemical interest: this is the formation of amides
from carboxylic acids in aqueous solution. Although the
equilibrium lies, for most systems, far to the left, some

$$RCOOH + R'NH_2 \rightleftharpoons RCONHR' + H_2O \qquad (13.72)$$

amide formation can be detected. Working with aliphatic carboxylic acids and aliphatic amines, Morawetz and Otaki [102] gave the rate equation for amide formation as $v = k[RCOO^-][R'NH_2]$, with typical values of the rate constant being of the order 10^{-8} M^{-1}-sec^{-1} at 75° and the equilibrium constant of the reaction, when formulated as Eq. (13.73), being 10^{-6} to 10^{-4} at 75° [103].

$$RCOO^- + R'NH_2 \rightleftharpoons RCONHR' + OH^- \qquad (13.73)$$

Higuchi and his co-workers have made extensive studies of reversible amide formation between amines and dicarboxylic and tricarboxylic acids in aqueous solution [104]. The key experimental factor in these studies was the use of aromatic amines, whose pK_a values are low enough so that the unprotonated amine can co-exist in significant concentrations with the undissociated acids in solutions of acidic pH. An important finding was that the rates of amide formation with monocarboxylic acids are two to three orders of magnitude smaller than with dicarboxylic acids. pH-rate profiles indicated that the free acid is kinetically important, with the rate being first order in free acid and first order in free amine; however, at high amine concentrations, it approaches a zero-order dependence on amine. The intermediate equilibrium formation of a cyclic anhydride was postulated, as shown for the succinic acid-aniline system in

$$(13.74)$$

With some acids the product anilic acid can cyclize to the imide.

The support for Eq. (13.74) comes primarily from indirect kinetic evidence, but this is strengthened by the advantageous capability of studying the several steps individually. At 65° these rate constant values are estimated for Eq. (13.74): $k_1 \approx 2.1 \times 10^{-7}$ sec^{-1}; $k_{-2} = 3.12 \times 10^{-5}$ sec^{-1}; $k_2/k_{-1} \approx 10^3$ M^{-1}. Although anhydride formation in aqueous solution may seem unlikely, it is calculated that about 2 parts per 10^7 at 25° and 3 parts per 10^5 at 95° of free succinic acid exists in the form of succinic anhydride [104], and this accounts for the rates of amide formation.

A rather special type of amide, a hydroxamic acid

RCONHOH, is usually formed by reaction of a carboxylic
acid derivative (ester, amide, anhydride, acid halide) with
hydroxylamine. Both O- and N-acylation of this α-effect
nucleophile can take place, and the O-acylhydroxylamine is
converted into N-acylhydroxylamine (hydroxamic acid) in a
subsequent step with hydroxylamine [40,105]. Hydroxamic
acid formation is base catalyzed.

It is generally asserted that hydroxamic acids are not
formed from carboxylic acids themselves, but Jencks et al.
[106] have shown that the equilibrium constant for acethydroxamic acid formation from acetic acid, Eq. (13.75), has
the value 1.46 at 25° and pH 7, where $[H_2O]$ is taken as
unity.

$$K = \frac{[CH_3CONHOH][H_2O]}{[CH_3COOH]_{total}[NH_2OH]_{total}} \tag{13.75}$$

The reaction is acid catalyzed, but the extent of conversion to the hydroxamic acid decreases with lower pH.

Lawlor has found that several metal ions catalyze hydroxamic acid formation from carboxylic acids [107]. Under
Lawlor's conditions [1 M acetic acid, 0.1 M hydroxylamine,
0.125 M Ni (II)], a bell-shaped pH-rate curve was observed,
which was attributed to a rate equation of the form
$k_{Ni}[CH_3COOH][NH_2OH][Ni^{2+}]$, the nickel presumably coordinating to the acid in much the same manner as the proton in
acid-catalyzed esterification. Figure 13.6 shows pH-rate
profiles for acethydroxamic acid formation from acetic acid
catalyzed by acid and by nickel (II) [108]. The latter
reaction is represented under markedly different conditions
from Lawlor's; for Fig. 13.6 the concentrations were acetic
acid 1 M, hydroxylamine 0.8 M, Ni(II) 0.04 M. A sigmoid
curve is obtained, and this can be accounted for with a
rate term $k'_{Ni}[CH_3COO^-][NH_2OH][Ni^{2+}]$. The pH behavior
therefore depends upon the hydroxylamine and nickel concentrations [109].

The formation of acethydroxamic acid in the absence
of nickel can be accounted for with rate equation (13.76).

$$v = k''[CH_3COOH][NH_2OH][H^+]^2 + k'[CH_3COOH][NH_2OH][H^+]$$
$$+ k[CH_3COOH][NH_2OH] \tag{13.76}$$

The individual contributions from these terms are shown by
dashed lines in Fig. 13.6, and the full line is their sum-
mation. The k term represents uncatalyzed nucleophilic
attack of neutral hydroxylamine on neutral acetic acid;
the k' term, which is kinetically equivalent to the product
$[CH_3COOH_2^+][NH_2OH]$, may be a simple acid catalysis term,

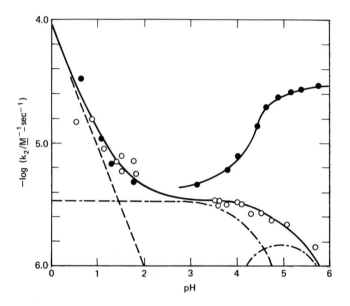

Fig. 13.6.   pH-rate profiles for acethydroxamic acid formation from acetic acid in aqueous solution at 90°.   Full
circles: 0.04 $\underline{M}$ NiCℓ$_2$; open circles: no NiCℓ$_2$.

with its contribution becoming pH independent at low pH
because of concomitant protonation of hydroxylamine.   The
k˝ term is less easy to account for, but it may indicate
that decomposition of the tetrahedral intermediate is rate
determining, and that this process can be acid catalyzed.
Equation (13.77) would therefore represent the kinetics
and mechanism over the pH range shown.

$$
\underset{\substack{k'_1[H^+] \\ }}{\overset{\substack{}}{}}
$$

$$
\begin{array}{c}
& & & \text{O} & & & k'_1[H^+] & & \text{OH} & & k'_2[H^+] & & \text{O} \\
& & & \parallel & & \xrightarrow{k_1} & & & | & & \xrightarrow{k_2} & & \parallel \\
\text{CH}_3\text{C-OH} & + & \text{NH}_2\text{OH} & & \underset{k_{-1}}{\rightleftharpoons} & & \text{CH}_3\text{-C-OH} & & \underset{k_{-2}}{\rightleftharpoons} & & \text{CH}_3\text{C-NHOH} & + & \text{H}_2\text{O} \\
& & & & & k'_{-1}[H^+] & & & \text{NHOH} & & k'_{-2}[H^+] & &
\end{array}
$$

$$\tag{13.77}$$

Fersht and Jencks [31] have constructed a Brønsted
plot for the attack of nucleophiles on acetic acid by combining equilibrium constant data with rates for the reverse reaction.   According to this plot and the results
of Fig. 13.6, hydroxylamine exhibits a positive deviation
of many orders of magnitude, which is why its reaction
with free acids is observable [110].   Concerning the

ability of a free carboxylic acid to undergo acyl transfer
reactions, it should be noted that the leaving group -OH
from RCOOH should be comparable with the leaving group
-OR´ from RCOOR´, where R´ is alkyl [111]. Acetic acid
behaves toward nucleophilic reagents like an ester with a
leaving group whose pK_a is 17 [31,112]. It is in fact observed that acethydroxamic acid is formed faster from acetic acid than from ethyl acetate in the pH range 3-5.5.

Turning briefly to acyl transfer reactions involving
carbon nucleophiles, we note a similarity to the reactions
of aldehydes and ketones with carbon nucleophiles (see
Chap. 12). In each case it is probable that a carbanion
(derived from an active methylene compound, a Grignard
reagent, etc.) adds to the carbonyl carbon giving a tetrahedral intermediate. From this point a major difference
is observed between carbonyl and carboxyl substrates, for
aldehydes and ketones possess very poor leaving groups,
but carboxylic acid derivatives possess good leaving groups.
The consequence is that the tetrahedral intermediate formed
from the carboxyl substrate decomposes [113]. To illustrate this sequence, the reaction of methyl propionate with
methyl benzoate is appropriate. The overall reaction is

$$\underset{\text{Ph-C-OMe}}{\overset{\overset{\textstyle O}{\|}}{}} + CH_3CH_2COOMe \xrightarrow{\text{NaOMe}} \underset{\underset{CH_3}{|}}{\overset{\overset{\textstyle O}{\|}}{Ph\text{-}C\text{-}CHCOOMe}} + MeOH \qquad (13.78)$$

The course of the reaction is as follows. The base abstracts a proton from the α-carbon of methyl propionate
and the resulting carbanion attacks the acyl carbon of
methyl benzoate. Loss of methoxide from the tetrahedral
intermediate gives the product, a β-keto ester. Since
this product possesses an active hydrogen, its ionization
draws the equilibrium to the right.

$$CH_3CH_2COOMe + \overset{-}{O}Me \rightleftharpoons CH_3\overset{-}{C}HCOOMe + MeOH \qquad (13.79)$$

$$\underset{\text{Ph-C-OMe}}{\overset{\overset{\textstyle O}{\|}}{}} + CH_3\overset{-}{C}HCOOMe \rightleftharpoons \underset{\underset{CH_3\text{-}CH\text{-}COOMe}{|}}{\overset{\overset{\textstyle O^-}{|}}{Ph\text{-}C\text{-}OMe}} \qquad (13.80)$$

$$\underset{\underset{CH_3\text{-}CH\text{-}COOMe}{|}}{\overset{\overset{\textstyle O^-}{|}}{Ph\text{-}C\text{-}OMe}} \rightleftharpoons \underset{\underset{CH_3}{|}}{\overset{\overset{\textstyle O}{\|}}{Ph\text{-}C\text{-}CHCOOMe}} + MeO^- \qquad (13.81)$$

This type of reaction, the base-catalyzed acylation of an active methylene compound by an ester, is called the Claisen condensation. Self-condensations can occur, as in the acetoacetic ester condensation of ethyl acetate,

$$2\,CH_3COOEt \xrightarrow{\text{NaOEt}} CH_3\overset{\overset{\displaystyle O}{\|}}{C}CH_2COOEt + EtOH \qquad (13.82)$$

When two different esters, both possessing α-hydrogens, are condensed, four products are possible, and the reaction is not useful. Self-condensations are synthetically practicable, and so are Claisen condensations, like that in Eq. (13.78), in which one ester has no α-hydrogens [114]. Some examples follow:

$$EtOOC\text{-}COOEt + CH_3CH_2COOEt \rightarrow EtOOC\text{-}\overset{\overset{\displaystyle O}{\|}}{C}\text{-}\underset{\underset{\displaystyle CH_3}{|}}{CH}COOEt + EtOH \qquad (13.83)$$

$$HCOOEt + PhCH_2COOEt \rightarrow H\text{-}\overset{\overset{\displaystyle O}{\|}}{C}\text{-}\underset{\underset{\displaystyle Ph}{|}}{CH}COOEt + EtOH \qquad (13.84)$$

$$EtOCOOEt + PhCH_2COOEt \rightarrow EtO\text{-}\overset{\overset{\displaystyle O}{\|}}{C}\text{-}\underset{\underset{\displaystyle Ph}{|}}{CH}COOEt + EtOH \qquad (13.85)$$

These reactions are written with the ester substrate given first and the active methylene compound second.

An intramolecular Claisen condensation will produce a cyclic ester; this reaction is known as the Dieckmann cyclization.

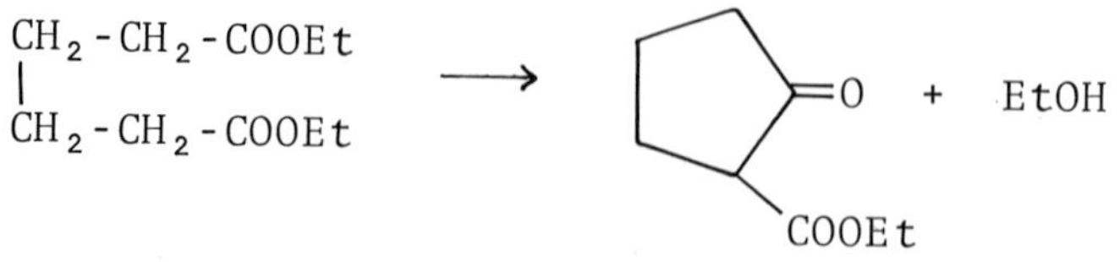

$$\begin{array}{l} CH_2\text{-}CH_2\text{-}COOEt \\ | \\ CH_2\text{-}CH_2\text{-}COOEt \end{array} \longrightarrow \text{[cyclopentanone ring]}{=}O + EtOH \qquad (13.86)$$

An intermolecular Claisen condensation may be followed by a Dieckmann cyclization, as in the reaction of diethyl oxalate with diethyl glutarate,

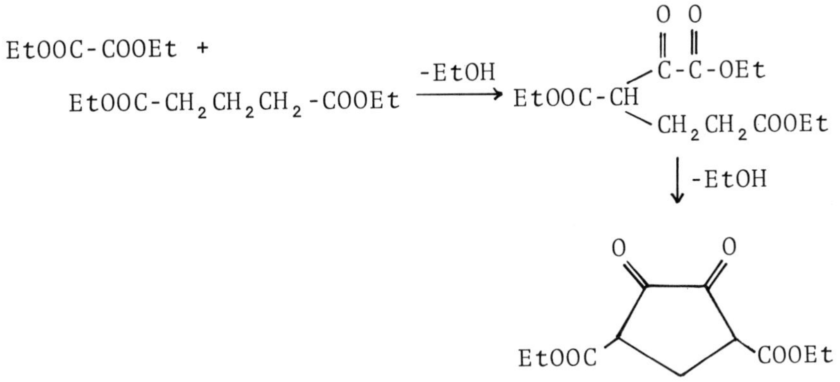

EtOOC-COOEt +

EtOOC-CH$_2$CH$_2$CH$_2$-COOEt $\xrightarrow{\text{-EtOH}}$

$$\text{EtOOC-CH}\begin{array}{l}\overset{\displaystyle O\ \ O}{\overset{\displaystyle \|\ \ \|}{\diagup\text{C-C-OEt}}} \\ \diagdown \text{CH}_2\text{CH}_2\text{COOEt}\end{array}$$

$\downarrow$ -EtOH

EtOOC   COOEt

(13.87)

As noted above, the position of equilibrium in the Claisen condensation is displaced in favor of the product by ionization of the remaining α-hydrogen; the product, in fact, is a stronger acid than is the reactant active methylene compound, because in the product the active hydrogen is flanked by two carbonyl groups. Since the product therefore exists primarily in the enolate form in the reaction mixture, acidification is necessary to recover the β-keto ester. If the reactant possesses but one active hydrogen, as in ethyl isobutyrate, Me$_2$CHCOOEt, the product contains no active hydrogen, and the final enolization equilibrium cannot take place. Thus sodium alkoxides are suitable bases for Claisen condensations of active methylene (-CH$_2$-) compounds, but are ineffective in condensing active methinyl (-CHR-) compounds. More powerful bases, such as triphenylmethylsodium, will effect the condensation of these compounds, presumably by increasing the concentration of the carbanion reactant sufficiently to drive the equilibrium, making the ionization of the product unnecessary [115].

13.2 Analytical Reactions

Acyl Compounds as Samples. Many reviews have been published on the analysis of carboxylic acids and their derivatives [116]. As in earlier chapters, the present concern is to give a sense of the kinds of reactions that are used, without detailed or comprehensive treatment of the available analytical methods. For convenience these reactions are dealt with in this order: (1) hydrolyses; (2) esterifications; and (3) reactions other than these.

The alkaline hydrolysis of a carboxylic acid derivative is the classical analytical method, especially for esters. This procedure is often called saponification (soap-making), because soaps are made by the alkaline hydrolysis of fatty acid esters of glycerine. The principle of the analysis is contained in the titration diagram of Fig. 1.1. It is based upon the consumption of alkali by a product of the hydrolysis, so that each mole of ester hydrolyzed results in a decrease in alkali of one mole. This reduction in alkali is determined by back titration with standard acid.

A practical difficulty with saponification methods is slowness of the reaction, especially for sterically hindered esters and for amides. The reaction is usually carried out at reflux to gain the advantage of the temperature effect on rate. Since the reaction is first order in hydroxide, its rate can be increased by increasing the hydroxide concentration, but a limit is imposed by the analytical requirement that a substantial fraction of alkali be consumed in the hydrolysis, because the quantitative finish measures this fraction. A common approach has been to utilize high-boiling solvents to achieve higher temperatures and higher rates; diethylene glycol is one of the solvents that has been recommended [117].

In selecting solvents for their higher boiling points, the isothermal effect of the solvent on reaction rate is sometimes overlooked. Considerable information has been collected, especially by Tommila and his co-workers [118]. Addition of methanol or ethanol to an aqueous medium decreases the rate of ester hydrolysis; in this connection recall the discussion, in Sec. 9.1, of the several factors contributing to solvent effects in alcohol-water mixture, and especially see Fig. 9.1, which pictures the alkoxide-hydroxide distribution for methanol-water and ethanol-water mixtures [119]. Bender and Glasson [120] found that the relative nucleophilicities of OH^-, OCH_3^-, and $OC_2H_5^-$, in hydrolysis and alcoholysis of esters, are 1:1.6:4.2, respectively; however, the specific rate constants for these three species all increase with increasing water content of the solvent. The net solvent effect is apt to be considerably different for methanol and for ethanol, especially in nearly nonaqueous solvents. Greive et al. [121] found that hydrolysis of methyl acetate by potassium hydroxide in essentially anhydrous methanol was incomplete; addition of a small amount of water led to complete hydrolysis. This effect is probably a manifestation of the alkoxide-hydroxide equilibrium.

For the quantitative saponification of highly hindered

esters, a medium containing 20 g of sodium hydroxide and
10 g of sodium perchlorate per liter of hexyl alcohol has
been recommended [122]. The role of the salt is obscure;
perchlorate is hardly likely to be a catalyst, as claimed.
It is possible that the salt exerts an effect on the alkoxide-hydroxide equilibrium, providing an optimum hydrox-
ide/alkoxide ratio at the recommended concentration. If
so, the salt might profitably be replaced with water.

Estimates of saponification time for esters can be
made on the basis of principles given in Sec. 4.2, or,
less rigorously, by comparison of the sample ester structure with known esters and the conditions required for
their determination. For example, a methyl ester always
hydrolyzes 2-3 times faster than does an ethyl ester of
the same acid (see Tables 4.IX and 13.X). Quantitative
predictions can sometimes be made with linear free energy
relationships such as the Hammett and Taft equations.
With an estimate of the second-order rate constant k_{OH}
the reaction time for complete hydrolysis can be calculated. An approximate calculation is appropriate, since
a check can always be made by carrying out the saponification for several reaction times. Suppose it is estimated that the rate constant for methyl benzoate hydroly-
sis is about 0.01 M^{-1}-sec^{-1}, and 0.5 $\underline{N}$ KOH is to be used
in the saponification. Sufficient sample will be taken
that the final hydroxide concentration will be significantly lower than this, so pseudo-first-order conditions
will not hold, but it is most convenient to assume first-
order behavior. Then $k_1 = k_{OH}[OH^-] \approx 0.005\ sec^{-1}$, so $t_{1/2}$
$= 0.693/k_1 \approx 140$ sec. The reaction is 99.9% complete
in 10 half-lives, or 23 min. Since the actual concentration of hydroxide will be less than 0.5 $\underline{M}$ during most of
the reaction, additional time should be allowed.

Several hydrolytic methods utilize a final measurement
of some product of the reaction rather than of the unreacted hydroxide. This overcomes the usual objection to
using a large excess of hydroxide in the saponification
medium. Amides, which as more resistant to alkaline hydrolysis than are esters, can be determined by hydrolysis
in alkali, distillation of the liberated ammonia or amine
into acid, and final titration [123]. The carboxylic acid
anion produced upon alkaline hydrolysis of an amide can be
converted to the free acid by passage through a cation-
exchange resin in the H^+ form; this also neutralizes the
excess alkali. The carboxylic acid in the eluate is then
titrated with standard base [124]. An ion-exchange resin
has been used to catalyze (by acid catalysis) the hydrolysis, as well as to convert the product to the free acid,
of esters and amides [125].

Among other applications of hydrolysis reactions, with

measurement of a product, are these: (1) esters can be
detected with an indicator spot test after being subjected
to the hydrolytic action of an enzyme [126]; (2) acetyl
groups can be hydrolyzed and the acetic acid produced
measured by gas-liquid chromatography [127]; (3) arylsulfonyl chlorides undergo pyridine-catalyzed hydrolysis,
$RSO_2Cl + H_2O \rightarrow RSO_3H + HCl$; the acidic products are titrated with standard base [128]; (4) carbonic acid derivatives, such as ureas, carbamates, and carbonate esters,
are hydrolyzed in concentrated phosphoric acid giving
quantitative yields of CO_2, which is determined gravimetrically [129]; (5) ethanolamides can be determined in
mixtures on the basis of their different saponification
rates. The reactions are followed by potentiometric titration of the weak bases liberated [130].

Both water and acetic anhydride have been determined
in acetic acid by reaction with an excess of the other
reactant. Spectrophotometric measurement of the acetic
anhydride is employed in one method [131]. Thermometric
titration is also possible, the reaction being catalyzed
by perchloric acid [132].

Determination of the amino acid sequence of a protein
requires hydrolysis of the peptide bonds. The classical
procedure is by acid catalysis in 6 $\underline{N}$ HCl at 110° for
24-72 hr in a sealed tube [133]. The amino acids are then
separated by ion-exchange chromatography and measured
colorimetrically, these steps usually being combined and
automated on apparatus designed for this analysis. The
hydrolysis procedure has drawbacks, some of the amino acids
being incompletely liberated during the course of the reaction, and others undergoing partial or complete degradation, so that corrections must be applied. Open flask
hydrolysis conditions give some improvement in amino acid
stability [134]. Hydrolysis in 6 $\underline{N}$ HCl (sealed tube) at
145° for 4 hr gives hydrolysis of ribonuclease equivalent
to that at 110° for 26 hr, and it is recommended that
amino acid yield be plotted against time for all amino
acids; suitable extrapolations are made to give the corrected yields [135]. Hydrolysis in the presence of oxalic
acid and hydrochloric acid is claimed to be more efficient
than the classical HCl procedure [136].

The haloform reaction is, in its final step, an acyl
transfer involving hydrolysis of a ketone. The reaction
is used as a qualitative test for methylketones CH_3COR,
where R is alkyl or aryl, or for compounds that can be
converted to this group, such as isopropyl alcohol [137].
When treated with iodine in basic medium, such compounds
yield iodoform according to the overall equation

$$CH_3COR + 3I_2 + 4OH^- \rightarrow RCOO^- + CHI_3 + 3I^- + 3H_2O \quad (13.00)$$

The reaction takes place via the enolate of the methyl
ketone, which is progressively iodinated to give the tri-
iodomethylketone. Successive iodination occurs on the
same carbon because each iodine makes ionization easier
and stabilizes the resulting carbanion. Written for acetone, these reactions are as follows:

$$CH_3-\overset{\overset{\displaystyle O}{\|}}{C}-CH_3 \underset{}{\overset{OH^-}{\rightleftharpoons}} CH_3-\overset{\overset{\displaystyle O}{\|}}{C}-CH_2 \overset{I_2}{\longrightarrow} CH_3-\overset{\overset{\displaystyle O}{\|}}{C}-CH_2I$$

$$\downarrow (etc.) \qquad (13.89)$$

$$CH_3-\overset{\overset{\displaystyle O}{\|}}{C}-CI_3$$

The triiodomethyl group is activating, and the methylketo
group in this molecule behaves like an acetyl group, alkaline cleavage occurring at the acyl carbon.

$$CH_3\overset{\overset{\displaystyle O}{\|}}{C}-CI_3 + OH^- \rightarrow CH_3\overset{\overset{\displaystyle O}{\|}}{C}-O^- + CHI_3 \qquad (13.90)$$

Several analytical methods are based on the iodoform reaction [138]. Most of these are titrimetric. Various sources
of inaccuracy, primarily side reactions, have been identified [139].
    Esterification and transesterification have given some
useful analytical methods. Free carboxylic acids are es-
terified by treatment with methanol and hydrochloric acid.
The methyl esters are amenable to gas chromatographic
separation and measurement, as in a method for the determination of pyruvate, lactate, fumarate, succinate, malate,
α-ketoglutarate, and citrate in urine [140]. After esterification of lactic acid by refluxing in methanolic HCℓ,
the methyl lactate has been determined by the ferric hydroxamate method [141], which is discussed later. Mixtures
of acetic, nitric, and hydrofluoric acids can be analyzed
by alkalimetric titration with the aid of an esterification [142]. The total acidity is found by titrating a
sample of the mixture. Another sample is refluxed with
methanol (the sample itself contains sufficient acid to
act as catalyst) to form methyl acetate; then this sample
is titrated potentiometrically. The HNO_3 and HF are found
by the differentiating potentiometric titration, and the
acetic acid is given by the difference in final end points
between the two titrations.
    Acids in gasoline fractions have been separated on the

basis of the differences in their esterification rates in
HCl-methanol. These rate differences have been attributed
to steric hindrance [143].

The boron trifluoride-catalyzed esterification of
acids, referred to earlier [93] as a recent synthetic
method, has been widely used in analytical chemistry.
Mitchell, Smith, and Bryant [144] reacted acids with excess methanol in the presence of boron trifluoride.

$$RCOOH + CH_3OH \underset{}{\overset{BF_3}{\rightleftharpoons}} RCOOCH_3 + H_2O \qquad (13.91)$$

The water liberated in the esterification was titrated
with Karl Fischer reagent. Aliphatic acids can be determined in the presence of aromatic acids because of their
greater esterification rates.

Lactic and succinic acids are determined in foods by
a gas chromatographic procedure utilizing a BF_3-n-propanol
reagent to form the propyl esters [145]. The propyl esters
were selected in preference to methyl, butyl, or silyl esters because they possess optimum properties for the two
requirements that they be extractable into an organic solvent and be well behaved in the chromatographic system.
A comparative study has been made of methyl ester preparation, for gas chromatography, by the methanol-HCl, methanol-
BF_3, and diazomethane techniques [146].

A transesterification by acidolysis (see Table 13.I)
yields solid derivatives of the alcohol portion of esters
[147]. 3,5-Dinitrobenzoic acid is refluxed with the ester
in the presence of sulfuric acid.

$$\qquad (13.92)$$

Transesterification by alcoholysis is used to convert esters of high-molecular weight alcohols into methyl esters,
which can subsequently be analyzed by gas chromatography.
In this way polyresin esters [148] and triglycerides [149]
have been analyzed. Usually base catalysis is employed,
the ester being refluxed in methanolic hydroxide or methoxide [150]. An ingenious modification incorporates 2,2-
dimethoxypropane in the transesterification mixture for
the preparation of methyl esters from triglycerides [151].
The role of the dimethoxypropane is to react with the

glycerol formed in the transesterification, forming iso-propylidene glycerol, thus displacing the transesterification equilibrium.

$$
\begin{array}{c}
CH_2\text{-}OCOR \\
|\\
CH_2\text{-}OCOR \\
|\\
CH_2\text{-}OCOR
\end{array}
+ MeOH \rightleftharpoons
\begin{array}{c}
CH_2OH \\
|\\
CH_2OH \\
|\\
CH_2OH
\end{array}
+ 3\ RCOOMe \qquad (13.93)
$$

$$
\begin{array}{c}
CH_2OH \\
|\\
CH_2OH \\
|\\
CH_2OH
\end{array}
+
\begin{array}{c}
OMe \\
|\\
CH_3\text{---}C\text{---}CH_3 \\
|\\
OMe
\end{array}
\rightleftharpoons
\begin{array}{c}
CH_2\text{---}O \\
|\quad\quad\ \ \ C \\
CH_2\text{---}O \\
|\\
CH_2OH
\end{array}
\begin{array}{c}
CH_3 \\
CH_3
\end{array}
+ 2\ MeOH \quad (13.94)
$$

Some peculiar kinetic behavior was observed in these systems, suggesting rather more complex chemistry than shown in these reactions, but quantitative yields of methyl esters were observed.

Transesterification of high molecular weight esters (for example, cetyl palmitate and cetyl stearate) has been effected by treatment with HBr and diethyl ether, the ethyl esters being produced [152]:

$$
RCOOR' \xrightarrow[Et_2O]{HBr} RCOOEt + R'OH
$$

The course of the reaction was not investigated.

With a sufficiently reactive acyl compound, esterification can be rapid enough to permit direct titration. Thus acid chlorides can be titrated with standard sodium methoxide in nonaqueous medium, just as if they were acids [153]. Naphthyl esters can be titrated with methoxide in pyridine solution; probably a pyridine-catalyzed transesterification occurs with production of the methyl ester and naphthol [154].

Analysis of acid halides and acid anhydrides can make use of several other acyl transfer reactions. Acids are converted to amide derivatives via their acid chlorides, which are prepared by treatment of the acid with thionyl chloride

$$
RCOOH + SOC\ell_2 \rightarrow RCOC\ell + SO_2 + HC\ell \qquad (13.95)
$$

$$
RCOC\ell + NH_3 \rightarrow RCONH_2 + HC\ell \qquad (13.96)
$$

The simple amides, as well as anilides, toluidides, and p-bromoanilides, are useful derivatives for characterizing acids [155].

    Quantitative analysis of acid anhydrides can be based upon amide formation. Aniline [156] and morpholine [157] are the usual reagents. The reaction with morpholine is shown below.

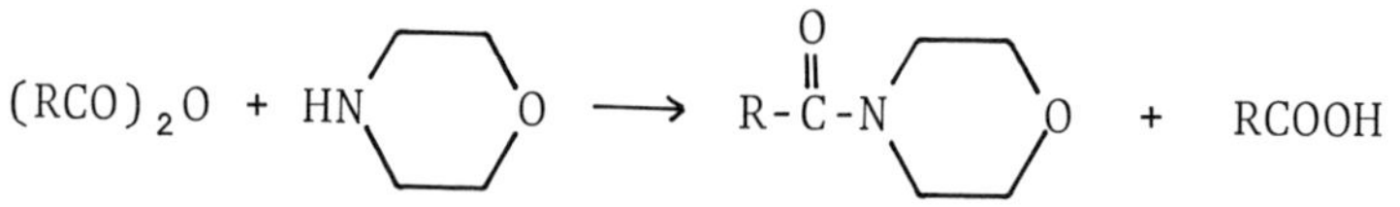

$$(RCO)_2O + HN\underset{}{\overset{}{\bigcirc}}O \longrightarrow R-\overset{\overset{\textstyle O}{\|}}{C}-N\underset{}{\overset{}{\bigcirc}}O + RCOOH \qquad (13.97)$$

After reaction is complete the excess amine is titrated with standard acid. It is interesting that succinic acid interferes in the determination of succinic anhydride by the aniline method, whereas monocarboxylic acids do not This interference is probably a manifestation of the type of reaction studied by Higuchi [104], in which acids capable of forming a cyclic anhydride yield anilides more readily than do monocarboxylic acids.

    Anilide formation is used in analyzing acid chloride samples, which may contain the hydrolytic products HCl and free carboxylic acid. The basis of the method, which combines anilide formation, hydrolysis, and halide determination, is that hydrolysis of an acid chloride yields two equivalents of acid, whereas anilide formation yields only one acid equivalent [158].

    A spot test detection of acid anhydrides employs o-nitrophenylhydrazine as a reagent [159]. Acylation to produce a blue-violet colored compound is believed responsible for the test.

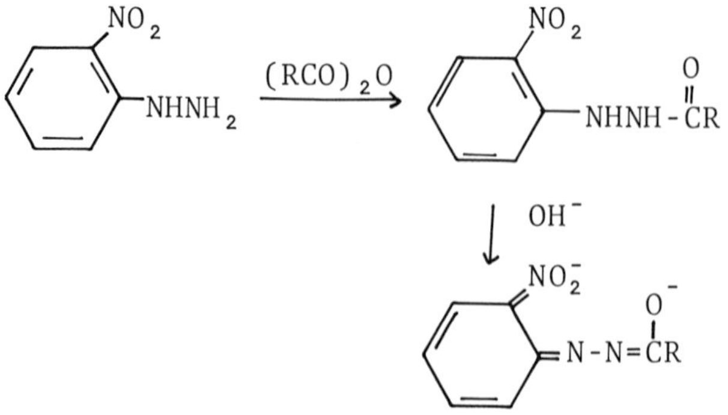

$$\underset{}{\overset{NO_2}{\bigcirc}}-NHNH_2 \xrightarrow{(RCO)_2O} \underset{}{\overset{NO_2}{\bigcirc}}-NHNH-\overset{\overset{\textstyle O}{\|}}{C}R \qquad (13.98)$$

$$\downarrow OH^-$$

$$\underset{}{\overset{NO_2^-}{\bigcirc}}=N-N=\overset{\overset{\textstyle O^-}{|}}{C}R$$

    One of the most generally applicable analytical methods for carboxylic acid derivatives is a spectrophotometric procedure utilizing preliminary formation of a hydroxamic

acid, which is then treated with iron (III) to form a
colored complex. This ferric hydroxamate method was introduced by Feigl as a spot test for carboxylic acid derivatives [160]. The first quantitative applications were
made by Hill [161]. The reaction, illustrated with an
ester substrate, is shown below.

$$\underset{\text{R-C-OR´}}{\overset{\overset{\displaystyle O}{\parallel}}{}} + NH_2OH \rightleftharpoons \underset{\text{R-C-NHOH}}{\overset{\overset{\displaystyle O}{\parallel}}{}} + R´OH \qquad (13.99)$$

The color produced on adding ferric ion appears to be due
to a 1:1 complex (at least under the usual acidic analytical conditions) [162].

$$\underset{\text{R-C-NHOH}}{\overset{\overset{\displaystyle O}{\parallel}}{}} + Fe^{3+} \rightleftharpoons \left[R-C \begin{matrix} O---Fe \\ \\ N-O \\ H \end{matrix} \right]^{2+} + H^+ \qquad (13.100)$$

The ferric hydroxamate method has been studied by many
workers. Its applicability depends upon two separate factors: (1) the rate and equilibrium of the hydroxamic acid
formation itself; (2) the yield and stability of the ferric
hydroxamate complex. Since hydroxamic acid formation is
base catalyzed, the reaction is usually carried out under
alkaline conditions. The color development, however, requires an acid medium. A solution of a hydroxamic acid
prepared during an analysis contains excess hydroxylamine,
which can lower the yield of the ferric hydroxamate by
reducing ferric ion to ferrous ion.

$$2NH_2OH + 2Fe^{3+} \rightarrow 2Fe^{2+} + N_2O + 2H^+ + 2H_2O \qquad (13.101)$$

In an unbuffered solution the hydrogen ion produced inhi-
bits further reduction, but concomitantly shifts equilibrium (13.100) to the left. The net result is that color
intensity should be maximal at some intermediate pH, which
is 1.4 for acethydroxamic and benzhydroxamic acids [162].
Color stability of ferric hydroxamate solutions has been
studied by Notari and Munson [163], who find that the
stability is controlled primarily by the ratio $[Fe^{3+}]/$
$[NH_2OH]$, which should be at least 5. If this ratio is
smaller than 5, good results can be obtained by letting
the solutions stand in the dark for an hour before making
the spectrophotometric measurement. If the ratio is less
than one, stable readings cannot be obtained.
    Esters, anhydrides, acid halides, amides, imides,

lactones, and thiolesters give hydroxamic acids with an
alkaline hydroxylamine reagent [161,164]. Under analytical conditions, the molar absorptivity of the acethydroxamic acid-ferric ion complex [108] at 530 nm is 1.15×10^3;
the method is therefore not highly sensitive, but its
generality (and the lack of a better method for low concentrations of these compounds) makes it valuable. An
important feature of the method is its freedom from interference by carboxylic acids, which do not form hydroxamic
acids in an alkaline hydroxylamine solution. Acids have
been detected and determined, however, after preliminary
conversion to esters [141,161] or acid chlorides [160,165].
Since alkaline conditions are usual for the hydroxamic
acid formation, concurrent hydrolysis of the carboxylic
acid derivative can contribute to low yields of the hydroxamic acid. For example, acetic anhydride gave only 81.5%
as much acethydroxamic acid as did an equal concentration
of ethyl acetate, probably because of the greater susceptibility of the anhydride to alkaline hydrolysis [166].
Moreover, even ethyl acetate gives less than the theoretical yield of acethydroxamic acid [105]. By using a neutral reagent of hydroxylamine, the more reactive carboxylic
acid derivatives such as anhydrides [164b], lactones, and
thiolesters [167] can be determined in the presence of
esters.

    Under alkaline conditions, amides react much slower
than do esters. It is therefore surprising to find that
amides react faster than esters in neutral solutions [168].
The rate of formohydroxamic acid formation from formamide
shows a maximum at pH 6.2-6.5, and the reaction is general
acid catalyzed by hydroxylammonium ion [169]. Reasonable
mechanisms can be written for this catalysis. The greater
susceptibility of amides, relative to esters, to hydroxamic
acid formation under conditions in which acid catalysis is
possible probably is a consequence of the greater basicity
of amides [170]; the pK_a of acetamide is about 0.0 and
that of ethyl acetate is about -6.5. Spot tests for amides
have been developed with propylene glycol as a solvent in
order to achieve higher reaction temperatures [171], but
the results seem to be explicable on the basis of the pH-
rate behavior for hydroxamic acid formation [169] and the
approximate pH of the several hydroxylamine reagents tested.

    The possibility of making analytical use of the direct
reaction between carboxylic acids and hydroxylamine is
suggested by Fig. 13.6, which shows the pH-rate curves for
acid-catalyzed and nickel (II)-catalyzed acethydroxamic
acid formation from acetic acid. The acid-catalyzed reaction is not analytically attractive because the position
of equilibrium is unfavorable in the low pH region [172],
but the nickel-catalyzed reaction is more feasible. A

spot test and a quantitative initial rate assay for carboxylic acids in aqueous solution have been based upon
nickel-catalyzed hydroxamic acid formation at pH 6 [108b].
Acid hydrazides also can be determined by the nickel-
catalyzed ferric hydroxamate method; this method is discussed later.

Acyl Compounds as Reagents. Acylation of a sample substrate is an important analytical process. If the sample
is a hydroxy compound, the reaction is an esterification;
if it is an amine, amide formation occurs. Thiols can
also be determined by acylation, and some C-acylations are
employed [173].
       Table 13.XI gives a good selection of the acylating
agents that have been introduced, with some indication of
reaction conditions and applicability. The principle of
those methods based on titration of the unreacted anhydride can be readily grasped by reference to the titration
diagram of Fig. 13.7, which shows the determination of an
alcohol by reaction with acetic anhydride [174].

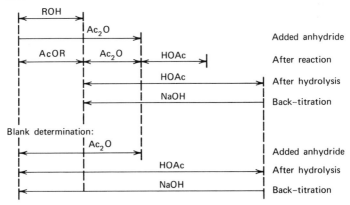

Fig. 13.7. Titration diagram of the determination of an
alcohol by acylation with acetic anhydride and back-
titration of unreacted reagent after hydrolysis. The
lengths of the lines represent numbers of moles or equivalents.

The procedure involves the addition of the sample to a
reagent containing the anhydride and a catalyst. After
acylation is complete, water is added to hydrolyze the

Table 13.XI   Acylating Agents Used in Analytical Chemistry

Reagent	Catalyst
Acetic acid, CH_3COOH	boron trifluoride
Acetic anhydride, $(CH_3CO)_2O$	pyridine
Acetic anhydride, $(CH_3CO)_2O$	perchloric acid
Acetyl chloride, $CH_3COC\ell$	zinc metal

Reagent	Catalyst
Phthalic anhydride, 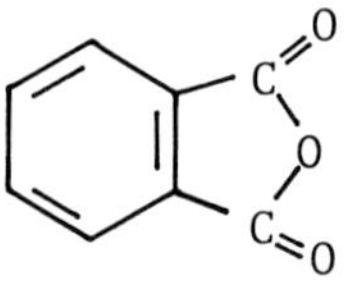	pyridine
Pyromellitic dianhydride, 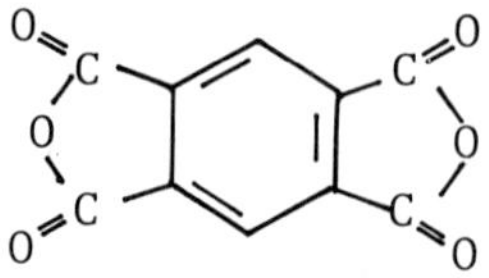	pyridine
3-Nitrophthalic anhydride, 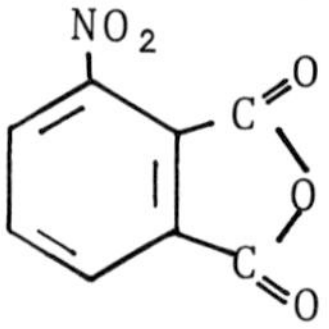	triethyl-amine
3,5-Dinitrobenzoyl chloride, 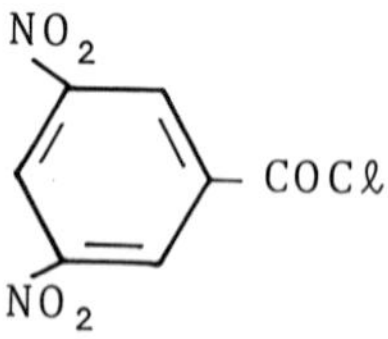	pyridine
p-Nitrobenzoyl chloride, 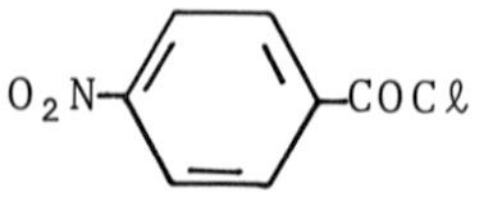 O_2N—COCℓ	pyridine
trans-Cinnamic anhydride, $(C_6H_5CH=CHCO)_2O$	none
Stearic anhydride, $[CH_3(CH_2)_{16}CO]_2O$	none

Sample	Finish	Ref.
alcohols	Karl Fischer titration of water produced	a
alcohols	titration of unreacted anhydride after hydrolysis	b
alcohols, phenols, amines, thiols	titration of unreacted anhydride after hydrolysis	c
alcohols	titration of unreacted reagent	d
alcohols, amines	titration of unreacted anhydride after hydrolysis	e
alcohols, amines	titration of unreacted anhydride after hydrolysis	f
alcohols	titration of unreacted anhydride after hydrolysis	g
alcohols	spectrophotometric measurement of ester in base	h
alcohols	spectrophotometric measurement of ester	i
amines	spectrophotometric measurement of amide	j
alcohols, phenols	titration of unreacted anhydride after hydrolysis	k

Table 13.XI continued

Reagent		Catalyst
Succinic anhydride		pyridine
N-Cinnamoylimidazole, $C_6H_5CH=CHCON$		none
1-Dimethylaminonaphthalene-5-sulfonyl chloride (dansyl chloride),	SO_2Cl / NMe_2	none
Pyruvic acid chloride 2,4-dinitrophenyl-hydrazone	$O_2N-\!\!\!\!-NHN=C-COCl$, NO_2 , CH_3	triethylene-diamine
Trifluoroacetic anhydride, $(CF_3CO)_2O$		none
Heptafluorobutyrylimidazole, C_3F_7CON		none

[a] W. M. D. Bryant, J. Mitchell, Jr., and D. M. Smith, J. Am. Chem. Soc., 62, 1 (1940).
[b] C. L. Ogg, W. L. Porter, and C. O. Willits, Ind. Eng. Chem., Anal. Ed., 17, 394 (1945).
[c] J. S. Fritz and G. H. Schenk, Anal. Chem., 31, 1808 (1959); G. H. Schenk and J. S. Fritz, ibid., 32, 987 (1960).
[d] D. Kyriacou, Anal. Chem., 32, 291 (1960).
[e] P. J. Elving and B. Warshowsky, Ind. Eng. Chem., Anal. Ed., 19, 1006 (1947).
[f] S. Siggia, J. G. Hanna, and R. Culmo, Anal. Chem., 33, 900 (1961).
[g] J. A. Floria, I. W. Dobratz, and J. H. McClure, Anal. Chem., 36, 2053 (1964).
[h] D. P. Johnson and F. E. Critchfield, Anal. Chem., 32, 865 (1960); G. R. Umbreit and R. L. Houtman, J. Pharm. Sci., 56, 349 (1967).

Sample	Finish	Ref.
alcohols, amines	titration of unreacted anhydride after hydrolysis	l
α-chymotrypsin	spectrophotometric titration (measurement of unreacted reagent)	m
amino acids	fluorimetric measurement of the sulfonamide	n
alcohols	spectrophotometric measurement of the ester	o
amines	gas chromatographic measurement of the amide	p
amines	gas chromatographic measurement of the amide	q

[i] M. W. Scoggins, Anal. Chem., _36_, 1152 (1964).
[j] W. H. Hong and K. A. Connors, Anal. Chem., _40_, 1273 (1968).
[k] B. D. Sully, Analyst, _87_, 940 (1962).
[l] C. K. Narang and N. K. Mathur, Indian J. Chem., _3_, 182 (1965); _4_, 263 (1966).
[m] G. R. Schonbaum, B. Zerner, and M. L. Bender, J. Biol. Chem., _236_, 2930 (1961).
[n] W. R. Gray, Methods in Enzymology, _11_, 139 (1967); J. P. Zanetta, G. Vincendon, P. Mandel, and G. Gombos, J. Chromatogr., _51_, 441 (1970).
[o] D. P. Schwartz, Anal. Biochem., _38_, 148 (1970).
[p] J. Hirtz and A. Gerardin, Ann. Pharm. Franc., _27_, 581 (1969); T. Green, Analyst, _95_, 168 (1970).
[q] J. Vessman, A. M. Moss, M. G. Horning, and E. C. Horning, Anal. Letters, _2_, 81 (1969); S. F. Sisenwine, J. A. Knowles, and H. W. Ruelius, ibid., _2_, 315 (1969).

unreacted anhydride. A blank determination is carried out
with the same quantity of anhydride. Since each mole of
unreacted anhydride yields two moles of acetic acid,
whereas each mole of reacted anhydride yields only one
mole of acetic acid, the difference between the blank and
sample back-titration volumes gives the amount of hydroxy
compound in the sample according to the relation: meq of
ROH = $N(V_B - V_S)$, where N is the normality of the alkali
used in the back-titration.

     If all of the anhydride were consumed in the acylation
of the sample, then V_S would be equal to $V_B/2$. This is an
undesirable result, because it shows that insufficient anhydride was present to react with all of the sample. If
V_S is not at least 60% of V_B, it is therefore advisable to
repeat the analysis with a smaller sample or more reagent.

     Considerable attention has been given to the variables
controlling the rate and extent of acylation in analytical
systems, though very little of this work has been systematic or quantitative (meaning that rate and equilibrium
constants are seldom obtained). These investigations can
be misleading, as a later example demonstrates. Nevertheless, much of practical value has been learned. Consider
first some of the catalysts used in analytical acylations.
Pyridine-catalyzed acetylation [175] and phthalation [176]
are standard methods; the mechanism of this nucleophilic
catalysis, which proceeds with the formation of an acyl-
pyridinium intermediate, was described in Sec. 13.1.
Schenk, Wines, and Mojzis [177] compared several tertiary
amines as catalysts in the acetylation of cyclohexanol
with acetic anhydride; they found that triethylenediamine,
<u>46</u>, is about 50 times more effective than pyridine.

<u>46</u>

This compound is also a better catalyst than is pyridine
in the acylation of alcohols with the acid chloride of the
2,4-dinitrophenylhydrazone of pyruvic acid [178]. Imidazole catalyzes acylations by pyromellitic dianhydride [179].
     Some complications can occur in pyridine-catalyzed
acetylations. Interference by aldehydes is common; in
Sec. 12.2 it was suggested that this might be caused by a
Perkin reaction. It has been reported [180] that the

acetic anhydride-pyridine reagent must contain 0.3-0.5%
of water to prevent a polymerization reaction, but this
is not the experience of all workers [180]. Though possibly irrelevant to this problem, but of potential analytical interest, is the observation that 47 is formed from
acetic anhydride and pyridine in the presence of an oxidized pyrrole mixture [181].

Although acid-catalyzed acylation methods have been
available for several decades, they did not find wide use
until the development of a rapid perchloric acid catalyzed
method by Fritz and Schenk [182]. A highly reactive reagent is composed of acetic anhydride, perchloric acid, and
ethyl acetate as the solvent. It is difficult to find
solvents for the potent acetic anhydride-perchloric acid
mixture; recall that even C-acetylation can occur with this
reagent [80,81]. The ethyl acetate solution of $HClO_4$ and
acetic anhydride tends to darken, this process not being
very reproducible from batch to batch. A discolored reagent obscures the visual end point. 1,2-Dichloroethane
has been recommended as a replacement for ethyl acetate
[183]. 2,4-Dinitrobenzenesulfonic acid is only slightly
less powerful as a catalyst than is perchloric acid, and
its solutions are said to show less tendency to discolor
[184].

Numerous interesting modifications and applications
of acylation, aside from those illustrated by Table 13.IX,
have been reported. Spectrophotometric titrations of
amines with anhydrides utilize the spectral properties of
the amine [185] or the anhydride [186]. Hydroxy groups
have been determined by acetylation followed by application of the ferric hydroxamate method to the ester [161,
187]. Similar in approach is the determination of hydroxy
compounds by acetylation followed by quantitative saponification of the ester, as in the "acetin" method for glycerol [188].

A reagent of formic acid and acetic anhydride reacts
much faster with amines than does acetic anhydride itself;

formamides are produced, and a mixed anhydride of formic and acetic acids is said to be the active acylating species [189]. This procedure is used to eliminate interference by primary and secondary amines in the titrimetric determination of tertiary amines.

The trifluoroacetyl group has achieved popularity as a "tag" for the identification and determination of amines and amino acids by gas chromatography and mass spectrometry. Amino acids are chromatographed as their N-trifluoroacetyl n-butyl esters [135,190]. Primary and secondary amines can be identified by GC and MS of the trifluoroacetamides [191] and by their rates of acylation with cinnamic anhydride [192].

The Hinsberg test is a classical method for distinguishing between primary, secondary, and tertiary amines [193]. The reagent, benzenesulfonyl chloride, of course does not react with tertiary amines. The sulfonamides derived from primary amines possess an acidic hydrogen and are soluble in alkali, whereas sulfonamides from secondary amines are insoluble.

$$C_6H_5SO_2C\ell + RNH_2 \xrightarrow{OH^-} C_6H_5SO_2NR^- \qquad (13.102)$$

$$C_6H_5SO_2C\ell + RR'NH \xrightarrow{OH^-} C_6H_5SO_2NRR' \qquad (13.103)$$

Indole-3-acetic acid reacts with acetic anhydride, in the presence of BF_3 or trifluoroacetic acid, to give compound 48, which has distinctive spectral characteristics [194].

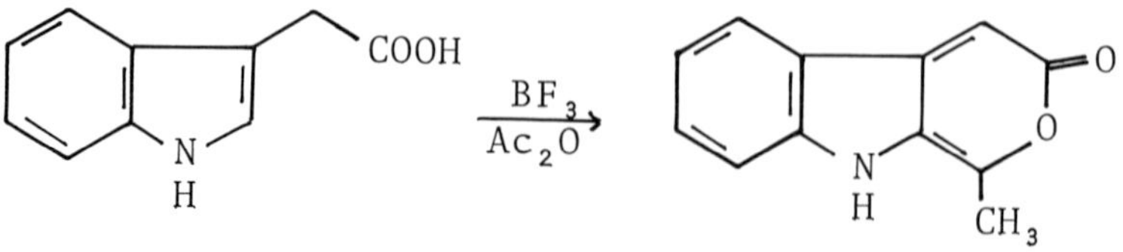

$$(13.104)$$

48

Indole-3-acetic acid has been determined in this way [195]. Presumably the initial step is similar to a Friedel-Crafts acylation to give 49, which undergoes dehydration to the lactone 48.

Several methods for the determination of water use an acyl compound as the reagent and measure the extent of

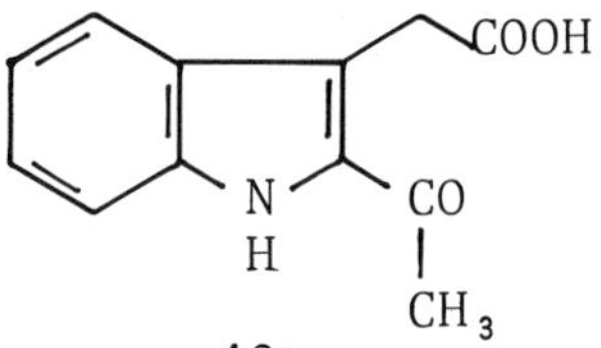

49

hydrolysis. These reagents include acetic anhydride [196],
dimethyl oxalate [197], acid chlorides [198], and ethyl
acetate [199]. The two ester procedures are very similar.
A solution of methanolic sodium methoxide in ethyl acetate
is stable with respect to total basicity (though transesterification can occur). Upon addition of water the
alkoxide-hydroxide equilibrium generates some hydroxide:

$$OR^- + H_2O \rightleftharpoons OH^- + ROH \qquad (13.105)$$

The hydroxide hydrolyzes the ester irreversibly.

$$OH^- + EtOAc \longrightarrow OAc^- + EtOH \qquad (13.106)$$

The net result is that each mole of water results in the
reaction of one mole of alkoxide in the reagent. This
decrease is determined by titration with a standard nonaqueous solution of acid [197,199].

Let us return now to Table 13.XI and the range of
applicability of several of the acylating agents. The
specifications given under the heading Sample do not sufficiently define the limits of the several methods, but
merely give the general functional groupings within which
the methods may be successfully applied. For example, the
cinnamic anhydride method is applicable to aliphatic amines
but not to aromatic amines, because the latter, being weakly nucleophilic, react very slowly [200]. Other reagents
may be capable of more extended use; thus dansyl chloride
has been used principally for amino end group determination in proteins, but it is a useful agent for tagging
other acylable compounds.

Acylation methods for hydroxy groups have variable
selectivity, depending partly on the acylating agent and
the catalyst. A pyridine-acetic anhydride reagent will
generally acylate primary and secondary alcohols, but
tertiary alcohols and highly branched secondary alcohols
are not quantitatively acylated. This reluctance is probably a consequence of steric hindrance, though dehydration of the alcohol may also take place. Perchloric acid-

catalyzed acetylations of these hindered alcohols often take place quantitatively, perhaps because the more reactive reagent is less selective, perhaps also because the acetylium ion, which is smaller than the acetylpyridinium ion, suffers less steric interaction with the substrate. Interestingly, tertiary alcohols are acylated by pyridine-3,5-dinitrobenzoyl chloride [201] and triethylenediamine-2,6-dinitrophenylhydrazone pyruvic acid chloride [178] reagents.

The analytical literature contains numerous statements to the effect that alcohols are acylated faster than phenols, this presumed rate difference being so great with some reagents as to permit the determination of alcohols in the presence of phenols. Let us examine these systems more closely. All of them employ the alkalimetric back-titration procedure diagrammed in Fig. 13.7. The success of this procedure depends upon three important reaction rates:

1. The rate of acylation of the sample ROH by the anhydride reagent.

2. The rate of hydrolysis of the unreacted anhydride prior to its titration. This is easily controlled, often by adding pyridine to catalyze the reaction, and it can be neglected as a factor in the present sense.

3. The rate of hydrolysis of the ester formed by the acylation of ROH.

In order to achieve quantitative analytical recovery, rate (1) must be relatively fast so that acylation is essentially complete, and rate (3) must be very slow (on the titration time scale), so that essentially no hydrolysis occurs during the back titration. In determinations with acetic anhydride, which is a fairly reactive reagent, it is feasible to acylate amines, most primary and secondary alcohols, and phenols quantitatively; moreover, most acetamides and acetates are sufficiently stable that they do not hydrolyze during the back titration of the acetic acid, which is usually carried to a phenolphthalein (pH 8-9) end point. It has been pointed out, however, that the acetates of p-nitrophenol and of catechol undergo some hydrolysis during the titration [202].

Very little quantitative information is available on the relative rates of acylation of alcohols and phenols under analytical conditions. Setting aside the obvious complication of highly sterically hindered secondary and tertiary alcohols and ortho-disubstituted phenols, we first note that alkoxide ions will be much more powerful nucleophiles than are phenoxide ions; however, under any given solution conditions, a larger fraction of a phenol than of an alcohol solute will exist in its conjugate base form, and the two opposing factors will largely compensate. Perhaps more to the point are the nucleophilicities of the neutral hydroxy compounds. Since the pK_a's (as bases) of alcohols are in the range -2 to -4, whereas the corresponding pK_a of phenol is -6.7, it may be concluded that alcohols are generally stronger bases than phenols [203], and therefore possibly also stronger nucleophiles. Opposing this factor is the steric influence of the alkyl group in even the simple alcohols. Substitution on the α carbon of primary alcohols, RCH_2OH, introduces significant steric effects toward alkaline hydrolysis of aliphatic esters [204], and it might be expected that similar effects will be manifested in acylation [205].

Reactions with highly reactive reagents cannot establish the relative reactivities of alcohols and phenols, because both types of compounds will be completely acylated, as with the perchloric acid-acetic anhydride reagent [182]. Observations with milder reagents give some helpful data. Thus, cyclohexanol is only partially acetylated with a pyridine-acetic anhydride reagent under conditions in which many phenols are completely acetylated [202]. A solution of acetic anhydride in ethyl acetate, with no catalyst, gave more extensive acylation with phenol than with ethanol [184]. Phenols are quantitatively acylated in about the same time as are alcohols by stearic anhydride in xylene [206]. By these observations and arguments it can be supposed that, as general classes, alcohols and phenols should have comparable susceptibilities toward acylation agents.

This conclusion appears to be contradicted by analytical results with the acylating agents phthalic anhydride [176] and pyromellitic dianhydride (PMDA) [207], which are reported not to react with phenols under conditions giving complete acylation of alcohols. One possible explanation for the unreactivity of phenols is based upon the tendency of large flat molecules to undergo noncovalent interaction with the equilibrium formation of molecular complexes [208]. Phenols, but not alcohols, would therefore be expected to complex with PMDA, which is known to form molecular complexes [209]. In such a plane-to-plane phenol-PMDA complex it is reasonable to expect the phenol to be inaccessible to attack by another PMDA molecule [210]. It has been found, however, that such complexes are not sufficiently stable to account for the apparent unreactivity of phenols, which would require most of the phenol to be in the complexed form [211].

These combined arguments show, therefore, that phenols probably are acylated by PMDA and phthalic anhydride. It follows that the resulting esters must hydrolyze during the back titration, and that this hydrolysis occurs under conditions such that (a) acylated alcohols do not hydrolyze, and (b) the corresponding acetate esters do not hydrolyze. Thus the effect is determined by the structures of the acyl function and the alkyl function of the ester.

The full range of observations is readily accounted for on the basis of Thanassi and Bruice's description of the hydrolysis kinetics of monoesters of phthalic acid [212]. Very briefly, these esters undergo intramolecular nucleophilically catalyzed hydrolysis, via the intermediate anhydride. If the pK_a of the conjugate acid of the leaving group is less than about 13.5, the ortho-carboxylic anion participates in the catalysis; if this pK_a is greater than 13.5, the un-ionized ortho-carboxylic acid is the catalytic function. The mechanisms can be illustrated with phenyl monohydrogen phthalate (pK_a of phenol = 10.0) and methyl monohydrogen phthalate (pK_a of methanol = 15.5) as examples.

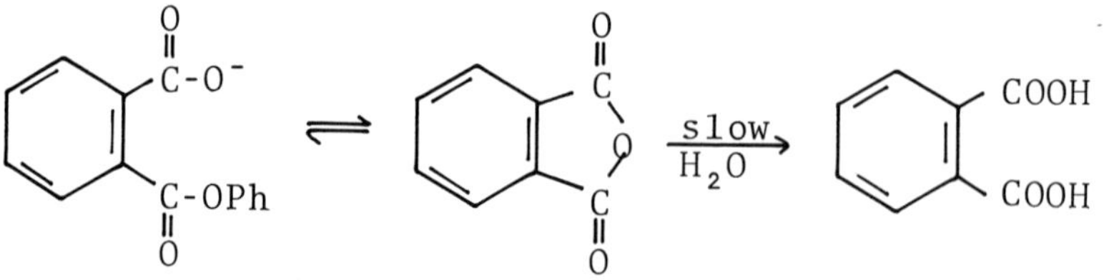

$$(13.107)$$

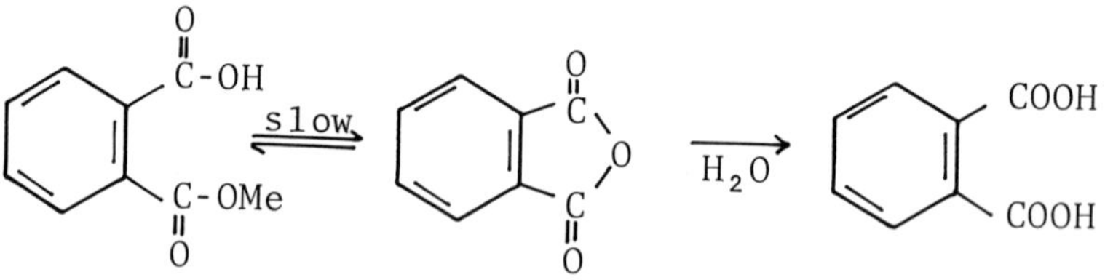

$$(13.108)$$

The evidence for this distinction, and its analytical implications, are clear from Fig. 13.8, showing pH-rate profiles for the hydrolysis of these two esters [212]. (Note that the two curves are on the same scale, but the phenyl ester hydrolysis is at 30° and the methyl ester at 100°.) Throughout nearly the whole of an acylation back titration, the product of a PMDA or phthalic anhydride acylation will be in the ionized form (the pK_a of the o-COOH group is 2.8-3.4). For most parent alcohols this is the most stable form of the ester, but for most parent phenols it is the most unstable form. Hydrolysis of the aromatic ester, therefore, occurs during the back-titration, with the result, apparent only, that phenols are not acyl-ated. Note that the critical factor is not whether the sample is an alcohol or a phenol, but its pK_a value.

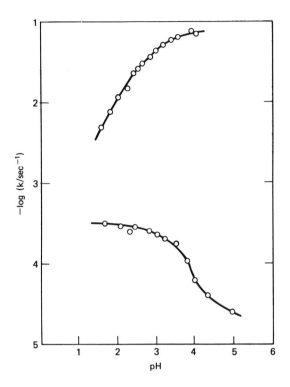

Fig. 13.8. pH-rate curves for the hydrolysis of phenyl hydrogen phthalate at 30° (upper curve) and methyl hydro-gen phthalate at 100° (lower curve) [212].

These arguments should apply to PMDA as well as to phthalic anhydride, and to other acylating agents (i.e., cyclic anhydrides) capable of similar mechanisms. Thus it is found that 3-nitrophthalic anhydride [213] and suc-cinic anhydride [214] cannot be used to determine phenols.

Comment. This chapter may suggest numerous areas, such as catalysis and solvent effects, in which analytical application of available data and concepts may be made. For example, the catalyst p-dimethylaminopyridine [92] warrants analytical investigation [215]. The effects of solvents on analytical saponifications, some complications

of which were discussed earlier, should be placed under
better control than presently exists. DMSO is a particularly interesting solvent [216]. Another area offering
room for improvement is the analytical finish of many acyl
transfer methods. Spectrophotometric and fluorimetric
detection of reaction products (rather than titrimetric
measurement of unreacted reagent) should be exploited to
give greater sensitivity and selectivity. The point here
is not that titrimetry is an inferior detection method,
but that direct measurement of a product is often preferable to measurement of a reactant.

The explanation given for the source of selectivity
in the cyclic anhydride methods for hydroxy group determinations should carry this meaning: The principle shown
should be examined for its possible utility in improving
and extending these methods and, more interestingly, in
designing new ones based on similar concepts [217].

Finally, an example will be given of analytical development based upon mechanistic studies. The reaction of
phenyl acetate with hydroxylamine to form acethydroxamic
acid was observed to be catalyzed by nickel (II). As
Jencks [40] has shown, the reaction occurs in two steps,
the first being formation of O-acetylhydroxylamine.

$$CH_3COOPh + NH_2OH \rightarrow CH_3COONH_2 + PhOH \qquad (13.109)$$

$$CH_3COONH_2 \xrightarrow{\quad NH_2OH \quad} CH_3CONHOH \qquad (13.110)$$

The rates of these reactions can be separately measured
by following the production of phenol (by ultraviolet
spectrophotometry) and of acethydroxamic acid (by color-
imetry with ferric ion). In this way it was shown that
nickel (II) catalyzes Reaction (13.110), the conversion
of O-acetylhydroxylamine to acethydroxamic acid. Formation of a five-membered chelate ring between nickel (II)
and O-acetylhydroxylamine, as in structure 50, was postulated to account for this catalysis.

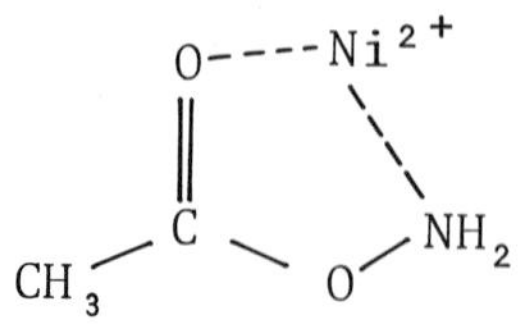

50 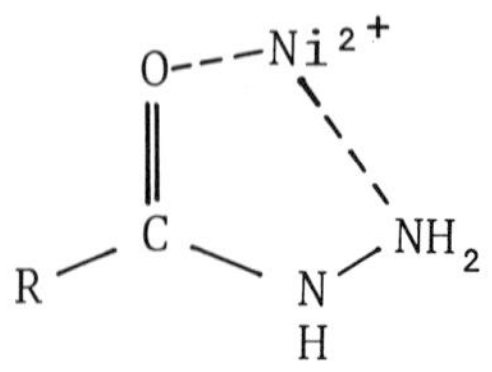   51

This hypothesis suggested that acid hydrazides (see 51),
which are structurally similar to O-acylhydroxylamines,
should also undergo hydroxylaminolysis catalyzed by nickel
(II). This predicted catalysis was observed, and a spectrophotometric method for acid hydrazides was developed
based upon their nickel(II)-catalyzed conversion to hydroxamic acids, followed by treatment with ferric ion to
give the colored ferric hydroxamate complex [218].

Problems

1. Consider the ester $CH_3COOCH_2CH_2C_6H_5$. In basic solu-
    tion these three reactions are concurrently possible:
    (a) B_{Ac} hydrolysis; (b) S_N hydrolysis; (c) β-elimina-
    tion. Devise experiments to measure quantitatively
    the extent of reaction by each pathway.

2. Compare oxime formation from a ketone with hydroxamic
    acid formation from an ester. Account for the products.

3. Why is the $B_{Ac}1$ mechanism an improbable reaction path
    for ester hydrolysis?

4. In Eq. (13.24), the term $k'[Im][OH^-]$ is kinetically
    equivalent to $k''[Im^-]$.
    (a) Find the relationship between k' and k''.
    (b) Offer mechanistic interpretations of these two
    rate terms.

5. Explain the observation that in the imidazole-catalyzed
    hydrolysis of an ester a term second order in imidazole
    is observed, but the N-methylimidazole-catalyzed hy-
    drolysis of the same ester shows no term second order
    in the catalyst.

    Answer: See Ref. [35].

6. Rationalize the instability of acylimidazoles relative
    to ordinary amides by sketching electron distributions
    showing the influence of the ring on resonance delo-
    calization.

7. List these bases in probable order of decreasing nu-
    cleophilic reactivity toward phenyl acetate: pyridine,
    imidazole, hydroxylamine, p-nitrophenolate, triethyl-
    amine, aniline, acetate.

8. Acetate ion catalyzes the hydrolysis of acetic anhy-
    dride. Is this an example of general base or of

nucleophilic catalysis?

9. p-Nitrophenyl acetate undergoes methanolysis, cata-
    lyzed by acetate ion, to give p-nitrophenol and methyl
    acetate. When this reaction was carried out with
    tritium-labeled acetate ion, the methyl acetate pro-
    duced was found to have a specific radioactivity
    0.56 ± 0.06 that of the catalyst. Give a mechanistic
    interpretation of this result.

    Answer: See R. L. Schowen and C. G. Behn, J. Am.
    Chem. Soc., 90, 5839 (1968).

10. (a) Calculate the difference in free energy of acti-
     vation between a glutarate and a succinate monoester,
     for their intramolecularly catalyzed hydrolysis,
     using the relative rate data of Table 13.VIII.
     (b) Assume that this rate effect is entirely the
     result of a change in the entropy of activation.
     Calculate this change.

     Answer: (a) $\delta_R \Delta G^{\ddagger}$ = 3.3 kcal/mole; (b) $\delta_R \Delta S^{\ddagger}$ =
     11 e.u.

11. Postulate a reaction pathway for nucleophilic cata-
     lysis of the hydrolysis of 8-acetoxyquinoline 5-
     sulfonate by zinc pyridinecarboxaldoxime anion through
     the mixed complex 32.

12. The following numbers are ratios of alkaline hydroly-
     sis rates. Give a qualitative account of these
     ratios:

     p-nitrobenzyl acetate
     ─────────────────────   2.0
       benzyl acetate

     p-nitrophenyl acetate
     ─────────────────────   14
       phenyl acetate

     methyl p-nitrobenzoate
     ──────────────────────   71
       methyl benzoate

     ethyl p-aminobenzoate
     ─────────────────────   0.03
       ethyl benzoate

13. From the data in Table 13.X, obtain ρ for the alkaline
     hydrolysis of p-fluorophenyl benzoates.

     Answer: ρ = 2.02

14. Benzamide and N-methylbenzamide undergo extensive
    oxygen exchange with the solvent during their alka-
    line hydrolysis, whereas N,N-dimethylbenzamide shows
    no detectable exchange. Account for these results.

    Answer: See C. A. Bunton, B. Nayak, and C. O'Connor,
    J. Org. Chem., 33, 572 (1968).

15. Postulate a reaction pathway for ester formation from
    an alcohol and a carboxylic acid promoted by N,N'-
    dicyclohexylcarbodiimide, 45, proceeding through an
    O-acylurea derivative.

16. Draw the structures of the four possible products
    from the Claisen condensation of ethyl acetate and
    ethyl propionate.

17. An Elcb mechanism has been proposed for the hydroly-
    sis of esters containing labile alpha hydrogens, for
    example $EtOOC-CH_2-COOR$, where R is $C_6H_4-NO_2$. Postu-
    late a reaction pathway.

    Answer: See T. C. Bruice and B. Holmquist, J. Am.
    Chem. Soc., 90, 7136 (1968); B. Holmquist and T. C.
    Bruice, ibid., 91, 2993 (1969); P. S. Tobias and F.
    J. Kézdy, ibid., 91, 5171 (1969); R. F. Pratt and T.
    C. Bruice, ibid., 92, 5956 (1970).

18. The rate constants for alkaline hydrolysis in water
    at 25° of ethyl benzoate and ethyl p-hydroxybenzoate
    are 293×10^{-4} $M^{-1}-sec^{-1}$ and 2.63×10^{-4} $M^{-1}-sec^{-1}$,
    respectively, as determined in solutions initially
    0.025 M in ester and in sodium hydroxide: E. Tommila,
    A. Nurro, R. Muren, S. Merenheimo, and E. Vuorinen,
    Suomen Kemistilehti, B32, 115 (1959). Account for
    this reactivity difference.

19. Using the literature, obtain estimates of the rates
    of acyl transfer reactions for acetyl compounds rela-
    tive to benzoyl compounds containing the same leaving
    group, i.e., $k(CH_3COX)/k(C_6H_5COX)$.

20. Postulate mechanisms for the base-catalyzed formation
    of hydroxamic acids from esters.

21. Devise a mechanism for the formation of 47 from pyri-
    dine and acetic anhydride.

22. Estimate an approximate half-life for the hydrolysis
    of phenyl hydrogen phthalate during the back-titration

of an attempted phenol determination by phthalation
(see Fig. 13.8).

23. It has been pointed out [202] that the acetates of
p-nitrophenol and catechol undergo hydrolysis during
back titration of an analytical reaction mixture with
alkali. Explain this high reactivity of catechol
monoacetate.

24. Develop analytical approaches to the quantitative
analysis of these mixtures:
(a) Ethanol and trifluoroethanol
(b) Acetyl chloride and benzoyl chloride
(c) n-Propylamine, di-n-propylamine, and tri-n-
propylamine.

25. Devise analytical methods to measure the rates of
these reactions:
(a) Hydrolysis of N,N-dimethylformamide
(b) Pyridine-catalyzed acylation of cyclohexanol
with benzoic anhydride
(c) Concurrent hydrolysis and n-butylaminolysis of
p-nitrophenyl benzoate.

References

1. One type of acyl transfer was treated in Chap. 7; the
Friedel-Crafts reaction is an acyl transfer to aromatic carbon, and it was discussed as an electrophilic aromatic substitution.
2. R. G. Pearson, D. H. Anderson, and L. L. Alt, J. Am.
Chem. Soc., 77, 527 (1955).
3. J. F. Bunnett, J. H. Miles, and K. V. Nahabedian, J.
Am. Chem. Soc., 83, 2512 (1961).
4. M. L. Bender and R. S. Dewey, J. Am. Chem. Soc., 78,
317 (1956).
5. O. R. Quayle and H. M. Norton, J. Am. Chem. Soc.,
62, 1170 (1940).
6. C. K. Ingold, Structure and Mechanism in Organic
Chemistry, Cornell Univ. Press, Ithaca, N. Y., 1953,
Chap. XIV.
7. M. L. Bender, Chem. Revs., 60, 53 (1960).
8. J. March, Advanced Organic Chemistry: Reactions,
Mechanisms, and Structure, McGraw-Hill, New York,
1968, pp. 309-313.
9. Many important reviews have dealt with the chemistry
of acyl transfer. Among these are the following:
T. C. Bruice and S. J. Benkovic, Bioorganic Mechanisms,
Vol. 1, Benjamin, New York, 1966; S. L. Johnson,

Advan. Phys. Org. Chem., 5, 237 (1967); W. P. Jencks,
Catalysis in Chemistry and Enzymology, McGraw-Hill,
New York, 1969; S. Patai (ed.), The Chemistry of
Carboxylic Acids and Esters, Wiley (Interscience),
New York, 1969, especially Chaps. 3 (J. Koskikallio),
9 (D. P. N. Satchell and R. S. Satchell), and 11
(E. K. Euranto); M. L. Bender, Mechanisms of Homo-
geneous Catalysis from Protons to Proteins, Wiley
(Interscience), New York, 1971.

10. L. P. Hammett, Physical Organic Chemistry, McGraw-
     Hill, New York, 1940, pp. 354-358.

11. M. L. Bender, J. Am. Chem. Soc., 73, 1626 (1951).

12. M. L. Bender and R. D. Ginger, Suomen Kemistilehti,
     B, 33, 25 (1960).

13. M. L. Bender and R. J. Thomas, J. Am. Chem. Soc.,
     83, 4189 (1961); M. L. Bender, H. Matsui, R. J.
     Thomas, and S. W. Tobey, ibid., 83, 4193 (1961).

14. S. A. Shain and J. F. Kirsch, J. Am. Chem. Soc., 90,
     5848 (1968).

15. M. L. Bender, R. D. Ginger, and K. C. Kemp, J. Am.
     Chem. Soc., 76, 3350 (1954); M. L. Bender and R. J.
     Thomas, ibid., 83, 4183 (1961).

16. M. L. Bender, R. D. Ginger, and J. P. Unik, J. Am.
     Chem. Soc., 80, 1044 (1958).

17. S. L. Johnson, Advan. Phys. Org. Chem., 5, 237 (1967).

18. W. P. Jencks, Catalysis in Chemistry and Enzymology,
     McGraw-Hill, New York, 1969, Chap. 10.

19. Thus sodium methoxide adds reversibly to ethyl tri-
     fluoroacetate with the position of equilibrium greatly
     favoring the addition product, as shown by infrared
     spectroscopy; M. L. Bender, J. Am. Chem. Soc., 75,
     5986 (1953).

20. E. S. Hand and W. P. Jencks, J. Am. Chem. Soc., 84,
     3505 (1962).

21. The reason for this behavior lies in the kinetic de-
     pendence upon the ionic states of the amine, the
     imido ester, and the tetrahedral intermediate. A
     steady-state treatment successfully describes the
     system [20].

22. W. P. Jencks and M. Gilchrist, J. Am. Chem. Soc.,
     86, 5616 (1964). The negative curvature is not due
     to ionization or complexation equilibria.

23. Not, however, by all; for alternate views see J. M.
     Briody and D. P. N. Satchell, J. Chem. Soc., 1965,
     1968; D. P. N. Satchell and R. S. Satchell, The
     Chemistry of Carboxylic Acids and Esters (S. Patai,
     ed.), Wiley (Interscience), New York, 1969, Chap. 9.

24. F. M. Menger and J. H. Smith, Tetrahedron Letters,
     1970, p. 4163.

25. (a) M. L. Bender and B. W. Turnquest, J. Am. Chem.

Soc., 79, 1656 (1957); (b) T. C. Bruice and G. L. Schmir, ibid., 79, 1663 (1957); (c) D. M. Brouwer, M. J. van der Blugt, and E. Havinga, Proc. Koninkl. Ned. Akad. Wetenschap., B60, 275 (1957).

26. J. F. Kirsch and W. P. Jencks, J. Am. Chem. Soc., 86, 837 (1964).

27. Each step of Eq. (13.22), the formation and the hydrolysis of acetylimidazole, proceeds through a tetrahedral intermediate as described earlier in this chapter.

28. W. P. Jencks and J. Carriuolo, J. Am. Chem. Soc., 83, 1743 (1961).

29. D. G. Oakenfull, T. Riley, and V. Gold, Chem. Commun., 1966, p. 385.

30. T. C. Bruice and R. Lapinski, J. Am. Chem. Soc., 80, 2265 (1958).

31. A. R. Fersht and W. P. Jencks, J. Am. Chem. Soc., 92, 5442 (1970).

32. J. F. Kirsch and W. P. Jencks, J. Am. Chem. Soc., 86, 837 (1964). For catalysis of ester hydrolysis by oxyanions, the borderline between nucleophilic and general base reactions is at a pK_a difference of 2.5 units between the leaving group and the nucleophile; see A. R. Fersht and J. Kirby, J. Am. Chem. Soc., 90, 5818 (1968).

33. J. O. Edwards and R. G. Pearson, J. Am. Chem. Soc., 84, 16 (1962).

34. Figure 4.3 shows further examples of this behavior.

35. For a thorough review, see T. C. Bruice and S. J. Benkovic, Bioorganic Mechanisms, Vol. I, Benjamin, New York, 1966, pp. 46-67.

36. Concurrent with the imidazole-catalyzed hydrolysis is the usual ester hydrolysis, which requires addition of the rate terms $k_H[H^+] + k_w + k_{OH}[OH^-]$.

37. (a) W. P. Jencks and J. Carriuolo, J. Biol. Chem., 234, 1272 (1280 (1959); (b) R. Wolfenden and W. P. Jencks, J. Am. Chem. Soc., 83, 4390 (1961); (c) J. F. Kirsch and W. P. Jencks, ibid., 86, 833 (1964); (d) D. G. Oakenfull, K. Salvesen, and W. P. Jencks, ibid. 93, 188 (1971).

38. W. P. Jencks, Ref. [18], p. 68-70.

39. (a) W. P. Jencks and J. Carriuolo, J. Am. Chem. Soc., 82, 1778 (1960); (b) W. P. Jencks, Ref. [18], pp. 107-111; (c) E. G. Sander and W. P. Jencks, J. Am. Chem. Soc., 90, 6154 (1968); (d) G. Klopman, K. Tsuda J. B. Louis, and R. E. Davis, Tetrahedron, 26, 4549 (1970); (e) J. E. Dixon and T. C. Bruice, J. Am. Chem. Soc., 93, 3248 (1971).

40. W. P. Jencks, J. Am. Chem. Soc., 80, 4581, 4585 (1958).

41. T. C. Bruice and S. J. Benkovic, Ref. [9], pp. 119-201.

42. T. C. Bruice and U. K. Pandit, Proc. Natl. Acad. Sci., 46, 402 (1960); T. C. Bruice and U. K. Pandit, J. Am. Chem. Soc., 82, 5858 (1960).

43. Perpendicular attack also occurs in nucleophilic intermolecular reactions at the carbonyl and carboxyl groups; see Bender [7]; R. G. Kadesch, J. Am. Chem. Soc., 66, 1207 (1944); and M. S. Newman, ibid., 72, 4783 (1950).

44. M. L. Bender, Y.-L. Chow, and F. Chloupek, J. Am. Chem. Soc., 80, 5380 (1958). The pH-rate profile, given in Fig. 3.10, shows that the free carboxylic acid group, or its kinetic equivalent, is the catalytically active form.

45. B. Zerner and M. L. Bender, J. Am. Chem. Soc., 83, 2267 (1961). Even in the general base or general acid mechanism the attack of the nucleophile water will take place perpendicular to the carboximide plane, but the geometrical constraints of this process are less severe than for direct intramolecular nucleophilic attack. On the other hand, the steric (space) requirement for general base catalysis, requiring as it does a molecule of solvent in the transition state, has its own restrictive features. Thus the hindered ester 2-carboxyphenyl mesitoate reacts by intramolecular nucleophilic catalysis, presumably because of crowding, in contrast with the general base catalysis observed with the unhindered analog aspirin; cf. H. D. Burrows and R. M. Topping, Chem. Commun., 1970, 1289.

46. A. R. Fersht and A. J. Kirby, J. Am. Chem. Soc., 90, 5818 (1968).

47. M. L. Bender, F. J. Kézdy, and B. Zerner, J. Am. Chem. Soc., 85, 3017 (1963).

48. M. L. Bender and M. C. Neveu, J. Am. Chem. Soc., 80, 5388 (1958).

49. W. P. Jencks, Catalysis in Chemistry and Enzymology, McGraw-Hill, New York, 1969, p. 11, 15.

50. Catalysis that can be ascribed to such a concentration increase is variously called a proximity effect, propinquity effect, or catalysis by approximation.

51. T. C. Bruice and S. J. Benkovic, J. Am. Chem. Soc., 85, 1 (1963).

52. Catalysis by proximity may occur within a molecular complex or micelle. An example that has been attributed to complex formation is described by J. R. Knowles and C. A. Parsons, Nature, 221, 53 (1969). p-Nitrophenyl decanoate hydrolyzes at 0.04 the rate of p-nitrophenyl acetate when N-ethylimidazole is the catalyst, but the decanoate reacts 19 times faster than the acetate in the presence of the catalyst

N-n-decylimidazole. The hydrophobic decanoate and N-decylimidazole are postulated to form a complex and thus to bring together the nucleophile and the ester.

53. D. R. Storm and D. E. Koshland, Jr., Proc. Natl. Acad. Sci. U.S., <u>66</u>, 445 (1970). For further discussion, and some criticism, of this concept see T. C. Bruice, A. Brown, and D. O. Harris, ibid., <u>68</u>, 658 (1971); B. Capon, J. Chem. Soc., B1971, 120<u>7</u>; G. N. J. Port and W. G. Richards, Nature, 231, 312 (1971); A. Dafforn and D. E. Koshland, Proc. Natl. Acad. Sci. U.S., <u>68</u>, 2463 (1971); T. C. Bruice, Nature, <u>237</u>, 335 (19<u>72</u>).

54. S. Milstein and L. A. Cohen, Proc. Natl. Acad. Sci. U.S., <u>67</u>, 1143 (1970).

55. (a) C. K. Ingold, Structure and Mechanism in Organic Chemistry, Cornell Univ. Press, Ithaca, N. Y., 1953, Chap. XIV; (b) E. K. Euranto, The Chemistry of Carboxylic Acids and Esters (S. Patai, ed.), Wiley (Interscience), New York, 1969, Chap. 11.

56. E. M. Arnett, Progr. Phys. Org. Chem., <u>1</u>, 223 (1963).

57. C. A. Lane, M. F. Cheung, and G. F. Dorsey, J. Am. Chem. Soc., <u>90</u>, 6492 (1968).

58. R. B. Martin, J. Am. Chem. Soc., <u>89</u>, 2501 (1967).

59. K. Yates and R. A. McClelland, J. Am. Chem. Soc., <u>89</u>, 2686 (1967); K. Yates, Accounts Chem. Res., <u>4</u>, 136 (1971).

60. C. A. Bunton, C. O'Connor, and T. A. Turney, Chem. Ind., 1967, p. 1835.

61. C. J. O'Connor, J. Chem. Educ., <u>46</u>, 686 (1969).

62. M. L. Bender and B. W. Turnquest, J. Am. Chem. Soc., <u>79</u>, 1889 (1957).

63. M. D. Alexander and D. R. Busch, J. Am. Chem. Soc., <u>88</u>, 1130 (1966). In these structures (en) represents ethylenediamine.

64. R. Breslow and D. Chipman, J. Am. Chem. Soc., <u>87</u>, 4195 (1965).

65. M. L. Bender, H. Ladenheim, and M. C. Chen, J. Am. Chem. Soc., <u>83</u>, 123 (1961).

66. Entropies of activation for A1 and A2 ester hydrolysis tend to fall in the ranges 0 to 10 e.u. for A1 and -15 to -30 e.u. for A2: L. L. Schaleger and F. A. Long, Advan. Phys. Org. Chem., <u>1</u>, 1 (1963).

67. M. S. Newman, J. Am. Chem. Soc., <u>63</u>, 2431 (1941).

68. M. S. Newman, H. G. Kuivila, and A. B. Garrett, J. Am. Chem. Soc., <u>67</u>, 704 (1945); H. P. Treffers and L. P. Hammett, ibid., <u>59</u>, 1708 (1937).

69. G. A. Olah, Science, <u>168</u>, 1298 (1970).

70. N. C. Deno, C. U. Pittman, Jr., and M. J. Wisotsky, J. Am. Chem. Soc., <u>86</u>, 4370 (1964). A solution of

SO_3 in H_2SO_4 is called fuming sulfuric acid or oleum.

71. C. W. L. Bevan and R. F. Hudson, J. Chem. Soc., 1953, 2187.

72. M. L. Bender and M. C. Chen, J. Am. Chem. Soc., 85, 30 (1963).

73. H. K. Hall and C. H. Lueck, J. Org. Chem., 28, 2818 (1963).

74. R. F. Hudson and G. E. Moss, J. Chem. Soc., 1962, 5157.

75. A thermodynamic treatment of specific solvation effects has been applied to the solvolysis of some substituted benzoyl chlorides in mixed solvents; T. F. Fagley, J. S. Bullock, and D. W. Dycus, J. Phys. Chem., 74, 1840 (1970). S. D. Ross, J. Am. Chem. Soc., 92, 5998 (1970), interpreted the reactions of benzoyl chlorides with ethanol in acetone and chloroform as hydrogen-bonded transition states leading to tetrahedral intermediates.

76. J. M. Briody and D. P. N. Satchell, J. Chem. Soc., 1965, 168.

77. R. A. Sneen and J. W. Larsen, J. Am. Chem. Soc., 91, 6031 (1969); see Sec. 9.1 for a discussion of the ion-pair hypothesis.

78. H. A. E. Mackenzie and E. R. S. Winter, Trans. Faraday Soc., 44, 159 (1948).

79. N. F. Hall and J. B. Conant, J. Am. Chem. Soc., 49, 3047 (1927); J. B. Conant and N. F. Hall, ibid., 49, 3062 (1927); N. F. Hall and T. H. Werner, ibid., 50, 2367 (1928); N. F. Hall, ibid., 52, 5115 (1930); J. B. Conant and T. H. Werner, ibid., 52, 4436 (1930).

80. H. Burton and P. F. G. Praill, J. Chem. Soc., 1950, 1203.

81. H. Burton and P. F. G. Praill, J. Chem. Soc., 1950, 2034.

82. Because of the low dielectric constant of acetic acid, even strong electrolytes are but slightly dissociated. For example, the overall dissociation constant (pK) of perchloric acid in acetic acid is 4.87 [83].

83. S. Bruckenstein and I. M. Kolthoff, J. Am. Chem. Soc., 78, 2974 (1956).

84. M. S. Newman, Steric Effects in Organic Chemistry (M. S. Newman, ed.), Wiley, New York, 1956, Chap. 4.

85. M. S. Newman, J. Am. Chem. Soc., 72, 4783 (1950). Newman [83] has emphasized that the rule-of-six is a paper-and-pencil substitute for scale molecular models.

86. R. W. Taft, Jr., Steric Effects in Organic Chemistry (M. S. Newman, ed.), Wiley, New York, 1956, Chap. 13.

87. M. L. Bender and R. J. Thomas, J. Am. Chem. Soc., 83, 4189 (1961). Actually the ratio k_h/k_e varied with structure, so, according to Eq. (13.14), k_2/k_3 varied.

But a change in R should affect k_2 and k_3 about equally. This dilemma was resolved by assuming that k_2/k_3 is constant, and invoking kinetically significant proton transfers in the tetrahedral intermediate to account for the variation in k_h/k_e.

88. R. J. Washkuhn, V. K. Patel, and J. R. Robinson, J. Pharm. Sci., 60, 736 (1971).

89. J. J. Ryan and A. A. Humffray, J. Chem. Soc., B1966, 842; 1967, 468.

90. For useful collections of synthetic procedures see R. B. Wagner and H. D. Zook, Synthetic Organic Chemistry, Wiley, New York, 1953; S. R. Sandler and W. Karo, Organic Functional Group Preparations, Academic, New York, 1968.

91. F. D. Chattaway, J. Chem. Soc., 1931, 2495.

92. W. Steglich and G. Höfle, Angew. Chem. Intern. Ed., 8, 981 (1969). For rates of nucleophilic attack by substituted pyridines on acetic anhydride see A. R. Fersht and W. P. Jencks, J. Am. Chem. Soc., 92, 5432 (1970).

93. J. L. Marshall, K. C. Erickson, and T. K. Folsom, Tetrahedron Letters, 1970, 4011.

94. R. B. Woodward, R. A. Olofson, and H. Mayer, J. Am. Chem. Soc., 83, 1010 (1961); R. B. Woodward and R. A. Olofson, Tetrahedron, Supple. No. 7, 415 (1966).

95. H. A. Staab, Angew. Chem. Intern. Ed., 1, 351 (1962).

96. O. O. Orazi and R. A. Corral, J. Am. Chem. Soc., 91, 2162 (1969).

97. B. Gutte and R. B. Merrifield, J. Am. Chem. Soc., 91, 501 (1969); R. Hirschmann, R. F. Nutt, D. F. Veber, R. A. Vitali, S. L. Varga, T. A. Jacob, F. W. Holly, and R. G. Denkewalter, ibid., 91, 507 (1969).

98. R. A. Boissonnas, Advan. Org. Chem., 3, 159 (1963), has reviewed amino protective groups and their deblocking reactions; J. F. W. McOmie, ibid., 3, 191 (1963) has surveyed protective groups for other functionalities.

99. For reviews of peptide synthesis see T. Wieland and H. Determann, Ann. Rev. Biochem., 35, 651 (1966); D. T. Elmore, Peptides and Proteins, Cambridge Univ. Press, Cambridge, 1968, Chap. 4; R. B. Merrifield, Advan. Enzymology, 32, 221 (1969).

100. J. March, Advanced Organic Chemistry: Reactions, Mechanisms, and Structure, McGraw-Hill, New York, 1968, pp. 321, 337.

101. R. Hirschmann, H. Schwam, R. G. Strachan, E. F. Schoenewaldt, H. Barkemeyer, S. M. Miller, John B. Conn, V. Garsky, D. F. Veber, and R. G. Denkewalter, J. Am. Chem. Soc., 93, 2746 (1971).

102. H. Morawetz and P. S. Otaki, J. Am. Chem. Soc., 85,

463 (1963).

103. A more realistic tentative rate equation would be
      $v = k_B[RCOO^-][R^\frown NH_2] + k_A[RCOOH][R^\frown NH_2]$. The data
      in Table I of Ref. [102], when tested against this
      equation, are too sparse to permit an estimate of
      k_A to be quoted, but chemical intuition suggests that
      $k_A \gg k_B$.

104. T. Higuchi and T. Miki, J. Am. Chem. Soc., 83, 3899
      (1961); T. Higuchi, T. Miki, A. C. Shah, and A. K.
      Herd, ibid., 85, 3655 (1963); T. Higuchi, S. O.
      Eriksson, H. Uno, and J. J. Windheuser, J. Pharm.
      Sci., 53, 280 (1964).

105. R. E. Notari, J. Pharm. Sci., 58, 1069 (1969); R.
      E. Notari and T. Baker, ibid., 61, 244 (1972).

106. W. P. Jencks, M. Caplow, M. Gilchrist, and R. G.
      Kallen, Biochemistry, 2, 1313 (1963).

107. J. M. Lawlor, Chem. Commun., 1967, 404.

108. (a) J. W. Munson and K. A. Connors, J. Am. Chem.
      Soc., 94, 1979 (1972); (b) K. A. Connors and J. W.
      Munson, Anal. Chem., 44, 336 (1972).

109. Spectral evidence shows that different nickel-
      hydroxylamine complexes are present under Lawlor's
      conditions and the conditions of Fig. 13.6.

110. An anhydride intermediate is not involved, as shown
      by first-order dependence on acetic acid, and by
      similar rates of hydroxamic acid formation from
      acetic and succinic acids.

111. T. C. Bruice and S. J. Benkovic, Bioorganic Mecha-
      nisms, Vol. I, Benjamin, New York, 1966, p. 4.

112. This analogy between an acid and an ester omits what
      may be a significant difference, namely that in an
      acid the two oxygen atoms are equivalent, but they
      are distinguishable in an ester.

113. For reviews see C. K. Ingold, Ref. [6], pp. 787-796;
      J. March, Ref. [100], pp. 362-370; E. E. Royals,
      Advanced Organic Chemistry, Prentice-Hall, New York,
      1954, pp. 802-829.

114. Acyl transfer reactions can also occur between active
      methylene compounds and carboxylic acid derivatives
      such as anhydrides and acid chlorides. These reac-
      tions are analogous to the reactions with esters
      (though possibly with different mechanisms), but the
      term Claisen condensation is reserved for reactions
      in which an ester is the substrate.

115. Possibly removal of a γ-hydrogen in the product con-
      tributes to the equilibrium displacement by the
      stronger base; see Ingold, Ref. [6], p. 795.

116. (a) R. T. Hall and W. E. Shaefer, Org. Anal., 2, 19
      (1954); (b) J. Mitchell, Jr., B. A. Montague, and
      R. H. Kinsey, ibid., 3, 1 (1956); (c) C. W. Hammond,

ibid., 3, 97 (1956); (d) S. Siggia, Quantitative
Organic Analysis via Functional Groups, 3rd ed.,
Wiley, New York, 1963, Chap. 3; (e) N. D. Cheronis
and T. S. Ma, Organic Functional Group Analysis by
Micro and Semimicro Methods, Wiley (Interscience),
New York, 1964, Chaps. 6 and 7; (f) F. E. Critch-
field, Organic Functional Group Analysis, Macmillan,
New York, 1963, Chaps. 8 and 9; (g) S. Veibel,
Treatise on Analytical Chemistry (I. M. Kolthoff
and P. J. Elving, eds.), Part II, Vol. 13, Wiley
(Interscience), New York, 1966, pp. 223-299; (h)
T. S. Ma, The Chemistry of Carboxylic Acids and
Esters (S. Patai, ed.), Wiley (Interscience), New
York, 1969, Chap. 17; (i) R. D. Tiwari and J. P.
Sharma, The Determination of Carboxylic Functional
Groups, Pergamon, Oxford, 1970; (j) F. T. Weiss,
Determination of Organic Compounds: Methods and
Procedures, Wiley (Interscience), New York, 1970,
Chaps. 5 and 6.

117.   Solvent effects on saponification analyses are discussed by Hall and Schaefer [116a] in their 1954
review on ester analysis. Alcohols are often used
because of the low solubility of esters in water.

118.   See, for example, E. Tommila, A. Nurro, R. Muren,
S. Merenheimo, and E. Vuorinen, Suomen Kemistilehti,
B32, 115 (1959); E. Tommila, ibid., B37, 117 (1964);
and the review by Euranto [55b].

119.   R. G. Burns and B. D. England, Tetrahedron Letters,
1960, No. 24, p. 1.

120.   M. L. Bender and W. A. Glasson, J. Am. Chem. Soc.,
81, 1590 (1959).

121.   W. H. Greive, K. F. Sporek, and M. K. Stinson, Anal.
Chem., 38, 1264 (1966).

122.   D. E. Jordan, Anal. Chem., 36, 2134 (1964).

123.   H. E. Hallam, Analyst, 80, 552 (1955); H. Roth and
P. Schuster, Mikrochim. Acta, 1957, p. 837.

124.   F. Kézdy and A. Bruylants, Bull. Soc. Chim. Belg.,
66, 565 (1957); T. M. Bednarski and D. N. Hume,
Anal. Chim. Acta, 30, 1 (1964).

125.   M. Qureshi, S. Qureshi, and S. C. Singhal, Anal.
Chem., 40, 1781 (1968).

126.   P. W. West and M. Qureshi, Anal. Chim. Acta, 26,
506 (1962).

127.   H. Jacin and J. M. Slanski, Anal. Chem., 42, 801
(1970).

128.   J. E. Barker, C. M. Payne, and J. Maulding, Anal.
Chem., 32, 831 (1960).

129.   W. Huber, Mikrochim. Acta, 1969, 897.

130.   F. H. Lohman and T. F. Mulligan, Anal. Chem., 41,
243 (1969).

131. S. Bruckenstein, Anal. Chem., 28, 1920 (1956).
132. L. H. Greathouse, H. J. Janssen, and C. H. Haydel,
      Anal. Chem., 28, 327 (1956).
133. S. Moore and W. H. Stein, Methods in Enzymology, 6,
      819 (1963); A. L. Light and E. L. Smith in The
      Proteins (H. Neurath, ed.), 2nd ed., Vol. I, Academic,
      New York, 1963, p. 1.
134. A. Mondino and G. Bongiovanni, J. Chromatogr., 52,
      405 (1970).
135. D. Roach and C. W. Gehrke, J. Chromatogr., 52, 393
      (1970).
136. N. Maravalhas, J. Chromatogr., 50, 413 (1970).
137. R. L. Shriner, R. C. Fuson, and D. Y. Curtin, The
      Systematic Identification of Organic Compounds, 5th
      ed., Wiley, New York, 1964, p. 137.
138. G. H. Schenk, Organic Functional Group Analysis,
      Pergamon, Oxford, 1968, pp. 17-19. The bromoform
      reaction has also been used; see K. C. Grover and
      R. C. Mehotra, Z. Anal. Chem., 160, 274 (1958).
139. C. F. Cullis and M. H. Hashmi, J. Chem. Soc., 1956,
      2512.
140. D. S. Zaura and J. Metcoff, Anal. Chem., 41, 1781
      (1969).
141. C. E. Tolbert and C. T. Kenner, J. Pharm. Sci., 60,
      596 (1971).
142. H. G. Griffin, Jr. and W. E. Sonia, Jr., Anal. Chem.,
      41, 1488 (1969).
143. K. Hancock and H. L. Lochte, J. Am. Chem. Soc., 61,
      2448 (1939); B. Shive, J. Horeczy, G. Wash, and H.
      L. Lochte, ibid., 64, 385 (1942).
144. J. Mitchell, Jr., D. M. Smith, and W. M. D. Bryant,
      J. Am. Chem. Soc., 62, 4 (1940).
145. H. Salwin and J. F. Bond, J. Assoc. Offic. Anal.
      Chem., 52, 41 (1969); W. F. Staruszkiewicz, Jr.,
      ibid., 52, 471 (1969).
146. M. L. Vorbeck, L. R. Mattick, F. A. Lee, and C. S.
      Pederson, Anal. Chem., 33, 1512 (1961). The diazo-
      methane reaction, CH_2N_2 + RCOOH → $RCOOCH_3$ + N_2, is
      an S_N reaction.
147. R. L. Shriner, R. C. Fuson, and D. Y. Curtin, Ref.
      [137], p. 269.
148. D. F. Percival, Anal. Chem., 35, 236 (1963).
149. I. G. Barr, R. J. Hamilton, and K. Simpson, Chem.
      Ind., 1970, 988.
150. L. Mazor and T. Meisel, Anal. Chim. Acta, 20, 130
      (1959).
151. M. E. Mason and G. R. Waller, Anal. Chem., 36, 583
      (1964); M. E. Mason, M. E. Eager, and G. R. Waller,
      ibid., 36, 587 (1964).
152. R. T. Coutts and K. K. Midha, J. Pharm. Sci., 58,

949 (1969).

153.   J. S. Fritz and N. M. Lisicki, Anal. Chem., _23_, 589 (1951).

154.   A. Groagova and V. Chromy, Analyst, _95_, 548 (1970). A related technique titrates aryl esters with potassium methoxide in ethylenediamine solvent. In this system aminolysis by the solvent can give an amide plus a phenol, which is then titrated; see A. R. Glenn and J. T. Peake, Anal. Chem., _27_, 205 (1955).

155.   R. L. Shriner, R. C. Fuson, and D. Y. Curtin, Ref. [137], pp. 235-236.

156.   S. Siggia and J. G. Hanna, Anal. Chem., _23_, 1717 (1951).

157.   J. B. Johnson and G. L. Funk, Anal. Chem., _27_, 1464 (1955).

158.   M. Pesez and R. Willemart, Bull. Soc. Chim. Franc., _15_, 479 (1948). For other determinations of acid chlorides see J. Simonyi and I. Kekesy, Z. Anal. Chem., _215_, 187 (1965); K. Stürzer, ibid., _216_, 409 (1966).

159.   L. Legradi, Mikrochim. Acta, 1970, 463.

160.   F. Feigl, Spot Tests in Organic Analysis, 6th ed., Elsevier, Amsterdam, 1960, pp. 249-254.

161.   U. T. Hill, Ind. Eng. Chem., Anal. Ed., _18_, 317 (1946); _19_, 932 (1947).

162.   G. Aksnes, Acta Chem. Scand., _11_, 710 (1957). At higher pH's, other complexes with the formulas $Fe(RCONHO)_2^+$ and $Fe(RCONHO)_3$ are formed.

163.   R. E. Notari and J. W. Munson, J. Pharm. Sci., _58_, 1060 (1969).

164.   (a) F. Bergman, Anal. Chem., _24_, 1367 (1952); (b) R. F. Goddu, N. F. LeBlanc, and C. M. Wright, ibid., _27_, 1251 (1955); (c) V. Goldenberg and P. E. Spoerri, ibid., _30_, 1327 (1958).

165.   D. W. Knight and G. H. Cleland, J. Chem. Educ., _47_, 781 (1970).

166.   W. M. Diggle and J. C. Gage, Analyst, _78_, 473 (1953).

167.   W. P. Jencks, S. Cordes, and J. Carriuolo, J. Biol. Chem., _235_, 3608 (1960). The unusually reactive β-lactam of penicillin also forms a hydroxamic acid in neutral solution; see J. H. Ford, Ind. Eng. Chem., Enal. Ed., _19_, 1004 (1947).

168.   F. Lipmann and L. C. Tuttle, J. Biol. Chem., _159_, 21 (1945).

169.   W. P. Jencks and M. Gilchrist, J. Am. Chem. Soc., _86_, 5616 (1964).

170.   E. M. Arnett, Progr. Phys. Org. Chem., _1_, 223 (1963).

171.   S. Soloway and A. Lipschitz, Anal. Chem., _24_, 898 (1952).

172.   R. E. Buckles and C. T. Thelen, Anal. Chem., _22_, 676

(1950) observed some positive tests for acids using an alkaline hydroxylamine reagent, but it is conceivable that these positive tests resulted from the acid-catalyzed reaction initiated upon acidification prior to color development.

173.   Hydroxy group determination is reviewed by V. C. Mehlenbacher, Org. Anal., 1, 1 (1953), and amine determination by E. F. Hillenbrand, Jr., and C. A. Pentz, ibid., 3, 129 (1956). See also several of the sources cited in Ref. [114]. Acylation methods have been discussed by N. K. Mathur, Talanta, 13, 1601 (1966) and G. H. Schenk, Organic Functional Group Analysis, Pergamon, Oxford, 1968, Chaps. 3 and 4.

174.   K. A. Connors, A Textbook of Pharmaceutical Analysis, Wiley, New York, 1967, p. 425.

175.   C. L. Ogg, W. L. Porter, and C. O. Willits, Ind. Eng. Chem., Anal. Ed., 17, 394 (1945).

176.   P. J. Elving and B. Warshowsky, Ind. Eng. Chem., Anal. Ed., 19, 1006 (1947).

177.   G. H. Schenk, P. Wines, and C. Mojzis, Anal. Chem., 36, 914 (1964).

178.   D. P. Schwartz, Anal. Biochem., 38, 148 (1970).

179.   B. H. M. Kingston, J. J. Garey, and W. B. Hellwig, Anal. Chem., 41, 86 (1969).

180.   H. N. Wilson and W. C. Hughes, J. Soc. Chem. Ind., 58, 74 (1939); V. C. Mehlenbacher, Ref. [173], pp. 23 and 28.

181.   I. Fleming and J. B. Mason, J. Chem. Soc., C1969, 2509.

182.   J. S. Fritz and G. H. Schenk, Anal. Chem., 31, 1808 (1959); G. H. Schenk and J. S. Fritz, ibid., 32, 987 (1960).

183.   J. A. Magnuson and R. J. Cerri, Anal. Chem., 38, 1088 (1966).

184.   D. J. Pietrzyk and J. Belisle, Anal. Chem., 38, 1508 (1966).

185.   C. A. Reynolds, F. H. Walker, and E. Cochran, Anal. Chem., 32, 983 (1960).

186.   K. A. Connors and W.-H. Hong, Anal. Chim. Acta, 43, 334 (1968).

187.   G. Gutnikov and G. H. Schenk, Anal. Chem., 34, 1316 (1962).

188.   V. C. Mehlenbacher, Ref. [173], p. 18.

189.   S. Goeroeg and G. Szepes, Z. Anal. Chem., 251, 303B (1970).

190.   N. Lachovitzki and B. Bjoerklund, Anal. Biochem., 38, 446 (1970).

191.   M. Pailer and J. W. Hübsch, Monatsh. Chem., 97, 1541 (1966); A. Zeman and I. P. G. Wirotama, Z. Anal.

Chem., 247, 158 (1969).

192.   W.-H. Hong and K. A. Connors, J. Pharm. Sci., 57, 1789 (1968).

193.   R. L. Shriner, R. C. Fuson, and D. Y. Curtin, Ref. [137], p. 119.

194.   H. Plieninger and W. Müller, Tetrahedron Letters, 1960, No. 11, p. 15; H. Plieninger, W. Müller, and K. Weinerth, Chem. Ber., 97, 667 (1964).

195.   A. Stoessl and M. A. Venis, Anal. Biochem., 34, 344 (1970).

196.   J. A. Barltrop and R. J. Morgan, Anal. Chim. Acta, 16, 520 (1957).

197.   J. Koskikallio, Suomen Kemistilehti, B, 30, 108 (1957).

198.   R. Belcher, L. Ottendorfer, and T. S. West, Talanta, 4, 116 (1960).

199.   R. L. Glass, Anal. Biochem., 37, 219 (1970).

200.   W.-H. Hong and K. A. Connors, Anal. Chem., 40, 1273 (1968).

201.   D. P. Johnson and F. E. Critchfield, Anal. Chem., 32, 865 (1960; W. T. Robinson, Jr., R. H. Cundiff, and P. C. Markunas, ibid., 33, 1030 (1961).

202.   C. W. DeWalt, Jr. and R. A. Glenn, Anal. Chem., 24, 1789 (1952).

203.   E. M. Arnett, Progr. Phys. Org. Chem., 1, 223 (1963).

204.   J. R. Robinson and L. E. Matheson, J. Org. Chem., 34, 3630 (1969).

205.   If the rate-determining step is formation of a reactive intermediate, then all nucleophiles will appear to have the same reactivity. Thus in the pyridine-catalyzed acetylation of toluidine and of anisidine by acetic anhydride, identical rate constants are observed for the two substrates and for the formation of the acetylpyridinium ion. This means that the acetylpyridinium ion formation must be rate determining. See A. R. Fersht and W. P. Jencks, J. Am. Chem. Soc., 92, 5432 (1970).

206.   B. D. Sully, Analyst, 87, 940 (1962).

207.   S. Siggia, J. G. Hanna, and R. Culmo, Anal. Chem., 33, 900 (1961).

208.   K. A. Connors, M. H. Infeld, and B. J. Kline, J. Am. Chem. Soc., 91, 3597, 5697 (1969); J. L. Cohen and K. A. Connors, J. Pharm. Sci., 59, 1271 (1970).

209.   L. L. Ferstandig, W. G. Toland, and C. D. Heaton, J. Am. Chem. Soc., 83, 1151 (1961).

210.   P. A. Kramer and K. A. Connors, J. Am. Chem. Soc., 91, 2600 (1969).

211.   P. A. Kramer, Purdue University, personal communication.

212.   J. W. Thanassi and T. C. Bruice, J. Am. Chem. Soc.,

88, 747 (1966).
213. J. A. Floria, I. W. Dobratz, and J. H. McClure,
      Anal. Chem., 36, 2053 (1964).
214. C. K. Narang and N. K. Mathur, Indian J. Chem., 4,
      263 (1966).
215. This compound has now been found to be a useful
      catalyst for hydroxy determinations by acetylation
      (K. A. Connors and K. S. Albert, this laboratory).
216. J. A. Vinson, J. S. Fritz, and C. A. Kingsbury,
      Talanta, 13, 1673 (1966), have developed a rapid
      saponification method using aqueous DMSO.
217. J. L. Cohen (University of Southern California,
      personal communication) has found that phenyl mono-
      hydrogen phthalate is hydrolyzed during the course
      of titration under usual analytical conditions.
218. J. W. Munson and K. A. Connors, J. Pharm. Sci., 61,
      211 (1972).

621